CONVERGING, EMERGING, INNOVATIVE, DISRUPTIVE, AND CRITICAL TECHNOLOGIES FOR MODERN AND FUTURE WARFARE

Prof. (Dr.) Jai Paul Dudeja

Made with ♥ on the Notion Press Platform
www.notionpress.com

Dedication

This book is most respectfully dedicated to our revered leader

(Late) Dr. APJ Abdul Kalam

Former President of India, and Former Chief of

Defence Research & Development Organisation (DRDO),

Under whose umbrella and guidance,

this author had the privilege of working as a scientist in DRDO.

Dr. Kalam was a Doyen and Beacon of Critical Defence Technologies,

a Friend, Philosopher, Guide and a Source of Inspiration for all of us.

TABLE OF CONTENTS

TESTIMONIALS

SECTION-I

INTRODUCTION AND OVERVIEW OF TECHNOLOGIES FOR MODERN AND FUTURE WARFARE

SECTION-II

CONVERGING TECHNOLOGIES FOR MODERN AND FUTURE WARFARE

SECTION-III

EMERGING TECHNOLOGIES FOR MODERN AND FUTURE WARFARE

SECTION-IV

INNOVATIVE TECHNOLOGIES FOR MODERN AND FUTURE WARFARE

SECTION-V

DISRUPTIVE TECHNOLOGIES FOR MODERN AND FUTURE WARFARE

SECTION-VI

CRITICAL TECHNOLOGIES FOR MODERN AND FUTURE WARFARE

SECTION-VI

NON-TECHNOLOGICAL WARFARE

TESTIMONIALS

TESTIMONIAL FROM DR. B K DAS,

DISTINGUISHED SCIENTIST & DIRECTOR GENERAL,
ELECTRONICS & COMMUNICATION SYSTEMS (ECS),
DEFENCE RESEARCH & DEVELOPMENT ORGANISATION (DRDO),
MINISTRY OF DEFENCE, GOVERNMENT OF INDIA, NEW DELHI.

I have glanced through the pre-published version of the manuscript titled, **"Converging, Emerging, Innovative, Disruptive, and Critical Technologies for Modern and Future Warfare",** authored by **Prof. (Dr.) Jai Paul Dudeja,** a former scientist in DRDO.

This comprehensive book is divided into seven sections, consisting of 60 chapters. Some of the technologies covered by Dr. Dudeja in this comprehensive book, relevant to modern and future warfare are: Smartphones, Network-Centric Warfare, Virtual Reality and Augmented Reality, Holography, and Drone Technology; Artificial Intelligence, 3D Printing, Nanotechnology, and Robotics; 5G and 6G Technologies, Cloud Computing, Renewable Energy Technologies, Internet of Things, and Wearable Devices; Generative AI, and Edge Computing; Advanced Composite Materials, Explosives and Energetic Materials, Advanced Magnets and Superconductors, Protective Clothing and Equipment, Application of Coatings, Smart Materials, Small satellites, Quantum Computing, Quantum Cryptography, Quantum Communication, Quantum Sensors, Free-Space Optical and Fiber-Optic Communication, RF & mm-Wave Communication, Undersea Wireless Communication, Electro-Optic Sensors, Fiber-Optic Sensors, Magnetic Field Sensors, Multispectral and Hyperspectral Imaging, Sonar Systems, Biotechnology, Chemical Warfare, Nuclear War, Environmental Warfare or Eco Terrorism, Cyber Warfare, Electronic Warfare, Space Warfare, Hybrid Warfare, Asymmetric Warfare, Irregular Warfare, Information Warfare, Stealth Technology, Integrated Missile Defence System, Radar Systems, Lidar Systems, Unmanned Aerial Vehicles, Unmanned Ground Vehicles, Unmanned Underwater Vehicles, and Directed Energy Weapons.

Besides these technologies, Dr. Dudeja has also covered the aspects of Cross-Border Terrorism and Proxy War, Psychological Warfare, Economic Warfare, Diplomatic Warfare, and Political Warfare.

After going through these topics in the manuscript, I must mention that Prof. Dudeja has covered almost all the conceivable technologies and other strategic aspects for the modern and future warfare, in a single volume of this comprehensive book.

I wish and hope that this guidebook is going to help all these readers and give inputs to them to implement some of these technologies and other aspects for the preparation of winning the modern and future warfare.

I congratulate Prof. Dudeja for completing this masterpiece book and wish that it will be appreciated by all the readers across the globe and the military-technology-related agencies.

TESTIMONIAL FROM PROF. HB SRIVASTAVA,

PROFESSOR, INDIAN INSTITUTE OF TECHNOLOGY, DELHI
Formerly, DIRECTOR GENERAL (technolohy management)
DEFENCE RESEARCH & DEVELOPMENT ORGANISATION (DRDO),
MINISTRY OF DEFENCE, GOVERNMENT OF INDIA, NEW DELHI.

'Converging, Emerging, Innovative, Disruptive and Critical Technologies for Modern and Future Warfare' by Prof JP Dudeja is an interesting read. From defining what constitutes converging to critical technologies, he takes the readers to dive deeper into each subject. He has covered a very vast spectrum of technologies ranging from cyber world and augmented/virtual reality to water/ underwater, air/aerospace and land warfare domains. All dimensions of modern warfare like robotics, unmanned systems, remote and non-conventional sensing, military communication, chem-bio warfare, missile shield, soldier protection and performance monitoring etc. find adequate space in the book. With this kind of content, the reader is expected to get full exposure into context and extent of defence technologies.

Some new age and deep technologies are set to revolutionize the world. While quantum and AI (and their variations) have found prominent place in engineering education, biotechnology is a new entrant in the field. While advances in biotechnology have potential to create weapons of mass destruction, the technology has immense defence and civilian applications. The technology has potential to save humans from chemical and biological warfare agents on one hand. On the other hand, biotechnology, in combination with AI, wearable sensors, block chain, robotics etc., will help the users produce fortified food, improve soil health, create climate resilient farming practices and result in regenerative agriculture. Biotechnology is just one example. Dr JP Dudeja has packed the book with several such dual use technologies having large potential for humanity, not just in war but also in peace and for promoting peace.

Finally, modern wars are not fought on borders only through killing of each other's soldiers and destruction of property; countries work on creating internal conflicts, low intensity engagements, 'salami-slices', breaking the will to fight wars, ruining the economy and isolating adversaries in the community of nations. The author has very effectively brought out these aspects in his book.

Overall, this comprehensive book has exhaustively covered all the essential aspects for modern and future warfare, and Dr. J P Dudeja deserves my congratulations for this wonderful work.

TESTIMONIAL FROM RAVI KUMAR GUPTA,

TV Commentator,
Author of 'Institutions That Shaped Modern India - DRDO'
Former Senior Defence Scientist
and Director, Public Interface, DRDO, GOVT. OF INDIA.

Modern technologies have often turned out to be consequential game-changers in wars. Writing a comprehensive volume on warfare-related technologies is not an easy task; a task that needs caliber, acumen, broad-based knowledge and experience along with exceptional writing skills of an author like Professor Jai Paul Dudeja, who has had a long exciting career as a defence scientist. I congratulate Prof Dudeja for bringing out a panoptic text on this very important and relevant aspect of warfare. The book exhaustively covers nearly every kind of military technology - conventional, contemporary, emerging and futuristic; beautifully presented by classifying them and discussing them in sixty chapters grouped in seven sections.

Additionally, the book also deals with non-technical, non-military kinds of warfare tactics as well as aspects such as economic and political warfare tactics, that govern short term and long-term strategies for winning wars. Modern warfare powered by such disruptive technologies has taken battles to homes and minds of the citizens, and the book accordingly covers related aspects like information, cyber and psychological warfare too.

Thus, from the three dimensions of space, and that of mind, the book comprehensively touches upon wide ranging topics of warfare. I'm sure the book will be handy for anyone dealing with defence and security matters, besides being an interesting read for a common citizen longing for general awareness.

TESTIMONIAL FROM DR. AT REGHUNATH,

Former Senior Defence Scientist, DRDO, Govt. Of India

About the great Indian epic Mahabharata, it is said, 'there is nothing under the sun that Veda Vyasa has not touched upon.' That is the feeling I got when I read this magnum opus on different technologies for warfare by Prof. (Dr.) Jai Paul Dudeja.

Prof. J P Dudeja has touched upon all technologies which are conceivable today. His teaching ability is in full display while presenting the complex technologies in a style which makes the subject easy for any interested student.

Every war is a cat and mouse game. However, technology has the ability to tilt the game, as the author says, 'a boy with a laptop is more dangerous than a division of soldiers.'

This book will serve as a handbook for ready reference to understand the basics of almost all technologies of this age.

I congratulate Prof. Dudeja for taking up and completing this masterpiece.

TESTIMONIAL FROM CM DHAWAN,

Former Senior Defence Scientist, Govt. of India

Modern and future warfare is increasingly shaped by a suite of critical technologies that redefine the battlefield and strategic dynamics. Among these, artificial intelligence (AI) and machine learning stand out as pivotal, driving autonomous systems, real-time decision-making, and predictive analytics. Cybersecurity and cyberwarfare capabilities are equally vital, protecting critical infrastructure and launching offensive operations in the digital realm. Advanced robotics and unmanned systems, such as drones and autonomous ground vehicles, offer unprecedented versatility in reconnaissance, logistics, and combat. Hypersonic weapons and precision-guided munitions enhance offensive capabilities with unparalleled speed and accuracy. Meanwhile, space technologies, including satellite networks and anti-satellite weapons, have made space a contested domain. Quantum computing and communication hold the potential to revolutionize encryption and data processing, while biotechnology and human augmentation are poised to impact soldier performance. Together, these technologies create a complex, multidimensional battlefield that demands adaptability and innovation from all actors.

I congratulate Dr. Dudeja who has taken lot of pain in consolidating most of the emerging and critical technologies in this book.

TESTIMONIAL FROM VINOD PANDE,

Former Senior Defence Scientist, Govt. of India

To enhance the chances of victory in the modern and future warfare, the upgradation of the equipment incorporating new and critical technologies is being developed by many countries. The future and modern warfare is likely to involve autonomous systems where less human involvement is a distinct possibility and caution must be exercised on this aspect. The artificial Intelligence (AI) is used for autonomous real time decision making and analyzing the data. Cyber security and cyber warfare capabilities are important to protect critical infrastructure and for offensive operations. Quantum technology is being used for detecting stealth aircraft and the advantage of quantum communications is that they can't be decrypted, and quantum radar can detect stealth aircraft.

In the present-day scenario the unmanned arial vehicle (UAV) / drone has the capability to carry missiles and other payload to deliver it at the desired target with very high precision. Hypersonic weapons are enhancing the offensive capability with missile travelling at speed greater than Mach5 with multiple warheads programmed to hit different targets; the Multiple Independent Targetable Reentry Vehicle (MIRV) technology is being used. Using high-power Microwave and laser sources, Directed Energy Weapon system is being developed where the electromagnetic waves itself acts as weapon travelling at a speed of light

These critical technologies are complex and multidimensional in nature and demand adaptability and innovation from all sectors.

I would like to thank and congratulate Dr Dudeja who has consolidated these new technologies of modern warfare at one place giving the shape of an excellent book.

I wish him and his book all the success.

PREFACE

Dear Readers,

I am extremely happy to see this book titled, **"Converging, Emerging, Innovative, Disruptive, and Critical Technologies for Modern and Future Warfare"**, in your hands. It is my firm belief that you have chosen to read this book with a specific aim in mind, and I assure you that you will not be disappointed with it.

Today's and Tomorrow's wars are not guaranteed to be won by yesterday's technologies. To enhance the chances of achieving victories in the modern and future wars, the nations have to embrace converging, emerging, innovative, disruptive, and critical technologies and new strategies. It is with this changed paradigm in view, that the current book is written.

This comprehensive book is divided into seven sections consisting of 60 chapters. These seven sections, with chapters in brackets, are: **I. Introduction and Overview**; **II. Converging Technologies** (Smartphones, Network-Centric Warfare, Virtual Reality and Augmented Reality, Holography, and Drone Technology); **III. Emerging Technologies** (Artificial Intelligence, 3D Printing, Nanotechnology, and Robotics); **IV. Innovative Technologies** (5G and 6G Technologies, Cloud Computing, Renewable Energy Technologies, Internet of Things, and Wearable Devices); **V. Disruptive Technologies** (Generative AI, and Edge Computing); **VI. Critical Technologies** (Advanced Composite Materials, Explosives and Energetic Materials, Advanced Magnets and Superconductors, Protective Clothing and Equipment, Application of Coatings, Smart Materials, Small satellites, Quantum Computing, Quantum Cryptography, Quantum Communication, Quantum Sensors, Free-Space Optical and Fiber-Optic Communication, RF & mm-Wave Communication, Undersea Wireless Communication, Electro-Optic Sensors, Fiber-Optic Sensors, Magnetic Field Sensors, Multispectral and Hyperspectral Imaging, Sonar Systems, Biotechnology, Chemical Warfare, Nuclear War, Environmental Warfare or Eco Terrorism, Cyber Warfare, Electronic Warfare, Space Warfare, Hybrid Warfare, Asymmetric Warfare, Irregular Warfare, Information Warfare, Stealth Technology, Integrated Missile Defence System, Radar Systems, Lidar Systems, Unmanned Aerial Vehicles, Unmanned Ground Vehicles, Unmanned Underwater Vehicles, and Directed Energy Weapons.); and **VII. Non-Technical Warfare** (Cross-Border Terrorism and Proxy War, Psychological Warfare, Economic Warfare, Diplomatic Warfare, and Political Warfare).

Besides the interested general readers across the globe, who wish to have a grasp of the converging, emerging, innovative, disruptive, and critical technologies, and new strategies for the modern and future warfare, this comprehensive book can also be used as a **'Reference Book in Warfare Technologies'** by the researchers, Goverments, and Military-technologies-related agencies.

This book is most respectfully dedicated to our revered leader **(Late) Dr. APJ Abdul Kalam**, Former President of India, and Former Chief of Defence Research & Development Organisation (DRDO), under whose umbrella and guidance, this author had the privilege of working as a scientist in DRDO.

The author would open heartedly love to receive any encouraging/critical comments as well as feedback from the dear readers at his Email ID: drjpdudeja@gmail.com

The value of this book has been enhanced by the generous 'TESTIMONIALS' from the present and former senior defence scientists of DRDO, in the Govt. of India.

Sincerely

2025 **Prof. (Dr.) Jai Paul Dudeja**

ACKNOWLEDGEMENTS

The seeds of my interest in the continual quest for new knowledge and truth were lovingly sown by my revered parents: **Late (Dr.) Shanti Sawrup Dudeja and (Late) Mrs. Jai Devi Dudeja**. I am sure that they are watching me every moment from wherever they are in the other world and continuously sending their blessings to me. I bow to them for their unconditional love, care, and guidance, and the human values nurtured by them in my life.

I have greatly benefitted in going through the books, reports, research papers and articles, referred to in the '**Bibliography'** in this Book. I gratefully acknowledge these authors for enhancing my understanding on the subject matter of this book.

Last but not the least, my greatest admiration is reserved for **Mrs. Rita Dudeja**, my wife and my best friend. She is the source of inspiration for me and a co-traveller on the path of trust and truth. I gratefully acknowledge the generous 'TESTIMONIALS' by the present and former senior defence scientists of DRDO, in the Govt. of India. It has certainly enhanced the value of this book.

2025 **Prof. (Dr.) Jai Paul Dudeja**

PROFILE OF THE AUTHOR

Born in June 1948, Prof. (Dr.) Jai Paul Dudeja holds a brilliant academic record. He did his Master's degree in Physics from Birla Institute of Technology and Science (BITS), Pilani (India), and Ph.D. degree in Physics from the Indian Institute of Technology (IIT), Delhi. He has been in regular employment as a Scientist, Professor, Dean, Director, Principal and a Senior Administrator in various educational institutions, universities, laboratories, public and private organizations. He superannuated as a Senior Scientist and Additional Director in May 2008 from the Defence Research and Development Organisation (DRDO), Government of India. After DRDO, he served for over 11 years, till his last posting as a Director at Amity University Gurgaon, from where he retired in Nov 2019.

Till date, Dr. Dudeja has published/presented about 90 research papers in various national and international journals/conferences. Besides this, he has authored the following **twenty-seven books** on science, technology, spirituality, mysticism, and consciousness, etc.:

1. Gayatri Mantra: A GPS to Enlightenment
2. Maha Mrityunjaya Mantra: An Invincible Armour for Conquering Death
3. Ajapa-Japa Sohum-Humsa Mantra: An Eternal Mantra for Inner Consciousness
4. The Third Eye: A Spiritual Laser for Stimulating Inner Awakening
5. Quantum Physics of Consciousness and Non-Duality in Eastern Philosophy
6. Quantum Science of Love, Healing, Happiness, and Bliss in Ancient Wisdom
7. Chakras Healing and Kundalini Awakening by Yogic Techniques,
8. Meditation Practices across the Globe and their Beneficial Effects.
9. Comparative Analysis of Hindu, Buddhist, and Jain Philosophies
10. Mantras for Happiness
11. Om Namah Shivaya: A Powerful Mantra for Mastering Five Elements
12. Trataka: A Concentrated Gazing Technique for Mystical Powers
13. Walking Meditation: Techniques and Benefits

14. Quantum Science of Ganesha Consciousness
15. Tantra Science
16. Quantum Brain, Mind, and Thinking
17. Profound Meditation Techniques in Tibetan Buddhism
18. REIKI: A Holistic Energy Healing Technique
19. Shaktipāt: Instant Transmission of Spiritual Energy from a Siddha Guru to the Disciple
20. Vāstu Shāstra: Ancient Indian Science of Architecture
21. Vedantic Thoughts on Māyā, Mithyā, and the Brahman
22. Siddhis (Supernatural Powers): A Guide for Understanding and Attaining These
23. Near-Death Experience: Scientific Interpretation (Inspired by True Story of a Friend)
24. Universe before the Big Bang: A Deeper Insight into Cosmology
25. Spiritual Intelligence: Significance, Applications, Measurement, and Development Techniques
26. Mind-Reading and Artificial Intelligence: Past, Present and Future (Science, Technology, Opportunity, Risks and Regulations)
27. Climate Change: A Global Challenge (Causes, Adverse Effects, Monitoring, Adaptation, And Mitigation).

Dr. Dudeja has delivered many invited lectures in the international conferences in India and abroad. He has been recognized as the "World's Who's Who in Science & Engineering".

SECTION-I

INTRODUCTION AND OVERVIEW OF TECHNOLOGIES FOR MODERN AND FUTURE WARFARE

CHAPTER 1

INTRODUCTION AND OVERVIEW OF TECHNOLOGIES FOR MODERN AND FUTURE WARFARE

Military theorists have researched the evolution of technology throughout history and analyzed the importance of new weapons developments that have dramatically altered the face of war. From the introduction of the chariot during Antiquity, which offered armies the power of mounted combat, to the development of the nuclear weapons during the Industrial Age, which provided nations the power to destroy whole cities with a single bomb, innovations have played a critical role in military success. In today's Information Age, our knowledge-based society operates on the promise of improved efficiency and personalization, and our news is constantly filled with announcements touting the latest technological advancements that will yield radical change for future armies. From power suits that turn soldiers into supermen, to swarm technology that will make autonomous drones a reality, a countless number of military innovations are being developed that seek to take advantage of the digital revolution. However, the single innovation that will truly reshape the modern battlefield in our lifetime is one that nearly every teenager in the U.S. already has in his or her possession—the portable smartphone.

1.1 WHAT IS CONVERGING TECHNOLOGY?

Convergence is a deep integration of knowledge, tools, and all relevant activities of human activity for a common goal, to allow society to answer new questions to change the respective physical or social ecosystem. Such changes in the respective ecosystem open new trends, pathways, and opportunities in the following divergent phases of the process.

Converging technology is a term that describes blending together, or integrating two or more wide range of single-function technologies, in a single device.

Converging technology (also known as 'Technological convergence') is the tendency for technologies that were originally unrelated to become more closely integrated and even unified as they develop and advance.

1.1.1 Some Examples of Converging Technologies

(i) Smartphones

Smartphones might be the best possible example of such a convergence. Prior to the widespread adoption of smartphones, consumers generally relied on a collection of single-purpose devices. These devices included (landline) telephones, wrist watches, digital cameras and global positioning system (GPS) navigators. Today, even low-end smartphones combine the functionality of all these separate devices, easily replacing them in a single device.

(ii) Smart TVs

A smart TV is an example of technological convergence because it combines multiple technologies, viz., a traditional television set, a computer that provides access to apps and other digital content, and digital media player that allows users to stream music and videos, into one device.

Smart TV is used to describe the current trend of integration of the Internet and Web 2.0 features into modern television sets and set-top boxes, as well as the technological convergence between computers and these television sets or set-top boxes. These new devices most often also have a much higher focus on online interactive media, Internet TV, over-the-top content, as well as on-demand streaming media, and less focus on traditional broadcast media like previous generations of television sets and set-top boxes always have had.

(iii) The Internet

The role of the Internet has changed from its original use as a communication tool to easier and faster access to information and services, mainly through a broadband connection. The television, radio and newspapers were the world's media for accessing news and entertainment; now, all three media have converged into one, and people all over the world can read and hear news and other information on the Internet.

(iv) Media

Media technological convergence is the tendency that, as technology changes, different technological systems sometimes evolve toward performing similar tasks. It is the interlinking of computing and other information technologies, media content, media companies and communication networks that have arisen as the result of the evolution and popularization of the Internet as well as the activities, products and services that have emerged in the digital media space.

(v) Digital Cameras

Digital cameras can take photos, shoot videos, access the internet, and store media.

(vi) WiFi-enabled Fridges

WiFi-enabled fridges allow users to order groceries directly from the fridge door.

(vii) LifeNome

LifeNome combines AI-powered tools, smart device health monitoring, and telehealth technology to support pregnancy health.

(viii) Drones

Drones continue to grow more agile and user-friendly every single year. They are what happens when engineers make rapid strides in aerial propulsion and dexterity and combine that with advancements in user experience applications and communication. Put it all together and you have remotely piloted aircraft that range from models easily controlled by hobbyists to state-of-the-art systems used by private businesses and the Armed Forces for reconnaissance, search and rescue missions, and more.

(ix) Virtual Reality and Augmented Reality

Virtual reality has existed in some form or another for a few decades, yet its capabilities never quite lived up to its promise until very recently.

The modern virtual reality headset, as exhibited by systems like the Oculus Quest, the Valve Index, PlayStation VR, and more, are made possible due to advancements in computer processing power, chip size, screen technology, graphics engines, and even lightweight construction materials.

It's the combination of all these factors together that have created a burgeoning virtual reality industry. On its own, each is impressive, but it's only upon convergence that something truly special and unique is created: the ability to seamlessly navigate virtual worlds like never before.

Augmented reality is yet another iteration of this, combining high-tech camera equipment with the incredible processing power of modern VR systems and cloud-based data retrieval to create a heads-up display detailing information about the world around you.

(x) Wireless Headphones

Wireless headphones are an interesting case wherein the tech convergence was actually a direct result of removing a benefit so commonplace that it was accepted as an absolute necessity.

Wireless headphones result from the convergence of headphone audio technology with Bluetooth technology. But the convergence of these two systems also coincides with the removal of headphone cords. Because the audio requires no external wiring system to get from the music player to the reception device, no cords are necessary.

1.2 WHAT IS EMERGING TECHNOLOGY?

Emerging technology is a term generally used to describe a new or developing technology, which may also include new applications of old technologies. Emerging technologies are often perceived as capable of changing the status quo.

1.2.1 Some Examples of Emerging Technologies

(i) Artificial Intelligence

Artificial intelligence (AI) is the sub intelligence exhibited by machines or software, and the branch of computer science that develops machines and software with animal-like intelligence. Major AI researchers and textbooks define the field as "the study and design of intelligent agents," where an intelligent agent is a system that perceives its environment and takes actions that maximize its chances of success. John McCarthy, who coined the term in 1956, defines it as "the study of making intelligent machines".

(ii) 3-D printing

3-D printing, also known as additive manufacturing, has been posited by Jeremy Rifkin and others as part of the third industrial revolution.

Combined with Internet technology, 3-D printing would allow for digital blueprints of virtually any material product to be sent instantly to another person to be reproduced on the spot, making purchasing a product online almost instantaneous.

Although this technology is still too crude to produce most products, it is rapidly developing and created a controversy in 2013 around the issue of 3D printed firearms.

(iii) Gene therapy

Gene therapy was first successfully demonstrated in late 1990/early 1991 for Adenosine deaminase (ADA) deficiency, a metabolic disorder that affects the immune system:, though the treatment was somatic – that is, did not affect the patient's germ line and thus was not heritable. This led the way to treatments for other genetic diseases and increased interest in germ line gene therapy – therapy affecting the gametes and descendants of patients.

Between September 1990 and January 2014, there were around 2,000 gene therapy trials conducted or approved.

(iv) Cancer vaccines

A cancer vaccine is a vaccine that treats existing cancer or prevents the development of cancer in certain high-risk individuals. Vaccines that treat existing cancer are known as therapeutic cancer vaccines. There are currently no vaccines able to prevent cancer in general.

On April 14, 2009, The Dendreon Corporation announced that. in their Phase III clinical trial of Provenge, a cancer vaccine designed to treat prostate cancer, had demonstrated an increased chance of survival. It received U.S. Food and Drug Administration (FDA) approval for use in the treatment of advanced prostate cancer patients on April 29, 2010. The approval of Provenge has stimulated interest in this type of therapy.

(v) Cultured meat

Cultured meat, also called in vitro meat, clean meat, cruelty-free meat, shmeat, and test-tube meat, is an animal-flesh product that has never been part of a living animal with exception of the fetal calf serum taken from a slaughtered cow. In the 21st century, several research projects have worked on in-vitro meat in the laboratory. The first in-vitro beefburger, created by a Dutch team, was eaten at a demonstration for the press in London in August 2013. There remain difficulties to be overcome before in vitro meat becomes commercially available. Cultured meat is prohibitively expensive, but it is expected that the cost could be reduced to compete with that of conventionally obtained meat as technology improves. In-vitro meat is also an ethical issue. Some argue that it is less objectionable than traditionally obtained meat because it does not involve killing and reduces the risk of animal cruelty, while others disagree with eating meat that has not developed naturally.

(vi) Nanotechnology

Nanotechnology (sometimes shortened to nanotech) is the manipulation of matter on an atomic, molecular, and supramolecular scale. The earliest widespread description of nanotechnology referred to the particular technological goal of precisely manipulating atoms and molecules for fabrication of macroscale products, also now referred to as molecular nanotechnology. A more generalized description of nanotechnology was subsequently established by the National Nanotechnology Initiative, which defines nanotechnology as the manipulation of matter with at least one dimension sized from 1 to 100 nanometers. This definition reflects the fact that quantum mechanical effects are important at this scale, and so the definition shifted from a particular technological goal to a research category inclusive of all types of research and technologies that deal with the special properties of matter that occur below the given size threshold.

(vii) Robotics

Robotics is the branch of technology that deals with the design, construction, operation, and application of robots, as well as computer systems for their control, sensory feedback, and information processing. These technologies deal with automated machines that can take the place of humans in dangerous environments or manufacturing processes, or resemble humans in appearance, behavior, and/or cognition. A good example of a robot that resembles humans is Sophia, a social humanoid robot developed by Hong Kong-based company Hanson Robotics which was activated on April 19, 2015. Many of today's robots are inspired by nature contributing to the field of bio-inspired robotics.

(viii) Stem-cell therapy

Stem cell therapy is an intervention strategy that introduces new adult stem cells into damaged tissue in order to treat disease or injury. Many medical researchers believe that stem cell treatments have the potential to change the face of human disease and alleviate suffering. The ability of stem cells to self-renew and give rise to subsequent generations with variable degrees of differentiation capacities offers significant potential for generation of tissues that can potentially replace diseased and damaged areas in the body, with minimal risk of rejection and side effects.

Chimeric Antigen Receptor (CAR)-modified T cells have raised among other immunotherapies for cancer treatment, being implemented against B-cell malignancies. Despite the promising outcomes of this innovative technology, CAR-T cells are not exempt from limitations that are yet to be overcome in order to provide reliable and more efficient treatments against other types of cancer.

(ix) Distributed Ledger technology (Blockchain Technology)

Distributed ledger or blockchain technology provides a transparent and immutable list of transactions. A wide range of uses has been proposed for where an open, decentralized database is required, ranging from supply chains to cryptocurrencies.

Smart contracts are self-executing transactions which occur when pre-defined conditions are met. The aim is to provide security that is superior to traditional contract law, and to reduce transaction costs and delays. The original idea was conceived by Nick Szabo in 1994, but remained unrealized until the development of blockchains.

1.3 WHAT IS INNOVATIVE TECHNOLOGY?

Innovative Technology is defined as the creation and application of new or improved technologies, tools, systems, and processes that bring about significant advancements or breakthroughs in various fields. It involves harnessing knowledge, expertise, and resources to develop innovative solutions that solve problems, improve efficiency, drive progress, and deliver value.

1.3.1 Some Examples of Innovative Technologies

(i) World-Wide Web

The development of the Internet and the World Wide Web revolutionized communication and information sharing. It enabled global connectivity, transformed how people access and share information, and laid the foundation for e-commerce, social media, and online services.

(ii) 5G and Advanced Connectivity

With speeds up to 100 times faster than 4G and ultra-low latency, 5G is set to enable a host of new technologies and applications. The technology is crucial for the widespread adoption of IoT devices, supporting millions of connected sensors and smart devices in dense urban areas. In healthcare, 5G could enable remote surgeries with haptic feedback, allowing surgeons to operate from thousands of miles away. For autonomous vehicles, 5G networks' low latency is essential for real-time communication between cars and infrastructure. In entertainment, 5G will facilitate high-quality streaming of virtual and augmented reality content. As 5G infrastructure continues to expand, it's paving the way for smart cities, industrial IoT and innovations we've yet to imagine.

(iii) Cloud Computing

Cloud computing, is the computing innovation that transformed the way computing resources are provisioned, accessed, and utilized. It provided scalable and flexible infrastructure, storage, and services on-demand, enabling businesses to leverage powerful computing capabilities without heavy upfront investments.

(iv) Renewable Energy Technologies

Technological innovations in renewable energy, such as solar power, wind power, and energy storage systems, have transformed the energy landscape. These innovations have increased the efficiency, affordability, and accessibility of renewable energy sources, leading to reduced dependence on fossil fuels and mitigating climate change impacts.

(v) Biotechnology

Advances in biotechnology and genetic engineering have revolutionized healthcare, agriculture, and environmental preservation. Innovations include gene editing technologies like CRISPR-Cas9, which enables precise modification of genes, and the development of biopharmaceuticals for personalized medicine.

CRISPR" (pronounced "crisper") stands for 'Clustered Regularly Interspaced Short Palindromic Repeats', which are the hallmark of a bacterial defense system that forms the basis for CRISPR-Cas9 genome editing technology. CRISPR-Cas9 was adapted from a naturally occurring genome editing system that bacteria use as an immune defense. When infected with viruses, bacteria capture small pieces of the viruses' DNA and insert them into their own DNA in a particular pattern to create segments known as CRISPR arrays.

(vi) Internet of Things (IoT)

The Internet of Things, the computing innovation, has resulted in the proliferation of interconnected devices and smart systems. Examples include:

(a) **Smart Homes:** IoT-enabled devices like thermostats, security systems, and appliances enable remote control, automation, and energy efficiency in homes.

(b) **Industrial IoT:** IoT technologies are used to monitor and optimize industrial processes, enabling predictive maintenance, supply chain optimization, and real-time monitoring of equipment.

(c) **Wearable Devices:** Innovations in wearable technology, such as fitness trackers and smartwatches, have enabled individuals to monitor their health, track physical activity, and receive personalized insights.

1.4 WHAT IS DISRUPTIVE TECHNOLOGY?

Disruptive technology is an innovation that significantly alters the established industries and markets, creating new sectors and business models. It is an innovation that radically changes the way the market is structured and how products and services are consumed.

Harvard Business School professor and business consultant, Clayton Christensen, coined the term "disruptive innovation" in the magazine Harvard Business Review back in 1995.

1.4.1 Some Examples of Disruptive Technologies

(i) Generative AI

Generative AI is a type of artificial intelligence that uses generative models to create content. Using massive amounts of data, these machines learn how to generate new content — spanning text, images, audio and video — by way of complex algorithms and neural networks. By identifying patterns and structures from an existing data set, these systems are able to answer prompts inputted by a user, predicting one word or pixel at a time.

Why is Generative AI disruptive?

Generative AI bots create content, music, art, trade stocks and perform administrative tasks. Industries that rely on manual, time-consuming creative work — like graphic design or writing — may be under fire; however, computer programmers, research analysts, paralegals and financial traders aren't exactly safe either. Generative AI offers unprecedented levels of personalized content delivery and does so at scale with nearly instant turnaround times. Despite ethical concerns around authorship and originality, generative AI content has been embraced by global press organizations and used to train facial recognition technology.

Autonomous fighting platforms, cyber warfare and Unmanned Aerial Vehicles have already begun to impact warfighting. Directed Energy Weapons, Big Data Analysis, and Internet of Things will have a major impact on warfighting in the future. We may not see large armies, instead, we will see fewer humans on the battlefield who would be more technologically advanced "augmented humans".

(ii) Edge Computing

Edge computing is a distributed computing framework that brings data processing and analysis closer to where it's generated, rather than relying solely on centralized cloud servers.

Why is Edge computing disruptive?

By processing data right at the "edge" (closer to where it's needed), edge computing reduces latency, which is crucial for applications that operate in real-time, like autonomous vehicles and augmented reality. Edge computing has a wide reach, as it is applicable to any sector that handles data or data analysis.

For example, edge computing allows for real-time patient monitoring in healthcare, making data more immediate and accessible. In manufacturing, it improves automation and quality control by minimizing delays. It's also playing a significant role in the development of smart cities, where it manages traffic, security, utilities and the distribution of energy in smart grids. In tandem with AI, edge computing enables the real-time processing required for tasks like facial recognition, language processing and object detection.

(iii) 'As-a-service' Models

'As-a-service' models refer to a paradigm shift in how products and services are delivered, where instead of traditional ownership, you access them on-demand, often through subscription-based or pay-as-you-go schemes. Cloud computing, connectivity and application programming interfaces, which allow different software systems to interact — make 'as-a-service' platforms possible.

Why it's disruptive?

No longer bogged down by closed systems and hardware, this disruptor creates ease and flexibility in the ways brands do business and how a customer experiences a product or service.

For example, in the software industry, software-as-a-service (SaaS) alters how businesses use and pay for software, reducing upfront costs and increasing accessibility. Other players in the tech sector now offer infrastructure-as-a-service (IaaS), where companies can outsource their IT infrastructure, alleviating data storage and cloud computing pain points. Platform-as-a-Service (PaaS) is another cloud-based service that lets developers focus on coding, while the platform manages the rest. Other iterations include data-as-a-service (DaaS), which leverages data sets for decision making, and knowledge-as-a-service (KaaS), which delivers data, information and experts on demand.

1.5 WHAT IS A CRITICAL TECHNOLOGY?

Critical technology is a technology that is considered important to a country's future economic growth, national security, and technological advancement. Critical technologies are often the result of cutting-edge research and innovation, and are strategically important.

Critical technologies can be defined as current and emerging technologies with the capacity to significantly enhance, or pose risk to, a country's national interests, including a nation's economic prosperity, social cohesion, and **national security**.

1.5.1 Some Examples of Critical Technologies

(i) Advanced composite materials

New materials created by combining two or more materials with different properties, without dissolving or blending them into each other. Advanced composite materials have strength, stiffness, or toughness greater than the base materials alone. Examples include carbon-fiber-reinforced plastics and laminated materials. Applications include vehicle protection, signature reducing materials, construction materials and wind turbine components.

(ii) Advanced Explosives and Energetic Materials

Materials with large amounts of stored or potential energy that can produce an explosion. Applications for advanced explosives and energetic materials include mining, civil engineering, manufacturing and defence.

(iii) Advanced Magnets and Superconductors

Advanced magnets are strong permanent magnets that require no or few critical minerals. Applications for advanced magnets include scientific research, smartphones, data storage, health care, power generation and electric motors.

Superconductors are materials that have no electrical resistance, ideally at room temperature and pressure. Applications for superconductors include creating strong magnetic fields for medical imaging, transferring electricity without loss, and hardware for quantum computers.

(iv) Advanced Protection of Clothing and Equipment

Clothing and equipment to protect defence, law enforcement and public safety personnel and defence platforms from physical injury and/or chemical or biological hazards. Examples include helmets, fire-retardant fabrics, respirators, and body armor.

(v) Coatings

Substances applied to the surface of an object to add a useful property. Examples include anti-biofouling coatings that prevent plants or animals growing on ships or buildings, super-hydrophobic coatings that repel water from solar panels or reduce drag on the hulls of ships, electromagnetic absorbing coatings that make airplanes and ships less visible to radar systems, thermal coatings that reduce heat loss and increase energy efficiency, and anti-corrosion coatings that prevent rust.

(vi) Critical minerals extraction and processing

Systems and processes to extract and process critical minerals safely, efficiently and sustainably. Applications for critical minerals extraction and processing include mining, concentrating minerals, and manufacturing battery-grade chemicals.

(vii) Nanoscale materials and manufacturing

Materials with essential features measuring less than 100 nm and the technologies for their synthesis. Applications for nanoscale materials include, paint, pharmaceuticals, wastewater treatment, data storage, communications, semiconductors, capturing carbon dioxide, and nanoscale tracking markers for critical materials.

(viii) Smart materials

Materials that have properties that change in response to external action. Examples include shape memory alloys that change shape when heated and self-healing materials that automatically repair themselves when damaged. Applications for smart materials include clothing, body armor, building materials and consumer electronics.

(ix) Wide and ultrawide bandgap semiconductors

Wide and ultrawide bandgap semiconductors are semiconducting materials that can operate at high temperatures and high power compared to conventional silicon devices. In addition, their higher switching speeds combined with their high power makes them ideal for making a complete radio system. Applications for wide and ultrawide bandgap semiconductors include improved electrical grid infrastructure (high power inverters and high electron mobility transistors with low switching losses), LEDs and lasers (operating in the blue region and frequencies above).

(x) Advanced Optical Communication

Devices and systems that use light to transfer information over optical fiber or free space (i.e. air or the vacuum of space) and use laser technologies, adaptive optics and optical routing to transfer information faster, more reliably and more efficiently. Applications for advanced optical communication include high-speed earth satellite communication, short-range visible light communication (i.e. 'Li-Fi'), narrow-beam laser communication and multi-gigabit broadband and corporate networks.

(xi) Advanced Radiofrequency Communication

Devices and systems that use radio waves to transfer information over free space (i.e. air or the vacuum of space) and use novel modulation techniques, advanced antenna designs and beamforming technologies to transfer information faster, more reliably and more efficiently. Applications for advanced radiofrequency communication include communication satellites, cellular networks (e.g. 5G and 6G), wireless local area networks (e.g. Wi-Fi), short-range wireless communication (e.g. Bluetooth), sensor networks, connected vehicles, implantable medical devices and mobile voice and data services for public safety and defence.

(xii) Advanced undersea wireless communication

Devices, sensors, and systems that enable untethered undersea data communication over longer distances and use novel techniques to achieve higher data transfer rates with acceptable error rates. Radiofrequency communications are absorbed by seawater making long-range communication a key challenge for both submarines and autonomous underwater vehicles. Extremely-low frequency (ELF) communication enables one-way communication to submarines at operational depth, but constructing transmitting antennas is a significant undertaking and only a handful of countries have them in operation.

(xiii) Protective cyber security technologies

Systems, algorithms and hardware that are designed to enable a cyber security benefit. Applications for cyber security technologies include operational technology security, trust and authentication infrastructures, protection of aggregated data sets, protection of AI systems and supply chain security.

(xiv) Adversarial AI

Methods to protect AI systems against malicious attacks that try to trick or exploit its vulnerabilities. Examples include poisoning the dataset the system is trained on to reduce the AI's accuracy or reverse engineering the system to extract the underlying model or recover the input training data. As AI systems become more widely used in a variety of sectors, they will be also become more vulnerable to actors who try to exploit their vulnerabilities. If systems are not resilient to reverse engineering, they cannot be trained on sensitive data. With less data available for training the systems will underperform.

(xv) Natural language processing (including speech and text recognition and analysis)

Systems that enable computers to recognize, understand and use written and/or spoken language in the same ways that people use language to communicate with each other. Natural language processing is a type of artificial intelligence. Applications for natural language processing include predictive text, language translation, virtual assistants and chat bots, summarizing long documents, sentiment analysis, and making technologies more accessible and inclusive.

(xvi) Synthetic biology

Designing and constructing biological systems and devices that have useful functions not found in nature. Applications for synthetic biology include creating microorganisms that can clean-up environmental pollutants and recycle plastics, manufacturing animal-free meat and dairy products, and biological computers.

(xvii) Nuclear medicine and radiation therapy

Nuclear medicine uses radioactive substances to diagnose or treat diseases. Applications for nuclear medicine include imaging internal organs and tissues, viewing biological processes and using radiopharmaceuticals to treat cancers and other diseases. Radiotherapy uses ionizing radiation to treat diseases by damaging the DNA in targeted cells, killing those cells. Applications for radiotherapy include treating some types of cancer and treating other diseases caused by overactive cells. While imaging and diagnosis techniques constitute a significant part of nuclear medicine, our publication dataset is focused on the application of these techniques to the diagnosis and treatment of diseases.

(xviii) Directed energy technologies

Systems and devices that transfer energy between two points in free space. Applications for directed energy technologies include powering consumer electronics, recharging electric vehicles, powering aerial drones, ground-space energy transfer, wireless sensor networks and internet of things devices, and advanced weapons.

(xix) Electric batteries

Devices that produce electricity from stored electrochemical energy and tolerate multiple charge and discharge cycles. Electric batteries utilize various materials and chemistries (e.g. lithium-ion (Li-ion), nickel metal hydride battery (Ni-MH)) and form factors (e.g. flow batteries for stationary grid storage, polymer electrolytes for vehicles and personal devices). Applications for electric batteries include electrified road and air transport, smartphones and personal electronic devices, medical devices and energy grid power storage.

(xx) Nuclear energy

Electricity generation using the energy released when the core of an atom (the atomic nucleus) splits into two or more lighter atomic nuclei. Applications include energy production for self-contained and/or remote uses, such as space travel, submarines, scientific research and medical isotope production.

(xxi) Supercapacitors

Electrochemical devices that can store large amounts of energy in small volumes. Supercapacitors store less energy and for shorter durations than rechargeable batteries (hours or days, rather than months or years), but can accept and deliver charge much faster than rechargeable batteries and tolerate many more charge and discharge cycles than rechargeable batteries before performance degrades. Applications for supercapacitors include regenerative braking, smartphones and personal electronic devices, grid energy storage and defence.

(xxii) Post-quantum cryptography

Mathematical techniques for ensuring that information stays private, or is authentic, that resist attacks by both quantum and non-quantum (i.e. classical) computers. The leading application for post-quantum cryptography is securing online communications against attacks using quantum computers.

(xxiii) Quantum communication (including quantum key distribution)

Devices and systems that communicate quantum information at a distance, including cryptographic keys. Applications for quantum communications include transferring information between quantum computers and sharing cryptographic keys (which are like secret passwords) between distant people in a way that means it is impossible for anyone else to copy.

(xxiv) Quantum computing

Computer systems and algorithms that depend directly on quantum mechanical properties and effects to perform computations. Quantum computers can solve particular types of problems much faster than existing 'classical' computers, including problems that are not practical to solve using even today's most powerful classical computers. Applications for quantum computing include accurately simulating chemical and biological processes, breaking currently widely used encryption methods, machine learning and efficiently optimizing very complex systems.

(xxv) Quantum sensors

Devices that depend directly on quantum mechanical properties and effects for high precision and high sensitivity measurements. Applications for quantum sensors include enhanced imaging, passive navigation, remote sensing, quantum radar and threat detection for defence.

(xxvi) Magnetic field sensors

Devices that can detect and measure the strength and/or direction of magnetic fields. Applications for magnetic field sensors include passive navigation, imaging for health, metallurgy, scientific research and threat detection for defence.

(xxvii) Multispectral and Hyperspectral Imaging sensors

Multispectral imaging sensors capture data across a few bands across the electromagnetic spectrum. Hyperspectral imaging sensors further this approach by capturing hundreds of bands continuously across the electromagnetic spectrum and map chemical content because of their specific spectral signatures. Applications for multispectral and hyperspectral imaging sensors include healthcare, defence, agriculture, minerals, forestry and machine vision for autonomous vehicles and robots.

(xxviii) Photonic sensors

Devices that use light to detect changes in the environment or in materials. Applications for photonic sensors are broad, ranging from mainstream photography, through to sensors for environments where electrical or chemical based sensors are impractical or unreliable, such as laser-based gas sensors to detect explosive materials or flexible photonic sensors embedded inside the human body to monitor bodily processes.

(xxix) Radar

Systems that listen for radio waves and microwaves reflected off objects and surfaces—such as people, buildings, aircraft and mountains—to 'see' how far away and how fast those objects are moving. Active radar systems send their own radio signals to reflect off (for example, ground penetrating radar) whereas passive radar systems listen for radio signals sent by targets or reflections of signals already present in the environment (for example, radio astronomy signals). Applications for radar include weather forecasting, situational awareness, connected and autonomous vehicles, virtual and augmented reality systems, and defence.

(xxx) Sonar and acoustic sensors

Systems that listen for sound waves created by, or reflected off, objects—such as boats, submarines, fish and underwater mountains—to identify those objects and/or 'see' how far away and how fast those objects are moving. Applications for sonar and acoustic sensors include monitoring marine wildlife, and threat detection, identification and targeting for defence.

(xxxi) Advanced aircraft engines (including hyper sonics)

Engine technologies that enable greater speed, range, and fuel-efficiency for aerial vehicles. Examples include hypersonic technologies such as ramjet and scramjet engines that allow aircraft and weapons to travel beyond Mach 5 (i.e. flying more than five times the speed of sound).

(xxxii) Advanced robotics

Robots capable of performing complex manual tasks usually performed by humans, including by teaming with humans and/or self-assembling to adapt to new or changed environments. Applications for advanced robotics include industry and manufacturing, defence and public safety, and healthcare and household tasks.

(xxxiii) Autonomous systems operation technologies

Self-governing machines that can independently perform tasks under limited direction or guidance by a human operator. Applications for autonomous systems operation technology include passenger and freight transport, uncrewed underwater vehicles, industrial robots, public safety and defence.

(xxxiv) Hypersonic detection and tracking

Systems that can quickly detect and accurately track the movement of object's moving at hypersonic speeds. As these objects move very quickly, and can maneuver mid-flight, predicting their trajectory is challenging. Due to the difficult in tracking, let alone intercepting, hypersonic-velocity missiles, these weapons may be very effective against high-value targets such as aircraft carriers. This technology category covers land, ship, air, and space-based sensors and algorithms for rapid trajectory analysis.

(xxxv) Drones, swarming and collaborative robots

Uncrewed air, ground, surface and underwater vehicles and robots that can achieve goals with limited or no human direction, or collaborate to achieve common goals in a self-organizing swarm. Applications for drones, swarming and collaborative robots include public safety, environmental monitoring, agriculture, logistics, and defence.

(xxxvi) Small satellites

Satellites with relatively low mass and size, usually mass under 500 kg and no larger than a domestic refrigerator or washing machine. Applications for small satellites include lower-cost earth observation constellations and wide area communications networks.

(xxxvii) Systems to transport payloads

Such as satellites or spacecraft—from the surface of the Earth to space safely, reliably and cost-effectively. Applications for space launch systems include launching defence, commercial, and scientific and research payloads into earth orbit.

(xxxviii) Stealth Technology

Stealth has become the single biggest game changer in warfare by changing the way aircraft, naval ships, and most other platforms are designed and used on contemporary battlefields. The aspects of stealth are very important because, through them, a mission remains a secret and there is a high chance of success.

(xxxix) Air-independent propulsion

Underwater propulsion systems that can operate without air intake. Underwater vehicles that use diesel engines for propulsion require periodic surfacing for air intake, which makes them less useful in stealth operations. Compact air-independent systems enable greater operational range and duration while maintaining stealth. Candidate systems include hydrogen fuel cells and Stirling engines.

(xL) Autonomous underwater vehicles

Underwater vehicles capable of conducting long-range missions without a remote operator. Route and mission are either pre-planned or calculated with minimal operator input. Applications include intelligence, surveillance, and reconnaissance missions in addition to anti-submarine warfare.

(xLi) Electronic warfare

The use of the electromagnetic spectrum to support military operations can be used in both an offensive and defensive capacity, to either deny, degrade, disrupt, deceive or destroy an adversary's electronic systems or protect one's own. Simultaneously applying tools, techniques, and technology to enable your forces to operate in contested and degraded environments (electronic protection). Also includes the collection and analysis of electromagnetic signals for intelligence, surveillance, reconnaissance purposes and to support targeting and operational planning (electronic support). It includes collection of foreign instrumentation signals intelligence (FISINT). Examples include GPS jamming, GPS spoofing, communication jamming, communication masking, and frequency-hopping communication. Electronic warfare techniques are also applied in hybrid warfare.

1.6 DIFFERENCE BETWEEN CONVERGING, EMERGING, INNOVATIVE, DISRUPTIVE, AND CRITICAL TECHNOLOGIES

1.6.1 Difference between Emerging and Converging Technologies

The main difference between emerging and converging technologies is that emerging technologies are new and rapidly growing, while converging technologies are the result of different technologies combining to create new efficiencies.

1.6.2 Difference between Emerging and Innovative Technologies

Emerging technologies are innovative technologies that are new or have recently been introduced into the market. Innovative technologies can refer to a new product, process, method of production, market, source of supply, or form of organization.

Emerging technologies are characterized by radical novelty, fast growth, coherence, prominent impact, and uncertainty and ambiguity. Innovative technologies can be characterized by relative advantage, compatibility, complexity, trialability, and observability.

Emerging technologies are typically nascent concepts gaining traction, whereas innovative technology actively introduces novel solutions to market challenges.

1.6.3 Difference between Innovative and Disruptive Technologies

Whereas innovative technologies actively introduce novel solutions to existing challenges, disruptive technologies creates new items, markets, and qualities to overthrow existing ones.

Disruptive technology is an innovation that significantly alters the established industries and markets, creating new sectors and business models.

1.6.4 Difference between Innovative and Critical Technologies

The main difference between innovative and critical technologies is that innovative technologies are a result of creativity and improvement, while critical technologies are the cutting-edge or state-of-the-art technologies that can impact national interests.

Innovative technologies are advancements that can improve business value and drive societal change through novel solutions. Critical technologies can be current or emerging and can pose a risk or enhance national interests. Critical technologies can be important for military or economic reasons.

1.7 MODERN WARFARE

Modern warfare is warfare that diverges notably from previous military concepts, methods, and technology, emphasizing how combatants must modernize to preserve their battle worthiness. As such, it is an evolving subject, seen differently in different times and places. In its narrowest sense, it is merely a synonym for contemporary warfare.

Modern warfare is tied to the introduction of total war, industrial warfare, mechanized warfare, nuclear warfare, counter-insurgency, or (more recently) the rise of asymmetric warfare. It is also a concept that can be interpreted in different ways, such as in the context of computer science, where it is also known as Fourth-Generation Warfare (4GW). In 4GW, the focus is on nonlinear tactics that target social, cultural, and economic objectives.

Typically, a modern warfare is characterized by several factors, including:

- **(i) Advanced technology:** The use of drones, cyber capabilities, and precision weaponry.
- **(ii) Globalization:** The increasing importance of non-state actors and irregular forces in conflicts.
- **(iii) Asymmetric strategies**: A focus on strategies that are not symmetrical.
- **(iv) Information:** The use of information and news media in conflict.
- **(v) Intelligence:** The use of intelligence in conflict.

1.7.1 Asymmetric Warfare

Asymmetric warfare is a type of warfare where the opposing forces have different military capabilities or methods of engagement. The weaker force must use their special advantages or exploit the enemy's weaknesses to win.

Asymmetric warfare can involve:

(i) Guerrilla tactics

A common form of asymmetric warfare where the weaker force uses tactics to weaken the enemy's will to fight.

(ii) Non-state entities

The weaker force may be a non-state entity like a terrorist group, criminal, or drug trafficker.

(iii) Surrogate warfare

When a non-state entity acts on behalf of another state.

(iv) Striking vulnerable points

The weaker force may strike at key points in the enemy's logistics network to render their military forces helpless.

Some examples of asymmetric warfare include: The French Indochina War, The Arab Revolt, The Mexican Revolution, and Global terrorism.

1.7.2 Asymmetric Warfare between India and Pakistan

'Bleed India with a Thousand Cuts' is a military doctrine followed by the Pakistani military against India. It consists of waging covert war against India using insurgents at multiple locations.

After the conclusion of the Soviet–Afghan War, the fighters of the Sunni Mujahideen and other Islamic militants had successfully removed the Soviet forces from Afghanistan. The military and civil government of Pakistan sought to utilize these militants in the Kashmir conflict against the Indian Armed Forces in accordance with the "thousand cuts" doctrine so as to "bleed India", using Pakistan's nuclear arsenal as a shield. In the 1980s cross-border terrorism started in the Kashmir region as armed and well-trained groups of terrorists were infiltrated into India through the border. Pakistan officially maintained that the terrorism in Kashmir was "freedom struggle" of Kashmiris and Pakistan only provided moral support to them. But this turned out to be inaccurate as Director-General of Inter-Services Intelligence (ISI) stated in the National Assembly of Pakistan that the ISI was sponsoring this support in Kashmir.

Pakistan has used the jihadist militias to conduct an asymmetric warfare with India. The militant groups have been used not just as proxies, but predominantly as "weapons" against India for Pakistan's "Bleed India" campaign.

In spite of grave provocations, there has been a lack of military retaliation by India. The nuclear deterrence has encouraged certain Pakistani elements to further provoke India. An "asymmetric nuclear escalation posture" of Pakistan has deterred conventional military power of India and in turn has enabled Pakistan's "aggressive strategy of bleeding India by a thousand cuts with little fear of significant retaliation".

1.8 FUTURE WARFARE

Future warfare is the evolution of military conflict in the coming decades, which is expected to be shaped by a number of factors, including:

(i) Information

Warfare will focus less on firepower and more on the power of information, including command, control, communications, computers, intelligence, surveillance, and reconnaissance (C^4ISR).

(ii) Technology

Critical technologies like cyber, virtual, and augmented reality (VR/AR), AI, and 3D printing will transform battlefields.

(iii) Geopolitical trends

Geopolitical trends like the rise of China, the realignment of Asia, and the emergence of an ambitious Russia will shape the future of warfare.

(iv) Other dynamics

Other dynamics that will shape the future of warfare include connectivity, lethality, autonomy, sustainability, and the risk of miscalculation and escalation.

Future warfare will be **Fifth-generation warfare** which aims to control the adversary's population by distorting their worldview and threat perceptions, even without knowledge of the target.

These shifting battlefields have transformed with critical technologies like cyber, virtual, and augmented reality (VR/AR), AI, and 3D printing. Several of these technologies have made the battlefield a complex and interconnected ecosystem with the convergence of the physical and virtual domains. They have allowed combatants to engage in fighting without resorting to kinetic means, which has been the hallmark of the battles of the previous centuries. At the same time, they have also spawned discussions about their ethical implications, legal frameworks, and potential unintended consequences. The use of lethal autonomous weapon systems, for instance, has been a controversial topic, with a growing chorus of voices from various sectors advocating for a global prohibition for their purported violation of international humanitarian law. Similarly, using cyber tools to target critical national infrastructure has raised concerns about potential widespread disruption and civilian casualties, which may have cascading effects on other essential infrastructure and services. Yet, several countries are pursuing these technologies to weaponize and deploy them as quickly as possible.

National security establishments and military planners worldwide are now faced with new challenges with this transformation in warfighting. It has made policy choices more complex and responses more challenging, like cyberwarfare, challenges of attribution, swarming drones, autonomous weapons, AI, and their impact on land warfare, blockchain, and warfighting while also looking at the impact of these technological advancements on nuclear weapons and space.

SECTION-II

CONVERGING TECHNOLOGIES FOR MODERN AND FUTURE WARFARE

CHAPTER 2

SMARTPHONES IN WARFARE

2.1 SMARTPHONES FOR CIVILIAN AND WARFARE APPLICATIONS

The smartphone already yields significant impact on civilian applications, providing personalized information and services at one's request and linking individuals into communities with shared interests. However, it is important to note that the smartphone technology is not radically new; it is a convergent technology which simply integrates existing technologies such as landline telephone, wireless sensor networks, embedded systems, machine-to-machine communications, cloud computing, and mobile applications on to a single device. Recent advances in each of these individual technologies is what has allowed such powerful synergies to take place, which, in turn, have enabled cost-efficiencies and relative ease of implementation of the smartphone's cyber-physical systems, including physical sensing, networking, data analysis, and tailored applications. The smartphone, therefore, has become so ubiquitous and prominent due to three critical and interrelated factors: (i) connectivity (its ability to connect and link into shared networks), (ii) adaptability (its ability to have tailored applications which users find most useful), and (iii) convergence (its ability to replace a number of single-purpose devices). These three factors also shape how future armies will leverage smartphone technology to gain decisive advantages in tomorrow's wars.

Digital Age soldiers leveraging military smartphones tailor them to their missions and rely upon a network of experts who serve as their reach-back support.

With connectivity, the smartphone permits armies to gain advanced persistent situational awareness and real-time intelligence. Historically, armies gained intelligence by employing their own limited number of sensors and scouts. With military smartphones, soldiers on the battlefield no longer rely exclusively on their military's own reconnaissance systems; soldiers are able to link into and leverage all existing surveillance systems on the battlefield, ideally tapping into even those that belong to the enemy. Consequently, soldiers gain access to the complete spectrum of intelligence feeds be they military-grade systems introduced on to the battlefield or available systems inherent to the area. Such an advantage would be akin to playing a game of Starcraft-II with an Orbital Command Vehicle that when ordered, clears the fog of war view and detects all clocked units within the target area. With such superior situational awareness and intelligence, the military smartphone enabled network-centric army possess a decided advantage.

Adaptability in the smartphone is what permits armies to conduct military operations with smaller footprints that take advantage of near real-time analysis and support. Through the acquisition of real-time intelligence, the military smartphone enables army consistently and expertly executes ad hoc objectives, and soldiers are able to request military strike packages tailored for their unique needs, circumstances, and missions. The smartphone's tailorable applications enable soldiers to objectively calculate what package is needed to successfully accomplish an objective as well as expertly request the whole host of military options available, including the delivery of a smart bomb, drone strike, naval gun fire, air force interdiction sortie, insertion of a military strike team, or humanitarian relief package. The smartphone provides soldiers with tailored military applications that not only assist with decision-making, but arguably more importantly, link them into the military analytical support centers capable of handling the large amounts of input feeds on the battlefield. Due to its adaptable interface, the military smartphone identifies and focuses soldiers on the set of most probable threats and is configured to address them—customizable for example with a suite of peripherals that detect nuclear, biological, or chemical hazards, sensors that detect armored equipment or armed personnel, biometric gadgets that pinpoint individuals on a wanted list, or a combination of such devices and applications. Digital Age soldiers leveraging military smartphones tailor them to their missions and rely upon a network of experts who serve as their reach-back support.

The convergence we have seen in today's smartphones similarly reduces the need for large standing armies, transforming them into smaller more nimble organizations that are networked for improved efficiencies. When military smartphone-enabled soldiers deploy for operations and encounter threats or situations that exceed their capabilities, their reach-back support has already either coordinated on their behalf a reinforcement package or other available aid in their immediate vicinity or queued them for long-range support that can assist from afar, such as from an aircraft, ship, or even space-based platform. As portrayed by the small away teams depicted on Star Trek, the need for large security forces becomes superfluous so long as teams maintain communication with the bridge, have access to the USS Enterprise's reach-back capabilities, and remain within the ship's protective footprint. By enabling soldiers to expertly perform multiple functions with the help of applications, the military smartphone makes army formations more efficient and leaner.

Throughout history, successful armies anticipated the future, adapted, and capitalized upon opportunities. Due to the advantages provided by their connectivity, adaptability, and convergence, smartphones today are already making significant changes in civilian applications. Armies of the future that leverage smartphone innovations will change the face of war and reshape the modern battlefield into one that is befitting of the Information Age.

2.2 SMART PHONES PLAYING PROMINENT ROLE IN RUSSIA-UKRAINE WAR

Smart phones have played a major role for several years now in strategic messaging and information operations. The U.S. Army previously forecast the critical importance of ubiquitous information technology in future competition as well as conflict. Russia's 2022-23 invasion of Ukraine reveals that future is now as smart phones have contributed directly to kinetic operations on a scale and to a degree never seen before on the battlefield.

The use of smart phones by Ukrainian and Russian combatants and noncombatants alike is transforming command, control, communications, computing, intelligence, surveillance, and reconnaissance (C4ISR) on the battlefield. Smart phones have crowd-sourced forward observation and turned entire populations into battlefield sensors. Smart phones have provided commanders at all echelons with capabilities historically available only at upper echelons, improving both situational awareness and agile response capability. Use of smart phones has also introduced new vulnerabilities that combatants have exploited.

Smart Phones Crowdsource Forward Observers and Listening Posts. Ukraine has used smart phone apps to mitigate Russian unmanned aerial system (UAS) attacks, particularly on civilian and strategic targets. UAS must travel long distances before striking because of Ukraine's large landmass. Forward-positioned teams, including civilians, with purpose-designed apps can observe and report on UAS movements. These reports are forwarded to a centralized system that plots likely targets and routes. Air defense forces then have enough lead time to plan interdiction. This use of spotters is reminiscent of British observer activities during WWII, when civilians would call in alerts about German air raids.

Civilians are also using smart phones to observe and report on enemy positions. These positions are shared through a variety of means, including social media posts, and this precise geolocated information can and has been used for fires targeting. Ukraine has capitalized on this technology by creating specialized apps that anyone can use to report enemy activity. Every citizen is now a sensor, and anyone with a smart phone can be a forward observer.

Smart Phones Deliver the Big Picture. Smart phones have enabled geospatial crisis monitoring, such as with the Live Universal Awareness Map (Liveuamap). Liveuamap has aggregated social and news media posts—many originating from smart phone apps—to provide real-time coverage of the Russian invasion of Ukraine and other global crises since 2014. Private citizens and soldiers positioned in the conflict zone post live updates to social media. Liveuamap uses artificial intelligence to collect and sort data, which feeds its geospatial platform. The result is a crowd-sourced, real-time operational picture. What is innovative is how Liveuamap has tweaked an established business concept designed to assess marketed product trends and public opinion and used it to generate what is an economical and efficient common intelligence picture-common operating picture (CIP-COP).

Russian soldiers are also using geo-spatial apps to build their own CIP-COP. The Ukrainian Armed Forces Eastern Group Commander reported in April 2023 that Russian forces had been using the off-road exploring app AlpineQuest GPS to locate Ukrainian equipment positions and units, as well as to identify objectives and attack routes. Russian forces can maintain an accurate, up-to-the-minute CIP-COP by using commercial, off-the-shelf tools.

2.2.1 Smart Phones Accelerate Counterbattery Fires

Soldiers can now use smart phones to respond to enemy fires. Russian military technology company VPK posted in January 2023 an account of Russian soldiers using self-created smart phone apps to triangulate Ukrainian artillery positions. Small teams of soldiers with smart phones or tablets disperse a few kilometers behind the line of contact. They use an app that detects the acoustic signature created when Ukrainian artillery guns fire. Their devices relay the information to another app that triangulates the position of the gun. A UAS confirms and refines the location. The command then targets the location with counterbattery fires. Artillery sound ranging using acoustic triangulation has been around for a long time but required time, specialized training, and advanced equipment.

The smart phone has pushed this capability down to the lowest echelons with minimal training and sustainment requirements. This results in a significantly reduced counterbattery fire response time.

2.2.2 Smart Phones Reveal Troop Assemblies

The efficiencies provided by battlefield smart phones come at a price, including at the cost of the users' lives. Widespread use of personal devices is an operational security (OPSEC) vulnerability exploited by combatants on both sides. Cellular devices constantly send signals to communication towers. Both Russia and Ukraine have intercepted these signals to geolocate enemy troop positions. Russia has unmanned aircraft dedicated to emulating cell phone towers to exploit this vulnerability. Russian forces launched a cruise missile strike on 13 March 2022 against Ukrainian Foreign Legion personnel at Yavoriv, killing more than 30 personnel. Russian forces targeted the site because Russian private military company Wagner personnel had detected a group of 12-14 smart phones with British country codes (+44). Wagner forwarded information to Russian intelligence, which assessed the site as a Foreign Legion activity. The surge of foreign fighters early in the war to support Ukraine was highly symbolic, so Russian commanders likely considered their base to be of strategic importance.

It may not be out-of-place to mention that Indians used the same technology of smartphones few years ago to locate some terrorist camps across the border and used the so-called 'surgical strike' in the precision air attack against Pakistan in Balakot.

2.2.3 Implications of Smart-phone based Technology in Training of Soldiers

Smart phones, with the wide range of capabilities they enable, are evolving conditions on the battlefield that contribute to increased transparency. Modern militaries are supplementing units' organic systems with civilian communications technology to achieve large-scale kinetic effects across the battlefield. Militaries are seeing improvements in the responsiveness and adaptability of lower echelons using C4ISR capabilities historically reserved for higher echelons. Improved precision means fires are regaining primacy among warfighting functions. Civilian noncombatants have an expanding role in warfare. The fog of war is continuing to evolve.

Ubiquitous information technology is a condition of the modern battlefield and should be replicated in training. Smart phones bring capabilities and vulnerabilities to both Red and Blue forces.

Conditions to replicate this dynamic could include the following:

- **Civilians' Growing Role in Hostilities.** Distinguishing between combatants and noncombatants—and responding appropriately—will be increasingly difficult. Red forces will employ civilians regularly and treat local populations with much harsher rules of engagement than would be acceptable for Blue.

- **Red Targeting of Civilian Networks.** Red forces targeting of civilian networks for their own security will alter the pattern of life and possibly increase movement of noncombatants on the battlefield.

- **Red Able To Be More Adaptable.** Mobile apps will democratize capabilities that allow Red forces to be more agile and adaptive to conditions on the ground than they historically would have been. Their fires will be commensurately more precise and responsive.
- **Expand Options for Communication.** Effective disintegration tactics against Red networks will be more difficult as civilian networks provide alternate communications channels.
- **Stress Additional OPSEC Measures**. Congregation of phones and even possession of a smartphone could be enough to reveal location to the enemy. Enemy forces will likely target clusters of troops with smart phones, even if these devices are not in active use. The location of Blue force operational centers, assembly areas, supply points, and other areas of congregation may be transparent to the enemy forces.

2.3 ABOUT SMARTPHONES IN THE BATTLEFIELD

Military planners want warfighters to have the same capability that civilian consumers get from their commercial smartphones and are testing different devices. However, they still have to overcome security hurdles and the short development cycles in the commercial market before full-scale deployment can happen.

The typical civilian smartphone – whether it is an iPhone 5, Samsung Galaxy III, or even a Blackberry – is easier to use and has more processing capability than any handheld device that soldiers, Marines, sailors, or airmen use in combat environments today. Modern cell phones have amazing technology, but are not seen as rugged or secure enough for military use on the battlefield. That is until recently. Different programs are in development in the Services to leverage commercial smartphones for battlefield use. One Army initiative – the Nett Warrior program, run by PEO Soldier is being used as a field device these days.

"No defense company in the world can beat the reliability and performance these small devices deliver," says Jason Regnier, Acting Program Manager for the Nett Warrior program at Ft. Belvoir, VA. "It is money well spent. What is enabling their use in part from a policy was a relaxing of the requirements about the environments they would be used in. For example, they don't have to survive a nuclear blast anymore. We are still looking at more ruggedized devices for underwater use and the like, but right now we are focused on commercial devices due to the tremendous cost savings and they are meeting all of our objectives so far."

"Smartphone development within the DoD is a testing environment," says Brett Kitchens, Senior Director, DoD Strategic Programs, U.S. Federal Government Markets at Motorola. "PEO Soldier wants a smartphone device at the edge running secret-level security, but it is not a program of record yet today. Some brigades are already testing different smartphone equipment and software. Eventually there will most likely be a pool of devices for the services to choose from based on their mission needs and user preference."

Right now Nett Warrior – an integrated, dismounted situational awareness and mission command system – is in the operational testing phase. However, the Army is in a hurry to get this technology out earlier and is fielding Motorola Atrix Android-enabled smartphones with certain brigades this year to improve situational awareness. They are still secure with strong encryption, but not certified by NSA for secret data. It is not under a program of record, but is more of an experimental requirement.

2.4 SMARTPHONES FOR SURVEILLANCE AND DECISION MAKING IN CONTEMPORARY WARFARE

Ukraine has shown how a mobilized society and a digital ecosystem can be orientated for the purposes of defending a country from invasion. These digital ecosystems form a constitutive part of an evolving 'geometry of power'. Built from a complex web of public and private organizations that needs careful disentanglement, these 'stacks' are shaping a new 'virtual sovereignty'. Much of this sovereignty is dependent on private internet capital and is exercised through digital platforms.

Nevertheless, prior to Russia's full-scale invasion of Ukraine in February 2022, and despite a history of civilian smartphone use in conflict that stretches back to at least the 2011 conflict in Libya, the role these devices played in battle rarely featured in thinking about war. It is clear, however, that smartphone technology is playing an increased role in shaping how we come to know and understand war. For example, one NGO documenting war crimes in Ukraine collected 2.8 million digital records between February and May 2022. By contrast, during eleven years of war in Syria, the Syrian Archive collected and preserved 5 million digital records. The number of records being produced reflects the extensive availability of the smartphone. At the same time, given this explosion of material, making sense out of it is beyond human comprehension. Instead, sophisticated algorithms must be employed to help analysts search and sift through the records. And yet the availability of AI is unevenly distributed among private sector businesses in different parts of the world.

Smartphone functionality means that users can post content more quickly than broadcast media, instantly gathering feedback on how well it is received. No editorial process is required, and so the speed at which people can engage and stories can 'go viral' has accelerated. This phenomenon is having an impact not just on how war is represented but also on how it is fought.

The technological means for globalized participative warfare has been the smartphone. But participative warfare is not limited to highly connected geographies. For example, WhatsApp, the end-to-end direct messaging app, has been used by US and Iraqi special forces to direct warfighting in Iraq. The Taliban used WhatsApp to help engage the Afghan public as they took over the country in 2021. In central and western Africa, mobile telephony has shaped insecurity across the region. Between 2007 and 2011, even before the widespread availability of the smartphone, mobile phones were important technologies for framing the way conflict unfolded in South Sudan.

Even before the full-scale Russian invasion of Ukraine in February 2022, smartphones made it possible for fighters in Donbas to remain connected with friends and family, to relax by playing a game or watching a film, or to engage in targeting, minefield-mapping and combat communication. At the same time, it has become increasingly clear that smartphones are an information source that can be geolocated either from the signal the phone emits or from images taken by civilians and soldiers. As a result, armed forces have made strenuous efforts to keep smartphones out of the hands of soldiers, even as the devices have proliferated across civil society.

In relation to mobilizing civil society and the armed forces for a full-scale invasion by Russia, Ukraine's IT sector has produced a sophisticated constellation of applications designed to work from mobile devices like tablets and smartphones. This includes Kropyva, also known as Nettle, a command-and-control platform developed by Army SOS, a Ukrainian IT volunteer organization. The app 'maps battle lines and targets and calculates artillery fire missions'. MilChat is a messaging app that allows communication between units and commanders and was developed with the support of Ukrainian NGO Noosphere. In use by an estimated 60,000 combatants, this app allows units to share geolocations, helping commanders keep track of troop movements and optimize resource allocation. The GIS Arta app, in use since 2014, is designed to take 'target information from drones, US and NATO intelligence feeds and conventional forward observers, and [convert] the information to precise coordinates for artillery'. The app, which is modelled on software developed for the Uber ride-hailing service, was created by a group of Ukrainian IT volunteers. Another company, UkropSoft, has designed an app freely available to download via Google Play called Ukrop/MyGun, which calculates target data for artillery units taking into account topographic, meteorological and ballistic training.

One way in which the Ukrainian approach differs, however, is the level of participation. Civilians participate in crowdsourcing the funding for the software now in use by Ukraine's armed forces. At the same time, civilians can directly involve themselves in warfighting through the apps that they have helped to crowdfund. These apps allow anyone with a smartphone to photograph and geotag the location of enemy formations and upload the relevant information to Ukrainian intelligence centres. Civilians gathered intelligence through an online chatbot called @stop_russian_war_bot, crowdfunded by the independent Come Back Alive Foundation for Ukraine's internal security agency, the Sluzhba Bezpeky Ukrainy (SBU) and initially available over the social media platform Telegram, through a Google web form or via Viber (a cross-platform voice over Internet Protocol smartphone app). Using the chatbot, 287,000 Ukrainians provided information on Russian movements and equipment in the first four months of the invasion.

Civilians are also able to help with bolstering Ukraine's air defenses. Another smartphone app, ePPO, allows citizens to report sightings of Russian aircraft, missiles and drones simply by 'point[ing] their device in the direction of the incoming object and press[ing] a single button for it to send a location report to the country's military'. This early warning application helps intelligence fusion cells to coordinate responses, whether the latter involve shooting down the object, tracing the original location from where it was fired or shaping responses by emergency services. Civilian sources of intelligence like e-Vorog and ePPO can be fused with other more traditional sources of intelligence and made available via Delta, a command, control and situational awareness management software system. Delta provides military units with real-time information about the battlefield via a web of connected devices and for convenience can be loaded 'on a laptop, tablet or mobile phone'. Although built and maintained by Ukraine's Air Intelligence Team, it was paid for out of volunteer funding.

A similar blurring of the details on the participation of individuals, activists and private companies can be seen in the annual report of the Come Back Alive Foundation. Raising around US $177 million in funds during 2022—292 times more than it raised in 2021—the foundation is crowdsourcing funds from 56 countries, making it possible for anyone to help Ukraine defend itself—not just Ukrainian taxpayers. In this respect, the approach being constructed in Ukraine is not just an open innovation ecosystem built to support professional armed forces along the lines being described by Eric Schmidt in relation to the United States. Rather, this is an ecosystem that is aimed at digitally mobilizing remote support from around the world; in the process, it is enabling the whole of Ukrainian society to participate in the defence of the state. As the Come Back Alive Foundation's funding partnership with Vodafone Ukraine states, what makes mass participation possible is the decision to use mundane connected devices like the smartphone to help transfer funds, while at the same time becoming the means by which society is mobilized to defend itself.

2.4.1 Participative warfare and its implications

A number of implications emerge when switching attention away from innovation and towards participation. In the first instance, it is important to note that the war in Ukraine has gone through several phases. Early in the invasion, civilian involvement in intelligence-gathering played a more prominent role than it did during periods when the front lines stabilized. It is also apparent that the precise gearing of interactions between civilian smartphone technology, social media and the intelligence collection/fusion cycle employed by Ukraine's armed forces needs further research. However, it is evident that Ukrainian citizens are directly participating in the process of finding, locating and targeting Russian armed forces. In effect, Ukraine has provided the tools to help its citizens resist invasion and occupation. The problem is that by participating in this way, civilians become combatants and become legal targets for Russian attacks. Indeed, Ukrainians with targeting apps loaded onto their smartphones have been rounded up and killed for behaving like combatants. This is principally because these citizens have become a part of the Ukrainian kill chain. This advances in three stages, starting with processing intelligence to understand enemy activities, followed by deciding on how to respond. It finishes by taking action to achieve a political effect. By providing information-gathering tools to ordinary citizens, the Ukrainian kill chain has become more resilient, but the forfeit for this has been the loss of civilian status in war.

For those thinking through what it means to fight in hybrid or sub-threshold warfare, smartphone-enabled participative warfare presents new opportunities for attacking and resisting adversaries. On the one hand, it creates new avenues for data collection. At the same time, it creates more resilient targeting processes. Controlling this type of activity will not be easy in this scenario. One might argue that platform providers like Google or Apple could prevent people from participating in conflict by denying access to apps intended for warfighting. However, as the internet splinters along geopolitical lines, the use of virtual private networks (VPNs) and a proliferation of providers make it unlikely that participative warfare can merely be 'switched off' at any future point.

CHAPTER 3

NETWOK-CENTRIC WARFARE

3.1 WHAT IS NETWORK-CENTRIC WARFARE?

Network-centric warfare, also called network-centric operations or net-centric warfare, is a military doctrine or theory of war that aims to translate an information advantage, enabled partly by information technology, into a competitive advantage through the computer networking of dispersed forces. It was pioneered by the United States Department of Defense in the 1990s.

Network-centric warfare (NCW) is defined as an information superiority-enabled concept of operations that generates increased combat power by networking sensors, decision makers, and shooters to achieve shared awareness, increased speed of command, higher tempo of operations, greater lethality, increased survivability, and a degree of self-synchronization. In essence, NCW translates information superiority into combat power by effectively linking knowledgeable entities in the battlespace.

NCW is about human and organizational behavior. NCW is based on adopting a new way of thinking— network-centric thinking—and applying it to military operations. NCW focuses on the combat power that can be generated from the effective linking or networking of the warfighting enterprise. It is characterized by the ability of geographically dispersed forces (consisting of entities) to create a high level of shared battlespace awareness that can be exploited via self-synchronization and other network-centric operations to achieve commanders' intent. NCW supports speed of command—the conversion of superior information position to action. NCW is transparent to mission, force size, and geography. Furthermore, NCW has the potential to contribute to the coalescence of the tactical, operational, and strategic levels of war. In brief, NCW is not narrowly about technology, but broadly about an emerging military response to the Information Age.

In 1996, Admiral William Owens introduced the concept of a 'system of systems' in a paper published by the Institute for National Security Studies in the United States. He described a system of intelligence sensors, command and control systems, and precision weapons that provided situational awareness, rapid target assessment, and distributed weapon assignment.

Also in 1996, the United States' Joint Chiefs of Staff released Joint Vision 2010, which introduced the military concept of full-spectrum dominance. Full Spectrum Dominance described the ability of the US military to dominate the battlespace from peace operations through to the outright application of military power that stemmed from the advantages of information superiority.

The term "network-centric warfare" and associated concepts first appeared in the United States Department of Navy's publication, "Copernicus: C4ISR for the 21st Century." The ideas of networking sensors, commanders, and shooters to flatten the hierarchy, reduce the operational pause, enhance precision, and increase speed of command were captured in this document. As a distinct concept, however, network-centric warfare first appeared publicly in a 1998-US Naval Institute Proceedings article by Vice Admiral Arthur K. Cebrowski and John Garstka. However, the first complete articulation of the idea was contained in the book "Network Centric Warfare: Developing and Leveraging Information Superiority" by David S. Alberts, John Garstka and Frederick Stein, published by the Command-and-Control Research Program (CCRP). This book derived a new theory of warfare from a series of case studies on how business was using information and communication technologies to improve situation analysis, accurately control inventory and production, as well as monitor customer relations.

NCW represents a powerful set of warfighting concepts and associated military capabilities that allow warfighters to take full advantage of all available information and bring all available assets to bear in a rapid and flexible manner.

The tenets of NCW are:

- A robustly networked force improves information sharing.
- Information sharing enhances the quality of information and shared situational awareness.
- Shared situational awareness enables collaboration and self-synchronization, and enhances sustainability and speed of command.
- These, in turn, dramatically increase mission effectiveness.

The goal of network-centric operations (NCO) is to enable forces to accomplish their objectives more efficiently: faster; with fewer troops in harm's way; and with fewer and lighter weapons and other equipment to bring to, sustain, and maneuver in the battlespace.

With timely and accurate intelligence, commanders can decide faster, deploy a force of the optimal size and characteristics, command and control that force better, and stay one step ahead of enemy forces. Network-centric operations can improve all of these functions.

In early October 2001, not quite a month after the 9/11 attacks on the World Trade Center, the Pentagon and in Pennsylvania, U.S. and British ships and aircraft introduced the 21st century (and the Taliban) to network-centric operations. With a striking blend of old and new technology, operating throughout the electromagnetic spectrum and across the range of operations, from ground forces to air and sea platforms and into space, U.S. forces in both conflicts used networked information to achieve huge efficiencies in combat. The "kill chain" against enemy targets was reduced in many cases from hours to minutes, and information about the location of enemy and friendly forces was relayed and tracked just as quickly. In Afghanistan, the deployment of American ground troops was minimal; in Iraq, a force one-quarter the size of the 1991 Desert Storm coalition defeated the Iraqi regime in 21 days, with only 161 troops killed in action. In both theaters, the incidence of civilian casualties and other collateral damage was minimal.

This information revolution has permeated the military world as well, with network-centric warfare replacing traditional combat methods. Technology is now at the forefront of battlefields, creating a new era of warfare - network-centric.

It's a new level of communication and coordination through what is known as **'tactical interoperability'**. From human soldiers to smart weapon systems, command & control systems, automatic sentry systems, and platforms on land, air, and space - all these elements are seamlessly connected in a single communication fabric, with encompass battle management systems for all services, catering to individuals from General HQs to soldiers on the field.

In Network Centric Warfare (NCW), the collaborative planning tools enable strike planners, potentially based on multiple ships or in units ashore, to plan and de-conflict multi-aircraft strike packages. This leads to improved capabilities for synchronization and de-confliction significantly decrease planning time and provide aircrews with the operational flexibility to rehearse or accelerate operational tempo. Net result is increased combat power.

3.2 TRACK RECORD OF NETWORK-CENTRIC WARFARE BY AMERICAN FORCES

Network-centric warfare (NCW) now has a track record. Practical application in Afghanistan and Iraq has given analysts enough data and experience to begin to evaluate its successes, weaknesses, and prospects for improvement. These first clashes in the war on terror have shown that NCW works. Albeit against markedly inferior military forces, American forces were able to integrate information and communications systems and procedures to accomplish more with less, and faster, than would have been possible even a decade ago. Sorties per target destroyed; speed of the "target observed-target destroyed" sequence; relatively little collateral damage; aggregate numbers of U.S. forces required — by these and other measures, both conflicts show that NCW is the right objective for American military planning.

By their very nature, and especially in the early going, major shifts in orientation are always more about challenges, risks, and shortfalls than about smooth, flawless implementation. But the promise of NCW is accompanied by a range of challenges. Understanding, resolving, and accounting for these challenges will enhance NCW as a tool in America's war on terror.

This ability to network real-time cueing from maneuvering ground forces and a range of other sensors to the aircraft and, in turn, to the bombs themselves, put massive, accurate, and close-in fire support at the disposal of U.S. and Coalition troops. Often the enemy was hit with lethal and accurate firepower, not from an armored division and thousands of U.S. troops, or massed formations of aircraft, but by handfuls of soldiers transmitting electrons to a single aircraft. Strikes that took hours to coordinate in Desert Storm a decade earlier were carried out in Afghanistan and Iraq as quickly as 45 minutes from the time a target was identified. Many carrier-based strike missions were not even targeted until the aircraft were underway. Long-range UAVs, controlled from afar strategic intelligence, surveillance and reconnaissance (ISR). Flying at altitudes of up to 60,000 feet, at high speed, the Air Force's experimental, high-flying, long-range Global Hawk unmanned aerial vehicle (UAV) provided constant coverage of wide areas of territory and added significantly to the imagery provided by satellites and other surveillance aircraft. With its synthetic aperture radar, electro-optical camera, and infrared (IR) and other sensors, Global Hawk saw through most environmental and battlefield conditions, and provided thousands of images for use in bomb damage assessment, fire support, and general reconnaissance.

From a networking standpoint, Global Hawk showed two especially unique capabilities: it was controlled through a communications network based from ground stations thousands of miles away, in Europe and California; and it flew not only pre-planned flight paths, but also in real time to locations as required by various missions. Global Hawk continues to fly responsive border surveillance missions in Iraq, providing closer follow-up reconnaissance when other assets point to suspicious activity.

3.3 OUTSTANDING ISSUES IN NETWORK-CENTRIC WARFARE

The term 'Network Centric Warfare' also carries some baggage. By mistake, some have focused on communication networks, not on warfare or operations where the focus should rightly be. Networks are merely a means to an end; they convey "stuff" from one place to another and they are the purview of technologists. NCW does not focus on network-centric computing and communications, but rather focuses on information flows, the nature and characteristics of battlespace entities, and how they need to interact. NCW is all about deriving combat power from distributed interacting entities with significantly improved access to information. NCW reflects and incorporates the characteristics necessary for success in the Information Age—the characteristics of agility and the ability to capitalize on opportunities revealed by developing an understanding of the battlespace that is superior to that developed by an adversary.

The network-centric paradigm is an enormous step forward in military thinking. But future enemies are going to find ways to defend against networked forces, and even the best network cannot solve or avoid every challenge forces will face. Growing reliance on network-centric operations should not blind the military to its inherent technical, operational, strategic and cultural limitations. Regardless of the particular weapons and other systems that emerge during this process, several broad objectives must be pursued.

Technical Limitations of Bandwidth is the information-carrying lifeblood of any network, and network-centric operations devour signal bandwidth. As technology proliferates and expectations for information "right now" increase, network-centric operations will constantly require more and more communications bandwidth. This means that the Services and industry must make constant efforts to manage bandwidth more efficiently, with better communications technology, and with command-and-control systems that are better able to prioritize and manage signal flow. Because networks begin with sensors, they will always be vulnerable to jamming by moderately sophisticated foes. Even the least capable enemy will be able sometimes to use deception or concealment to foil sensors, particularly with the coverage gaps that currently exist. An adept opponent may find ways to attack the networks themselves. Because the technology is always vulnerable, and frequently fragile, networks must be tough, flexible and redundant.

3.3.1 Operational Limitations

From a commander's perspective, more information generally is better than less. However, a somewhat ironic difficulty can arise when commanders at different levels become inundated with information from different sensors and sources. Information may become intoxicating, turning tactical challenges into quantitative equations and distracting commanders from such basic military principles as initiative and decisiveness. Too much information may cause commanders to 'tune out.' Ultimately, the appropriate information — not just data — must be matched to the differing requirements of tactical commanders and theater commanders. Networks allow for collaborative planning throughout the chain of command, which can develop more effective plans faster. But the same collaboration allows senior level commanders to micromanage. Military operations rely on a properly functioning chain of command, where commanders at each level have a manageable span of control and can focus on operations at their appropriate level. As real-time battlefield information passes before senior commanders, there will be a temptation to over-direct small units and lose focus on broader objectives. One of the military's greatest strengths is the initiative of small-unit commanders; if these commanders grow accustomed to centralized control from above, they may grow hesitant and indecisive. As forces become more accustomed to high-quality, real-time information, there is a risk that they will be hesitant or even paralyzed without it. For example, networking enabled some remarkably responsive close air support in Afghanistan and Iraq, and raises expectations for conflicts in the future. But because technology can and will fail, our military must have an ongoing commitment to mastering basic soldiering skills — such as map-and-compass navigation, communications and the accurate verbal call for fire support — even when the optimal 'network' is not there to facilitate them.

3.4 SENSORS IN NETWORK-CENTRIC WARFARE

The operational performance of a sensor network (a collection of networked sensor entities) in generating battlespace awareness depends upon a number of factors including:

(i) the performance of component sensors; sensor geometry:

(ii) the locations of the sensors with respect to each other and the objects of interest;

(iii) the velocity of information;

(iv) fusion capabilities; and

(v) tasking capabilities.

In the fundamental shift to network-centric operations, sensor networks emerge as a key enabler of increased combat power. The operational value or benefit of sensor networks is derived from their enhanced ability to generate more complete, accurate, and timely information than can be generated by sensors operating in stand-alone mode. The performance advantage that emerges from the enabling of sensor networks is a function of the type of sensors being employed (e.g., active, passive) and the class of objects of interest (e.g., missiles, aircraft, tanks, submarines, etc.). Sensor networks can generate significantly increased battlespace awareness of objects in the battlespace. Sensor networks provide significant performance advantages over stand-alone sensors in key mission spaces by overcoming the fundamental performance limitations (e.g., coverage, accuracy, and target identification properties) of individual stand-alone sensors. The value-adding processes of data fusion and sensor tasking can partially overcome these limitations. This does not imply that the level of awareness generated against all targets will be 100 percent in all mission areas, but rather that almost all mission areas can benefit to some degree from the shift to network-centric operations.

NCW provides opportunities to improve both C2 and execution at each echelon in the context of particular missions and tasks.

These opportunities will come about because:

(i) decision entities or elements will be more knowledgeable;

(ii) actor entities will be more knowledgeable;

(iii) actor and decision entities will be better connected;

(iv) sensor entities will be more responsive; and

(v) the footprint of all entities will be much smaller.

Each of these improvements makes it possible for us to do things differently. It is important to stress that these properties of NCW offer opportunities to better match our approach to each set of battlespace circumstances and conditions to achieve greater levels of both effectiveness and efficiency.

Associated with the employment of actor entities are certain characteristics that determine their effectiveness and efficiency.

Included are:

(i) the targets they can engage or their engagement envelope;

(ii) their exposure to enemy attacks or their risk profile;

(iii) the speed of command and rate of engagement they can sustain (or their tempo);

(iv) the responsiveness of forces or support units;

(v) their ability to move (or their maneuverability);

(vi) their lethality (or the probability of kill); and

(vii) the extent to which their activities can be synchronized.

Adoption of NCW provides us with the ability to enlarge the engagement envelope, reduce risk profiles, increase operating tempo and responsiveness, improve maneuverability, and achieve higher kill probabilities.83 A number of examples follow which illustrate these points.

3.5 COOPERATIVE ENGAGEMENT CAPABILITY (CEC) IN THE AIR DEFENSE

The Cooperative Engagement Capability (CEC) improves the ability to conduct Air Defense. In this mission area, time is a key factor since there is a limited amount of time available to detect, track, classify, and engage targets. Engagement time is further compressed for high-speed or low-observable targets. This stresses all elements of the combat power value chain: sensors, command and control, and weapons.

The CEC increases combat power by changing the relationships between battlespace and battle-time. The CEC component forces currently consist of surface combatants (e.g., AEGIS Cruisers) and early warning aircraft (e.g., E-2 Hawkeye). Concepts will emerge enabling other elements, such as fighter aircraft, and ground-based missiles (e.g., Patriot Missiles or Hawk Missiles) to be employed as part of the CEC, serving to further increase combat power.

The CEC is enabled by the close coupling of an integrated communications capability in the form of the Data Distribution System (DDS), with a computational capability, in the form of the Cooperative Engagement Processor (CEP). This infostructure provides a high-performance backplane which is key to increasing the velocity of information among sensor, fire control nodes. The netting of sensors generates a level of battlespace awareness that far surpasses that which could be generated by sensors operating in stand-alone mode.

Shared engagement quality information is provided directly to the cognizant air defense commander, as well as to all other warfighters that have access to the CEC infostructure. The actor entities that are linked to the CEC infostructure give the air defense commander the capability to employ forces in multiple modes. In the first mode, the netting of command and control and fire control capabilities provides the commander with automated decision support capabilities that help him identify the locations and weapons status of linked shooters. This information is combined with other battlespace information to identify the shooters that can engage each incoming target. The commander is then able to make effective force employment decisions: when to engage each target and what weapon to engage with. In this mode, the commander has centralized operational control over all connected weapons systems. Because of the short timelines involved, and the large number of decisions that potentially need to be made, a second mode has been created which automates the weapon target assignment process.

The value added by the CEC is a result of its ability to extend the engagement envelope, enabling incoming targets to be engaged in depth with multiple shooters with increased probability of kill. Furthermore, the inherent capability to engage adversary missiles by aircraft using engagement quality information generated by sensors not organic to the ship can increase the survivability of one's own target by enabling it to engage without generating an electrical signature. The net result is the ability of the CEC to successfully engage and defeat threats capable of defeating a platform-centric defense. The whole is clearly greater than the sum of the parts.

CHAPTER 4

VIRTUAL REALITY AND AUGMENTED REALITY IN WARFARE

Virtual Reality (VR) and Augmented Reality (AR) are being increasingly used by the military for training and operations. VR allows soldiers to experience realistic combat scenarios without the risk of actual harm, while AR enhances their perception of the battlefield by overlaying vital information onto their view. The Indian Armed Forces are also increasingly utilizing VR and AR technologies for training, wargaming, and enhancing the capabilities of their equipment. However, challenges such as the potential adverse effects of VR on young soldiers and the high installation costs need to be addressed before these technologies can be fully integrated into the Indian military.

4.1 INTRODUCTION TO VIRTUAL REALITY, AUGMENTED REALITY, AND WARFARE

Towards the end of the 1999 cult classic film, The Matrix, Neo, after transforming into the all-powerful 'The One', says, "I can see everything clearly now." The camera then shifts to Neo's point of view, displaying a cascade of ones and zeros against a green screen. The filmmaker wants the viewer to see that Neo can now access the duality behind the Matrix—a simulation created by machines to keep humans in a state of stupor while their bodies are used as bio-electric fuel for the machine civilization. The simulation immerses human beings into the world as it was in 1999, and life inside the Matrix is designed to be as normal as possible to create a sense of presence for the users while precluding them from ever thinking of the greater reality beyond the simulation.

The aim of virtual reality (VR)-based applications and hardware is to create such user immersion inside a synthetic environment, though for purely benign and educative purposes. However, the current peculiarities of the hardware, software, and user requirements have created a number of devices and programmes that combine elements of both physical and virtual reality. This mixed reality (MR) can be thought of as forming part of a reality–virtuality continuum with the physical and virtual environments (VE) acting as extreme bounds. Two intermediate states of augmented reality (AR), which comprises the use of certain equipment and data to accentuate the user's perception of the physical world, and augmented virtuality (AV), which is the augmentation of VE with real or unmodelled imaging data, fall between the two bounds.

VR and AR have increased relevance for warfighting, as VR models worlds and conditions that are impossible to create in a laboratory or on training grounds for safety reasons, and adds to a quantitative and systematized approach towards training. AR, meanwhile, adds a layer of additional information (audio, visual, haptic) between the soldier and their physical world and heightens certain sensory performances required for the battlefield. Whether these are cost-effective and adequately ruggedized needs to be studied carefully.

4.2 AUGMENTED REALITY VS. VIRTUAL REALITY: WHAT'S THE DIFFERENCE?

Augmented reality (AR) and virtual reality (VR) are both technological experiences that change how digital technology interacts with the physical world. AR and VR are often lumped together, but each has its way of interacting with the virtual environment.

(i) Augmented Reality (AR)

AR overlays video or images onto a display of the physical world, typically through a smartphone. The overlaid image creates an interaction between the user, the digital, and the physical worlds, allowing them to make connections or have new experiences. AR and CGI (Computer-generated imagery) allow for the visualization of objects in the real world.

To create an AR experience, you need a device with a camera and software that uses the 3D elements with the AR application. Some popular examples of AR include entertainment-related options like Snapchat filters and the mobile video game Pokemon Go, but it also has implications for business training.

(ii) Virtual Reality

Unlike AR, virtual reality creates a fully immersive experience using a headset and computer-generated images (CGI) to put the user in a virtual world. In VR, the user interacts with a fully virtual world using a headset and a controller. Virtual reality creates a sensory experience by stimulating and tricking the sensory organs into interacting with the virtual world as they would the physical world.

To sum up:

- AR uses a real-world setting while VR is completely virtual;
- AR users can control their presence in the real world; VR users are controlled by the system;
- VR requires a headset device, but AR can be accessed with a smartphone; and
- AR enhances both the virtual and real world while VR only enhances a fictional reality.

4.3 VIRTUAL REALITY (VR) AND THE MILITARY

The mainstreaming of VR began with the entertainment industry in the US with the intention of immersing the user within a movie using all of their five senses. Morton Heilig, a professional cinematographer, developed the "Sensorama" in 1962 as a rudimentary VR device. Ivan Sutherland wrote about the 'ultimate display' in 1965 that would include interactive graphics, force-feedback devices, audio, smell, and taste. Jaron Lanier, a computer scientist and founder of VPL Research, coined the term 'virtual reality' in 1987.

In terms of the military, the introduction of VR was in the form of flight simulators for training pilots. From thereon, the utility of VR has involved the increasing use of VE, i.e., digitally created worlds either mediated through real-world inputs or totally insulated from it. VR as a military utility has grown to encompass computer simulations, computer games, flight simulators, networked simulators for small teams, formation simulators, joint simulators, and even multi-domain simulators for combat tasks.

Before diving into the findings of the studies, however, it is important to define certain terms specific to the AR and VR fields.

(a) Immersion: Refers to the tracking and display that a VR/AR system delivers to the user. This can be measured objectively. As per Mel Slater, the more a system delivers displays and tracking that preserves fidelity to their equivalent real-world sensory modalities, the more immersive it is.

(b) Presence: Presence, which cannot be measured yet, is the subjective experience of the user inside a VR world. In other words, presence is a human reaction to the immersion. Given the same levels of immersion, different humans can experience different levels of presence. One of the primary reasons to achieve presence through a VR system is to elicit human physiological responses that would be commensurate with that of the real world. There are two ways to achieve presence. One is to mirror reality within the VR to such fidelity that there is no distinction between the virtual and physical worlds. The second is to extract the relevant sensory stimuli through knowledge of the perceptual system, i.e., to find out what is important in a human's representation of reality and deliver presence without a high level of immersion.

(c) Tracking: The tracking sensors and devices are the main components of the VR system. They interact with the system's processing unit and relay the user's orientation to the system. These include electromagnetic, acoustic, mechanical, and optical tracking systems.

(d) Registration: This is a term used for Augmented Reality (AR) systems and can be defined as a process that merges virtual objects generated by a computer with real-world images caught by a camera to create an accurate alignment of the two. Without accurate registration, the issue of the virtual object 'dangling' in the overlaid graphic without context cannot be ruled out, defeating the purpose of using AR.

The Office of Naval Research in USA, based on user feedback, developed the Virtual Iraq application with a 'virtual Afghanistan' scenario as an addendum in an initial open clinical trial with 20 soldiers, positive clinical outcomes, such as improvement in neuro-cognitive functioning, memory and learning, spatial cognition, and executive functioning. One of the more practical uses of VR, also emphasized by the US Department of Defense in their Comprehensive Soldier Fitness program, is stress resilience testing. VR exposure therapy uses a mix of cognitive–behavioral treatment with prolonged exposure (PE) that is delivered through multi-sensory and context-relevant cues that evoke the trauma, the intensity of which can be calibrated by the clinician. The use of PE as a psychotherapeutic tool is based on the emotional processing theory that states that PTSD involves "pathological fear structures" when information represented in the structures is encountered. Treatment using this theory calls for the emotional processing of the fear structures to modify their pathological elements so that the stimuli do not provoke fear.

Terrain familiarization is another area that can be modelled using VR and troops practiced on the same. India, one of the largest contributors of soldiers to multiple United Nations (UN) missions across the globe, can benefit from these technologies, and the Centre for UN Peacekeeping can take a lead in test bedding VR technologies for training Indian troops. The US Army has had success in the use of VR-based applications for training their troops for overseas deployments. A 2003-report by the North Atlantic Treaty Organization's Research and Technology Organisation lists the use of VR in military operations other than war where Intelligent Virtual Agents (IVAs) can simulate indigenous personnel, communicate non-verbal cues associated with foreign cultures, and function as coaches and mentors for trainees.

4.4 VIRTUAL REALITY (VR) IN THE MILITARY TRAINING

(i) Air Force Pilot Training

The use of VR for Air Force pilot training is transforming the traditional cockpit learning environment. Programmes such as the AR and VR in Military Combat provide highly realistic flight simulations, enabling pilots to experience airborne scenarios without the risks and costs associated with actual flight. It also allows for more frequent training sessions and the ability to simulate a wide range of challenging conditions and situations.

(ii) Battlefield Simulations

Battlefield simulations via VR prepare soldiers for real-world combat situations by providing a versatile and controlled environment. The British Army has integrated VR technology into their training, as highlighted in efforts like the Virtual Reality Training Technology trial.

This allows troops to train in detailed, simulated terrains that mirror potential combat zones, greatly improving their spatial awareness and decision-making under pressure.

(iii) Medical and First Aid Preparedness

In the realm of Medical and First Aid Preparedness, VR is an invaluable tool for medics. Virtual environments offer the opportunity to practice life-saving procedures in a variety of trauma scenarios. This technological approach is essential for enhancing the readiness of military medical personnel, ensuring that they can respond with confidence and precision in any situation.

(iv) Mission Rehearsal and Strategy

When it comes to Mission Rehearsal and Strategy, VR enables military units to visualize and rehearse operations before deployment. These immersive simulations provide an interactive platform for practicing various strategies, allowing troops to analyze and refine tactics in a dynamic, risk-free setting.

Through repeated rehearsals of mission-critical tasks and procedures, units can identify potential challenges and find solutions in advance of real-world execution.

4.4.1 Advantages of VR Training for the Military

Virtual reality (VR) training offers the military cutting-edge solutions to simulate real-world environments and scenarios. Through immersive experiences, it enhances skill sets while ensuring safety and reducing costs.

(i) Enhanced Realism

VR training allows military personnel to experience highly realistic battlefield scenarios. Systems like the British Army's Virtual Reality In-Land Training (VRLT) replicate complex environments and situations with great precision, enabling soldiers to train for actual combat without the real-world risks.

(ii) Improved Safety

VR establishes safe training grounds. There is a significant reduction in risk for injury or death as service members can learn to navigate dangerous situations within a controlled, virtual space.

(iii) Cost-Effectiveness

The use of VR also presents an opportunity for cost savings. Virtual environments mean less reliance on physical assets and locations, which can be expensive to maintain and secure. Investing in VR can lead to long-term savings.

(iv) Measurable Competence

VR training provides quantifiable evidence of a soldier's skills and adaptability. It allows for the collection of data on performance, measuring and demonstrating competence in a variety of training modules, which can be essential for targeted improvements.

4.4.2 Challenges and Limitations of VR

While virtual reality (VR) provides dynamic training opportunities for the military, several challenges and limitations can affect its effectiveness and wide-scale implementation.

(i) Technical Barriers

In the realm of VR military training, technical hurdles are significant. Hardware issues such as limited battery life, image resolution, and latency can hamper the realism and immersion crucial for effective training.

Content creation is complex and resource-intensive, often requiring extensive programming and 3D modelling. AR and VR systems must also be robust enough to withstand diverse and harsh operational environments faced by military personnel.

(ii) Physical and Psychological Adaptation

Trainees may experience physical side effects such as motion sickness, eye strain, and discomfort due to prolonged use of VR equipment. Furthermore, there's a risk that prolonged immersion might lead to a blurring of lines between virtual and real combat scenarios. This necessitates careful consideration of how VR is introduced and the frequency of its use, to prevent any negative psychological impact.

(iii) Integration with Traditional Training

Combining VR with conventional methods poses its challenges; finding the correct balance is crucial. Traditional training has been honed for decades and integrates nuances of real-world physics and social dynamics that VR hasn't fully replicated.

Moreover, skepticism amongst military trainers towards the effectiveness of VR means that it must demonstrate tangible benefits to be embraced as a complementary training tool.

(iv) Development of VR Training Content

In the realm of military training, virtual reality (VR) content creation is pivotal for the delivery of immersive and effective training programmes. The successful development of this content hinges on meticulous scenario design, the incorporation of interactive elements, and the integration of analytics and feedback systems to measure performance and outcomes.

(v) Scenario Design

VR training scenarios are crafted with precision to mimic real-world situations as closely as possible. Subject matter experts employ their extensive field knowledge to infuse each scenario with realistic challenges, relevant to military training objectives. Every environment, from arctic warfare to rapid jet landings, is methodically produced to provide authentic experiences.

(vi) Interactive Elements

Interactive features play a crucial role in VR training, engaging the user and simulating true-to-life operations. These elements range from basic navigational controls to complex, decision-making tasks that provide military personnel with hands-on experience. For instance, the British Army's VR training technology includes interactive tools to aid soldiers in learning critical skills required in the field.

(vii) Analytics and Feedback Systems

To quantitatively assess trainees' performance, developers implement sophisticated analytics and feedback systems into VR training programmes. These systems track progress, pinpoint areas of improvement, and deliver tailored feedback. The data collected is instrumental for continuous improvement of both the training content and the trainee's performance, thereby enhancing the overall quality of military training.

4.4.3 Future Trends in Military Training with VR

The landscape of military training is being transformed by Virtual Reality (VR), focusing on increased immersion, collaboration, and realism for service members.

(i) Adoption of AI Technologies

Virtual Reality is being increasingly combined with Artificial Intelligence (AI) to create more dynamic and responsive training scenarios. For instance, soldiers can experience realistic and unpredictable combat situations where the AI adapts the mission and environment in real-time, providing a tailored training experience that improves decision-making skills under stress.

(ii) Multi-User Virtual Environments

Multi-User Virtual Environments (MUVEs) are becoming a crucial trend in military training, allowing multiple participants to train simultaneously in a shared digital space. Teams can practice complex collaborative tasks, such as tactical maneuvers or coordinated assaults, fostering unit cohesion and operational preparedness.

(iii) Hardware Advancements

Advancements in VR hardware are providing soldiers with more lightweight and comfortable headsets, coupled with better haptic feedback systems. These developments aim to enhance physical fidelity and reduce training-related fatigue, enabling longer and more frequent training sessions without compromising on the quality or intensity of the experience.

(iv) Case Studies and Real-World Applications

Virtual Reality (VR) training within the military context has been instrumental in enhancing the preparedness of forces across various scenarios. The following subsections detail specific ways in which VR has been integrated into military training programmes.

(v) Joint Military Exercises

VR technology enables the creation of complex, multinational joint exercise scenarios that facilitate interaction and coordination among different military units. These exercises can simulate environments and conditions, such as airborne operations or urban warfare, in a controlled and repeatable manner, allowing forces to gain valuable experience without the associated risks and costs of live exercises.

(vi) Special Forces Training

Special Forces units employ VR training to hone their skills in high-stakes environments. VR allows for the repetition of specific scenarios, like hostage rescue or covert operations, that require precision and adherence to strict protocols. The immersive nature of VR provides realistic stress inoculation, which is crucial for decision-making under pressure.

(vii) Vehicle and Weapons Systems Operation

Through VR simulations, military personnel can gain proficiency in the operation of various vehicles and weapons systems without consuming physical resources. For instance, pilots can practice complex maneuvers and engage in dogfights, and tank operators can navigate difficult terrain.

These simulations also enable immediate feedback and detailed debriefing, significantly accelerating the learning process.

4.4.4 Ethical Considerations with VR

Virtual reality (VR) in military training carries unique ethical implications that must be scrutinized. As the technology advances, it is paramount to address concerns relating to data privacy and the psychological impact on service members.

(i) Data Privacy and Security

VR military training often involves the collection and processing of sensitive personal data. Security protocols must be stringent to protect against unauthorized data breaches. Explicit consent from participants should be acquired before data collection, and they should be informed about how their data will be used and the measures taken to safeguard it. It is critical to ensure that their personal data is not misused or exposed to external threats.

(ii) Psychological Impact and Desensitization

Utilizing VR for military training can lead to potential psychological effects. A question arises about the impact of immersing service members in virtual scenarios that mimic real-life combat.

Prolonged exposure to VR training environments may result in desensitization to violence. It is vital to monitor the psychological well-being of personnel and provide support to mitigate adverse effects.

4.5 AUGMENTED REALITY

Augmented reality (AR) merges the virtual and the physical worlds. Although it has become synonymous with a helmet-mounted display (HMD) or see-through glasses, AR has also been consumerized in the form of mobile applications like Pokemon Go. It is a blend of technologies that accentuate the user's perception of the physical reality. One of the more prominent examples of AR is the series of space telescopes launched by the US National Aeronautics and Space Administration (NASA), the latest iteration being the James Webb Space Telescope (JWST). Though JWST 'sees' in the infrared range, the images are translated into a form that is visible to the human user. AR can also extend into the auditory and haptic domains, using devices, technologies, and processes to increase human perceptivity of a particular strand of the physical reality.

For the military, AR can be used to augment situational awareness through the merging of multiple intelligence, surveillance, and reconnaissance streams, generating a common operational picture and then disseminating them to the soldiers on the ground through an HMD. However, this will also require an intelligent and context-aware AR application to cater to the chaotic and dynamic battlefield. In closely contested areas with multiple agents and installations, there is a danger of information overload on the soldier, which may retard rather than enhance their sense of the combat zone.

An urban terrain is considered to be the ideal setting for an AR-based device due to two reasons: rapid urbanization of areas previously considered and suited for mechanized warfare, and the issue of clutter necessitating the use of AR in the first place. An urban warfare AR has three objectives: transparent battlefield, intuitional perception, and natural interaction. For this, better hardware, powerful software, and a much more integrated process of combining geographical information systems with a virtual geographic environment are essential. In peacetime, AR has multiple uses, such as enhancing the level of detail in table-top and sand-model wargames, maintenance training for military equipment, and even merging AR with the web—a superb tool that can be used for augmenting open-source intelligence data for effective debunking of disinformation operations.

4.6 APPLICATIONS OF AR IN WARFARE

Far before Snapchat released its filters (which is the simplest form of AR) and augmented reality mobile app development wasn't a thing, the army had already implemented the technology of real-time overlaying for their fighter-jet pilots.

Because warfare is constantly evolving, armies have to keep up with the newest military "trends" and look for opportunities to get ahead in the technological war. And, with the expanding possibilities of data and graphics processing, the number of uses of augmented reality in military grows exponentially.

Here-below, we cover the most recent developments as wells the applications in the area of augmented reality (AR) that can revolutionize the warfare.

4.6.1 Three Solutions to Warfare Augmentation

(i) Tactical Augmented Reality (TAR)

All the crucial information (spatial orientation data, weapons targeting, etc.) is being superimposed onto the pilot's visor, so they do not have to look down at their panels all the time and have much better situational awareness.

Something like that has been developed by the U.S. Army Research, Development and Engineering Command's Communications-Electronics Research, Development and Engineering Center (CERDEC) that are actively researching the potential of augmented reality technology.

TAR looks like the night-vision goggles (NVG), but it can offer much more possibilities. It can show a soldier their exact location, and the positions of the allied and enemy forces. The system is mounted to the helmet the same way the goggles are and can operate during both night and day. So, TAR basically replaces the typical handheld GPS device and goggles. As a result, a soldier would not have to look down whenever they want to check their GPS location. Moreover, there is a thermal site on the weapon that is wirelessly connected to the tactical augmented reality eyepiece and a tablet on the soldier's waist. Such a system allows soldiers to see the target they're aiming at and the distance to it. Also, the display can be split in two so that you can see where your gun is pointing at and the view from your frontal camera mounted on the helmet at the same time. For instance, a soldier can see around a corner or over the wall without any risk of getting a headshot.

To add to it, tactical augmented reality has its own wireless network that allows soldiers to share information among their squad members or input data whenever the situation changes.

Tactical augmented reality heads-up display can improve the soldiers' battlefield awareness, reduce the number of devices that must be carried, and help beat the tar out of your enemies more efficiently.

(ii) Synthetic Training Environment (STE)

Training is nothing compared to the real combat. When you're in the heart of a hot zone where gunfire is all around you, it is difficult to stay calm and make the right calls. Spending time in barracks and shooting at 2D cardboard models won't prepare you for the real actions to the full extent. However, there might be a looming solution provided by augmented reality.

Synthetic Training Environment, an AR system that should help train soldiers in a more immersive way, putting them into more physically and mentally stressing operational environments.

One of the key objectives pursued by the STE developers is to create such a training option that would allow commanders to establish adaptive units with a higher readiness level.

STE is a mixture of all three realities – virtual, augmented and physical. It is supposed to be flexible enough to allow for mission rehearsals of most types and be intuitive enough to make training effective. One of the main advantages of such a platform would be the number of iterations you can go through during one training and the ability to adjust the system in real time.

(iii) ARES – Augmented Reality Sand table

Sand tables have been used to plan military operations and train cadets for many years. Usually, it is a table with a scaled physical model of the involved terrain and some kind of plastic figures to position the allied and enemy troops.

Army Research Laboratory had finished developing an AR version of a military sand table in 2015. They used an off-the-shelf projector, a simple LCD monitor, laptop and Microsoft Kinect. ARES provides improved battlefield visual representation, decreases the time required to model the terrain and scenarios, and provides a much higher level of engagement for students.

The main features of ARES are:

- Wide selection of terrain types;
- Control with gestures;
- Real-time terrain generation (using hands);
- Intuitive interface;
- Network for distant collaboration;
- Scalability.

An AR headset can do that and overlay combat information for even better awareness. And, the gunner can operate a turret without showing on the top of the tank thus risking his life.

4.6.2 Pros of AR

(i) Operational and situational awareness (SA)

To make better judgments, operation commanders are usually looking for the ways to gain and exploit SA. Augmented reality can give them such an opportunity by providing tools for better monitoring of operational space, time and forces.

These AR instruments benefit both leaders that manage the operations and field soldiers. The former can get a better perspective on large-scale missions, and the latter is provided with real-time information superimposed on their visors by augmented reality systems.

Less costly combat training.

The ability to conduct training without using expensive and often fragile equipment (often using mixed reality). Instead of getting real vehicles out in the field and expend consumables, AR can offer a flexible training platform that can be even more effective when it comes to getting a better feel of what an actual war is like.

Other advantages of augmented reality:

- Safer training environments;
- More accessed mission rehearsals;
- Terrain diversity and customization;
- Real-time targeting aid;
- Enhanced spatial awareness;
- Engaging mission planning.

4.6.3 Cons of AR

(i) Information overload

Such a phenomenon might occur when there's too much data available, and it is not appropriately structured. The continuous streams of information provided by augmented reality systems might be more of a hinder than an enhancement. Therefore, it is vital to filter the data you receive through AR devices to avoid overload and prevent distraction.

(ii) Dependence on AR technologies

There are concerns that people responsible for making key decisions might rely too much on AR enhancements. Should that happen, then any type of disruption or destruction of augmented reality systems will allow the enemy to gain a significant advantage. Therefore, besides using AR technologies, basic decision-making tools (like maps) must be maintained.

(iii) Security

The way of handling the communication and data storing by AR and VR systems has always been one of the main concerns in the military along with the policy of AR applications and equipment and their accreditation.

Although all those futuristic scenes of warfare are becoming more feasible, AR technology is quite immature and requires further research and development. Especially in terms of visual registration, overlapping, robustness and the convenience of head-mounted AR eyewear.

4.7 USE OF AR AND VR IN THE INDIAN ARMED FORCES

The Indian Armed Forces have slowly started utilizing VR to simulate and immerse soldiers into virtual training grounds, while using VR-based war games to practice operational strategies at the same time. Whereas simulators have been a part of the forces' inventory since the 1970s, the induction of VR-based systems has taken place only in the last few years. Indeed, in the Army Commanders' Conference of April 2023, the Raksha Mantri (Minister for Defence) reviewed an equipment display that also comprised VR-based systems. The Indian Army has a wargaming centre that is in the process of employing VR and AR technologies in conjunction with AI and data analytics to create metaverse-enabled gameplay. Incidentally, "metaverse for mental health" has been adjudged as one of the top ten emerging technologies of 2023 by the World Economic Forum. A custom-built CBT tool will be used for teaching strategies to student officers and extrinsic factors will also be included and/or modified.

Some Indian defence start-ups have also developed VR applications. HoloSuit, a motion capture haptic suit, has nine haptic feedback devices fitted across the body. When connected to the HoloSuit Engine on any mobile or desktop operating system, the application creates an instant 3D avatar that faithfully reproduces the user's movements, in effect providing a wealth of stance, posture, and reactivity data for in-depth training. The suit is reportedly being used by the Indian Armed Forces. Another start-up, Mumbai-based Parallax Labs, has developed a VR-based personal flight simulator. The first prototype is installed at a naval aviation unit in Goa while talks are on for a second one in Nashik. Certain army units are also using VR systems for terrain familiarization along the Line of Control. Missile simulators are being used by personnel of the Corps of Army Air Defence to provide realistic targeting practice without expending precious surface-to-air missiles.

Meanwhile, AR is being recognized as a breakthrough for mechanized operations, a fact that has been acknowledged in the updated request for information (RFI) for see-through armor in the future-ready combat vehicle (FRCV), which is slated to start domestic production in 2030. A see-through armor combines data and footage from multiple sources, such as drones and nearby vehicles and overlays the same onto a tank gunner's sight, enabling increased field of view within the safety of the tank. The FRCV is also supposed to be integrated with assets on land and air, ensuring a longer detection range and enhancing situational awareness—it is another reason why a tethered drone system has also been included in the FRCV RFI.

CHAPTER 5

HOLOGRAPHY IN WARFARE

5.1 INTRODUCTION TO HOLOGRAPHY

The word holography comes from the Greek words (holos; "whole") and (graphē; "writing" or "drawing").

Holography is a technique that enables a wavefront to be recorded and later reconstructed. It is best known as a method of generating 3-D (three-dimensional) images, and has a wide range of other uses, including data storage, microscopy, and interferometry. In principle, it is possible to make a hologram for any type of wave.

A **hologram** is a recording of an interference pattern that can reproduce a 3-D light field using diffraction. In general usage, a hologram is a recording of any type of wavefront in the form of an interference pattern. It can be created by capturing light from a real scene, or it can be generated by a computer, in which case it is known as a computer-generated hologram, which can show virtual objects or scenes. Optical holography needs a laser light to record the light field. The reproduced light field can generate an image that has the depth and parallax of the original scene. A hologram is usually unintelligible when viewed under diffuse ambient light. When suitably lit, the interference pattern diffracts the light into an accurate reproduction of the original light field, and the objects that were in it exhibit visual depth cues such as parallax and perspective that change realistically with the different angles of viewing. That is, the view of the image from different angles shows the subject viewed from similar angles.

A hologram is traditionally generated by overlaying a second wavefront, known as the reference beam, onto a wavefront of interest. This generates an interference pattern, which is then captured on a physical medium. When the recorded interference pattern is later illuminated by the second wavefront, it is diffracted to recreate the original wavefront. The 3-D image from a hologram can often be viewed with non-laser light. However, in common practice, major image quality compromises ca be removed by laser illumination to view the hologram.

A computer-generated hologram is created by digitally modeling and combining two wavefronts to generate an interference pattern image. This image can then be printed onto a mask or film and illuminated with an appropriate light source to reconstruct the desired wavefront. Alternatively, the interference pattern image can be directly displayed on a dynamic holographic display.

Holographic portraiture often resorts to a non-holographic intermediate imaging procedure, to avoid the dangerous high-powered pulsed lasers which would be needed to optically "freeze" moving subjects as perfectly as the extremely motion-intolerant holographic recording process requires.

Early holography required high-power and expensive lasers. Currently, mass-produced low-cost laser diodes, such as those found on DVD recorders and used in other common applications, can be used to make holograms. They have made holography much more accessible to low-budget researchers, artists, and dedicated hobbyists.

Most holograms produced are of static objects, but systems for displaying changing scenes on dynamic holographic displays are now being developed.

5.2 ORIGINS OF HOLOGRAMS

The first holograms had no connection with military goals or contexts, at least not directly. The principle of the hologram was, however, conceived by the Jewish Hungarian engineer Dennis Gabor in England, who had felt the need to emigrate from Germany in 1933 following the rise of Adolph Hitler to power. Ironically, Gabor's development projects at British Thomson-Houston (BTH), a major military contractor during the Second World War, were determined by his status as a potential enemy alien: he was excluded from BTH's war work on projects such as radar and infrared detection, and was segregated in a building outside the secure area of the company premises. This physical and intellectual exclusion may have contributed to Gabor's attention to innovative commercial concepts.

In 1947, Gabor conceived the hologram process, a form of two-step imaging. First, light of a single wavelength and point of origin (i.e. coherent radiation), would cast a shadow of an opaque object, a shadow ringed by bright and dark fringes owing to the diffraction of light by the object's edges, and subsequent constructive and destructive interference. This physical shadow or hologram would be recorded on photographic film. Second, the processed film would be situated in a beam of coherent light, and the fringes would diffract the light to reconstruct an image of the original object. This complicated and seemingly pointless procedure had some advantages in principle. Gabor imagined it solely with reference to the electron microscopes being developed by a sister company (AEI) during the 1940s. Such microscopes were intrinsically limited by poor-quality electron lenses; Gabor hoped that by recording the physical shadow from the coherent electron beam and then reconstructing with visible light, it would be possible to correct for the optical aberrations using high quality optical lenses, and so yield a higher-resolution image. He also suggested that this technique would allow three-dimensional imaging, or at least an image having a large depth of field. In practice, however, Gabor's ideas were stillborn. Between 1948 and about 1955, he collaborated with AEI colleagues to generate electron microscope holograms; at BTH, and later with his doctoral students at Imperial College, London, where he became Reader in Electronics, he attempted optical recording and reconstruction of holograms. Results were poor and did not impress either his principal audience – electron microscopists – or influential optical physicists such as Sir Lawrence Bragg and Max Born.

In the late 1963, lasers permitted their most spectacular achievement in holography, which was reconstruction of three-dimensional images of solid objects.

5.3 APPLICATIONS OF HOLOGRAPHY IN WARFARE

Early holograms in the military paved the way for advanced applications, showcasing their potential to revolutionize training simulations, enhance situational awareness, and transform communication methods on the battlefield.

Holograms can also be used for strategic planning and situational awareness.

Holographic technology has many potential applications in modern and future warfare, including:

(i) 3-D visualizations

Holograms can create 3D models of terrain, assets, and personnel, which can help with decision-making and coordination.

(ii) Improved reconnaissance

Holographic maps can help soldiers view terrain in 3D, look around corners, and train for missions.

(iii) Virtual simulations

Holograms can be used to simulate realistic aircraft and equipment maintenance environments, and for advanced mission rehearsals.

(iv) Visual deception

Holograms could revolutionize visual deception and transform military tactics.

(v) Remote briefings and teleconferencing

Holograms can be used for remote briefings and teleconferencing.

Holographic technology is also used in other fields, such as medicine, weather forecasting, virtual reality, digital art, and security.

By using holograms to augment traditional planning tools, military leaders can make more informed decisions, reduce risk and increase the success of their operations. By visualizing complex environments, mission parameters, and potential scenarios, holograms provide a comprehensive understanding of the operational landscape with holographic maps, aiding in effective mission planning, risk assessment, and resource allocation.

If the holographic technology advances to where holographic systems could be portable, the display large enough, and the resolution high enough, holograms could be used as camouflage or to augment reality for deception.

5.4 HOLOGRAMS IN TACTICAL OPERATIONS

Holograms have transformed tactical operations within military contexts, offering a novel approach to strategy and communication. Their capacity to project three-dimensional images allows commanders and soldiers to visualize complex scenarios in real-time, thus enhancing collaborative efforts.

In operations, holograms can serve multiple functions, such as:

- Visualizing enemy movements and terrain dynamics;
- Communicating detailed mission plans to ground troops;
- Conducting reconnaissance through virtual representations.

The integration of holographic technology into tactical frameworks fosters a more immersive decision-making process. As military personnel engage with dynamic holograms, they gain a deeper understanding of mission parameters and immediate surroundings, improving situational responsiveness.

Moreover, holographic displays can deliver critical intelligence, ensuring that all units have synchronized information. This real-time data sharing significantly mitigates the risk of miscommunication during high-stakes operations, reinforcing the efficacy of tactical maneuvers under pressure.

5.5 HOLOGRAPHIC DISPLAY AND MILITARY OPERATIONS

The utilization of holograms in military communications represents a groundbreaking innovation. Holographic technology enables interactive displays that convey complex information clearly and efficiently. As the military explores the use of holograms in operations, it stands to revolutionize how information is shared and interpreted in critical situations.

Following are some ways to show that holographic display technology is changing the landscape of next-generation warfare:

(i) Improved Contextual Awareness

On the battlefield, holographic displays are essential for improving situational awareness. The Holographic Display provides soldiers a perspective of the battlefield that is unmatched. A lidar scanner and holographic display technology make sure that the augmented data blends in perfectly with the surrounding surroundings. Holographic situational awareness gives you the ability to see past walls, maneuver through dangerous terrain with ease, and predict opponent movements.

Key advantages of holograms in enhancing situational awareness include:

- **Realism:** Holograms offer a lifelike representation of various scenarios, helping soldiers anticipate real-world challenges.
- **Integration:** Seamlessly integrating various data sources enables clearer analysis of complex situations.
- **Collaboration:** Teams can interact with shared holographic images, fostering cooperative strategies and quick responses.

(ii) Planning a mission and accurate targeting

For the purpose of limiting collateral damage and guaranteeing mission success, accurate targeting is essential. Commanders now have a new way to organize and visualize missions in three dimensions: holographic screens. Through the use of holographic depictions of the battlefield, soldiers can engage in various engagement scenarios, launch virtual forces, and identify targets with extreme precision.

(iii) Improved Cooperation and Communication

In a combat zone, a breakdown in communication can have catastrophic consequences. Effective communication is critical in the military. Soldiers may exchange vital information, give orders, and plan actions in real time without having to worry about enemy eavesdropping when they use holographic communication.

No matter where units are physically located, holographic display technology enables improved communication and collaboration among them. Commanders may view and plan in real-time via holographic projections, which facilitates quicker decision-making. Military leaders may communicate 3D holographic models of topography, mission goals, and tactical plans.

(iv) Advanced Training and Simulations

Soldiers must receive quality training in order to be prepared for the reality of combat. Conventional training techniques frequently lack realism and are unable to accurately simulate the confusion and stress of combat.

With the use of holographic simulations, soldiers can train in an incredibly realistic setting while facing hypothetical dangers, practicing tactical maneuvers, and refining their decision-making abilities. These realistic scenarios offer soldiers priceless experience that boosts their self-assurance and gets them ready for any eventuality in the real world.

(v) Psychological Warfare and Deception

The use of holographic projection technology in military operations creates new avenues for strategic deception and misdirection. One is able to produce holographic decoys and illusions that can fool and distract foes. There are difficulties in incorporating holographic display technology into military operations. It is necessary to address concerns regarding costs, data security, and ethical factors. Nonetheless, there is no denying this technology's prospective advantages.

5.6 CHALLENGES AND LIMITATIONS OF HOLOGRAPHIC TECHNOLOGY

Holographic technology, while promising, presents several challenges and limitations when applied in military contexts. One significant concern is the high cost associated with developing and implementing advanced holographic systems. The initial investment in equipment, as well as the continuous updates needed for maintenance, can strain military budgets.

Another limitation lies in the technological constraints of current holographic displays. Many systems have issues with resolution and clarity, particularly in outdoor environments, where light interference can distort images. This can hinder the effectiveness of holograms in critical military communications.

Additionally, the complexity of operating holographic technology requires specialized training for personnel. Without adequate training, the potential benefits of holographics in military operations could be undermined, leaving soldiers ill-prepared to leverage these innovations effectively.

Traditional communication methods still dominate military operations, as these established systems are often more reliable. Thus, despite the potential advantages of the use of holograms in military applications, overcoming these challenges is essential for successful integration into military communications.

5.7 FUTURE PROSPECTS OF HOLOGRAMS IN MILITARY APPLICATIONS

What will military holographic displays look like in 10 years? Some think we will see deployed battlespace sand tables with a field of view of 100-to 120-degrees, HD resolution and several feet of depth volume by generating 10 billion plus rays.

Others think these sand tables will be here as well, but not yet operational with the military still doing human factors testing to better determine how best to visualize the battlespace and to quantify the collaborative and cognitive benefits it delivers. This is a necessary step before actual deployment in live military activities.

It seems clear that there is wide interest in advanced 3-D displays in the military and government agencies. But there are also many other uses for these holographic and light-field displays, so consumer and commercial activities may be enough to help with the military advancements. If not, government agencies and the armed services may have to step up to further fund or speed up development for their more urgent use cases.

The future of holograms in military applications appears promising as technological advancements continue to reshape military communications. Enhanced holographic systems could significantly improve the quality of real-time communications, allowing for more effective decision-making during operations.

Next-generation holograms may facilitate advanced training programs, providing soldiers with unparalleled exposure to realistic scenarios without the need for extensive physical setups. By integrating augmented reality, these simulations can adapt to various combat environments, enhancing preparedness.

Moreover, the use of holograms in command centers could revolutionize strategic planning. Real-time, three-dimensional visuals of battlefield dynamics would enable commanders to devise more accurate and effective strategies, fostering collaboration among units in diverse locations.

Military requires holograms for battlefield intelligence, military planning and explosives disposal purposes. Situational awareness as some critical battlefield information would be better understood when viewed in three dimensions rather than two. The desire to enable the U.S. military to view such information via holograms is the driving force behind a number of research projects under way at some of the nation's top universities and imaging companies. Military also interested in holography as a psychological warfare tool. The "Face of Allah" weapon would beam a massive, lifelike hologram over a battlefield, projecting the image of some deity "to incite fear in soldiers on a battlefield," according to one researcher.

Scientists at the University of Arizona have demonstrated "actual moving holograms that are filmed in one spot and then projected and viewed in another spot." It is like the holograms to the tiny image of Princess Leia that R2D2 showed Luke Skywalker in the beginning of Star Wars, only "a lot more haltingly, as the display changes only every two seconds."

This hologram is created by a suite of 16 cameras that use lasers to record data on "smart" plastic some distance away that, when hit by a special light, project the image in solid-looking 3-D. A partner team at Columbia University is studying ways to beam the holo-data via the Internet, to allow 3-D video chats or instantaneous transmission of holographic maps, blueprints or medical scans. It might take a decade for the technology to become affordable and widespread. Weaponization would be much further behind (though we wouldn't bet on today's cash-strapped military to invest in a Face-of-Allah gun). Cost aside, it's just not very PC.

Early holograms are already a fixture in military headquarters. A company called Zebra Imaging in Texas has been selling 2-by-3-foot plastic holographic maps to the Pentagon – its "main customer" – for $1,000 to $3,000 a pop. The military "sends data in computer files to the company. Zebra then renders holographic displays of, for example, battlefields in Iraq and Afghanistan." No goofy 3D glasses required, just a custom-made LED flashlight that "activates" the image encoded in the plastic.

Zebra's technology has other military applications, such as post-blast IED forensics, according to the company's Website. "Analysts trying to understand the nature and construction of an explosive device ... are able to understand the scene in 3D far better than the classic 2D 'bird's-eye view.'"

As research progresses, the **integration of artificial intelligence with holographic technologies** is likely to emerge. This synergy could lead to predictive analytics, thereby improving threat assessments and operational readiness, ultimately transforming military communications.

5.8 HOLOGRAPHIC HEAD-UP DISPLAYS

A head-up display, or heads-up display, also known as a HUD, is any transparent display that presents data without requiring users to look away from their usual viewpoints. The origin of the name stems from a pilot being able to view information with the head positioned "up" and looking forward, instead of angled down looking at lower instruments. A HUD also has the advantage that the pilot's eyes do not need to refocus to view the outside after looking at the optically nearer instruments.

Although they were initially developed for military aviation, HUDs are now used in commercial aircraft, automobiles, and other (mostly professional) applications.

Head-up displays were a precursor technology to augmented reality (AR), incorporating a subset of the features needed for the full AR experience, but lacking the necessary registration and tracking between the virtual content and the user's real-world environment.

HUD systems have been developed for military fighter aircraft in the 1950's. Still in use, they provide critical information to the pilot without them having to look at their instrumentation – a big deal when travelling faster than the speed of sound where a millisecond could truly be the difference between life and death.

On aircraft avionics systems, HUDs typically operate from dual independent redundant computer systems. They receive input directly from the sensors (pitot-static, gyroscopic, navigation, etc.) aboard the aircraft and perform their own computations rather than receiving previously computed data from the flight computers. On other aircraft (the Boeing 787, for example) the HUD guidance computation for Low Visibility Take-off (LVTO) and low visibility approach comes from the same flight guidance computer that drives the autopilot. Computers are integrated with the aircraft's systems and allow connectivity onto several different data buses such as the ARINC 429, ARINC 629, and MIL-STD-1553.

5.8.1 Displayed data symbology of a head-up display

Typical aircraft HUDs display airspeed, altitude, a horizon line, heading, turn/bank and slip/skid indicators. These instruments are the minimum required by 14 CFR Part 91. *Title 14 CFR – Aeronautics and Space is one of the fifty titles that make up the United States Code of Federal Regulations (CFR).*

Other symbols and data are also available in some HUDs:

(i) **Boresi**ght or waterline symbol — is fixed on the display and shows where the nose of the aircraft is actually pointing.

(ii) Flight path vector (FPV) or velocity vector symbol — shows where the aircraft is actually going, as opposed to merely where it is pointed as with the boresight. For example, if the aircraft is pitched up but descending as may occur in high angle of attack flight or in flight through descending air, then the FPV symbol will be below the horizon even though the boresight symbol is above the horizon. During approach and landing, a pilot can fly the approach by keeping the acceleration indicator or energy cue — typically to the left of the FPV symbol, it is above it if the aircraft is accelerating, and below the FPV symbol if decelerating.

(iii) Angle of attack indicator — shows the wing's angle relative to the airflow, often displayed as "α".

(iv) Navigation data and symbols — for approaches and landings, the flight guidance systems can provide visual cues based on navigation aids such as an Instrument Landing System (ILS) or augmented Global Positioning System such as the Wide Area Augmentation System. Typically this is a circle which fits inside the flight path vector symbol. Pilots can fly along the correct flight path by "flying to" the guidance cue.

Since being introduced on HUDs, both the FPV and acceleration symbols are becoming standard on head-down displays (HDD.) The actual form of the FPV symbol on an HDD is not standardized but is usually a simple aircraft drawing, such as a circle with two short angled lines, (180 ± 30 degrees) and "wings" on the ends of the descending line. Keeping the FPV on the horizon allows the pilot to fly level turns in various angles of bank.

5.9 MILITARY AIRCRAFT-SPECIFIC APPLICATIONS OF HUDS

In addition to the generic information described above, military applications include weapons system and sensor data such as:

(i) **Target designation (TD) indicator** — places a cue over an air or ground target (which is typically derived from radar or inertial navigation system data.)

(ii) V_c — closing velocity with target.

(iii **Range** — to target, waypoint, etc.

(iv) **Weapon seeker or sensor line of sight — shows where a seeker or sensor is pointing.**

(v) **Weapon status — includes type and number of weapons selected, available, arming, etc.**

5.9.1 VTOL/STOL approaches and landings

During the 1980s, the United States military tested the use of HUDs in vertical take-off and landing (VTOL) and short-take-off and landing (STOL) aircraft. A HUD format was developed at NASA Ames Research Center to provide pilots of VTOL and STOL aircraft with complete flight guidance and control information for Category III C terminal-area flight operations. This includes a large variety of flight operations, from STOL flights on land-based runways to VTOL operations on aircraft carriers. The principal features of this display format are the integration of the flightpath and pursuit guidance information into a narrow field of view, easily assimilated by the pilot with a single glance, and the superposition of vertical and horizontal situation information. The display is a derivative of a successful design developed for conventional transport aircraft.

5.9.2 Holographic Head-up Displays

In recent years, it has been argued that conventional HUDs will be replaced by holographic AR technologies, that use Holographic Optical Elements (HOE.) The HOE allows for a wider field of view while reducing the size of the device and making the solution customizable.

The advantage of holographic HUDs is the depth information and the possibility to project objects as augmented and mixed reality in 3D in the pilot's field of view. Hence, holographic HUDs are paramount to explore the possibilities of augmented and mixed reality for air safety.

What is the key phase that military end users and their holographic display development partners point to as the key application? That phase is battlespace visualization. Situational awareness (SA) and establishing a common operating picture (COP) are some of the most critical operational objectives for warfighters today — especially those responsible for making strategic and tactical decisions. Holographic or light field displays represent next generation 3D displays that these military users want. Given the 2022 Russian invasion of Ukraine, such visualizations may have taken on new urgency.

The Ukraine situation presents another interesting scenario — tracking Russian and Ukrainian forces and providing intelligence as appropriate. Advanced 3-D display systems would offer a clear advantage in conflicts like this.

Holographic displays should enable a more intuitive visualization of the battlespace, say industry officials. The most obvious scenario is visualizing all of the assets and personnel in the battlespace, by their 3D positions and motion vectors. But there are many other data problems that better visuals can help solve. For example, it is often necessary to model the coverage area of military radios to understand when signals will be blocked and/or reflected off of surfaces. With a holographic display, a field operator can quickly see if he can communicate with another person or asset.

In air operations, telemetry data may show the altitude and directional vector of an asset, but it can be hard to visualize if it is ascending or descending, and at what rate. Again, a 3-D holographic display makes such information more intuitively obvious.

For naval operations, it takes years of training to understand sonar signals and learn how to mentally map underwater objects in 3D space. Turning such signals into a visual 3-D volume representation reduces the cognitive load and allows less-trained operators to have the same skill sets as experienced ones.

CHAPTER 6

DRONE TECHNOLOGY IN WARFARE

6.1 WHAT IS A DRONE?

An unmanned aerial vehicle (UAV), commonly known as a drone, is an aircraft with no human pilot, crew, or passengers on board. UAVs were originally developed through the twentieth century for military missions too "dull, dirty or dangerous for humans, and by the twenty-first century, they had become essential assets to most militaries. As control technologies improved and costs fell, their use expanded to many non-military applications. These include aerial photography, area coverage, precision agriculture, forest fire monitoring, river monitoring, environmental monitoring, policing and surveillance, infrastructure inspections, smuggling, product deliveries, entertainment, and drone racing.

An unmanned aerial vehicle (UAV) is defined as a "powered, aerial vehicle that does not carry a human operator, uses aerodynamic forces to provide vehicle lift, can fly autonomously or be piloted remotely, can be expendable or recoverable, and can carry a lethal or nonlethal payload". UAV is a term that is commonly applied to military use cases. Missiles with warheads are generally not considered UAVs because the vehicle itself is a munition, but certain types of propeller-based missile are often called "kamikaze drones" by the public and media. Also, the relation of UAVs to remote controlled model aircraft is unclear, UAVs may or may not include remote-controlled model aircraft. Some jurisdictions base their definition on size or weight; however, the US FAA (Federal Aviation Administration) defines any unmanned flying craft as a UAV regardless of size. A similar term is remotely piloted aerial vehicle (RPAV).

UAVs or RPAVs can also be seen as a component of an unmanned aircraft system (UAS), which also includes a ground-based controller and a system of communications with the aircraft. The term UAS was adopted by the United States Department of Defense (DoD) and the United States Federal Aviation Administration (FAA) in 2005 according to their Unmanned Aircraft System Roadmap 2005–2030. The International Civil Aviation Organization (ICAO) and the British Civil Aviation Authority adopted this term, also used in the European Union's Single European Sky (SES) Air Traffic Management (ATM) Research (SESAR Joint Undertaking) roadmap for 2020. This term emphasizes the importance of elements other than the aircraft. It includes elements such as ground control stations, data links and other support equipment. Similar terms are unmanned aircraft vehicle system (UAVS) and remotely piloted aircraft system (RPAS). Many similar terms are in use. Under new regulations which came into effect 1 June 2019, the term RPAS has been adopted by the Canadian Government to mean "a set of configurable elements consisting of a remotely piloted aircraft, its control station, the command-and-control links and any other system elements required during flight operation".

The term drone has been used from the early days of aviation, some being applied to remotely flown target aircraft used for practice firing of a battleship's guns, such as the 1920s Fairey Queen and 1930s de Havilland Queen Bee. Later examples included the Airspeed Queen Wasp and Miles Queen Martinet, before the ultimate replacement by the GAF Jindivik. The term remains in common use. In addition to the software, autonomous drones also employ a host of advanced technologies that allow them to carry out their missions without human intervention, such as cloud computing, computer vision, artificial intelligence, machine learning, deep learning, and thermal sensors. For recreational uses, an aerial photography drone is an aircraft that has first-person video, autonomous capabilities, or both.

6.2 WHO INVENTED THE DRONE AND WHEN WAS IT?

The invention of the first drone is often credited to **Karem**, an aerospace engineer and innovator whose work in the 1970s and 1980s laid the foundation for modern UAVs. Karem, born in Baghdad and raised in Israel, was passionate about aviation and engineering from a young age.

His most notable achievement came after he moved to the United States in the 1970s, where he founded his companies, 'Leading Systems, Inc.', and 'Karem Aircraft'. He began developing an unmanned aerial vehicle, and he created a drone prototype called the **Albatross**.

While this prototype was a far cry from today's lightweight drones — it weighed 200 pounds! The technology caught the attention of the US military and the US Defense Advanced Research Projects Agency (DARPA).

Under contract with DARPA, Karem helped evolve the Albatross technology into what eventually became the Predator drone, one of the most successful UAVs in military history. The Predator became a key asset in reconnaissance and combat operations, showcasing the potential of UAVs in modern warfare and significantly influencing modern drone warfare and surveillance.

Karem's innovations didn't stop with the Predator. His contributions to drone technology continued to influence the development of more advanced drones, both in military and civilian applications. His work laid the groundwork for the proliferation of drones we see today, making him a pivotal figure in the history of unmanned aerial vehicles.

6.3 HOW WAS THE DRONE INVENTED?

While Karem's work was key in shaping the drone industry, the invention of the drone was not the result of a single event or individual but rather a series of technological advancements and innovations that spanned centuries.

The idea of creating a pilotless aircraft dates back to the 19th century when inventors and engineers began experimenting with remote-controlled and unmanned vehicles.

One of the earliest attempts to create an unmanned aircraft occurred in 1783 when French inventors Joseph-Michel and Jacques-Étienne Montgolfier launched the first unmanned hot air balloon. While not a drone in the modern sense, this event marked the beginning of humanity's exploration of unmanned flight. In the years that followed, inventors continued to push the boundaries of aviation, leading to the development of the first radio-controlled aircraft in 1898. This innovation, created by Nikola Tesla, allowed for the remote control of vehicles without the need for direct human intervention, a concept that would later become a cornerstone of drone technology.

6.4 DRONE-HISTORY TIMELINE

The history of drones is a fascinating journey through time, marked by significant milestones and technological breakthroughs. Below is a timeline that highlights key events in the evolution of drone technology, starting with unmanned air balloons, model aircraft, and eventually evolving into what we know as a modern drone today:

(i) 1783 – First Unmanned Aircraft

French inventors Joseph-Michel and Jacques-Étienne Montgolfier launched the first unmanned hot air balloon. Their linen and silk balloon, fueled by a stove burning wool and straw, ascended roughly 6,000 feet and traveled over a mile in 10 minutes. This event marked the beginning of humanity's exploration of unmanned flight and the first time in recorded history that humans have used armed drones.

(ii) 1849 – First Military Use of UAVs

In the earliest recorded military use of drones, the Austrian military used unmanned balloons loaded with explosives to attack Venice, Italy. However, the incendiary balloons were reported to be largely ineffective after the wind shifted and the majority missed their target.

(iii) 1898 – First Radio-Controlled Craft

Nikola Tesla, Serbian-American inventor and engineer, demonstrated the first radio-controlled craft with his invention of a radio-controlled boat. The boat was only about three feet long and While this craft was on water, it showcased the potential of remote-controlled vehicles. This innovation laid the groundwork for the development of remotely-controlled aircraft.

(iv) 1917 – The First Aerial Torpedo

In 1917, American Charles F. Kettering of Dayton, Ohio, created the first unmanned aerial torpedo for the US Army. It was known as the Kettering Aerial Torpedo, or "Bug" and had a maximum speed of approximately 50 mph and a maximum range of approximately 75 miles. Test flights gave mixed results, and the Bug was never used in combat. Fewer than 50 Bugs were produced, and a full-size reconstruction is on display at the National Museum of the United States Air Force in Dayton, Ohio.

(v) 1935 – The First Modern Drone

The British Royal Navy introduced the de Havilland DH82B "Queen Bee," a remotely controlled aircraft used for target practice. 412 of the DH82B were built. The Queen Bee is considered one of the first modern drones and marked the beginning of the drone technology program in the United Kingdom.

The metaphor of a "bee" is thought to contribute to the use of the word "drone" to describe UAVs. However, this term has been in popular use throughout history.

(vi) 1936 – The US Drone Program Starts

The US began its own drone program in 1936, developing the "Radioplane" also called the Curtiss N2C-2, a remote-controlled propeller aircraft used for training anti-aircraft gunners in the military during World War II. More than 9,400 of the Radioplane OQ-3 were produced during World War II. After the war, more than 60,000 of an updated Radioplane BTT model were produced. The Radioplane program marked the United States' entry into the field of unmanned aerial vehicles.

(vii) 1943 – The First First-Person View Aircraft

In 1943, the U.S. Air Force and Boeing collaborated to develop the BQ-7, an early attempt at a first-person view (FPV) flight system. This crude FPV technology was integrated into modified bombers, which were stripped of non-essential equipment and loaded with explosives. A human pilot would manually fly the aircraft toward a target, and once it was in sight, they would engage the autopilot and bail out. The BQ-7 would then continue on its path to the target autonomously.

However, the BQ-7 proved largely ineffective in warfare, with many of the pilots who bailed out either dying or being captured. Despite its challenges, this early effort at FPV flight highlighted both the potential and the significant challenges of remotely controlled aerial systems.

During World War II, the German military also developed the "V-1 flying bomb," an early cruise missile with a first-person view (FPV) system. This technology allowed operators to guide the missile to its target using a live video feed.

(viii) 1964 – First Large-Scale Military Deployment of Drones in War

The US first deployed reconnaissance UAVs at a large scale during the Vietnam War. The models they used were called Ryan Model 147s or Lightning Bugs. Over 3,400 drones were deployed, each one completing an average of three missions before being lost. During this time, the military also started using drones for additional tasks such as combat decoys, missile launches, and item distribution.

(ix) 1986 – The RQ2 Pioneer Drone

The United States Navy and Marine Corps introduced the RQ2 Pioneer, a UAV used for reconnaissance and surveillance during the Gulf War. The Pioneer became one of the first widely used drones in modern warfare.

(x) 1996 – The Predator Drone

Based on the drone technology developed by Abraham Karem, General Atomics released the MQ-1 Predator in 1996. The Predator has become one of the most iconic military drones, known for its role in reconnaissance and targeted strikes.

(xi) 2006 – UAVs Permitted in US Civilian Airspace

In 2006, the United States Federal Aviation Administration (FAA) began allowing the use of UAVs in civilian airspace under specific regulations. This marks the beginning of the widespread use of drones in the United States.

(xii) 2013 – DJI Phantom Drone

In 2013, DJI, a Chinese technology company, released the Phantom drone, an extremely popular consumer-friendly UAV that revolutionized the drone market. The Phantom's ease of use, affordability, and advanced features make it a popular choice for hobbyists and professionals alike. Now, these drones and similar consumer models can be seen everywhere—used for everything from travel vlogs to real estate listings and more.

6.5 COMMON TYPES OF DRONES

Drones come in various shapes and sizes, each designed with different features for specific applications. Here are some of the most common types of drones:

(i) **Military Drones:** These are the most advanced and sophisticated drones, used primarily for reconnaissance, surveillance, and combat operations. Examples include the Predator and the RQ-4 Global Hawk.

(ii) **Consumer Drones:** These are drones designed for recreational use, photography, and videography. The DJI Phantom and Mavic series are some of the most popular examples of consumer drones used for tasks like photography and videography.

(iii) **Commercial Drones:** These drones are used for industrial applications such as agriculture, geological surveying, and building inspection. They often feature advanced sensors and cameras for specialized tasks.

(iv) **Agricultural Drones:** This subset of commercial drones is designed with special features to perform agricultural tasks like crop monitoring, irrigation, and soil analysis.

(v) **Delivery Drones:** Another common subset of commercial drones, companies like Amazon and Google are developing drones that can deliver packages directly to customers. These drones are designed for efficiency and precision and can navigate complex environments like cities and neighborhoods.

(vi) **Racing Drones:** These small, high-speed drones are designed for competitive racing. They are equipped with FPV systems that allow pilots to precisely control them.

(vii) **Search and Rescue Drones:** These drones are used in emergency situations to locate and assist people in distress, such as hikers, skiers, and natural disaster survivors. Search and rescue drones are often equipped with thermal imaging cameras and other sensors to detect heat signatures from body heat.

6.6 THE FUTURE OF DRONES

As drone technology continues to evolve, the future of UAVs looks promising.

The integration of artificial intelligence and machine learning into drones is expected to enhance their capabilities, allowing for more autonomous operations and advanced decision-making.

One of the most exciting developments in drone technology is the potential for urban air mobility (UAM). Companies like Uber and Airbus are exploring the possibility of using drones for passenger transport, creating a new mode of urban transportation that could alleviate traffic congestion and reduce travel times.

In the military domain, drones are expected to play an even more significant role in national security and warfare, with advancements in stealth technology, endurance, and payload capacity.

The development of **"swarms" of drones**, capable of working together to complete complex tasks, is also on the horizon. (see picture on the cover of this book)

On the consumer side, drones are becoming more accessible and affordable, with improvements in battery life, camera quality, and ease of use. The rise of 5G technology is expected to enhance real-time communication between drones and operators, enabling more precise control and data transmission.

As UAVs become more prevalent in civilian airspace, governments worldwide are working to create frameworks that ensure safety, privacy, and security. Balancing innovation with regulation will be key to unlocking the full potential of drone technology.

6.7 COMBAT DRONE

An unmanned combat aerial vehicle (UCAV), also known as a combat drone, fighter drone or battlefield UAV, is an unmanned aerial vehicle (UAV) that is used for intelligence, surveillance, target acquisition, and reconnaissance and carries aircraft ordnance such as missiles, anti-tank guided missiles (ATGMs), and/or bombs in hardpoints for drone strikes. These drones are usually under real-time human control, with varying levels of autonomy. Combat drones are used for reconnaissance, attacking targets and returning to base; unlike kamikaze drones which are only made to explode on impact, or surveillance drones which are only for gathering intelligence.

As the operator runs the vehicle from a remote terminal, equipment necessary for a human pilot is not needed, resulting in a lower weight and a smaller size than a manned aircraft. Many countries have operational domestic drones, and many more have imported fighter drones or are in the process of developing them.

6.8 THE RACE FOR KILLER DRONES

Russian President Vladimir Putin has stated that the nation that leads in AI "will become the ruler of the world." The advancement of AI in modern warfare will forever alter the relationships between great powers like the United States, China, and Russia as well as the private technology industry. For this reason, China has committed 150 billion dollars to become the world leader in AI technology, compared to Russian spending of 181 million dollars from 2021 to 2023 and U.S. spending of 4.6 billion dollars. In 2019, Jane's stated that more than 80,000 surveillance drones and almost 2,000 attack drones will be purchased around the world in the next decade. The UK operates missile-bearing drones and plans to spend 415 million pounds on Protector drones by 2024. Saudi Arabia also cannot be underestimated as a newer entrant in the drone marketplace, having invested 69 billion dollars in 2023—23 percent of its national budget—on defense. Additionally, Saudi Arabia plans to create a 40-billion-dollar fund to invest in AI, which would make it the world's largest AI investor.

With spending on drone and AI development increasing so rapidly, advancements in technology might eventually enable drones to make decisions instantaneously without human input during conflict. This may potentially eliminate peaceful negotiation in conflict, as drones' reactions will consist purely of retaliatory violence. Drone technology has already advanced from its use by NATO to identify hidden Serbian strategic positions during the Kosovo War in 1999 to its use by the United States in the immediate aftermath of the September 11 terrorist attacks. After an ISR drone successfully located Osama Bin Laden, the U.S. military increasingly used and outfitted drones with lethal payloads, carrying out 14,000 drone strikes in Afghanistan alone from 2010 to 2020.

The United States, the United Kingdom, and Israel remain the largest users of drones, and their arsenals continue to grow. The United States and the United Kingdom have used weaponized drones for over a decade, including the Predator and the Reaper, both made by the California-based company General Atomics. According to Drone Wars, in four years of conflict in Syria from 2014 to 2018, the United Kingdom used Reaper drones more than 2,400 times during strategic missions, the equivalent of two per day. The Pentagon estimates that by 2035, remotely piloted aircraft will make up 70 percent of the U.S. Air Force. Meanwhile, Israel has been developing its own weaponized drones, and it has deployed drones in Gaza to conduct surveillance, deliver explosives, and more.

Furthermore, drone technology is spreading rapidly to militaries around the world. Nearly every NATO member state now has the capability to use drones in conflict. In the last five years, both Turkey and Pakistan have also created drone manufacturing programs. China currently supplies several states with its Wing Loong and CH-series drones, including the UAE, Egypt, Saudi Arabia, Nigeria, and Iraq. Even non-state actors are using drones. Hezbollah has used Iranian-built reconnaissance drones to violate Israeli airspace, while Hamas has been using drones against Israel since October 2023.

AI use in warfare is also spreading rapidly. Reports suggest that Ukraine has equipped its long-range drones with AI that can autonomously identify terrain and military targets, using them to launch successful attacks against Russian refineries. Israel has also used the "Lavender" AI system in the conflict in Gaza to identify 37,000 Hamas targets. Accordingly, the current conflict between Israel and Hamas has been dubbed the first "AI war." However, no evidence indicates that an AWS, a system without significant human control, has been used in conflict yet.

As anxieties grow over the emergence of so-called "killer robots," AI use in warfare raises increasingly salient ethical and legal questions. In particular, drones may not be able to distinguish the difference between combatants and civilians. Thankfully, many AI technologies are still in development. The "killer robots" analogy refers to unmanned aircraft that can operate autonomously;

however, most current AI only functions well in a narrow, predetermined set of circumstances, with input by human operators.

Nonetheless, the increasing integration of AI into drones and other AWS creates very real dangers of conflict being decided without meaningful human control. The use of violence during conflict may be determined by the instincts of machines incapable of navigating the moral ambiguities of war and making ethical decisions. It is impossible to predict how the law will keep up with or even stop such technological advances, but the current legal framework certainly lacks clarity and foresight.

6.9 SWARM DRONES - NEW FRONTIER OF WARFARE

Swarm drones are multiple unmanned aerial flying platforms integrated as a single networked system self-contained for communication, reconnaissance and weapons/munitions to strike an enemy target.

The concept of swarm drone operation envisages integrating and flying a large number of UAVs carrying sensors, weapons/munitions and communication equipment on board and obtaining inputs, building a comprehensive battle picture and communicating to multiple users in a simple manner and in real-time. Such a concept increases the probability of success of a mission as there are built in redundancies in the concept. For example, if an UAV is shot down or jammed its functions get taken over by another UAV. Swarm drones are multiple unmanned aerial flying platforms integrated as a single networked system self-contained for communication, reconnaissance and weapons/munitions to strike an enemy ground-based target. The successful use of drones by Azerbaijan which was supported by Turkey and Israel has comprehensively validated the employment of drones in a conventional setting beyond doubt. This once again has upheld the supremacy of air power in a future battlefield against a legacy large ground forces with relatively poor air defence forces.

Drone Warfare is being extensively and effectively used by many countries to conduct hybrid warfare against their adversaries. For example, the US has employed drones with telling effect in the Afghanistan war to target the fighters and leaders of Al-Qaeda and Taliban. Recently it targeted the Iranian Major General Qasem Soleimani. Other countries that have used drones are Turkey against the Kurdistan Workers' Party, Nigeria against Boko Haram, Iraq against the Islamic State in Iraq and Syria (ISIS), and Saudi Arabia in Libya and Yemen. Swarm drone attacks with loiter capability has made drone warfare deadlier and even more lethal. A classic example of swarm drone attack was conducted by Yemeni Houthi rebels on two Aramco installations of Saudi Oil Companies by 18 drones and three low-flying missiles which many believe was actually conducted by Iran given the sophistication of the attack. However, drone warfare shot into limelight during the recent Azerbaijan - Armenian conflict over Nagorno-Karabakh. During the conflict drones were used by Azeri defence forces against the Armenian land forces with telling effect.

6.10 SWARM DRONES SCENARIO IN THE WORLD

The United States, China, Russia, and Britain are in advance stages of developing this technology. In 2017, the US carried out trials with 103 'Perdix Quadcopter Drones' **functioning as a swarm**. As per Lt General Bali Pawar, a well-known aviation expert, the US research agency DARPA has developed micro-drones of the size and shape of missiles designed to be dropped from planes. The US Navy is also working on 'Low-Cost Unmanned Aerial Vehicle Swarming Technology' (LOCUST). Russia is also working on the concept of drone swarming and is probably trying to integrate drones with its 'Sixth Generation Fighter Aircraft'. Similarly, the UK is also working on **Swarm attack technology**. While qualitatively the US is still world leader in drone technology, it is the Chinese who have taken the lead in manufacturing and proliferating drones in the world. As per an article published in November 2020 by a New York based Foreign Affairs article that, prior to 2011 only three countries possessed armed drones but between 2011 to 2020; there are now18 countries that have armed drones of which 11 countries have been supplied drones by China. What is of even greater worry is that most of the countries who have procured these drones are non-democracies. Therefore, its employment for hybrid warfare and by militants or terrorist organizations is likely to increase. The Iran backed Houthi swarm drone attack on Saudi Oil Company facility is a recent example.

China is working towards developing a **swarm drone** operational concept of war fighting. Towards this end it has started testing swarm drone attack technology for warfighting since 2017 on a regular basis as reported by Global Times a Chinese Daily.

Swarm drones are of two types:

- "Fixed Wing Swarms" and
- Quadcopter Swarms.

China holds the world record in both categories beating the US in the process. In the fixed wing UAV swarms it has flown 119 UAVs in June 2017 by Chinese Electronic Technology Corporation (CETC) beating the US record of 103 fixed wing UAVs in January 2017.

The fixed wing UAV swarm is divided into different groups, with each group circling over its intended target and thereafter, maneuvering as a group to ascertain the viability of carrying out coordinated reconnaissance, distributed surveillance and saturated strikes over ground targets. Similarly, the Chinese company Ehang demonstrated the world largest quadcopter UAV swarms comprising 1,374 UAVs on April 29, 2018. China recently conducted a test of Stealth attack Swarm tech drone. The state-of-the-art drone is capable of being launched from trucks, helicopters, and Sea platforms. It is about 1.2 meters long, maximum speed 150 kmph and an operating radius of 15 km. These type of drones function like normal drones during recce mode but transform into a cruise missile and launch suicide attacks when they receive the order. Swarm drones attack the target by integrating two critical technologies: machine and artificial intelligence.

At present air defence guns and missiles are being used to counter drone threat however, they are practically ineffective against hundreds of Swarm drones. But as it happens with any path-breaking technology, in the initial stages it pays rich dividends, for example, dynamite, guns, tanks, aircrafts but soon counter measures are also invented to neutralize or minimize the impact of such technologies.

Research is on to counter swarm drone threats. An area which looks most promising to counter such threat may lie in the electronic and communication domain - it is just a matter of time when an effective counter-UAV-system would be developed. Even in the physical destruction domain efforts are on to find suitable weapon systems. The number of flaws that were noticed in the Chinese demonstration of the world's largest swarm display on May 1, 2018 justifies this conclusion. 496

drones in the May 1, 2018 demonstration deviated from their path and few of them just returned back after taking off. This was attributed to GPS jamming. The other limitation of the Chinese drones is their limited scope in terms of range and GPS denied area operability. On the other hand, the US drones are air launched and can be recovered by aircrafts, operate upto 300 nautical miles and continue on their mission even in a GPS denied area of operations. Although the Chinese are actively pursuing research in making their drones jamming resistant and capable of operating in GPS denied areas there is a scope to disrupt this disruptive new warfare capability.

6.11 ADVANTAGES OFFERED BY SWARM DRONES

(i) **Greater Survivability** as the drones are difficult to detect hence chances of their survival is far greater as compared to manned aircraft. Similarly, being a network of weapon systems of multiple drones, even if they get detected and a few of them shot down, the function of the destroyed is taken over by other UAVs in the Swarm. However, if they get detected they are easy to shoot down. Houthi rebels have shot down a few drones of the US. An important conclusion that can be drawn from this is that the system may not survive against a contested battlefield environment with robust AD systems and offensive cyber-attack capabilities.

(ii) **Cheaper Options** to Armed Forces especially that do not enjoy conventional weapon superiority over their adversaries such as Pakistan over India or India over China. The drone technology is gaining popularity rapidly over countermeasure systems to such threats such as air defence or electronic countermeasure systems. As discussed weapon systems to counter drone technologies will be developed in the short term. Swarm drones are likely to rule the roost and if India is not to be left behind then at least in this technology it must quickly catch up with the world leaders.

(iii) **Effectively beats legacy AD Systems** as proved in the Nagorno-Karabakh conflict especially when faced with swarm drone attacks. However, their chances of success against integrated AD systems of leading powers like the US, Russia or China is highly debatable which are networked and layered comprising long, medium and short-range weapon systems to defend their air and land space. Without a networked and layered AD system even the US systems failed to detect the Houthi drone attack on Saudi Oil Company facilities. The Air Defence systems today must be supported by electronic warfare and specialized counter unmanned systems (C-UAS) to defend against Swarm Drone attacks.

(iv) **Disruptive Effect against Ground based Weapon Systems.** During the Nagorno-Karabakh conflict the Armenian tanks, Mechanized Infantry vehicles, artillery guns and limited Air Defence weapon systems were picked up and destroyed like ninepins by the Azeri drones supplied by Turkey and Israel thus proving a point that efficacy of such weapons in a limited AD environment will do wonders. It also gives an opportunity to countries like India to redefine their force structuring and military capability development. The Nagorno-Karabakh conflict has proven that a small conventional force led by persistent drone attacks is not only a lethal and potent option but also a cheaper option. Although critics might say that we can operate with impunity after achieving air superiority or favorable air situation but in a swarm drone battle space this may just be partially true. The primary weapon platform of the ground forces, the tank, also came under fire from the drones thus suggesting that probably tanks will now have to carry a C-UAS system with them to ensure their security like an anti-missile system.

(v) **Limited effect of Terrain, Camouflage and Dispersion** was distinctly visible, as highlighted by Michael Kofman writing in Moscow Times. The drones were able to overcome the terrain and camouflage advantages of the ground forces through onboard sensors and reconnaissance equipment of the drones. Similarly, with loiter capability and mass drone employment the dispersion tactics was also neutralized. The Armenians Armed Forces had no chance in hell to fight the Azeris who were larger in numbers (Azerbaijan is ranked 68th to Armenia being 100th in Global Firepower Index) and better in quality due to strong backing by Turkey and even supported to a limited extent by the Israelis.

6.12 MILITARY DRONE SWARM INTELLIGENCE

A critical element of modern warfare. A military drone swarm can efficiently carry out missions while minimizing the danger to military personnel.

Swarm intelligence refers to the collective behavior observed in decentralized, self-organized systems. In the context of unmanned technologies like drones, it deals with large numbers of simple drones that together accomplish complex tasks. Each of the drones follows simple rules but through their interactions, the swarm exhibits emergent intelligence beyond the abilities of the individual parts. Potential applications include surveillance and combat. Drone swarms take inspiration from social insects like ants and bees, leveraging swarm intelligence to create a powerful collective entity out of many simple agents.

Swarm intelligence is the principle that a group of simple intelligences operating in concert can operate as a single, collective intelligence with superior capabilities to any of the individuals. Swarm intelligence is defined as the idea of a decentralized "brain" where there is no single leader or decision maker, and all the required tasks are accomplished by the overall collective behavior of the individuals. Swarm intelligence is a type of emergent property, a concept in biology in which the interactions of individual parts of a system acting together produce an overall capability that exceeds that of the individuals. Essentially, swarm intelligence is a situation in which the whole is more than the sum of its parts. Examples of swarm intelligence exist in nature, in formations of migratory birds and in swarms of insects, from which swarm intelligence derives its name.

Swarm intelligence as observed in nature refers to the emergent collective behaviors that arise from decentralized coordination between organisms. Essentially, this means that when each member of a swarm individually follows local rules, the combined action of these members creates complex and "intelligent" group behaviors. As an example, a "V" formation of geese is made up of individual birds who each follow simple rules to stay in formation. Such self-organized and adaptive group behaviors can accomplish things together that would be impossible for individuals on their own, which, when applied to the military context, is a key benefit of swarms of drones beyond the effectiveness of individual drones.

Swarm intelligence is usually applied to systems where the individuals composing the system have very limited ability to perceive their environment or accomplish needed tasks alone. The behavior of bees provides a useful metaphor for understanding how military drone swarm intelligence works. The drones are like the individual bees (incidentally, also known as drones) that complete their tasks such as locating a food source or a new place to live together more efficiently as a collective whole. However, in this case there is no "queen bee;" no individual drone outranks the others in the sense of command or decision-making ability. Instead, a bee can use what scientists term a "dance" to communicate the goodness of a location with resources to the rest of the hive, and the swarm collectively converges on that location. No one bee instructs the swarm to move; the swarm simply moves together; thus the swarm itself behaves in a way that simulates intelligence.

6.12.1 How can a Military Drone S3warm use Swarm Intelligence in Battle?

Applying the principles of swarm intelligence learned in nature to the military context means that a drone operator does not have to precisely control every movement or every drone within a swarm. Instead, the drone operator can send the swarm where it needs to be, and then the swarm intelligence primarily takes over to get the job done.

The most important aspect of swarm intelligence is the adaptability of the individuals in the system, to be assigned any required task as the need for that task increases in priority. Mission-critical operations can be performed by the drone most able to perform them at a given moment, rather than any hierarchy within the swarm dictating which drone should do so. Additionally, the swarm intelligence plays a major role in "military swarming" which is the military tactic of overwhelming the enemy by saturating that enemy's defenses.

6.13 IMPORTANCE OF HUMAN AND AI COLLABORATION IN DRONE TECHNOLOGY

While drones are designed to operate with a degree of autonomy, they, like any artificial intelligence, require human supervision. While AI can perform complex data analysis and change its behavior accordingly, it lacks empathy and an understanding of nuances that are critical to human decision making.

By pairing AI technologies with human experts, military drone swarms can operate more effectively. AI can process huge data sets quickly and efficiently, while humans can aid in decision making that better takes into account the context and complexities of a given situation. Together, this produces a synergistic command and control structure, limiting the risks and improving the success of military drone missions.

Human oversight should be incorporated into AI uses from the start and at every step in the process. This not only improves strategic outcomes for AI, but also ensures that it adheres to any ethical, legal, or compliance requirements. For example, AI technologies should be set up to identify situations that require review by human workers. This can include unexpected situations, borderline cases, or critical decisions. Professional AI operators should have a system in place to review and respond to these situations, as well as monitor AI operations and data in real time for the greatest possible oversight.

Critical decisions in particular are important to flag for human intervention and decision making. This includes engaging a target or any other action with major potential consequences. Without human oversight, it is impossible to be assured that AI will make the right decision. Balancing strategic, ethical, and legal needs in these cases requires human involvement.

It is also necessary to create fail-safes for AI that bring in human intervention when needed. If a system fails, for instance, safety protocols should go into effect that can circumvent mission parameters. This not only helps to maintain the drone's integrity, but also reduces the risk of collateral damage. Human workers should be on standby to provide oversight in these instances, ensuring safety, success, and working within ethical, legal, or compliance boundaries.

For the best possible results, AI technologies should operate with some degree of autonomy while still submitting to human oversight in the command and control (C2) structures of drone swarms.

From the outset, human actors should be responsible for setting mission objectives. Prior to a mission, an AI professional can input these goals, operational parameters, and rules of engagement into the swarm's control system. From there, AI can create an overarching plan, including optimally dividing tasks among specific drones based on their location, function, and abilities. AI can autonomously design and perform these actions faster than their human counterparts, creating operational efficiencies. As such, these tasks can be easily scaled. AI can manage complex swarm coordination and task delegation on a large scale, while a smaller team of AI operators oversee these swarms through high-level commands.

When needed, AI operators can also alter the mission objectives, operational parameters, and rules of engagement in real time. As the human team gains more information, the situation may change, requiring them to intervene. With AI's efficient data processing and strategic capabilities, it can reallocate tasks appropriately within the swarm to account for these changes, moving faster and more efficiently than if humans handled these changes. Instead, human teams can simply provide oversight to AI.

6.13.1 Real-Time Decision-Making by AI

Using AI to power drone swarms enables much more efficient decision making in real time. With humans providing the mission objectives, operational parameters, and rules of engagement, AI can work autonomously to optimally meet its goals. AI is able to use sensors to continuously collect information, and then review that data to spot any changes in the environment that could affect its operations. In real time, AI can autonomously respond to these changes, changing the formation, tactics, or individual tasks of the swarm for optimal results.

Not only is AI effective at identifying and responding to environmental changes, but it also can anticipate the strategy, movement, and actions of the enemy using machine learning capabilities. Depending on the circumstances, AI can either adjust its plan autonomously or flag possible threats or important decisions for human review.

While human operators supply pre-defined rules to AI, it can leverage machine learning to take past missions into consideration and improve its strategy and performance for future missions. This powerful combination allows AI to recognize targets and objects, maneuver around obstacles, and coordinate swarm activities autonomously, even in complex environments.

One of the areas in which AI has recently improved drones is autonomous navigation. By leveraging AI algorithms, drones can avoid obstacles, move with autonomy, and otherwise make changes in real time based on their environment. As a result, they can operate more safely and efficiently for the tasks for which drones are well suited, which include everything from delivering packages and surveying areas aerially, to conducting search and rescue missions. When it comes to surveillance and monitoring tasks especially, AI can make drones more successful than ever before. AI technology can make a drone's flight path more-efficient and safe, as well as plan out its mission optimally. This allows drones to finish missions faster while reserving more energy.

AI excels in rapid data processing, so it can analyze the significant data input that drones collect and quickly produce actionable insights. In agriculture, for instance, AI can quickly survey the aerial imagery collected by a drone and flag potential issues in crop health or yield. There are also growing applications for AI-powered drones in inspecting infrastructure, monitoring wildlife, and supporting event security. With the use of AI-powered computer vision, drones can identify infrastructure, wildlife, and possible threats with greater accuracy than before, as well as track moving objects to monitor them.

Improved payload capabilities for AI-powered drones also bring significant benefits, especially in conjunction with rapid data processing. Looking again at the agriculture industry and its need for analysis of crops and farming practices, advanced payload capabilities in AI can allow drones to more effectively transport and use cameras, sensors, and other payloads.

These recent developments in AI make drones work more efficiently as well as create new potential use cases for them. Drones are increasingly capable of tasks like delivering critical medical supplies as part of emergency response procedures, for example. AI will inevitably continue to advance over time, paving the way for drones to become even more integral to operations in a wide variety of fields. Not only will drones become safer and more efficient over time, but also they will continue to find new and innovative applications.

6.14 WHERE DOES INDIA STAND IN COMBAT DRONE TECHNOLOGY?

India recently demonstrated a swarm drone technology demonstration with Quadcopters on Army Day 2021. However small it may be but it has definitely indicated the intent of the Indian Armed Forces to move in this direction. It is the next frontier of warfare and we better be amongst the leaders in this domain. While there are a number of strategic lessons to be learnt from the Nagorno-Karabakh conflict, it is military lessons highlighted above that should be of concern for us. India is deadlocked into perpetual conflict with Pakistan which greatly resembles the setting of Nagorno-Karabakh conflict. Moreover, in recent times India is dealing with an aggressive and expansionist China and locked into a prolonged eyeball to eyeball conflict in Ladakh whose prognosis remains uncertain. We must also be cognizant of the fact that China today enjoys a significant edge in drone technology in fact it is almost in league with the world leader the US. So, where does India stand? Sadly, while a number of startups have sprung up in this sector, we have a long way to go. While interacting with a few players and experts it was realized that a decisive public-private partnership backed strongly by the Ministry of Defence in collaboration with the world leaders (USA, Israel or UK) in this technology is the only way to catch up with China.

6.15 INDIAN MILITARY HAS NOW NARROWED THE GAP WITH CHINA IN DRONE WARFARE

As an update, the development in 2024 is that the Indian Military has now narrowed the gap with China in drone warfare.

Hardly a day goes by without news of various types of drones being inducted into the armed forces. A large number of military drone manufacturers have jumped into the fray from corporations such as Adani Group, Tata, Reliance, and a large number of small and medium scale enterprises, and startups. Prime Minister Narendra Modi stated that India has the potential to become a global drone hub by 2030. As per some estimates, about 2,000 to 2,500 drones have been inducted so far and many more are in the pipeline. Approximately Rs. 3,000 to 3,500 crore has been spent on drones, spare parts and maintenance contracts under emergency and special financial powers delegated to the armed forces.

So, have our armed forces caught up with China in the field of unmanned aerial vehicles (UAV)? In terms of numbers, the gap has certainly been narrowed but in terms of effectiveness as a force multiplier system, we are still miles behind.

India's drone technology for warfare in 2024 is advancing through events, demonstrations, and the development of new weapons:

- **HIM-DRONE-A-THON 2**

This event took place in September 2024 at Wari La Pass in Ladakh, where over 25 indigenous drone manufacturers demonstrated their capabilities in high-altitude conditions. The Indian Army plans to select some drones for procurement based on their performance.

- **Vajra-Shot**

This handheld anti-drone gun was developed by Big Bang Boom Solutions and showcased at the Indian Navy's Swavlamban 2024 exhibition. The gun has a range of 4 km.

- **Indian Navy's anti-swarm drones**

The Indian Navy has developed anti-swarm drones to protect against enemy attacks.

- **Swarm drone deal**

In August 2023, a Delhi-based startup won a $36 million swarm drone deal from the Indian Air Force.

- **Drone ecosystem**

India has over 200 startups dedicated to drone technology, which could help the country quickly replace lost drones.

Drones are becoming a critical component of military operations around the world, and India is investing in indigenous drone technology to bolster its defense capabilities.

6.15.1 India Unveils Lethal '*Swadeshi*' (Indigenous) Kamikaze Drones with 1,000 Km Range

Here is another update (2024) about the drone technology in India:

As India prepares to celebrate the 78th Independence Day, the National Aerospace Laboratories (NAL) unveiled that it is making potent swadeshi (indigenous) Kamikaze Drones, do-and-die unmanned aerial vehicles with home-built engines that can power them to fly up to 1,000 kilometres.

Loitering munitions are do-and-die machines. They have been widely used in the ongoing Russia-Ukraine war and the Israel-Hamas conflict in Gaza. These unmanned aerial vehicles have been used extensively by the Ukrainians to target Russian infantry and armored vehicles.

They loiter in the general area of interest for an extended period, carry explosives and ram the target when commanded by a human controller sitting far away. They can be sent in swarms, i.e. multiple drones and attack enemy installations by overwhelming the radars and enemy defenses.

The Kamikaze suicide missions were first seen towards the end of World War II when the pilots of a depleted Japanese air force would ram their fighter planes on Allied aircraft and ships.

Dr. Abhay Pashilkar, Director of the National Aerospace Laboratories, who is spearheading the research says, "India is developing these fully indigenous kamikaze drones, they are a game-changing 21st century new age war machine".

The Indian kamikaze drone will be around 2.8 meters long with a wingspan of 3.5 meters, weigh about 120 kg and carry an explosive charge of 25 kilograms.

Dr. Pashilkar told NDTV, a TV news channel, that the Indian loitering munition will have an endurance of about nine hours meaning, that once launched, it can continuously hover in the area of interest. After the target is identified, the controller after authorization can send the do-and-die drone on its suicide mission.

The Council of Scientific and Industrial Research (CSIR) has given in-principle approval for launching a project on Loitering Munitions or kamikaze drones with CSIR-NAL as the nodal laboratory and participation from all the major engineering laboratories of CSIR. This capability will address our national security needs.

The Indian kamikaze drone will use a 30-horsepower Wankel Engine designed and developed by the National Aerospace Laboratories and can fly continuously for 1,000 kilometres with a maximum speed of 180 kilometres per hour.

The Indian version will be able to work in GPS-denied scenarios and can use the Indian NAViC for navigating and homing on to the target. Such drones deployed by other nations have shown great potential in the modern ongoing wars elsewhere.

SECTION-III

EMERGING TECHNOLOGIES FOR MODERN AND FUTURE WARFARE

CHAPTER 7

ARTIFICIAL INTELLIGENCE (AI) IN WARFARE

7.1 ARTIFICIAL INTELLIGENCE IN MODERN AND FUTURE WARFARE

Artificial intelligence (AI) is now influencing every area of human life. The past decade has seen a drastic increase in the use of AI, including facial recognition software, self-driving vehicles, search engines, and translation software. These accepted uses of AI in modern society have also coincided with an increased presence of AI in **modern warfare**. The escalating weaponization of AI parallels the nuclear arms race of the Cold War, with nuclear weapons being replaced with **automated weapons systems.** However, the international community, the United Nations, and international law have been struggling to adapt to and regulate the use of automated weapons, which are rapidly changing the landscape of modern warfare.

The international community started to take notice of AI and its influence on modern warfare in 2012, with a series of documents outlining the use of automated weapons systems. These documents included policy directives by the U.S. Department of Defense (DoD) on autonomy in weapons systems and a report from Human Rights Watch and the Harvard Law School's International Human Rights Clinic (2012 HRW-IHRC report) calling for an outright ban on automated weapons.

The development and use of weapons that can undertake autonomous functions during conflict is becoming the focus of states and tech companies. In 2017, an open letter from the Future Life Institute to the United Nations (UN) signed by 126 CEOs and founders of artificial intelligence and robotics companies "implored" states to prevent an arms race for autonomous weapons systems (AWS). However, no international legal regulatory framework exists to address these concerns around the use of AI, particularly in the context of conflict. The only legal framework for AI that does exist, established by Article 26 of the International Covenant on Civil and Political Rights, only relates AI use to the right to privacy.

7.2 WHAT IS AN AUTOMATED WEAPON?

There are competing definitions of what constitutes Automated Weapon System (AWS), although the UK Ministry of Defence (MoD) and the U.S. Department of Defense (DoD) have developed the two main definitions. In 2011, the UK MoD defined AWS as "systems capable of understanding higher level intent and direction, namely of achieving the same level of situational understanding as a human and able to take appropriate action to bring about the desired state." Comparably, the U.S. DoD in 2023 proposed a different approach and defined "AWS as being capable of once activated, to select and engage targets without further intervention from a human operator." The 2012 HRW-IHRC report advanced a similar definition for the international community, defining AWS as "fully autonomous weapons that could select and engage targets without human intervention." The NATO Joint Air Power Competence Centre (JAPCC) also extends the notion of automation to "consciousness and self-determination." Examples of automated weapons include defensive systems like the **Israeli Iron Dome** and the German MANTIS as well as active protective vehicles like the Swedish LEDS-150. A new definition would also need to include automated weapons used in non-conflict situations, like the South Korean Super aEgis II, which is used as a peacetime surveillance device along the South and North Korean border.

However, the real problem lies in future-proofing definitions of AWS. Definitions must not only include systems not already accounted for, such as the Super aEgis II, but also anticipate AWS that may emerge in the future. In particular, the international community must agree on a definition that can encompass AI human cognitive inputting algorithms, which have humanlike decision-making capabilities.

Despite this need, the international community has yet to agree on regulations of AWS. The UN Convention on Conventional Weapons (CCW) has a special Amended Protocol (1986) governed by the Group of Governmental Experts (GGE), who meet annually to discuss the implementation of the Protocols of the CCW and related weapons issues. The latest GGE meeting, held in May 2023, ended without any substantial progress on AWS, as the GGE did not agree on any regulatory safeguards. Their draft report also failed to advance a legal framework. However, the report did introduce prohibitions centered on the need for human control of AWS as well as regulations centered around the development of AWS. Fifty-two states issued a joint statement of support for the draft report, but they also stated that the draft was a minimum standard and emphasized the need for a much more robust and ambitious legal framework. In its May 2023 meeting, the GGE resolved to organize longer discussions on emerging lethal AWS technologies in March and August 2024. While more meaningful developments may occur in the latest GGE discussions, a legal framework to regulate the development and deployment of AWS has yet to emerge.

7.3 APPLICATION OF AI IN WARFARE

Advances in Artificial Intelligence (AI) have been in the headlines lately. However, AI has been used for a variety of military and defense purposes for decades and will continue to play a major role in the way wars are waged in the future. As a result, the centralized data centers and embedded computing systems that enable AI will be required to provide unprecedented levels of computing power and bandwidth as new military applications of AI are developed and deployed.

A groundbreaking milestone in military history is a precursor to AI as we know it today: mathematician Alan Turing broke the German Enigma encryption codes in World War II. He then went on to develop the Turing Test, which is still widely used to determine if a computer can match human intelligence.

The U.S. military also has a long history of experimenting with and using AI. In 1991, an AI program called the Dynamic Analysis and Replanning Tool (DART) was used to schedule the transportation of supplies and personnel and to solve other logistical problems, saving millions of dollars. Today, the military is exploring new ways to use AI to help improve effectiveness and efficiency, including the following examples of AI's use in warfare.

Following are some of the **applications of AI** in modern warfare:

(i) Identifying and Neutralizing Threats

Locating the unique signature of a radio or radar signal that's emitted by a particular aircraft within the entire electromagnetic spectrum is like trying to find a needle in a haystack. In the past, sensors were recording broad frequency ranges. These recordings were then sent to a lab where trained technicians scoured the recording to isolate and identify any signals of interest before designing jamming systems to address each signal.

Today, AI systems are being employed to help detect and identify signals of interest and to implement the appropriate technology to jam the signal or to help intercept the threat as needed.

AI can also be used to help intercept incoming threats. When a ballistic missile launch is detected, it's possible to determine the missile's expected trajectory and intervene before the missile hits its intended target. Predictive AI can be used to identify the missile's electromagnetic signature and to either jam the signal and redirect the missile or direct interceptors to destroy the missile before it reaches its target. An AI-based system may also be able to decipher any encrypted communications, providing valuable information about the opposition's capabilities or intent.

(ii) Guiding Manned and Unmanned Aircraft and Vehicles

Unmanned vehicles play an important role in modern warfare, but most require some level of human control. Vehicles without a human crew or guidance have some kind of AI capabilities or local logic to avoid catastrophe, especially if communication with the unmanned vehicle is lost even for a moment.

Future generations of fighter jets, submarines, and other military vehicles, currently on the drawing board, are likely to operate with a swarm of unmanned vehicles. AI will be used to control the swarm of unmanned aerial vehicles (UAVs) while the pilot focuses on executing his mission. Military forces around the world are experimenting with this emerging technology as engineers work to mature artificial intelligence technology and its integration with various platforms.

Swarm technology will likely be incorporated into the next generation of fighter jets being developed today so that the autonomous crafts will coordinate seamlessly with their manned counterparts.

(iii) Gathering Intelligence

The quality of AI-based transcription and translation programs improves dramatically with each software update. For instance, it can allow audio communications to be transcribed and translated in real time.

(iv) Preparing for Battle

Simulators and training programs that employ AI are replacing textbooks as the preferred media for training. Lessons delivered via tablet or virtual/augmented reality can be updated immediately to reflect the latest developments and to better simulate real-life battle situations and conditions.

(v) Empowering an Integrated Defense System

The U.S. Department of Defense (DoD) is developing a centralized system called Joint All-Domain Command and Control (JADC2) that uses AI to link sensors from all branches of the armed forces into a unified network. The goal of this effort is to improve access to strategic and tactical data across all branches of the U.S. military. JADC2 aims to achieve this by identifying and eliminating physical and technological barriers that limit the sharing of mission-critical data and information. If successful, the result will be unprecedented levels of interconnectivity among military personnel and equipment.

(vi) Restrictions on the Use of AI in Warfare

As the race to find new military applications of AI goes on, nations and government organizations are working to establish the guidelines and standards to help ensure that AI is applied ethically.

In January of 2023, the DoD announced an update to Directive 3000.09 Autonomy in Weapon Systems that applies to autonomous and semi-autonomous weapon systems, including those that incorporate AI. The updated requirements state: "Autonomous and semi-autonomous weapon systems will be designed to allow commanders and operators to exercise appropriate levels of human judgment over the use of force." So, while AI may be used to help leaders and commanders make more informed decisions, it's ultimately a person — not a machine — that determines the final action.

(vii) Importance of Embedded Systems in Electronic Warfare

Regardless of how they are being used, AI applications require significant amounts of data that must be processed quickly. High-speed connections are needed to keep the data flowing between the sensors and the black boxes used to implement the AI.

Most industries that use AI rely on ground-based, climate-controlled data centers where space for processing equipment is virtually unlimited. Defense and military applications of AI will also require ruggedized embedded systems that must function reliably and consistently in confined and potentially hostile environments on the ground, in the water, in the air, or in space.

Local processing is required to minimize the constraints of securely sending the vast amount of information that can be collected. Embedded systems on board each vehicle will require extraordinary computing power to handle the demands of AI applications.

(viii) TE Connectivity (TE) Enables Military Applications of AI

When designing embedded electronic systems, the two prime constraints are the physical layout of the circuit board and the thermal management within the system. The amount of data that can be collected and stored, and how quickly that data can be sent from point A to point B, are also extremely important. For the design to be efficient, a balance between the computing power that's available today and the rates at which data will be able to be transmitted in the near future is needed.

Current systems are capable of transmitting data at rates of more than 32 gigabytes per second, but that's not enough. Designers and engineers will always need or want to move more data, faster.

7.3.1 TE's VITA- and SOSA-aligned Portfolio

TE offers a broad portfolio of VITA- and SOSA-aligned interconnect components that can handle speeds of up to 112 gigabits per second (Gb/s), even under harsh battlefield conditions, including the:

(i) **STRADA Whisper** backplane family was designed for high-performing, high-bandwidth systems. Its revolutionary design transfers data up to 112 Gb/s—allowing for efficient future system upgrades without costly backplane or midplane redesigns.

(ii) **MULTIGIG** connectors support speeds of more than 32 Gb/s, making them among the fastest rugged backplane connectors for embedded computing or VPX systems.

(iii) **DEUTSCH CeeLok FAS-X Connectors** offer high-speed solutions for Ethernet data delivery using rugged, mil-spec components.

(iv) **Raychem** brand cable assemblies, harnessing, heat shrink tubing and tape, joints, terminations, wraparound sleeves, terminal, splices, wires, and cables are designed for a wide range of applications in the most demanding environments.

7.4 ADVANTAGES OF ARTIFICIAL INTELLIGENCE IN WARFARE

Warfare systems such as weapons, sensors, navigation, aviation support, and surveillance can employ AI in order to make operations more efficient and less reliant on human input. This additional efficiency means that these systems may require less maintenance. Taking away the need for full human control of warfare systems reduces the impact of human error and frees up humans' bandwidth for other essential tasks.

Specifically regarding weapons, the Pentagon recently updated its autonomous weapons policy to take into account recent advances in AI. Since the policy's original creation in 2012, a number of technological leaps forward have been made that necessitated this update. The update provides guidance for the safe and ethical development and use of autonomous weapons, one of the most useful military applications of AI. In addition to review and testing requirements, the policy creates a working group focused on autonomous weapons systems to advise the DoD.

Following are some of the benefits of using AI in warfare:

(i) Military Drone Swarm

AI-controlled swarms of drones are actually programmed to act in the same manner that swarms of insects act in nature. For example, when a bee finds something that could benefit the rest of the hive, it will report that information in detail to other bees. The drones can do the same. They are able to communicate the distance, direction, and elevation of a target, as well as any potential dangers, just as a bee does. The ability to use AI-powered drone swarms to put this powerful collective intelligence to work towards military objectives represents a critical frontier in the military applications of AI.

(ii) Strategic Decision Making

One of the best benefits of artificial intelligence in the military is in an area that military commanders might feel hesitant to let AI contribute. That is helping with strategic decision making. AI's algorithms are able to collect and process data from numerous different sources to aid in decision making, especially in high-stress situations. In many circumstances AI systems can quickly and efficiently analyze a situation and make the best decision in a critical situation. It is also able to neutralize prejudices that may come with human input, with the caveat that AI may not yet have a fully developed understanding of human ethical concerns and there is a danger of AI learning from the biases that may exist in materials in its database. However, decision making under pressure is a critical part of being a service member, and AI and humans can work together to make this process easier. The combination of humans' ethical understanding and AI's quick analytical abilities can speed up the decision-making process.

Generative AI can contribute to the decision-making process in military settings. Rapidly sorting through large amounts of data, generative models can show connections, patterns, and potential implications that humans alone would take a longer time to find. This information can be presented to human decision makers not only as reports but also in a conversational format, facilitating human-AI collaboration. AI can also create simulations to test out possible scenarios, allowing for more informed decision making. Upon receiving this information from AI, humans must fill in the gaps, using their understanding of ethical principles, national security interests, and situational nuances to create optimal outcomes.

With close human supervision, generative AI has a lot of potential to enhance military leaders' strategic thinking. Among the considerations in implementing AI to assist decision making include counteracting harmful biases, accounting for real-world conditions that may be beyond models' understanding, safeguarding classified data, not using AI as a replacement for human judgment, and ensuring alignment with regulations, ethics, and more. The key takeaway here is that the capabilities of AI for decision making are those of assisting humans, rather than taking this function out of their hands.

(iii) Data Processing and Research

In many cases, processing large volumes of data can be extremely time consuming, but AI's capabilities can really add value in this area. AI can be helpful for quickly filtering through data and selecting the most valuable information. It can also aid in grouping information from various datasets. This can allow military personnel to identify patterns more efficiently, draw more accurate conclusions, and create plans of action based on a more complete picture of the situation. Generative AI's analysis capabilities mean that it can find connections in large volumes of information that might escape humans' notice, or can find them faster than a human would. Thanks to their NLP abilities, generative AI models can also communicate this information to humans in a conversational manner and engage in a dialogue to explain it.

AI can also be used in order to filter through large amounts of content from news and social media outlets in order to aid in identifying new information. This allows analysts to save time when tasked with large quantities of content. AI systems can also eliminate repetitive information, as well as inaccurate information. This can optimize the research process, helping analysts finish a job faster, and more accurately, as well as, again, reducing human error.

Generative AI can speed up the analysis process when it's critical to understand information as quickly as possible. Models can bring order to chaos when handling massive datasets, uncovering connections between seemingly unrelated data points. This allows military leaders to formulate strategy based on a deeper understanding of conditions. AI can also rapidly generate and compare thousands of scenarios by making small changes to variables; by understanding a wide range of permutations of a problem or situation, military commanders can prepare for many contingencies. Furthermore, generative models can quickly compare intelligence with existing knowledge and research, and make useful suggestions, enabling better predictions. Humans, rather than AI, will still need to make final strategic decisions, due to their ability to take into account context that may elude AI. However, by collaborating with AI, military leaders can have a more detailed understanding of what is happening around them and what may happen in the future.

(iv) Combat Simulation and Training

Military training simulation software has been used in the U.S. Army for quite some time. It combines systems engineering, software engineering, and computer science in order to build digitized models that prepare soldiers with combat systems deployed during operations. In simpler terms, military training simulation software is essentially a virtual "wargame" that is used in order to train soldiers.

This software can be used for just about anything from mathematical models to simulating strategies used in non-combative environments. In turn, this will better prepare soldiers for real-life situations. These simulations are able to provide realistic missions and tasks to soldiers, to ensure they gain the most experience possible before applying their skills to real-life situations.

Generative AI can improve military training and educational programs. AI-powered language models can read training manuals and other sources and use them to create new training materials, including notes, quizzes, and study guides. AI can also help evaluate students' current abilities and tailor training to their specific needs. With NLP, generative AI can answer students' questions and explain concepts similar to the way that a human instructor would. By analyzing large amounts of intelligence data, records of previous combat experiences, and more, AI can craft more comprehensive training, including detailed military simulations. Conversational AI can also provide customized feedback to help students build their skills and help commanding officers know where a particular student may be struggling.

While AI holds a lot of potential for military training applications, it should never entirely replace human instructors. To avoid issues like bias or misinformation, leadership should always review AI-generated materials and be in charge of the ultimate analysis of students' skills. Human instructors should determine the overall syllabus, while AI can craft individualized lessons that human instructors can then review for accuracy and other issues. However, with AI assistance, instructors can create and administer more effective training programs, because of individualized attention to students that humans may not be able to always provide, and do so more quickly due to AI's processing speed.

One of the most important details about combat simulation is that it is far safer than reality. Many casualties can occur from training with real weapons and situations. This allows soldiers to experience the best simulation of the realities of warfare, without being endangered. These virtual realities can aid soldiers in understanding how to handle clones of weapons just like their real-life counterparts, make decisions in stressful situations, as well as work with their teammates. The training software can prepare soldiers for just about anything, and can save them in the long run. Not only can AI-based simulation train soldiers, but it can personalize training programs, as well as make fair assessments in order to make future adjustments to the programs. Combat simulation can also save time and money due to being more efficient at certain tasks than humans are. Check out our innovative AI model, Strat Agent, which acts as a modern-day battlefield commander that can be used in combat simulation.

(v) Target Recognition

Artificial intelligence can aid in making target recognition more accurate in combat environments. AI can improve the ability for systems like this to identify the position of their targets. It can also allow defense forces to acquire a detailed understanding of an operation area by examining reports, documents, news, and other forms of information, aggregating and analyzing these sources much more quickly than humans would be able to do so. With generative AI's conversational abilities, there can be a two-way discussion about this information, so military decision makers can ask questions to make sure the most relevant information comes to the surface. AI systems have the ability to predict enemy behavior, anticipate vulnerabilities, weather and environmental conditions, assess mission strategies, and suggest alleviation plans. This can save time and human resources, putting soldiers a step ahead of their targets, but as always requires humans to make the ultimate decision.

(vi) Threat Monitoring

Threat monitoring, as well as situation awareness uses operations that gain and analyze information to aid in many different military activities. There are unmanned systems that can be remotely controlled or sent on a pre-calculated route. These systems use AI in order to aid defense personnel in monitoring threats, and thus leveraging their situational awareness. Drones with AI can also be used in these situations. They can monitor border areas, recognize threats, and alert response teams. Additionally, they can strengthen the security of military bases, as well as increase the safety of soldiers in combat.

(vii) Cybersecurity

Even highly secure military systems can be vulnerable to cyber-attacks, which is where AI can be of great help. Attacks can put classified information at risk, as well as damage a system altogether, which can endanger military personnel and jeopardize the mission. AI has the ability to protect programs, data, networks, and computers from persons not authorized to access them. AI also has the skills to study patterns of cyber-attacks and form protective strategies in order to fight against them. These systems can recognize the smallest behaviors of malware attacks far before they enter a network.

Generative AI's analysis, scenario generation, and communication capabilities can also improve cybersecurity in military settings. With the ability to analyze large quantities of data and find patterns, generative AI can detect potential threats and use predictive analytics to help predict future attacks. At the same time, generative AI also poses its own threats in the wrong hands, such as the concern that attackers can leverage the power of generative models for social engineering. The military will also have to take care to counteract this possibility as well, with consistently updated training and mitigation plans. When applied with care and close supervision, generative AI can enhance cyber defense, even for critically important military applications.

As it does in many other areas, advanced AI has a mixed impact on cybersecurity. Functions such as the ability to write malware may make AI dangerous in the hands of bad actors, but AI can also help to detect and mitigate these threats. In essence, the military applies AI to counter adversaries who may also have access to AI. This means that it is critical for the military to have access to the most advanced and tailored AI cybersecurity solutions in order to stay safe amid a constantly evolving landscape of AI-driven cybersecurity risks.

(viii) Transportation

AI is able to play a role in the transportation of ammunition, goods, armaments, and troops. The logistics and transportation of these things is obviously vital to the success of military operations. AI can lower transportation costs and reduce the need for human input by, for example, plotting the most efficient route to travel under current conditions. It can also pre-identify problems for military fleets in order to increase efficiency of their performance. As the combination of innovation in computer vision and autonomous decision making over time also continues to bring self-driving vehicles closer to common use in the commercial space, this technology may also prove useful in the military context.

(ix) Casualty and Evacuation

Because soldiers and medics have to make decisions in high-stress situations, AI is able to aid them when a fellow service member may need help. The dangerous environment of the battlefield presents a lot of obstacles to providing medical treatment to the wounded. The assistance of AI in a highly emotional environment can help with analyzing the situation and suggesting the best course of action. This advice can help humans to base their sometimes, split-second decisions on information provided by an analytical rather than emotional mind. However, obviously, AI cannot make these decisions for humans, precisely because AI does not have an understanding of the emotional and contextual factors involved in a life-or-death situation.

This type of AI uses an algorithm and large medical database that is able to access data containing medical trauma cases, which include diagnoses, vital sign sets, medications given, treatments, and outcomes. It then takes this data, combined with manually entered information in order to provide indications, warnings, and suggestions for treatment. This is another situation in which AI needs human guidance in order to operate effectively; while the AI will make recommendations without emotional considerations as a hindrance, humans must use their emotional abilities to make appropriate decisions that take into account these recommendations. AI is not qualified to make medical decisions but it can provide rapid analysis to give humans more information on which to base their decisions.

7.5 LOOKING AHEAD TO THE FUTURE OF AI IN WARFARE

The accelerated progress in software and AI technology virtually guarantees that the defense industry will see tremendous advances and continue to find new applications for AI in the military. Expertise in various commercial, industrial, medical, aerospace, and defense applications that also employ AI tech allows TE to adapt existing products to work in harsh environments on the ground, on the water, and in space.

For example, TE's STRADA Whisper high-speed connectors are based on connectors developed for use in ground-based data centers but have been ruggedized to withstand the extreme conditions on the battlefield and in space.

7.6 IMPACT OF AI IN THE INDIAN ARMY

The impact of AI is transforming industries worldwide. The worldwide expenditure on AI reached $118 billion in 2022 and is expected to exceed $300 billion by 2026. Like other sectors, the influence of AI has resulted in a growing trend of global militaries using AI in their combat systems.

The defence sector uses AI-based technologies for training, surveillance, logistics, cybersecurity, UAV, advanced military weaponry like LAWS, autonomous combat vehicles, and robotics. Rajnath Singh, the Indian Defence Minister, unveiled 75 recently created AI technologies on July 11, 2022. This event occurred at the inaugural 'AI in Defence' (AIDef) symposium and exhibition organized by the Ministry of Defence in New Delhi.

Furthermore, the minister emphasized the significance of promptly incorporating advanced technologies such as AI and Big Data in the defense sector. It is crucial to ensure we remain up-to-date with technical advancements and fully leverage technology for our services.

7.6.1 Applications of AI in Defence

Border monitoring can be enhanced by integrating cameras, radar feeds, sensors, and other technologies aided by AI-based solutions. These advanced technologies aid in identifying border breaches, categorizing targets, and improving the precision of defense operations.

- **Unmanned Aerial Vehicles (UAVs)** - Drones outfitted with AI-based aircraft technology are highly proficient in conducting day and night surveillance operations, encompassing border control and comprehensive surveillance.
- **Lethal Autonomous Weapon Systems (LAWS)** are equipped with integrated sensors and pre-programmed algorithms that assist in identifying, selecting, and tracking hostile targets. These weapons can engage targets independently and thus reduce the need for personnel.

- **Autonomous armored vehicles and robots** perform unattended, real-time surveillance, transport injured individuals, and carry supplies under challenging locations, including deserts and mountainous areas. Robots demonstrate superior performance in hazardous and high-pressure environments, surpassing the skills of humans.
- **Data management** - AI can analyze and utilize unutilized or underutilized data, generating more practical and valuable insights for the Indian armed forces. It will improve the capabilities of Intelligence, Surveillance, and Reconnaissance (ISR).
- **Pattern recognition** - AI can analyze data from many sources and discern patterns.

  The purpose of this technology is to forecast possible terrorist attacks and insurgency activity and propose proactive measures.
- **Training and simulation** encompass a range of disciplines that utilize system and software engineering principles to develop models that aid soldiers in training for various combat **systems employed in real-world military missions.**

7.6.2 AI Adoption in the Indian Military

The Ministry of Defence's Department of Defence Production established a task force in February 2018 to examine the prospective implementation of AI in defence contexts. Its report was submitted in June 2018 under the "Strategic Implementation of AI for National Security and Defense" task force. In 2019, a Defence AI Project Agency (DAIPA) and a Defence AI Council (DAIC) were established by the task force's recommendations.

The DAIC comprises the three service commanders, the defence secretary, the national cyber security coordinator, and representatives from the DRDO, industry, and academia. The Defence Minister chairs it. The DAIC is tasked with convening biannually to deliver essential guidance that facilitates and executes policy-level modifications, operational framework development or customization, and structural support.

The ex-officio President of the DAIPA is the Secretary (Defense Production), while members are selected from academia, industry partners, the services, Defense Public Sector Undertakings, DRDO, and the DAIPA. The DAIPA shall establish and implement benchmarks for AI initiatives' technology development and delivery process and consult user groups regarding the adoption strategy for AI-led and AI-enabled systems and processes.

At the level of the Defence Ministry, these are plausible measures; however, they must be supplemented with modifications to organizational structure, planning, and processes at the level of the end users, namely the military. Adoption of artificial intelligence will be inconsistent and suboptimal unless the military is adequately prepared to assimilate this technology.

7.7 ETHICAL, LEGAL, AND PRACTICAL IMPLICATIONS OF AI IN MODERN WARFARE

In an era defined by rapid technological advancements and escalating global conflicts, the integration of artificial intelligence (AI) into warfare has emerged as both a revolutionary opportunity and a profound ethical challenge. As AI and machine learning technologies evolve, they are reshaping numerous aspects of armed conflict—from the development of autonomous weapon systems to their application in strategic decision-making. These advancements promise to transform the conduct of warfare, offering benefits such as increased precision in targeting and enhanced decision-making capabilities. However, they also raise critical ethical and legal questions, particularly concerning the preservation of human control and accountability, and the potential violation of international humanitarian principles such as distinction and proportionality.

AI's role in warfare spans a range of applications, including the automation of military hardware, cyber warfare capabilities, and information warfare strategies. Autonomous weapon systems (AWS), capable of selecting and engaging targets without human intervention, are at the forefront of these concerns. The potential for these systems to operate with minimal human oversight poses significant risks, including unintended consequences and violations of **International Humanitarian Law (IHL)**. Moreover, AI-driven systems in cyber operations and information warfare introduce new dimensions of conflict, potentially leading to escalations that could severely impact civilian populations.

7.7.1 Examining AI-Enabled Weapons and Autonomous Decision-Making

(i) Lethal Autonomous Weapon Systems (LAWS)

The definition and use of lethal autonomous weapon systems (LAWS) are among the most contentious issues in modern warfare. Despite the lack of a universally accepted definition at the state and international levels, it is clear that many nations are developing and deploying these systems, which could revolutionize the future of warfare. The International Committee of the Red Cross (ICRC) offers a comprehensive definition: "After initial activation or launch by a person, an autonomous weapon system self-initiates or triggers a strike in response to information from the environment received through sensors and based on a generalized target profile."

The autonomy of weapon systems can be categorized into three levels: semi-autonomous, supervised autonomous, and fully autonomous operations. In semi-autonomous operations, machines perform tasks but require human approval to proceed. Supervised autonomous operations allow machines to continue tasks until stopped by a human operator. Fully autonomous operations, however, are beyond human control once initiated. While autonomous systems are used in navigation, intelligence, surveillance, and reconnaissance, here we focus on weapons that autonomously select and engage targets.

According to Autonomous Weapons Watch, 17 weapon systems can operate autonomously, with China, Germany, Israel, South Korea, Russia, Turkey, Ukraine, and the USA leading their development. Thirteen of these systems are unmanned aerial systems, one is an unmanned surface system, and three are unmanned ground systems. Some systems, like Anduril's Altuis 600, explicitly state their autonomous capabilities, while others, such as Turkey's Kargu-2, have downplayed their autonomy in response to international concerns.

Despite the lack of consensus on regulation, the UN General Assembly took a historic step on December 22, 2023, by adopting the first resolution against LAWS with 152 countries in favour. The resolution affirms that international humanitarian law applies to LAWS and supports the efforts of the UN Group of Governmental Experts to introduce restrictions or prohibitions on such systems. However, more comprehensive regulations are needed.

A two-pronged approach to LAWS regulation, as proposed by several countries, is necessary: prohibiting systems without human control over target selection and engagement, and regulating other systems in compliance with IHL. The Human Rights Council's resolution of October 7, 2022, highlighted the risks of using representative data sets, algorithm-based programming, and machine learning, which may perpetuate structural discrimination, marginalization, and unpredictability. These practices conflict with IHL principles and pose challenges to state responsibility and human accountability.

7.7.2 AI in Decision-Making Processes

The integration of AI and machine learning into decision-making processes in armed conflict represents one of the most transformative developments in modern warfare. These technologies enable the extensive collection and analysis of diverse data sources to identify individuals or objects, assess patterns of behavior, recommend military strategies, and predict future actions or situations. Such systems extend traditional intelligence, surveillance, and reconnaissance capabilities by automating large data set processing and providing recommendations to human operators or autonomously initiating actions based on analyses.

AI-driven decision-making systems can influence critical decisions, such as targeting, detention, military strategies, and even the potential use of nuclear weapons. However, these systems also introduce significant risks, particularly given the limitations of current AI technologies, such as unpredictability, lack of transparency, and inherent biases. From a humanitarian perspective, the deployment of AI in conflict must be scrutinized to ensure compliance with IHL. Decisions impacted by AI—especially those affecting human life or property—must adhere to IHL rules governing hostilities.

The ethical and legal concerns extend to detention decisions, mirroring debates in the civilian sector about human oversight and the accuracy of risk assessment algorithms. Moreover, AI tools could personalize warfare by integrating personally identifiable information from various sources to form algorithmic assessments of individuals, leading to targeted violence, wrongful detention, identity theft, and other humanitarian risks.

7.7.3 Evaluating Ethical Concerns and International Norms

While inter-state efforts to limit AI use, particularly within the framework of LAWS, have yet to conclude, international law remains relevant in this area. Article 36 of Additional Protocol I requires states to assess the legality of new weapons, means, or methods of warfare before deployment. This obligation ensures that new technologies, including LAWS, comply with IHL principles. The Martens

Clause further emphasizes the need to adhere to humanity and public conscience in warfare, complementing Article 36 by embedding a moral dimension into weapon assessments.

The fundamental principles of IHL—distinction, proportionality, and precautions in attack—are directed toward human actors, who bear responsibility for adhering to and implementing these norms. Consequently, accountability for violations rests with humans, not machines or algorithms. Combatants must make nuanced, context-specific judgments related to these principles, ensuring that AI systems used in warfare do not undermine human responsibility.

Discussions under the Convention on Certain Conventional Weapons have affirmed the need for "human responsibility" in deploying weapon systems and using force. This consensus among states, international organizations, and civil society groups underscores the importance of maintaining human control to ensure compliance with IHL and ethical standards.

To address the challenges posed by AI in warfare, we propose the following recommendations:

(i) **Prohibit Autonomous Weapon Systems Without Human Control:** Weapon systems that do not allow for sufficient human oversight in target selection and engagement should be prohibited.

(ii) **Establish Positive Obligations for Human Control:** For systems that are not prohibited, establish clear obligations for human control over weapon parameters (e.g., type of target), the environment of use, and human-machine interaction during use.

(iii) **Ensure Human Command and Control:** Any use of weapon systems with autonomous functionalities must be guided and overseen by a responsible chain of human command and control.

(iv) **Preserve Human Judgment in the Use of Force:** Actions that may result in the loss of human life through the use of force should remain under human intent and judgment. Once a human initiates a sequence of actions intended to end with lethal force, autonomous systems may complete the sequence only with ongoing human oversight.

7.8 MILITARIZATION OF AI HAS SEVERE IMPLICATIONS FOR GLOBAL SECURITY AND WARFARE

AI systems, especially those employing machine learning, have the potential to have a profound effect on society. They can make decisions or predictions that affect individuals and communities, raising ethical concerns.

How, for example, can we guarantee that AI systems are impartial and do not perpetuate existing biases? How can we ensure that AI decision-making mechanisms are transparent? How can we safeguard privacy when AI systems frequently rely on vast data? How do we get people who understand the technical and regulatory frameworks? How do we bridge the gap between the AI haves and the AI have-nots?

These considerations are an essential aspect of AI governance. The sober news is that fixing a defective AI system is far easier than fixing a broken human.

Regulating AI is challenging due to its technical complexity, rapid evolution, and widespread applicability across multiple industries. Regulatory frameworks must balance fostering innovation and preventing potential harm to society. They should ensure accountability, transparency and impartiality in AI systems while promoting competition and preventing misuse.

7.8.1 Adaptive AI Governance

Due to the global nature of AI development and the rapidity with which it evolves compared to traditional legislative procedures, it is difficult to create such frameworks. Therefore, an adaptive AI governance framework is essential, and we can draw many lessons from genetic algorithms, an AI method based on the principles of natural evolution.

AI governance is a global issue, not just a national one. Effective governance of AI systems necessitates international cooperation, as AI systems and their effects do not respect national boundaries. Countries must collaborate to establish global AI usage standards and norms.

International organizations, such as the United Nations, can play a vital role in facilitating dialogue and cooperation regarding AI governance. Cooperation between governments, private sector companies and civil society is essential to guarantee a comprehensive approach to AI governance.

What are some of the essential elements that should be taken into account to create a global AI governance?

(i) We need to standardize AI

The need for standardization increases as AI evolves and becomes more integrated into society. Standardization can ensure AI systems' consistency, dependability and fairness while fostering innovation and competition.

The swift evolution of AI technology is one of the most significant obstacles to standardization.

Standardization in AI is essential for a variety of reasons. It can assure the interoperability of AI systems, allowing them to communicate and collaborate effectively. This is crucial as AI becomes more prevalent in vital infrastructure such as healthcare and transportation systems.

Second, standardization can promote AI system transparency and confidence. By adhering to recognized standards, developers can demonstrate the predictability and dependability of their AI systems. Standardization can assist in addressing ethical and societal issues associated with AI, such as bias, privacy, and accountability.

The swift evolution of AI technology is one of the most significant obstacles to standardization. Frequently, the rate of AI development outpaces the rate at which standards can be developed and implemented, resulting in a never-ending game of catch-up. In addition, the complexity and variety of AI technologies make it challenging to develop universal standards. Additionally, there are concerns that excessively rigid standards could stifle innovation and competition in the AI industry.

Despite these obstacles, the prospects for standardizing AI are promising. Several groups, including the International Organization for Standardization (ISO) and the Institute of Electrical and Electronics

Engineers (IEEE), are actively developing AI standards. These efforts concentrate on terminology, ethical considerations, and technical specifications for AI systems.

In addition, there is a growing recognition of the significance of involving a broad range of stakeholders in the standardization process, including AI developers, consumers, regulators, and civil society.

(ii) Machine Learning Governance

The dominant version of AI technology that urgently requires governance is machine learning. AI systems, especially those based on machine learning, rely significantly on data. This data's quality, diversity and quantity can considerably affect the performance and behavior of AI systems. Consequently, regulating data utilized for AI training is a crucial concern.

Striking a balance between the need for data to train AI systems and the need to safeguard privacy remains a formidable challenge.

Privacy protection is a primary challenge in regulating AI training data. Data protection and privacy concerns are raised because AI systems frequently require large quantities of personal data. To safeguard individuals' data privacy, regulations such as the General Data Protection Regulation have been implemented in the European Union. How do we do the same in the global South?

However, striking a balance between the need for data to train AI systems and the need to safeguard privacy remains a formidable challenge. Data bias is a further crucial concern. If the data used to train AI systems is prejudiced, the AI systems can become biased and produce unfair or discriminatory results.

Consequently, it is crucial to regulate data to ensure it is representative and bias-free. Nevertheless, identifying and eliminating bias in data can be difficult and complex. Transparency in data collection, storage and use for AI training is an additional crucial aspect of regulation.

Transparency can aid in developing confidence in AI systems and ensuring accountability. However, providing transparency without compromising privacy or confidential information can be difficult.

There are significant opportunities for regulating data for AI training despite these obstacles. Regulations can aid in establishing data privacy, bias and transparency standards, thereby fostering the responsible and ethical use of AI. Additionally, they can promote innovation by levelling the playing field and fostering competition.

As AI continues to advance, the need to effectively regulate these algorithms becomes more crucial. Regulation of AI training algorithms is crucial for multiple reasons. It can first assure the fairness and openness of AI systems. Inadvertently, algorithms can perpetuate or amplify biases present in training data, resulting in unjust outcomes.

Regulation can aid in ensuring that algorithms are created and utilized in a manner that mitigates these biases. Yes, the traditional training, testing and validation process, even though necessary, is not enough.

There is a growing recognition of the need for the participation of multiple stakeholders in algorithm regulation, including AI developers, consumers, regulators and affected communities.

Second, regulation can aid in establishing accountability. As AI systems become more complex, it can be challenging to comprehend how they make decisions. Regulation can help ensure that algorithms are transparent and interpretable, allowing their decisions to be held accountable.

The technical complexity of these algorithms is one of the primary challenges. The regulatory experts often treat this as a black box and have little understanding of the interworkings of these algorithms with devastating consequences. The need to understand the complex networks of neural networks, optimization and backpropagation is required to have effective governance of algorithms. A high level of technical expertise is required to comprehend how they operate and how they must be regulated.

In addition, the swift development of AI can make it challenging for regulations to keep up. There is also the possibility that excessively restrictive regulations will inhibit AI innovation.

There are promising prospective directions for algorithm regulation despite these obstacles. Creating technical standards for AI algorithms can guide their design and implementation. Third-party audits are another method for evaluating the impartiality and transparency of algorithms.

In addition, there is a growing recognition of the need for the participation of multiple stakeholders in algorithm regulation, including AI developers, consumers, regulators and affected communities.

(iii) Use and activity

Finally, we need to regulate the use of AI. Each sector will have to develop its regulatory standards. For example, the World Health Organization has developed ethical guidelines for using AI in health. However, one area that requires global regulation is the weaponization of AI.

In addition to its numerous benefits, AI poses potential dangers, especially when weaponized. The weaponization of AI refers to using AI in military and warfare contexts, such as autonomous weapons and cyber warfare.

The militarization of AI has profound implications for global security and warfare. AI can improve military capabilities by allowing quicker decision-making, more accurate targeting and more efficient resource allocation. AI-powered autonomous weapons can operate without human intervention, potentially reducing the danger to human soldiers.

However, these developments raise concerns regarding the escalation of conflicts, the possibility of autonomous weapons being compromised or misused and the possibility of an AI arms race.

The regulation of AI weaponization presents significant difficulties. The rapid development of AI and its technical complexity makes it challenging for regulations to keep up. In addition, international cooperation is difficult; effective regulation requires consensus among nations, which can be difficult due to divergent national interests. In addition, the dual-use nature of AI technology (i.e. its use for civilian and military purposes) complicates regulation.

CHAPTER 8

3D PRINTING IN WARFARE

8.1 INTRODUCTION TO 3D PRINTING

3D printing or **additive manufacturing** is the construction of a three-dimensional object from a CAD model or a digital 3D model. It can be done in a variety of processes in which the material is deposited, joined or solidified under computer control, with the material being added together (such as plastics, liquids or powder grains being fused), typically layer by layer.

In the 1980s, 3D printing techniques were considered suitable only for the production of functional or aesthetic prototypes, and a more appropriate term for it at the time was rapid prototyping. As of 2019, the precision, repeatability, and material range of 3D printing have increased to the point that some 3D printing processes are considered viable as an industrial-production technology; in this context, the term additive manufacturing can be used synonymously with 3D printing. One of the key advantages of 3D printing is the ability to produce very complex shapes or geometries that would be otherwise infeasible to construct by hand, including hollow parts or parts with internal truss structures to reduce weight while creating less material waste. Fused deposition modeling (FDM), which uses a continuous filament of a thermoplastic material, is the most common 3D printing process in use as of 2020.

The umbrella term **additive manufacturing (AM)** gained popularity in the 2000s, inspired by the theme of material being added together (in any of various ways). In contrast, the term **subtractive manufacturing** appeared as a retronym for the large family of machining processes with material removal as their common process. The term 3D printing still referred only to the polymer technologies in most minds, and the term AM was more likely to be used in metalworking and end-use part production contexts than among polymer, inkjet, or stereolithography enthusiasts.

The **3D printing process is the opposite of subtractive manufacturing**, in which, instead of carving out a particular shape from a solid block of material, as done with a milling machine, an object is built by adding layers of material. 3D printing allows the creation of intricate shapes using less material compared to traditional manufacturing methods.

3D printing is widely used across industries, from creating such consumer products as eyewear and furniture to industrial tools, prototypes and functional parts. It is also revolutionizing fields, such as dentistry and prosthetics by enabling efficient production of custom items.

By the early 2010s, the terms 3D printing and additive manufacturing evolved senses in which they were alternate umbrella terms for additive technologies, one being used in popular language by consumer-maker communities and the media, and the other used more formally by industrial end-use part producers, machine manufacturers, and global technical standards organizations. Until recently, the term 3D printing has been associated with machines low in price or capability. 3D printing and additive manufacturing reflect that the technologies share the theme of material addition or joining throughout a 3D work envelope under automated control. These terms are still often synonymous in casual usage, but some manufacturing industry experts are trying to make a distinction whereby additive manufacturing comprises 3D printing plus other technologies or other aspects of a manufacturing process.

Other terms that have been used as synonyms or hypernyms have included desktop manufacturing, rapid manufacturing (as the logical production-level successor to rapid prototyping), and on-demand manufacturing (which echoes on-demand printing in the 2D sense of printing). The fact that the application of the adjectives rapid and on-demand to the noun manufacturing was novel in the 2000s reveals the long-prevailing mental model of the previous industrial era during which almost all production manufacturing had involved long lead times for laborious tooling development. Today, the term subtractive has not replaced the term machining, instead complementing it when a term that covers any removal method is needed. Agile tooling is the use of modular means to design tooling that is produced by additive manufacturing or 3D printing methods to enable quick prototyping and responses to tooling and fixture needs. Agile tooling uses a cost-effective and high-quality method to quickly respond to customer and market needs, and it can be used in hydro-forming, stamping, injection molding and other manufacturing processes.

8.2 HOW DOES 3D PRINTING FUNCTION?

The process begins with a 3D model. You can either design it yourself or download one from a 3D model library. Since the late-1970s, companies have been using 3D printers in their design process to make prototypes. This practice is known as rapid prototyping. It is fast and cost-effective. You can take an idea, turn it into a 3D model and have a physical prototype in hand within days, not weeks.

Adjustments are simpler and cheaper, without requiring expensive molds or equipment. Besides rapid prototyping, 3D printing is used for rapid manufacturing. This is a modern manufacturing method where companies use 3D printers to produce small batches or custom items in limited quantities quickly and efficiently.

8.3 IS 3D PRINTING THE FUTURE OF WAR?

As 3D printing continues to revolutionize the manufacturing industry by enabling complex products to be printed in a matter of hours, the considerable benefits of additive manufacturing are appealing.

One industry, in particular, that stands to reap these benefits is the arms industry. In 2020, global military spending was estimated to be around $1.83 trillion—and this trend is predicted to continue. As defense forces around the world continue to need more ammunition, tools and machine parts, additive manufacturing is an effective recourse to bolstering this vast, complex supply chain.

Titomic and Repkon recently signed an agreement to operate a 3D printing facility **in Australia.** They aim to combine Repkon's cylinder-creating expertise with Titomic's 3D printing technology to enhance Australia's defense manufacturing processes. Turkey-based Repkon has nearly four decades of experience in metal-building techniques like hot spinning, flow-forming and shear forming that it uses to produce warheads, gun barrels, bullet casings and missile bodies, among other pieces of equipment. The defense industry relies heavily on highly efficient, perfectly calibrated cylindrical metal components, be it gun barrels or bullet shells.

Guns, bullets and artillery rounds are the most fundamental aspect of any military's arsenal. In 2017, around 133 million firearms were estimated to be in circulation in the world's militaries with an additional 100,000 artillery weapons. Moreover, around 10 billion bullets are manufactured every single year. To put this number into perspective, one could shoot every single man, woman and child on the planet today and still have nearly 2.5 billion bullets to spare.

For guns and artillery to function, they require precise and durable construction. Gun barrels, bullet casings and artillery rounds need to be perfectly calibrated to the millimeter for ballistic purposes. Additionally, they require high tensile strength to prevent breakage or cracks while fulfilling heat resistance requirements. Bullet and artillery casings in particular need to be constructed from cheap, sturdy material since thousands of them are fired routinely.

Titomic, an Australia-based company, has made a name for itself through its patented additive manufacturing process. Called Titomic Kinetic Fusion (TKF), the process involves using pressurized gas to deposit metal powder at supersonic speed onto a build tray. At such speeds, the kinetic energy released upon impact is high enough to plastically deform the metal particles and bond them in an extremely dense layer. As more metal powder is deposited, it forms layers of tightly bonded metal parts into a desired shape that needs little post-processing. TKF's additive manufacturing style also reduces material loss by 90 percent **while reducing greenhouse gas (GHG) emissions by 60 percent**.

In Israel's West Bank, an Israeli patrol discovered several **3D-printed firearms on the West Bank**; these included several Glock-like pistols, as well as what appeared to be a submachine gun. It was the first time 3D-printed weapons have been discovered in the West Bank and has caused concern for Israeli forces.

In the war in Ukraine, **Ukrainian forces are using 3D printers to print the bodies of grenades** that are filled with shrapnel and explosive material. They're also printing stabilization fins for drone-dropped grenades.

You might not think that making small plastic widgets was very relevant, but a 3D printer can produce a lot of items which are extremely useful on the front line, from munitions to lifesaving gadgets.

In the early days of the war, Eman, founder of volunteer group WildBees Poland 3D printing group, bought everything he could and drove it into Ukraine from Poland. He still makes those runs, but now his efforts are focused on items that can be easily 3D printed and which are the most valuable to the troops. Demand for everything is high.

"We have printed more than a million items," says Eman. "Several thousand different designs."

The variety of items highlights a major advantage of 3D printing, that it can quickly and easily pivot from one product to another. Traditional manufacturing methods require machinery to be at the very least reconfigured for a new product, but with a 3D printer can switch over at the click of a mouse.

3D printing makes iterative development easy. Prototypes are made quickly and sent out for evaluation, feedback from users in the field is quickly incorporated into the next version. And if users want a design in three different sizes, or left-handed and right-handed versions, accommodating them is easy.

Some of their most popular products are munition casings, similar to those Ukraine's Steel Hornets which functions like an Amazon for drone bombs, sending munitions (minus explosive and detonators) via commercial couriers.

Most modern Western military forces are also adopting the technology. For example, the French military has taken the same route and in the civil war in Mali, French forces experimented with 3D printers to produce replacement parts for broken equipment; and the British military wants to use 3D printers to manufacture parts and equipment.

Guns, bullets and artillery rounds are not the only critical equipment the military requires. Countless machine parts, vehicular parts, construction tools and other equipment are pivotal for a fully operationally, successful military. As such, downtime, logistical errors and delays, and replacement or repair costs are a serious detriment to the military. This is especially true given the time-sensitive nature of many of their missions.

Various defense sectors have begun exploring how feasible it would be to incorporate on-site, easily deployable additive manufacturing. For example, the U.S. Navy is already working with Xerox to 3D print machine parts onboard their ships. As such, 3D printing companies have an opportunity to offer cost-effective solutions that curtail downtime and maintain a high-volume build rate.

SPEE3D is an additive manufacturing company that, like Titomic, uses a gas chamber to eject powdered metal at Mach 3 speeds, resulting in the metal plastically deforming into densely packed layers. All parts printed this way are durable and affordable, costing between $40 to $100 per kilogram. SPEE3D's printers mainly use copper, stainless steel, aluminum alloys and aluminum bronze, which have the tensile strength, cost-effectiveness and corrosion resistance necessary for numerous military applications.

Last year, the Australian Army entered a $1.5 million partnership with SPEE3D to administer a series of field tests on the company's industrial 3D printer, WarpSPEE3D. The printer has a build rate of 100g per minute and can print objects as heavy as 40 kg.

8.4 OTHER MILITARY APPLICATIONS OF 3D PRINTING

Additive manufacturing has applications beyond printing tools or machine parts. As 3D printing software grows more sophisticated, designing and printing increasingly complex parts paves the way for innovations that were not possible before. For example, the standard design of **the night-vision goggles** used by the U.S. Army has a selector switch that is used to turns the goggles on or off. If damaged, the part cannot be replaced and the entire night-vision device must be discarded. Additive manufacturing has enabled the U.S. Army to **redesign the switch** in a more sophisticated, durable way that not only extends the life cycle of the night-vision goggles, but also ensures that the part can be reprinted and replaced easily.

Another application of additive manufacturing involves increasing the service life of entire machines. In May 2020, researchers at Wichita State University (WSU) announced their collaboration with the U.S. Army on a Black Hawk tear-down project. A tear-down is when a large machine, like a vehicle, is disassembled down to its smallest part for documentation purposes. In this case, the researchers will not just be documenting every one of the Black Hawk's 20,000 components, but will also be pursuing a goal of creating the aircraft's digital twin.

Given the age of the helicopter, some models of the aircraft have been out of production for almost 15 years. As such, the only way to find replacements is through additive manufacturing. Incorporating 3D printing in the process also makes it possible to redesign some of these components to improve the helicopter's performance.

When a part in your equipment breaks and it takes typically 12 weeks for the manufacturer to get you a new one, you have a problem. However, if you can replicate it, print it, or find a fix on the battlefield, then you're ahead of the curb. To do just that, the United States Army has established Expeditionary Labs. These self-contained labs occupy a simple 20-foot container. Inside, Soldiers utilize 3D printers, CNC machines, routers, welders, and more to rapidly produce military equipment parts. Expeditionary Labs were used in Afghanistan and are being championed for future potential conflicts with Russia and China.

The Marine Corps is experimenting with the same technology and has used 3D printers of all sizes to produce everything from protective gear to bunkers, a reinforced concrete bridge, and a barracks room. The Corps has even equipped helicopters with 3D printers and in 2023 it successfully printed a medical cast aboard an airborne Marine Corps Osprey.

The Black Hawk has been in service for the U.S. Army for nearly four decades, and now—through 3D printing—the army can expect to see the Black Hawk in service for much longer instead of phasing it out like it has had to do with other once-reliable equipment. The effective harnessing of additive manufacturing could possibly even allow the reintroduction of equipment that has long been out of service.

8.5 US NAVY 3D-PRINTING SUBMARINES

It usually takes about 5 months to build a small submarine. Now, thanks to 3D printing, US Navy SEALS can be transported to and from combat zones in small 3D printed submersible hulls that take merely 4 weeks to build.

The hull in the submarine is 30 feet long, and made of six carbon fiber composite sections. The project only took four weeks to complete, and the proof-of-concept also cut production costs by 90%.

According to the Department of Energy, a traditional SEAL Delivery Vehicle costs between $600,000 and $800,000, and it takes three to five months to manufacture. So, that means that they made this sub for as low as $60,000 and it was printed in a number of days — total development time took four weeks, but it only took a few days to print the six sections.

While US Marines are used to having to "MacGyver" things out in the field, they now have the option in some locations to use 3D printers to help them replace belt buckles, parts for various equipment as well as drones.

Once drones are in use, Marines can work on customizing and otherwise altering the initial design to fit their needs. Then, Marines share their ideas with others and version 2.0 can be immediately pushed out to the rest of the corps.

The US Army is looking to use as many 3D printed titanium parts on their vehicles as possible. The Army's Next-Generation Combat Vehicle (NGCV) is a family of new combat machines now being developed for future warfare. The plan involves building new infantry carriers, tanks, and robotic vehicles networked together as part of an integrated tactical maneuver strategy.

The Army Research Laboratory is now working to engineer new, lightweight yet survivable vehicle parts such as brackets, turret components, propulsion systems, and weapons, using 3D printing technology. In particular, they are exploring the use of lightweight metals such as titanium, titanium alloys, and hybrid ceramic tile/polymer-matrix composites.

The Department of Defense (DoD) wants to find a way to turn discarded battlefield plastic waste, such as water bottles, milk jugs, and yogurt containers into useful replacement parts for US soldiers.

Researchers at the Army Research Laboratory have taken recycled polyethylene terephthalate (PET) from bottles and plastics without any chemical modifications or additives and used it as the material in a fused filament fabrication (FFF filament), one of the many types of 3D printing.

The research will let U.S. forces 3D print replacement parts on demand. This will increase the readiness of their equipment, but also let them make mission-specific devices in the field.

Each unit carries a stockpile of replacement parts for emergencies, but it can be costly to store and ship them. Having the ability to 3D print on demand reduces costs and frees up space.

It will be interesting to watch how 3D printing will continue to find its way into every branch of military service, especially during these unprecedented times while we struggle to keep factories and businesses open during a pandemic that would normally be supplying our military with parts they need to maintain their equipment.

One thing we know for sure is that 3D printing technology opens up a lot of doors for our military personnel that will allow them to continue to provide their selfless service to our country and the world with a little less stress and struggle.

8.6 INDIA MULLS '3D PRINTING' OF WEAPONS

India, which faces the potential challenge of two fronts, is also aware of this development. In future conflicts, 3D printing, known as Additive Manufacturing, can help the Indian forces plug this critical gap in supply.

This technology is not futuristic, but it is already here. In the hands of rebels and non-state actors, it has provided an endless supply of weapons. In Myanmar, images of 3D-printed drones made by rebels have been catalyzing the civil war in the country.

The 3D printers are producing in-house weapons for the rebels, including drones, mortar stabilizers, and other munitions for the Karenni Nationalities Defence Force (KNDF), which was outgunned by the Myanmar Junta. The use of 3D printers in Myanmar has further shown that wars can be prolonged on a shoestring budget.

For India, 3D technology can help produce specific missile components, particularly rocket motors. Additive manufacturing (3D printing) can surely speed up missile key component manufacturing. For example, the inertial navigation system platform design and implementation, internal component design, and mounting assemblies can be expedited through additive manufacturing.

In 2024, Indian space startup *'Agnikul Cosmos'* successfully launched the nation's first 3D-printed rocket engine, paving the way for reduced time and costs associated with building rockets and boosting the country's spacefaring capabilities.

8.7 INTRODUCTION TO 4D PRINTING

Using 3D printing and multi-material structures in additive manufacturing has allowed for the design and creation of what is called 4D printing. 4D printing is an additive manufacturing process in which the **printed object changes shape with time, temperature, or some other type of stimulation.** 4D printing allows for the creation of dynamic structures with adjustable shapes, properties or functionality. The **smart/stimulus-responsive materials that are created using 4D printing can be activated to create calculated responses such as self-assembly, self-repair, multi-functionality,** reconfiguration and shape-shifting. This allows for customized printing of shape-changing and shape-memory materials.

4D printing technology is the process that allows the printed objects to self-transform over time. It is also known as additive manufacturing technology that allows nanoscale manipulation and programming during the production process to ultimately create products designed to adapt to their environment. The printed object can change shape due to many factors such as air, heat, pressure and magnetism. It is an advancement to the 3D printing technology, which can be created by using smart material and software program in the 3D printing machine. Hence, it is a novel technology and is on the verge of commercialization.

4D printing has the potential to find new applications and uses for materials (plastics, composites, metals, etc.) and has the potential to create new alloys and composites that were not viable before. The versatility of this technology and materials can lead to advances in multiple fields of industry, including space, commercial and medical fields. The repeatability, precision, and material range for 4D printing must increase to allow the process to become more practical throughout these industries.

4D Printing is the technological mixture of additive manufacturing, nanotechnology, and geometric programming. From flight suits that self-adapt during times of increased g-forces, to adaptable aircraft struts, to nanotechnology munitions, the applications for 4D Printing intersect multiple disciplines and myriad applications. Skylar Tibbits, one of the 4D Printing industry pioneers, defines 4D Printing as creating "multi-material prints with the capability to transform over time, or a customized material that can change from one shape to another, directly off the print bed." Essentially, 4D Printing combines the field of "programmable matter" with the additive manufacturing capabilities of 3D Printing to generate products in the fourth dimension—time.

To become a viable industrial production option, there are a few challenges that 4D printing must overcome. The challenges of 4D printing include the fact that the microstructures of these printed smart materials must be close to or better than the parts obtained through traditional machining processes. New and customizable materials need to be developed that have the ability to consistently respond to varying external stimuli and change to their desired shape. There is also a need to design new software for the various technique types of 4D printing. The 4D printing software will need to take into consideration the base smart material, printing technique, and structural and geometric requirements of the design.

The global military 4D printing market size is projected to be valued at $16.1 million in 2030, and is expected to reach $673.4 million by 2040, registering a CAGR of 45.2% from 2030 to 2040.

8.7.1 Military 4D Printing Market

Currently, the defense industry is adopting the 4D printing technologies to simplify the process of manufacturing of parts and components, as well as weapons and equipment. With the changes in battlefield requirements, the equipment that changes (shape or properties) in accordance with the environment is likely to help armies gain an upper hand. To have this tactical advantage, investments in 4D printed weapons are anticipated to grow rapidly in the future. For instance, the U.S. Army Research Center is coming up with new technologies, such as a soldier's uniform that can offer protection against toxic gases effectively or alter its own camouflage while traveling through different climates. Such robust development in the defense sector is anticipated to boost the growth for the global military 4D printing market in the forecast period.

8.8 4D PRINTING FOR THE US AIR FORCE

Over the past 25 years there has been a revolution in American military engineering, design, and manufacturing (EDM)—the revolution is 4D Printing. The US Air Force now operates over 1000 different specialty aircraft and thousands more prototype aircraft, creating the most robust, layered, and globally capable Air Force in history—despite the increased global instability and transnational unrest. Total development costs of this revolutionized Air Force are less than the sequestration budgets of 2015, because 4D Printing changed the acquisition paradigm of burdensome manufacturing overhead, decades-long-phased development, and economies of scale. By transitioning to 4D Printing at the industrial level and throughout the defense enterprise, the USAF has denied adversaries the ability to gain significant military benefit by cybertheft, because hardware can be continuously improved at the small batch level. The USAF now can produce aircraft capable of adapting to their environment and dominating the newest warfighting domain—the Nano-Dimension. 4D Printing is an additive manufacturing technology that allows nanoscale manipulation and programming during the production process to ultimately create products designed to adapt to their environment. While still an emerging technology and process, 4D Printing is based on current capabilities, and is poised to revolutionize much of the world's EDM over the next twenty-five years. 4D Printing is a game-changer for airpower technology, but the United States must purposefully invest in this technology and EDM infrastructure to both exploit the benefits and prepare for 4D Printing adversaries. Ultimately, to retain national airpower dominance (and by extension, American hegemony) into the future, the United States must lead the world in developing operational 4D Printing capability and capacity.

8.9 5D PRINTING TECHNOLOGY

3D printing is an additive process which was invented by Japanese researcher **Hideo Kodama** in 1981. 3D printing works on stereo lithographic process. It has three-dimensional x, y and z axis. 3D printing requires lot of time to print the components and it is slow in process which is a drawback of this technology. These drawbacks of 3D printing can be overcome by 4D printing. In February 2013, **Skylar Tibbits** invented 4D printing. In this technology, smart materials are incorporated because the 4th dimension is time and it has self-repair, self-assembly and it will save 70% to 90% of printing time. It is applicable in all fields like fashion, medical fields, and automobile, industrial polymer components, and the modern warfare. The drawback of 4D printing is that it is not able to print complex shapes having curved surface. These drawbacks of 4D printing can be overcome by 5D printing. Recent manufacturing industries focus on 4D and 5D printing due to more advantages in present technology. The concept behind 5-dimensional additive manufacturing is rotation of extruder head and rotation of print bed in order to print in 5 different axes. 5D printing is 3 to 4 times stronger than 3D and 4D printing. In 5D printing technology, the curved complex surfaces can be produced and its applications are mainly used in different fields like biomedical, automobile and aerospace components.

5D printing is a new technology in additive manufacturing in which prints head and printable object rotate along with x, y and z axis altogether with five degrees of freedom. It produces curved layer or concave shapes very accurately as per design constraints. In this process, printed part move while the printer head is printing in five axis printing simultaneously print bed also moves forward and backward along with x, y and z axis which allows the object to be printed from all 5 axes instead of form one point of printing.

CHAPTER 9

NANOTECHNOLOGY IN WARFARE

Nanotechnology in warfare is a branch of nano-science in which molecular systems are designed, produced and created to fit a nano-scale (1-100 nm). The application of such technology, specifically in the area of warfare and defence, has paved the way for future research in the context of weaponization. Nanotechnology unites a variety of scientific fields including material science, chemistry, physics, biology and engineering.

Advancements in this area, have led to the categorized development of such nano-weapons with classifications varying from; small robotic machines, hyper-reactive explosives, and electromagnetic super-materials. With this technological growth, has emerged implications of associated risks and repercussions, as well as regulation to combat these effects. Its impacts give rise to issues concerning global security, the safety of society, and the environment. Legislation may need to be constantly monitored to keep up with the dynamic growth and development of nano-science, due to the potential benefits or dangers of its use. Anticipation of such impacts through regulation, would 'prevent irreversible damages' of implementing defence-related nanotechnology in warfare.

9.1 ORIGINS OF NANOTECHNOLOGY IN WARFARE

Historical use of nanotechnology in the area of warfare and defence has been rapid and expansive. Over the past two decades, numerous countries have funded military applications of this technology including; China, United Kingdom, Russia, and most notably the United States. The US government has been considered a national leader of research and development in this area, however now rivalled by international competition as appreciation of nanotechnology's eminence increases. Therefore, the growth of this area in the use of its power has a dominant platform in the front line of military interests.

9.1.1 U.S. National Nanotechnology Initiative

In 2000, the United States government developed a National Nanotechnology Initiative to focus funding towards the development of nano-science and its technology, with a heavy focus on utilizing the potential of nano-weapons. This initial US proposal has now grown to coordinate application of nanotechnology in numerous defence programs, as well as all military factions including Air Force, Army and Navy. From the financial year 2001 through to 2014, the US government contributed around $19.4 billion to nano-science, on the development and manufacturing of nano-weapons for military defence. The 21st Century Nanotechnology Research and Development Act (2003), envisions the United States continuing its leadership in the field of nanotechnology through national collaboration, productivity and competitiveness, to maintain this dominance.

9.1.2 Developments Nanotechnology into Defence

Successful transitions of nanotechnology into defence products in USA has the following features:

- Lifetime of material coatings increased from hours to years, however further development continuing.
- Nano-structured silicate manipulation reducing insulation weight by 980 lbs.
- High Power Microwave (HPM) devices with reduced weight, shape and power consumption.

The United States government has had military-purposed development of nanotechnology at the forefront of its national budget and policy throughout various administrations, with the Department of Defense planning to continue with this priority throughout the 21st century. In response to America›s assertive public funding of defence-purposed nanotechnology, numerous global actors have since created similar programmes.

9.1.3 China's Nanotechnology Initiative

In the sub-category of nano materials China secures second place behind the United States in the amount of research publications they have released. Conjecture stands over the purpose of China's quick development to rival the U.S., with 1/5 of their government budget spent on research (US $ 337 million). In 2018, Tsinghua University, Beijing, released their findings where they have enhanced carbon nanotubes to now withstand the weight of over 800 tonnes, requiring just 1 cm cm3of material. The scientific nanotechnology team hinted at aerospace, and armor boosting applications, showing promise for defence related nano-weapons. The Chinese Academy of Science's Vice President has stated the need to focus on closing the gap between "basic research and application," in order for China to advance its global competitiveness in nanotechnology.

Between 2001 and 2004, approximately 60 countries globally implemented national nanotechnology programmes. Research and development in this area has now become a socio-economic target, an area of intense international collaboration and competition. As of 2017, data showed 4725 patents published in USPTO by the USA alone, maintaining their position as a leader in nanotechnology for over 20 years.

9.2 CURRENT RESEARCH INTO MILITARY NANOTECHNOLOGICAL WEAPONS

Most recent research into military nanotechnological weapons includes production of defensive military apparatus, with objectives of enhancing existing designs of lightweight, flexible and durable materials. These innovative designs are equipped with features to also enhance offensive strategy through sensing devices and manipulation of electromechanical properties.

(i) Soldier battle suit

The Institute for Soldier Nanotechnologies (ISN), deriving from a partnership between the United States Army and MIT, provided an opportunity to focus funding and research activities purely on developing armor to increase soldier survival. Each of seven teams produces innovative enhancements for different aspects of a future U.S. soldier bodysuit. These additional characteristics include energy-absorbing material protecting from blasts or ammunition shocks, engineered sensors to detect chemicals and toxins, as well as built in nano devices to identify personal medical issues such as hemorrhages and fractures. This suit would be made possible with advanced nano-materials such as carbon nanotubes woven into fibers, allowing strengthened structural capacities and flexibility, however preparation becomes an issue due to inability to use automated manufacturing.

Improved body armor is a major focus for military nanotechnology research. Several different technologies have been explored, some of which will be operational in just a few years:

- Si or TiO_2 nanoparticles embedded in epoxy matrix;
- SiO_2 nanoparticles in a liquid polymer which hardens on ballistic impact (Shear Thickening Fluid)
- Iron nanoparticles in inert oil which hardens on stimulation with an electrical pulse (Magnetorheological Fluid)

(ii) Enhanced materials

Creation of sol-gel ceramic coatings has protected metals from; wear, fractures and moisture, allowing adjustability to numerous shapes and sizes, as well as aiding materials that cannot withstand high temperature. Current research focuses on resolving durability issues, where stress cracks between the coating and material set limitations on its use and longevity. The drive for this research is finding more efficient and cost-effective uses in application of nanotechnology for Airforce and Navy military groups. Integration of fiber-reinforced nano-materials in structural features, such as missile casings, can limit overheating, increase reliability, strength and ductility of the materials used for such nanotechnology.

(iii) Communication devices

Nanotechnology designed for advanced communication is expected to equip soldiers and vehicles with micro antenna rays, tags for remote identification, acoustic arrays, micro GPS receivers and wireless communication. Nanotech facilitates easier defence related communications due to lower energy consumption, light weight, efficiency of power, as well as smaller and cheaper to manufacture. Specific military uses of this technology include aerospace applications such as; solid oxide fuel cells to provide three times the energy, surveillance cameras on microchips, performance monitors, and cameras as light as 18g.

(iv) Mini-nukes

The United States, along with countries such as Russia and Germany, are utilizing the convenience of small nanotechnologies, adhering it to nuclear "mini-nuke" explosive devices. This weapon would weigh 5 lbs, with the force of 100 tonnes of TNT, giving it the possibility to annihilate and threaten humanity. The structural integrity would remain the same as nuclear bombs, however manufactured with nano-materials to allow production to a smaller scale.

Engineers and scientists alike, realise some of these proposed developments may not be feasible within the next two decades as more research needs to be undertaken, improving models to be quicker and more efficient. Particularly molecular nanotechnology, requires further understanding of manipulation and reaction, in order to adapt it to a military arena.

(v) Integrating soldiers

Imagine the soldier at the next-generation platform. The Star analogy is this modular soldier of the future, plugged into virtual combat zones born from the dreams of science fiction, currently evolves in military programs all over the world, including Germany's IdZ (Infanterist der Zukunft, or 'Infantryman of the Future'), India's F-INSAS (Futuristic Infantry Soldier as a System) and the United Kingdom's FIST (Future Integrated Soldier Technology).

Similarly, France's FELIN (Fantassin a Equipement et Liaisons Integrees, or 'Foot Soldiers with Integrated Equipment and Links') and BOA (Bulle Operationnelle Aeroterrestre, the network-enabled 'Aeroterrestrial Operational Bubble' combat system), Singapore's ACMS (Advanced Combat Man System), Norway's NORMANS (Norwegian Modular Network Soldier), Australia's Land, the United States' Future Combat Systems and Future Warrior.

The digital battlefield, involving the total integration of bits of matter with bits of code, necessarily extends from the macro scale down to the nano scale.

Nanotechnology would seem to provide revolutionary solutions for integrating soldiers and information systems in the battle spaces of the future.

(vi) Nano-Enhanced Sensors

Many sensors have already been developed which take advantage of the unique properties of nanomaterials to become smaller and more sensitive, compared to conventional technology. Portable, efficient sensors will be highly valuable to military field operatives, for example:

- Highly sensitive infrared thermal sensors;
- Small, lightweight accelerometers and GPS for motion and position sensing;
- Miniature high performance camera systems;
- Biochemical sensors; and
- Health-monitoring sensors and drug/nutrition delivery systems.

(vii) Radar and Sonar Invisibility

Stealth ships and aircraft are being improved with the use of nanomaterials which can help 'hide' military hardware, such as submarines, from detection by radar and sonar systems.

9.3 INDIA AND NANOTECHNOLOGY

As regards India, the potential was recognized during the 9th Five Year Plan. The 9th Five Year Plan covered the period from 1998 to 2002. It was for the first time that national facilities and core groups were set up to promote research in frontier areas of Science and Technology to include nano materials. In actuality, the thrust came with the launch of the Programme on Nano Materials, Sciences and Devices, by the Department of Science and Technology (DST). The National Nano-science and Nano-technology Initiative (NSTI) was launched in October, 2001 and the aim was to create research infrastructure and promote basic research in nano science and nano-technology. It focused on various issues relating to infrastructure development, application-oriented programmes in nano material including drugs, gene targeting and DNA chips. Nano-technology was heralded for revolutionary technology with applications in almost every aspect of life. Gradually, mission-oriented projects were initiated in this field.

It is interesting to note that the DST has launched bilateral joint research projects with more than 25 countries and multilateral regional projects with multi-lateral bodies such as the European Union, BRICS and other organizations. These international collaborative projects cover a range of nano-science and nano-technology basic and applied research in areas having diverse applications as enumerated below:

- Healthcare/medicine/drug delivery.
- Sensors.
- Energy storage in batteries.
- Solar energy.
- Fuel cells.
- Other areas of use of nano materials such as thin films, carbon nano tubes and nano composites.

It is also important to know that the Department of Bio-technology (DBT) is extremely active in the area of nano bio-technology research & development. The Department has been actively engaged in promoting inter-disciplinary research, fostering innovations and promoting development of translational research in various areas of nano bio-technologies including new therapeutics, diagnostics, for early diagnostics. This program could be applied also in the field of Army Medical Research and Veterinary Research. The various aspects which they focused are as under:

- **Nanotechnology for food/agriculture:** weed utility nano-sensor for crop protection; pesticide delivery vehicles; smart packaging; sensors for detecting pathogens; and chemicals in food. Militarily, these sensors could be utilized for surveillance as also getting early warning of tunneling activity and inspection of ammunition and missiles.
- **Nano-technology in other allied areas such as bi-engineering**, water filtration and variety of other tasks. This could be also used for military applications.
- **Activity and inspection of ammunition and missiles** for correct chemical composition.

The Department of Electronics and Industry has also been taking a few baby steps and are doing extremely well in the manufacturing of semi-conductors and chips for Indian Space Research Organisation. All these could be used for the development of high-speed communications.

9.3.1 Indian Army and Nanotechnology

Nano-technology has the potential to influence warfare technology in number of ways. Lighter, stronger, heat-resistant nano material could be used in manufacturing all types of war-like equipment, weapon platforms, and missiles enabling mobility over all types of terrain including mountains. With regard to boots on the ground, the military use of nano-technology will ensure better protection, more lethality, enhanced endurance and better capabilities of repair in the battle space.

Areas Impacted

The first aspect that nanotechnology would directly impact would be the all class of vehicles. The aim would be to reduce the overall weight, improve its fuel efficiency and improve the ergonomics. This would be applicable to tanks, infantry combat vehicles, gun towers, engineer vehicles and load carriers. Nano-technology makes the vehicle stronger in the event of a crash. Overall, nanotechnology can concentrate for five areas for the Indian Army. These are as under:

- Focus on the combat soldier,
- Ability to introduce nano technology in information dominance,
- Weapons,
- Vehicles and platforms,
- Logistics.

The combat soldier is keen that the weight of his equipment is reduced. Further, the sensors he carries would need lesser electrical output for the same functional capability. The soldier of the future of the Indian Army should have light protective clothing, Battlefield Management System in the form of a wrist watch which would provide him all details including the Common Operating Picture. Further, his weapon would be lighter and he would be camouflaged against Infra-Red (IR) devices. The future soldier will have an all-impact suit enabled by nano-materials combined with micro or macro fiber to offer protection against bullets, grenade fragments, bio agents, chemical agents. His back pack and weapon would be light with precise ammunition. His smart helmet would be light-weight and he will have a light solar system for preservation of food and water.

Further, nano-electronics will basically lead to improved equipment and quality for sensors and other communication equipment. In the field of electronics it will lead to lower power consumption with regards to microchips. It would also lead to less noise and higher processing speeds. In times to come, equipment with nano-electronics will take on cognitive functions leading to higher human applications by machines. Gradually these will increase leading to better command functions being performed by machines. The Indian Army, which has a wide surveillance network, will be deeply impacted by these changes.

Nano-materials will assist in creating a smoother control of energy release and shorter diffusion paths for ammunition blasts of High Explosive. This will improve the overall impact at the target end. Further, nano-particles on ammunition can create greater penetration of kinetic shells particularly against armor. The construction flexibility can be improved so that light falling on these particles is scattered, thereby improving the stealth capability of the material. The impact of Nano-technology would in the next decade result in light weight structures for guns, rifles and automatic firing systems. Other armies are using nano-weapons for correcting the bullet path against body vibration while firing.

Future platforms, be it a tank or a land vehicle, would be lighter to fit in greater numbers into an IL 76, C 130 or a Globe Master transport aircrafts. The tank would be faster and more lethal. The overall aim would be to develop light weight nano-composite plates to cover critical parts of the tank, infantry combat vehicle or any other vehicle.

An essential element of future warfare is logistics. In the Indian Army, supply chain management forms an important component of logistics. This would help operations in the field to ensure that food, water, ammunition and fuel supply are simpler to move to the right place, on time and in desired quantities. In order to undertake such tasks, there is the need for modular containers. These could be of various sizes which can be moved to the desired destinations for the troops. All these containers would have RFID tags which would assist in collection, transit and distribution. Nanotechnology would make these containers light and easy to move by transport aircraft and utility helicopters. All these aspects, combined with Artificial Intelligence, would make easy the delivery of items in the battle space.

DRDO is carrying out extensive work in the field of nanotechnology to enhance its application in defense sector. Major focus areas have been NBC (Nuclear, biological and Chemical) attack protection devices, stealth and camouflage, sensors, high-energy applications, nanoelectronics, structural applications.

DRDO has also set up nano research and production facility in various parts of India. A Bengaluru based Materials startup is also collaborating with the defense industry to help it build various products and applications while conserving energy. However, the progress made by the country is not enough and this work needs to be accelerated.

9.4 OTHER APPLICATIONS OF NANOTECHNOLOGY

In the longer term, it seems likely that most military technology will be dependent on nanomaterials. Some of the more speculative applications in this area include:

- Nano-machines to mimic human muscle action in an exoskeleton;
- Stealth coatings;
- Self-healing (self-repair) material;
- Smart skin materials;
- Adaptive camouflage; and
- Adaptive structures

9.5 IMPLICATIONS OF NANOTECHNOLOGY USE IN WARFARE

Nanotechnology and its use in warfare promises economic growth however comes with the increased threat to international security and peacekeeping. The rapid emergence of new nanotechnologies have sparked discussion surrounding the impacts such developments will have on geo-politics, ethics, and the environment.

(i) Geo-political

Difficulty in categorization of nano-weapons, and their intended purposes (defensive or offensive) compromises the balance of stability and trust in the global environment. A lack of transparency about an emerging technology not only negatively affects the public perception but also negatively impacts the perceived balance of powers in the existing security environment. The peace and cohesion of the international structure may possibly be negatively affected with a continuing military-focused development of nanotechnology in warfare.

Ambiguity and a lack of transparency in research increases difficulty of regulation in this area. Similarly, arguments put forward from a scientific standpoint, highlight the limited information known, concerning the implications of creating such powerful technology, in regards to reaction of the nano-particles themselves. Although great scientific and technological progress has been made, many questions about the behavior of matter at the nanoscale level remain, and considerable scientific knowledge has yet to be learned.

(ii) Environmental

The introduction of nanotechnology into everyday life enables potential benefits of use, yet carries the possibility of unknown consequences for the environment and safety. Possible positive developments include creation of nano-devices to decrease remaining radio-activity in areas, as well as sensors to detect pollutants and adjust fuel-air mixtures. Associated risks may involve: military personnel inhaling nanoparticles added to fuel, possible absorption of nanoparticles from sensors into the skin, water, air or soil, dispersion of particles from blasts through the environment (via wind), alongside disposal of nano-tech batteries potentially affecting ecosystems. Applications for materials or explosive devices, allow a greater volume of nano-powders to be packed into a smaller weapon, resulting in a stronger and possibly lethal toxic effect.

(iii) Social and Ethical

The full extent of consequences that may arise in social and ethical areas is unknown. Estimates can be made on the associated impacts as they may mirror similar progression of technological developments and affect all areas. The main ethical uncertainties entail the degree to which modern nanotechnology will threaten privacy, global equity and fairness, while giving rise to patent and property right disputes. An overarching social and humanitarian issue, branches from the creative intention of these developments. 'The power to kill or capture debate', highlights the unethical purpose and function of destruction these nanotechnological weapons supply to the user.

Controversy surrounding the innovation and application of nanotechnology in warfare highlights dangers of not pre-determining risks, or accounting for possible impacts of such technology. The threat of nuclear weapons led to the cold war. The same trend is foreseen with nanotechnology, which may lead to the so-called nano-wars, a new age of destruction. Similarly a report released by Oxford University, warns of the pre-eminent extinction of the human race with a 5% risk of this occurring due to development of 'molecular nanotech weapons'.

9.6 REGULATION OF NANOTECHNOLOGY USE IN WARFARE

International regulation for such concerns surrounding issues of nanotechnology and its military application, are non-existent. There is currently no framework to enforce or support international cooperation to limit production or monitor research and development of nanotechnology for defensive use. Even if a transnational regulatory framework is established, it is impossible to determine if a nation is non-compliant if one is unable to determine the entire scope of research, development, or manufacturing.

Producing legislation to keep-up with the rapid development of products and new materials in the scientific spheres, would pose as a hindrance to constructing working and relevant regulation. Productive regulation should assure public health and safety, account for environmental and international concerns, yet not restrict innovation of emerging ideas and applications for nanotechnology.

9.6.1 Proposed regulation

Approaches to development of legislation, possibly include progression towards classified non-disclosive information pertaining to military use of nanotechnology. A paper written by Harvard Journal of Law and Technology, discusses laws that would revolve around specific export controls and discourage civilian or private research into nano-materials. This proposal suggests mimicking the U.S. Atomic Energy Act of 1954, restricting any distribution of information regarding the properties and features of the nanotechnology at creation.

CHAPTER 10

ROBOTICS IN WARFARE

10.1 INTRODUCTION TO ROBOTS IN WARFARE

Military robots are autonomous robots or remote-controlled mobile robots designed for military applications, from transport to search & rescue and attack. Some such systems are currently in use, and many are under development.

Broadly defined, military robots date back to World War II and the Cold War in the form of the German Goliath tracked mines and the Soviet teletanks. The introduction of the MQ-1 Predator drone was: when CIA officers began to see the first practical returns on their decade-old fantasy of using aerial robots to collect intelligence.

The use of **robots in warfare**, although traditionally a topic for science fiction, is being researched as a possible future means of fighting wars. Already several military robots have been developed by various armies. Some believe the future of modern warfare will be fought by automated weapons systems. The U.S. military is investing heavily in the RQ-1 Predator, which can be armed with air-to-ground missiles and remotely operated from a command center in reconnaissance roles. DARPA has hosted competitions in 2004 & 2005 to involve private companies and universities to develop unmanned ground vehicles to navigate through rough terrain in the Mojave Desert for a final prize of 2 million dollars.

Artillery has seen promising research with an experimental weapons system named "Dragon Fire II" which automates loading and ballistics calculations required for accurate and predicted fire, providing a 12-second response time to fire support requests. However, military weapons are prevented from being fully autonomous; they require human input at certain intervention points to ensure that targets are not within restricted fire areas as defined by Geneva Conventions for the laws of war.

There have been some developments towards developing autonomous fighter jets and bombers. The use of autonomous fighters and bombers to destroy enemy targets is especially promising because of the lack of training required for robotic pilots, autonomous planes are capable of performing maneuvers which could not otherwise be done with human pilots (due to high amount of G-force), plane designs do not require a life support system, and a loss of a plane does not mean a loss of a pilot. However, the largest drawback to robotics is their inability to accommodate for non-standard conditions. Advances in artificial intelligence in the near future may help to rectify this.

In 2020, a Kargu 2 drone hunted down and attacked a human target in Libya, according to a report from the UN Security Council's Panel of Experts on Libya, published in March 2021. This may have been the first time an autonomous killer robot armed with lethal weaponry attacked human beings.

10.2 EXAMPLES OF ROBOTS IN WARFARE

10.2.1 Robots in Warfare In Current Use

(i) D9T Panda, Israel

The IDF Caterpillar D9 — nicknamed Doobi (for teddy bear) — is a Caterpillar D9 armored bulldozer used by the Israel Defense Forces (IDF). It is supplied by Caterpillar Inc. and modified by the Israel Defense Forces, Israeli Military Industries and Israel Aerospace Industries to increase the survivability of the bulldozer in hostile environments and enable it to withstand attack.

In the 1980s the IDF began modifying D9 bulldozers to incorporate armor. The bulldozers can also be fitted with weaponry: machine guns and grenade launchers. There are various models, including a remote-controlled version.

The Office of the UN High Commissioner on Human Rights has advised Caterpillar Inc. that by supplying the bulldozers to the IDF it may be complicit in human rights violations.

The IDF Caterpillar D9 has been used to cause civilian deaths, such as the killing of activist Rachel Corrie in 2003 or civilians sheltering outside the Kamal Adwan Hospital during the 2023 Israel–Hamas war. The IDF Caterpillar D9 is operated by the Israel Defense Forces (IDF) Combat Engineering Corps for combat engineering and counter-terrorism operations.

(ii) Elbit Hermes 450, Israel

The Elbit Hermes 450 is an Israeli medium-sized multi-payload unmanned aerial vehicle (UAV) designed for tactical long endurance missions. It has an endurance of over 20 hours, with a primary mission of reconnaissance, surveillance and communications relay. Payload options include electro-optical/infrared sensors, communications and electronic intelligence, synthetic-aperture radar/ground-moving target indication, electronic warfare, and hyperspectral sensors.

(iii) Ghost Robotics Vision 60, Israel

In an effort to avoid harming soldiers and dogs, the IDF has been experimenting with the use of robots and remote-controlled dogs in the Gaza War. Most of the tests have been with a "robot dog," which is also equipped with a drone and can replace or reinforce the Oketz Unit's dogs in certain situations.

(iv) Goalkeeper CIWS

The Goalkeeper CIWS is a Dutch close-in weapon system (CIWS) introduced in 1979. It is an autonomous and completely automatic weapon system for short-range defence of ships against highly maneuverable missiles, aircraft and fast-maneuvering surface vessels. Once activated the system automatically undertakes the entire air defence process from surveillance and detection to destruction, including the selection of the next priority target.

(v) Guardium

Guardium, developed by G-NIUS, is an Israeli unmanned ground vehicle (UGV) used by the Israel Defense Forces along Gaza's border. It was jointly developed by Israel Aerospace Industries and Elbit Industries. It can be used in either tele-operated or autonomous mode. Neither of these modes require human interaction. The more unmanned ground vehicles patrolling the area the less human resources needed while guaranteeing deterrence. The joint program was terminated in April 2016, but the vehicle has remained in service with the IDF.

(vi) IAIO Fotros, Iran

The HESA Fotros is an Iranian reconnaissance, surveillance, and combat unmanned aerial vehicle built by Iran Aircraft Manufacturing Industries Corporation and unveiled in November 2013. It was the largest Iranian drone at its unveiling. It has an operational range of 1,700 km to 2,000 km with flight endurance of 16 to 30 hours depending on armament. The name refers to a fallen angel in Shia mythology which was redeemed by Imam Husayn ibn Ali. The Fotros carries up to six missiles or bombs.

(vii) PackBot

PackBot is a series of military robots by the USA's Endeavor Robotics (previously by iRobot), an international robotics company founded in 2016, created from iRobot, that previously produced military robots since 1990. More than 2000 were used in Iraq and Afghanistan. They were also used to aid searching through the debris of the World Trade Center after 9/11 In 2001. Another instance of the PackBot technology being implemented was to the damaged Fukushima nuclear plant after the 2011 Tōhoku earthquake and tsunami where they were the first to assess the site. The PackBot technology is also used in collaboration with NASA for their rovers and probes.

(viii) MQ-9 Reaper

The General Atomics MQ-9 Reaper (sometimes called Predator B) is an unmanned aerial vehicle (UAV, one component of an unmanned aircraft system (UAS)) capable of remotely controlled or autonomous flight operations, developed by General Atomics Aeronautical Systems (GA-ASI) primarily for the United States Air Force (USAF). The MQ-9 and other UAVs are referred to as Remotely Piloted Vehicles/Aircraft (RPV/RPA) by the USAF to indicate ground control by humans.

(ix) MQ-1 Predator

The General Atomics MQ-1 Predator (often referred to as the Predator drone) is an American remotely piloted aircraft (RPA) built by General Atomics that was used primarily by the United States Air Force (USAF) and Central Intelligence Agency (CIA). Conceived in the early 1990s for aerial reconnaissance and forward observation roles, the Predator carries cameras and other sensors. It was modified and upgraded to carry and fire two AGM-114 Hellfire missiles or other munitions. The aircraft entered service in 1995, and saw combat in the war in Afghanistan, Pakistan, the NATO intervention in Bosnia, the NATO bombing of Yugoslavia, the Iraq War, Yemen, the 2011 Libyan civil war, the 2014 intervention in Syria, and Somalia.

(x) TALON

The Foster-Miller TALON is a remotely operated, tracked military robot designed for missions ranging from reconnaissance to combat. It is made by the American robotics company QinetiQ-NA, a subsidiary of QinetiQ.

(xi) Samsung SGR-A1

The SGR-A1 is a type of autonomous sentry gun that was jointly developed by Samsung Techwin (now Hanwha Aerospace) and Korea University to assist South Korean troops in the Korean Demilitarized Zone. It is widely considered as the first unit of its kind to have an integrated system that includes surveillance, tracking, firing, and voice recognition. While units of the SGR-A1 have been reportedly deployed, their number is unknown due to the project being "highly classified".

(xii) Shahed 129, Iran

The Shahed 129 is an Iranian single-engine medium-altitude long-endurance unmanned combat aerial vehicle (UCAV) designed by Shahed Aviation Industries for the Islamic Revolutionary Guard Corps (IRGC). The Shahed 129 is capable of combat and reconnaissance missions and has an endurance of 24 hours; it is similar in size, shape and role to the American MQ-1 Predator and is widely considered as one of the most capable drones in Iranian service.

(xiii) Baykar Bayraktar TB2, Turkey

The Bayraktar TB2 is a medium-altitude long-endurance (MALE) unmanned combat aerial vehicle (UCAV) capable of remotely controlled or autonomous flight operations. It is manufactured by the Turkish company Baykar Makina Sanayi ve Ticaret A.Ş., primarily for the Turkish Armed Forces. The aircraft are monitored and controlled by an aircrew in a ground control station, including weapons employment. The development of this UAV has been largely credited to Selçuk Bayraktar, a former MIT graduate student.

(xiv) Albatross, Taiwan

The Albatross, also known as the Chung Xiang II, is a medium unmanned aerial vehicle made by National Chung-Shan Institute of Science and Technology. It is in service with the Republic of China Navy.

(xv) THeMIS, Estonia

THeMIS (Tracked Hybrid Modular Infantry System), unmanned ground vehicle (UGV), is a ground-based armed drone vehicle designed largely for military applications, and is built by Milrem Robotics in Estonia. The vehicle is intended to provide support for dismounted troops by serving as a transport platform, remote weapon station, IED detection and disposal unit etc.

(xvi) PLA's robot soldiers, deployed on the China-India border

China is deploying robots equipped with machine guns along the border with India. A report claims that China is replacing soldiers with these machines - as the Chinese troops are unable to cope with the high altitudes.

10.2.2 Robots in Warfare In development

(i) MIDARS

MIDARS is a four-wheeled robot outfitted with several cameras, radar, and possibly a firearm, that automatically performs random or preprogrammed patrols around a military base or other government installation. It alerts a human overseer when it detects movement in unauthorized areas, or other programmed conditions. The operator can then instruct the robot to ignore the event, or take over remote control to deal with an intruder, or to get better camera views of an emergency. The robot would also regularly scan radio frequency identification tags (RFID) placed on stored inventory as it passed and report any missing items.

(ii) Tactical Autonomous Combatant (TAC) units, described in Project Alpha study:

The United States Joint Forces Command (USJFCOM) was a Unified Combatant Command of the United States Department of Defense. USJFCOM was a functional command that provided specific services to the military. The last commander was Army Gen. Ray Odierno and the Command Senior Enlisted was Marine Sergeant Major Bryan B. Battaglia. As directed by the President to identify opportunities to cut costs and rebalance priorities, Defense Secretary Robert Gates recommended that USJFCOM be disestablished and its essential functions reassigned to other unified combatant commands. Formal disestablishment occurred on 4 August 2011.

Among the command's many directorates and departments was **Project Alpha**, a JFCOM rapid idea analysis group created to identify high-impact ideas from industry, academia and the defense community that could transform the United States Department of Defense into an organization better equipped to deal with the uncertain landscape of the future. Project Alpha was discontinued as part of an internal reorganization of U.S. Joint Forces Command's Joint Experimentation Directorate.

(iii) Autonomous Rotorcraft Sniper System

Autonomous Rotorcraft Sniper System is an experimental robotic weapons system being developed by the U.S. Army since 2005. It consists of a remotely operated sniper rifle attached to an unmanned autonomous helicopter. It is intended for use in urban combat or for several other missions requiring snipers.

(iv) The "Mobile Autonomous Robot Software" research program was started in December 2003 by the Pentagon who purchased 15 Segways in an attempt to develop more advanced military robots. The program was part of a $26 million Pentagon program to develop software for autonomous systems.

(v) ACER

The Armored Combat Engineer Robot (ACER) is a military robot created by Mesa Robotics. Roughly the size of a small bulldozer and weighing 2.25 tons, ACER is among the larger military robots. ACER is able to reach speeds of 6.3 mph, using treads for movement. Uses for this robot include clearing obstacles, removing explosives, hauling cargo and disabled vehicles, and serving as a platform for various other tasks, such as clearing buildings and disarming landmines and laser mines.

(vi) Atlas (robot)

Atlas is a bipedal humanoid robot primarily developed by the American robotics company Boston Dynamics with funding and oversight from the U.S. Defense Advanced Research Projects Agency (DARPA). The robot was initially designed for a variety of search and rescue tasks, and was unveiled to the public on July 11, 2013. In April of 2024, the hydraulic Atlas (HD Atlas) was retired from service. A new fully electric version was announced the following day.

(vii) Battlefield Extraction-Assist Robot

The Battlefield Extraction-Assist Robot (BEAR) is a remotely controlled robot developed by Vecna Robotics for use in the extraction of wounded soldiers from the battlefield with no risk to human life. The humanoid robot uses a powerful hydraulics system to carry humans and other heavy objects over long distances and rough terrain, such as stairs.

Work on the robot commenced in 2005 and it was featured in Time Magazine's Best Inventions of 2006. Vecna Robotics wrapped up development and testing for applications on and off of the battlefield in 2011.

(viii) Dassault nEUROn (French UCAV)

The Dassault nEUROn is an experimental unmanned combat aerial vehicle (UCAV) being developed with international cooperation, led by the French company Dassault Aviation. Countries involved in this project include France, Greece, Italy, Spain, Sweden and Switzerland. The design goal is to create a stealthy, autonomous UAV that can function in medium-to-high threat combat zones.

(ix) Dragon Runner

Dragon Runner is a military robot built for urban combat. At 20 pounds (9 kg) it is light enough to be carried and thrown. The original project was funded by the United States Marine Corps Warfighting Laboratory in conjunction with Carnegie Mellon University. It was designed at Carnegie Mellon University while the electronics and thermoplastic shell is developed and made by QinetiQ, Inc. Early development was conducted by the United States Naval Research Laboratory, including initial design, production and field-testing.

(x) MATILDA

Mesa Associates' Tactical Integrated Light-Force Deployment Assembly (MATILDA) is a remote-controlled surveillance and reconnaissance robot, created and designed by the 'Mesa Robotics Corporation'. It is available in many different models such as the Urban Warrior, Block II, and Scout, which feature different combinations of components for increased utility. These options include a sensor mount, manipulator arm, weapon mount, fiber optic reel, remote trailer release, and disrupter mount. When purchased the basic system includes the platform, the control unit, and battery charger.

(xi) MULE (US UGV)

The Multi-Mission Unmanned Ground Vehicle, previously known as the Multifunction Utility/Logistics and Equipment vehicle (MULE), was an autonomous unmanned ground combat vehicle developed by Lockheed Martin Missiles and Fire Control for the United States Army's Future Combat Systems and BCT Modernization programs.

(xii) R-Gator

The iRobot R-Gator is an unmanned robotic platform from iRobot Corporation and John Deere.

The 1,450 pounds (660 kg) robot is built upon Deere's M-Gator currently in use by the US Military. The R-Gator can operate autonomously, performing perimeter patrol and other missions while keeping personnel out of harm's way. It can operate autonomously by following a map or choosing its own waypoints to reach a pre-determined destination. It can also do "follow the leader" operations to keep up with troops, be tele-operated, or driven manually if needed. In military exercises, the R-Gator has shown an ability to carry gear for soldiers to lighten their loads. It also demonstrated its capacity to carry and drop off explosive ordnance disposal robots weighing more than 100 pounds (45 kg). An R-Gator could deploy the smaller machine and provide unmanned perimeter security while the EOD robot dismantles the bomb. The first R-Gator sale to the military was to the U.S. Navy Space and Naval Warfare Systems Command for autonomous perimeter security.

(xiii) Ripsaw MS1

The Ripsaw is a series of developmental unmanned ground combat vehicles designed by Howe & Howe Technologies (now part of Textron Systems) for evaluation by the United States Army.

(xiv) SUGV

The XM1216 Small Unmanned Ground Vehicle (SUGV) is a Future Combat Systems specific, man packable (< 30 pounds (14 kg)) version of the iRobot's PackBot.

(xv) Syrano

SYRANO (Système Robotisé d'Acquisition pour la Neutralisation d'Objectifs, "Robotic acquisition system for neutralization of targets") is the first operational battlefield robot of the French military. SYRANO was designed by a consortium of CGEY, THALES, GIAT Industries and SAGEM to collect information in combat zones, especially, in urban combat conditions. It is based on the Wiesel AWC of the German Army.

(xvi) iRobot Warrior

The iRobot Warrior (also described as the Warrior 700 or X700) is an unmanned robotic platform from iRobot Corporation.

The 285 lb (129 kg) robot can traverse land at up to 9.3 mph (15 km/h) and is capable of carrying up to 500 lb (227 kg), including 150 lb (68 kg) in its manipulator. Able to climb steps and slopes at up to 45°, the next generation of remote-control robotic vehicles is bigger, faster, and more capable than their smaller counterparts. Initial intended uses are Explosive Ordnance Disposal, route-clearance, surveillance, enhanced security measures, scouting, reconnaissance, casualty extraction, firefighting, manipulating and welding.

(xvii) PETMAN

PETMAN (Protection Ensemble Test Mannequin) is a bipedal device constructed for testing chemical protection suits. It is the first anthropomorphic robot that moves dynamically like a person.

(xviii) Excalibur unmanned aerial vehicle

The Aurora Excalibur was an unmanned aerial vehicle (UAV) developed by Aurora Flight Sciences between 2005 and 2010 capable of vertical takeoff and landing (VTOL). The design combined ducted fans and hybrid drive. A smaller scale model with a 13-foot (4.0 m) wingspan was successfully tested on June 24, 2009. A full-scale version was to be capable of carrying four AGM-114 Hellfire missiles and traveling at 460 mph (740 km/h). Aurora used the experience of developing this vehicle to inform their submission to the Vertical Take-Off and Landing Experimental Aircraft (VTOL X-Plane) program funded by the Defense Advanced Research Projects Agency (DARPA).

(xix) Teng Yun medium size reconnaissance UAV program, Taiwan

The Teng Yun is a UAV under development by the National Chung-Shan Institute of Science and Technology (NCSIST) of Taiwan. It was said to be able to carry armaments to conduct combat missions.

10.3 RISE OF AUTONOMOUS ROBOTS IN WARFARE

Autonomous robots are among the latest innovations bringing immense military benefits, including improved intelligence, reduced casualties, and increased speed and efficiency. The use of autonomous robots in warfare has been a topic of debate for quite some time now. While some believe that robots represent the future of warfare, others argue that they pose a significant threat to humanity. Regardless of where you stand on the matter, it is essential to be informed about the various examples of autonomous robots for warfare that exist today and the potential implications they may have. From drones, unmanned ground vehicles, and humanoid military robots, the military is deploying these machines for different roles, including surveillance, reconnaissance, logistics, and combat. With increasing advancements in robotics and AI, the future of warfare is full of promises and uncertainties. As the deployment of advanced robotics solutions globally is a foregone conclusion, the need for reliable connectivity will play a critical role in ensuring the operability of these warfighting machines.

The first autonomous robot in warfare was a sea vessel developed by Leonardo da Vinci in the late 15th century. Since then, the military has used various robots to save lives, reduce risks, and enhance performance. One of the most exciting examples today is Boston Dynamics' robot dog, the **'Spot'**. This four-legged robot can now speak thanks to AI, navigate complex terrain, climb stairs, and perform various field tasks, including carrying equipment. The US Armed Forces have deployed the

Spot in combat zones, where it collects data and provides situational awareness, among other roles. The robot can also communicate with other systems and work collaboratively with human soldiers. Keeping robots like 'Spot' reliably connected around the world will be critical to future missions.

Perhaps the most advanced and sophisticated example of an autonomous robot for warfare is the **humanoid robot**. These robots are designed to be humanoid in appearance and are capable of performing a wide range of tasks, from carrying heavy equipment to engaging in firefighting missions. Humanoid robots can be programmed to operate independently, without human control, and can be outfitted with advanced AI systems that allow them to make complex decisions on the battlefield.

One area where autonomous robots have shown tremendous promise is in the field of unmanned aerial vehicles, or drones. These drones can be outfitted with cutting-edge sensors and cameras, allowing them to gather intelligence and reconnaissance information from the air. This information can then be used to inform military decision-making, without putting human pilots at risk of harm. Drones can also be deployed for reconnaissance missions in areas that are too dangerous or inaccessible for human soldiers, making them a valuable tool for gathering information in hostile environments. Beyond visual line of sight (BVLOS) drones are extending the reach of drones thanks to advanced global connectivity technology.

When it comes to military drones, we often think of fixed wing drones like the Predator drone, but Small Unmanned Aircraft Systems (sUAS) like FLIR's Black Hornet, the Military's stealthy pocket-sized recon drone is proving to have an impact in theater.

DARPA announced the Nano Air Vehicle (NAV) program and its goal to develop agile systems that could fit in one hand. One example that started development almost 2 decades ago is the Nano Hummingbird or Nano Air Vehicle (NAV) is a tiny, remote-controlled aircraft built to resemble and fly like a hummingbird, developed in the United States by AeroVironment, Inc. to specifications provided by the Defense Advanced Research Projects Agency (DARPA).

Another example of an autonomous robot for warfare is the **unmanned ground vehicle, or UGV**. These machines are designed to operate on the ground, and can be used for a variety of tasks, from carrying supplies and equipment to patrolling dangerous areas. UGVs can be equipped with advanced sensors and cameras, enabling them to navigate complex terrain and gather valuable intelligence information. In addition to their intelligence-gathering capabilities, UGVs can also be equipped with weapons systems, making them a potential threat to enemy forces.

Perhaps the most controversial example of an autonomous robot for warfare is the autonomous weapon system, which is capable of making decisions to engage targets without human intervention. While no fully autonomous weapon system has been developed yet, several countries, including the United States, Russia, and China, are working on them. There is considerable concern surrounding the development of these systems, as their potential use could lead to catastrophic consequences.

Despite their benefits, autonomous robots in warfare raise serious ethical questions. Some experts worry that these machines could lead to uncontrolled, unstoppable killing machines. You guessed it, like the Terminator. There is also a concern that robots could malfunction or fall into the wrong hands, leading to disastrous consequences. Moreover, robots could reduce the value of human life in warfare and normalize violence.

Still, there's no doubt that autonomous robots in warfare are a game-changer, offering significant benefits to the military and solving complex challenges. These machines can reduce the risks for human soldiers, reduce costs, and offer better security for nations. Governments and tech companies need to work together to establish clear guidelines for developing, deploying, and using autonomous robots in warfare. As the military continues to embrace technology, we must keep all ethical considerations at the forefront of the conversation.

10.4 TYPES OF MILITARY ROBOTS AND THEIR USES

One of the first military robots used in combat dates back to the 1930s. Believe it or not, the Soviet Union initially used radio signals to operate their "teletanks" in World War II.

Since then, there have been substantial technological advancements in robots in the military such as firefighting, mine clearance, and even backpackable robots. Nearly 100 years later, militaries around the world use robots on a daily basis to assist them on the battlefield.

Today, there are several types of military robots used globally with various functions and roles. While there are a wide variety of features, uses, and applications for military robots, the most popular military robots are unmanned vehicles. These can be broken down into three primary classes:

- Unmanned Ground Vehicle (UGV),
- Unmanned Aerial Vehicle (UAV), and
- Unmanned Underwater Vehicle (UUV)

Following are typically the types of Military Robots:

(i) Transportation Robots

One of the most common types of military robots is transportation robots. This robotic technology can help soldiers transport different supplies like artillery, bombs, and other supplies.

They can also be used to transport humans, including picking up casualties from the battlefield.

These military robots are typically unmanned ground vehicle (UGV) robots that come equipped with wheels or legs. However, they could also be used in the air as an unmanned aerial vehicle (UAV), or unmanned underwater vehicle (UUV).

(ii) Search and Rescue Robots

One of the most important roles of military robots is to assist in search and rescue missions. These military robots provide critical support in finding missing or captured personnel.

Search and rescue robots are advantageous since they can often go where humans can't—whether underwater, through floods, wildfires, or over mountains.

These types of robots are especially useful in disaster relief such as supporting local search and rescue after a tsunami, earthquake, or other natural disasters. They've also been crucial in providing support to human-caused disasters like 9/11 and Chernobyl.

(iii) Mine Clearance Robots

Minefields can be incredibly lethal in war. However, what's worse is that they're incredibly challenging to clear after a war zone is cleared and life returns to normal. Oftentimes, uncleared mines kill locals years after a war has ended.

Mine clearance robots are unmanned ground vehicles (UGVs) that are created specifically to detect and clear land mines. They can help identify the exact location of land mines and deactivate them. They're often controlled by a human operator who is clearing the land mines remotely from miles away.

(iv) Firefighting Robots

Fires are commonplace in combat. While the military trains up human firefighters, it can be incredibly dangerous and challenging to only use human skills to put out fires. A firefighting robot is a futuristic type of military robot that can assist in extinguishing fires in the military.

These military robots can be linked up to a fire hydrant to help investigate a fire site as well as put out fires to save victims' and firefighters' lives.

(v) Surveillance and Reconnaissance Robots

Surveillance and reconnaissance robots are essential technologies that assist the military to survey and spy on the enemy or potential threats. These robots are powerful in surveillance and investigative military operations.

(vi) Armed Robots

This type of military robot is pretty self-explanatory. Armed with weapons to eliminate threats in combat and help in defense systems, these robots are quite commonplace in helping the military improve their operations.

(vii) Training Robots

A new, futuristic military robot is a training robot. For instance, Polytronic International AG develops robotic target (RT) systems for cutting-edge, live-fire training.

One of their top infantry training solutions offers an autonomous robotic target giving combat training leaders an interactive training solution that simulates live combat.

10.5 MILITARY ROBOTICS DEVELOPMENTS IN INDIA

Realizing the potential of robotic technology for military purposes, India is actively pursuing the development of robots through its Defence Research and Development Organisation (DRDO) and select public sector enterprises. Centre for Artificial Intelligence and Robotics (CAIR) was established in Oct 1986. Its research focus was initially in the areas of Artificial Intelligence (AI), Robotics, and Control systems. In 2000, R & D groups working in the areas of Command Control Communication and Intelligence (C^3I) systems, Communication and Networking, and Communication Secrecy in Electronics and Radar Development Establishment (LRDE) were merged with CAIR.

Since the last few decades, DRDO, Hindustan Aeronautics Limited (HAL) and Aeronautical Development Agency (ADA) are involved in the development of a range of UAVs/UCA. 'Lakshya', the indigenously-developed pilotless target aircraft was inducted into the IAF in 2005. DRDO is also involved in developing the know-how for a swept wing, stealth design and composite construction technical demonstrator that will demonstrate the technical feasibility, military utility and operational value for a networked system of high-performance weaponized UCAVs. DRDO's Research and Development Establishment (Engineers) in Pune, has developed a Remotely Operated Vehicle (ROV) called '*Daksh*', manufactured by a consortium of firms and in use with the army. 'Daksh' is an electrically powered and remotely controlled robot used for locating, handling and destroying hazardous objects safely. It is a battery-operated robot on wheels and its primary role is to recover bombs. It locates bombs with an X-Ray machine, picks them up with a gripper-arm and defuses them with a jet of water. It has a shotgun, which can break open locked doors, and it can scan cars for explosives. 'Daksh' can also climb staircases, negotiate steep slopes, navigate narrow corridors and tow vehicles.

DRDO is experimenting with robot mules to carry arms and equipment in difficult terrain and high altitude. Autonomous underwater vehicles capable of carrying out multiple tasks are also being developed. A UGV for nuclear biological and chemical (NBC) surveillance operations is under development at VRDE, Ahmednagar. 'Netra' UAV is being developed for surveillance and recognizance operations for counter-terrorist operations in urban as well as jungle terrain. 'Netra' is set to enter into the production phase following successful user and field trials, including those in high-altitude. R&DE (Engineers) is also developing a gun-mounted robot (light machine gun and grenade launcher), which can be deployed in anti-terrorist situations. Collaboration of DRDO with PSUs, academic institutions (like IITs) and private sector firms is being encouraged for this mission.

10.5.1 Indian Army Unveils 'Robotic Buddy' for Battlefield Missions

The Indian Army unveiled, in 2023, a robotic system capable of serving a wide variety of battlefield needs.

Called the "Robotic Buddy," the cutting-edge system was developed by army engineers at the Military College of Electronics and Mechanical Engineering and unveiled during the Artificial Intelligence for Military Applications seminar in southern India. It can reportedly detect humans, track specific areas, measure distances, and transmit intelligence, surveillance, and target acquisition data to aid an attack.

The tech features a robotic arm and a platform equipped with two cameras each to give an advantageous view of the surroundings.

Operated remotely from a ground control station, the Robotic Buddy's arm can lift up to 15 kilograms, allowing it to retrieve unexploded shells and save soldiers the dangerous task of collecting them manually. The arm can also be detached to accommodate other functions, such as becoming a battlefield stretcher to evacuate wounded troops.

This battery-powered platform is built to withstand rugged terrains and measures one meter by one meter. The development of the Robotic Buddy is a testament to the country's increasing investment in technological innovation to address evolving threats.

Apart from the robotic system, New Delhi is set to unveil other state-of-the-art defense items for ground, air, and naval use.

This week, the Indian Navy will introduce its Autonomous Weaponized Boat Swarms and the Autonomous Vessel Underwater for the first time.

According to developer Sagar Defence Engineering, the systems are equipped with various types of weaponry and sensors for remote or autonomous operation. They can execute littoral patrols, high-speed interdiction, coastal surveillance, and constabulary operations.

10.6 ADVANTAGES AND CONCERNS OF ROBOTICS IN WARFARE

10.6.1 Advantages of Robotics in Warfare

Efficiency: In contrast to humans, robots do not require rest, sustenance, or recreational breaks during work and have high concentration.

Robots need to take breaks only for **recharging or refueling**, and the repetitive nature of complex, dangerous activities does not affect their efficiency or efficacy.

They can be employed on the battlefield for tasks requiring incredibly high concentration, and difficult for humans to sustain for long periods.

Robotic systems can operate in environments contaminated by biological, chemical or radiological weapons, where a human would have to wear a bulky suit and protective gear.

They can sustain extremes of operating situations, like high 'G' turns (gravity force when you change direction and velocity due to inertia) that can render aircraft pilots inoperative. Unmanned systems can fly faster and turn harder.

In underwater operations, small robotic boats can operate in the rough ocean where waves are eighteen feet high or more, and human sailors would suffer serious physical injury from all the tossing about.

Aggressive attacking: Humans can only react to incoming mortar rounds by taking cover at the last second, whereas the Counter Rocket Artillery and Mortar (CRAM), a robotic system deployed in Iraq by the US, could detect and shoot them down before they arrived at the target.

10.6.2 Concerns of Robotics in Warfare

Ethical concerns: It raises ethical concerns as they may lack the ability to make moral judgments, potentially violating human rights and international humanitarian laws.

No accountability: The lack of clear accountability for actions taken by autonomous weapons poses a significant challenge.

Unpredictability: The unpredictable nature of autonomous systems poses a risk as to how they will respond in battlefield scenarios.

Potential arms race: It can lead to an arms race, with nations competing to enhance their robotic capabilities.

Lack of empathy: The removal of humans from decision-making eliminates the capacity for empathy and nuanced judgement.

Fear of unemployment: It could lead to a reduction in the demand for human military personnel, potentially leading to job losses.

SECTION-IV

INNOVATIVE TECHNOLOGIES FOR MODERN AND FUTURE WARFARE

CHAPTER 11

5G AND 6G TECHNOLOGIES IN WARFARE

11.1 INTRODUCTION TO 5G TECHNOLOGY

5G stands for fifth-generation wireless, and it is the latest iteration of cellular technology. 5G aims to improve upon the current 4G LTE (Long-Term Evolution) standard with faster data download and upload speeds, more reliable connections, and the ability to connect more devices at once. This is achieved through the use of new, higher frequency bands known as millimeter waves.

11.1.1 Advantages of 5G Technology

(i) Better Capabilities than 4G and 4G LTE Technologies

There are considerable differences between the 5G standard and the previous 4G and 4G LTE standards in terms of network capabilities and performance. The fifth-generation standard has a wider spectrum allocation and uses higher frequencies of electromagnetic radiation within the upper limits of radio waves and the range of microwaves.

It is important to underscore that data transmission speed is partly dependent on frequency levels and wavelengths. Higher frequencies and shorter wavelengths translate to better data transmission speeds. This comes from the fact that a higher frequency is associated with a faster movement of signal-bearing electromagnetic waves.

An advantage of 5G is that it runs on two different network specifications. These are the upper limits of the sub-6 GHz specification and the mm Wave specification. A particular 5G network can use frequencies between 3.3 GHz and 4.2 GHz or within the frequencies above 24 GHz or the UHF and EHF areas of the electromagnetic spectrum.

This standard can provide a 1 Gbps bandwidth compared to the 200 Mbps bandwidth of 4G. It can also demonstrate a network latency of fewer than 10 milliseconds compared to the 20 to 30 milliseconds latency of its predecessor. Data transmission is between 50 Mbps and 200 to 400 Mbps compared to the 25 Mbps average speed of 4G LTE.

(ii) Faster and Reliable Connectivity Means Better Streaming

Bandwidth, in digital communication, is the maximum amount of data transmitted over a network in a given amount of time while network throughput pertains to how much data was transferred from a source at a particular given time. Network latency represents the time it takes for data to be transmitted between its source and its destination.

The terms above determine overall network quality. Wired digital communication has better values than wireless digital communication. However, because of the capabilities of the 5G standard, it could rival fiber-based wired internet services in these three areas and support novel and more advanced wireless communication applications. It can support a range of streaming applications. These include the consumption of content such as high-definition video streaming from platforms and services such as Netflix, online-enabled or cloud gaming services such as Xbox Cloud Gaming, and video conferencing using platforms such as Zoom Meetings, Microsoft Teams, and Google Meet.

(iii) Support for Advance and Future Internet Applications

Another benefit of 5G centers on advancing the scope of wireless digital communication. Take note that several network service providers have rolled out 5G-enabled wireless internet service or wireless broadband service for residential customers. It is positioned as a counterpart to wired broadband internet service based on fiber optics technology.

(iv) It also supports the deployment and adoption of autonomous vehicles. Some of the important features of these vehicles depend on an internet connection. Examples include obtaining and processing real-time traffic information and downloading maps for navigation. 5G enables better vehicle-to-vehicle and vehicle-to-traffic communications

(v) Support for the implementation of smart cities could also be made possible under a fifth-generation cellular network. It can enable the integration of smart devices in public infrastructure to allow autonomous and wireless operations, data gathering, and assimilation of public infrastructures with other wireless communication devices

(vi) The technology can also enhance other existing technologies or expand further their adaptation. These include better native support for mixed reality or virtual reality and augmented reality, various on-device applications of the metaverse, remote medical diagnostics and remote surgery, telecommuting or remote work, and cloud computing services.

11.1.2 Disadvantages and Key Limitations of 5G Technology

(i) Expensive Initial Rollout and High Public Adoption Costs

Costs represent one of the disadvantages of 5G. Network deployment would require cellular network carriers or operators to both upgrade existing network infrastructures and build new ones to meet the 3GPP standard. These also require purchasing new equipment, securing new licenses, and leasing public spaces and private properties.

Take note that cellular network operators must also integrate and use other relevant wireless technologies such as Multi-User MIMO (Multiple-Input Multiple-Output), Massive MIMO, and beamforming, The barrier to entry for local network operators remains in developing and underdeveloped countries high because of the high initial cost, knowledge, and other resource requirements.

Adding to this is the fact that the market in these countries might not be able to afford equipped mobile devices. Even most flagship smartphones from 2019 and mid-range smartphones from 2020 and earlier will not run on fifth-generation networks. Consumers need to purchase capable devices to experience the advantages of 5G technology.

(ii) Sub-6 and mm-Wave Difference and Limitations of mm-Wave

Another problem with 5G is that it is based on two different specifications. Sub-6 uses different technologies and works on distinctive principles than mm-Wave. A Sub-6 modem would not be able to connect to a mm-Wave network. Some devices have modems for both specifications but most are compatible with either one of the two and cannot utilize both

The C-Band 5G also falls under the Sub-6 5G standard. Not all 5G networks and 5G devices are the same. Consumers would need to know if the devices they are planning to purchase are compatible with the 5G network in their area. It is better to purchase one that supports both considering the differences between these two specifications.

It is also important to note that mm-Wave 5G is superior to sub-6 5G in terms of bandwidth, network latency, and data transmission speed. It is the closest rival to wired broadband internet and fiber-based wired communication. But there is one drawback. The mm-Wave specification has a limited range and coverage because it uses higher frequencies.

Network operators would need to build and place hundreds to thousands of smaller cells to cover an entire area. This has cost implications. Some have expressed concern over the unappealing placements of these cells. These issues and limitations of mm-Wave make it ideal for dense urban areas or particular target spots like stadiums and airports.

(iii) Limited Real-World Performance and High-Power Draw

Another disadvantage of 5G is that the theoretical transmission speeds do not translate to actual performance. There is also a mismatch between upload and download speeds. Upstream transmission is slower than downstream transmission. Factors such as congestion level, cell site proximity, and device capabilities affect the network performance.

It is also important to reiterate the fact that the Sub-6 specification is slower than the mm-Wave specification. There are instances in which an LTE Advanced network can provide the same level of overall network performance as a Sub-6 5G network. This is true in dense areas or certain spots within a particular area with high network traffic.

Another disadvantage of 5G is that it consumes a lot of power. The battery runtime of a mobile device would be shorter while connected to this network than a connection to 4G LTE and 3G networks. This is a problem for mobile devices running on batteries with small charge capacities. Heating issues have also been observed in some devices.

11.1.3 How 5G Technology Works

5G technology operates using a combination of existing 4G LTE networks and new, high-frequency bands. These higher frequencies are able to transmit data more quickly, but they have a shorter range and can be more easily blocked by obstacles. To overcome this, 5G networks use small cell sites that can be placed every few hundred meters to provide continuous coverage.

11.1.4 Future of Connectivity with 5G Technology

The introduction of 5G technology is set to have a profound impact on the future of connectivity. With 5G, we can expect to see a surge in the **Internet of Things (IoT),** with more devices than ever before being able to connect to the internet. This could lead to advancements in everything from smart homes to autonomous vehicles. However, with these advancements come new challenges. As more devices connect to the internet, cybersecurity will become increasingly important.

11.2 5G TECHNOLOGY IN THE WARFARE

That fact that 5G is a largely commercially developed solution that is now on track for adaptation to military use has enabled defense electronics manufacturers to operate in collaborative environments. This reality will prove to be valuable, as one of the driving factors behind widespread 5G adoption in the military is not the advent of the waveform itself, but the set of network-management standards that will follow.

Multidomain operations, as seen in military programs such as Joint All Domain Command and Control (JADC2), could greatly benefit from the interoperability enabled through 5G use. One way is through 5G technology's ability to bring disparate solutions together under a unified network management system.

With 5G adoption occurring all around the world, electronic warfare (EW) and spectrum-management applications could also see an evolution following the implementation of 5G architectures. The use of 5G and its incorporation into intelligence, surveillance, and reconnaissance (ISR) systems could indelibly alter how the military gathers actionable intelligence, ranging from how the DoD detects/is detected to further opening vast avenues of data collection.

The transition from Long Term Evolution 4G (4G LTE) to standalone 5G deployment will be gradual, just as it was from the early days of 1G (launched in Japan in 1979, adopted in 1983 in the U.S.) all the way to December 1, 2018, when South Korea became the first country to offer 5G. The process will take even longer to reach the military, with networking standards, current DoD networking architectures, and measured acquisition timelines among the current hurdles bogging down immediate fielding.

11.2.1 Difference between 4G and 5G

Following the introduction and implementation of 4G LTE in consumer and military markets, the unprecedented streaming ability that it offered was regarded as revolutionary. While patchiness in the 4G LTE network did leave users in certain regions unsatisfied, it's a fact that 5G was already in the planning stages.

5G, as it actually came to be, was thought of as more than just a generational change in how we communicate through a cellular environment with cell phones. It was actually pretty far-reaching in its genesis. It is more than a comms system, as it enables a system of different systems. 5G also envisions how that might affect IoT [Internet of Things], or even satellite communications. There's terrestrial communication, there's people, there's data, there's IoT, and then there's the satellite piece of it.

Often referred to by the industry professionals as the IoT-era, 5G dawned amid the understanding that data was being created far more quickly and in volumes greater than what current networks could handle. The vast quantities of unstructured sensor data and the military's need to make sense of it also produces processing challenges that 5G could likely handle.

11.2.2 5G and how the military interoperates

There's sort of a historical precedent for this. If you look at how successful interoperability has been achieved in the past, it's usually by doing it at the network level or sometimes the application level. But the internet is an example of doing it at the network level where you have a TCP/IP (internet protocol suite) network and a bunch of different networks that aren't even always compatible. But you do have the networking standards that allow them to interoperate through either the management or the gateway functions, and 5G can do this.

Joint All-Domain Command and Control is key to enabling multi-domain operations. JADC2 and global operations, including space-based 5G terrestrial networks, will require the elimination of silos to ensure seamless communications between nodes. For the military, 5G technologies allow for the operation of several potential applications to include C2 [command and control], logistics, maintenance, training, AI [artificial intelligence], augmented and virtual reality, and ISR systems – all of which can benefit from improved data speeds and lower latency.

Fielding a network as extensible, indiscriminate, and interoperable as 5G, however, will present security concerns, especially in multidomain defense arenas. Officials claim that the security question will influence how 5G networks and 5G network management systems are architected.

As you bring more users together, there are more paths to give to any individual user. Being able to understand the data and metadata on your network and create a cyber-defense around that is one thing we've put a lot of time and energy into in the past decade. This will be really important on 5G networks, especially in the JADC2 construct as you bring these networks together because you're going to have to move beyond cyber hygiene and checklist-based and boundary-based cyber-defense and instead move to behavioral analysis and watching how users interact.

11.2.3 Military Advantages of 5G Technology

(i) 5G-powered Augmented Reality for Training

One of 5G's capabilities is powering Augmented Reality (AR). Armed forces on the field can utilize AR technology to simulate various training scenarios, thus improving situational awareness and training effectiveness.

(ii) Real-time Enemy Surveillance

The high-speed data transmission and low latency of 5G technology allow armed forces to deploy surveillance systems that provide real-time intelligence on enemy movements and activities. This, in turn, enhances intelligence and tactical decision-making.

(iii) Intersection of NTN and 5G in defence

5G can also enable Satellite Communication. It can extend coverage to remote areas and provide critical connectivity for tactical military users and first responders. Non-terrestrial networks (NTNs) leverage space-borne assets to deliver ubiquitous cellular coverage, empowering military operations in previously inaccessible regions.

11.2.4 Electronic warfare and 5G

There is a possibility of 5G network-management systems designed to support various protected waveforms and to be ideal for EW environments. Pairing a native 5G network with supported waveforms that are built for low probability of detection could enable warfighters to roam between available networks, depending on the mission.

The part of the 5G spec and waveform that's operating in the millimeter band – those very high frequencies that tend to not propagate as far – create some capabilities that are a little more interesting in terms of the EW environment. This is because it's harder for the adversary to see them, simply because they don't propagate as far. So, these little bubbles of operation are more difficult for adversaries to reach into. And a hybrid networking case where you have a 5G network with different physical layer waveforms, that also gives you room to play in an EW environment.

Moreover, 5G could also fare well in EW arenas because it's designed with the inherent ability to listen only to what it needs to, avoiding disruption from both accidental and purposeful signals that may get in the way. Even so, in the instance that disruption on the electromagnetic spectrum did occur, corresponding 5G-powered technologies could aid in identifying it.

If an adversary is using 5G technology in the area, then EW systems will need to be able to understand if there's 5G communication because this technology is not just owned by any specific country. It's worldwide, so adversaries could very well use 5G to communicate as well. EW systems will need to start being engineered to identify 5G signals and understand what adversaries could potentially be trying to communicate. That's one side of it, and the other would be the fact that we use 5G technology and we don't want unwelcome users to be able to disrupt or see what we're doing. That's where EW and 5G technology environments will have to coexist.

Data collection and interpretation are so important to EW and are major pieces of the advantage 5G could provide on the battlefield. With the seemingly insurmountable quantities of data brought in by both military and commercial sensors, 5G-powered computing could better enable processing capabilities.

5G allows for multiaccess edge computing (MEC), where data is processed locally near a device to speed the completion of computing tasks. MEC allows users to aggregate mission-critical data on site and offers the flexibility of delivering cloud services closer to the edge with localized compute functionality and optimized 5G cloud services. 5G networks are designed with robust, integrated cybersecurity protections to help secure user data.

While edge computing is undoubtedly a start when it comes to military data analysis, there is still far too much of it for humans alone to process. Whether it's anomalous network behavior, signals on the spectrum, or information from the cloud, automated 5G network-management systems powered by AI could be pivotal in rendering out actionable details.

11.2.5 Pairing 5G with AI

5G technologies could also be incorporated into ISR (Integrated Services Router) systems. These require increasingly high bandwidth to process, exploit, and disseminate intelligence data from a network of terrestrial and airborne sensors. This could enhance C2 by providing commanders with timely access to key intelligence and actionable information that can improve decision making in split seconds and allow commanders a better understanding of an adversary's decision cycle.

The faster and more efficient transmission of data that 5G could enable would thereby allow AI-powered analysis engines to run more effectively. Because AI and machine learning (ML) algorithms must be trained, enhanced low-latency access to data could result in smarter AI/ML systems.

What 5G does bring is this somewhat open architecture where you can plug all kinds of different things in that could allow you to plug in different networks at the right level and offer the ability to command and control using the 5G network. Historically, with all of the different sensors and systems out there, it's been very difficult to get actionable, multi-sensor data from all of the different deployed technologies to give you a cohesive, one-point location for that information. 5G makes that more likely because it can plug into different systems, assuming the necessary security is there.

11.3 INTRODUCTION TO 6G TECHNOLOGY

In telecommunications, 6G is the designation for a future technical standard of a sixth-generation technology for wireless communications.

It is the planned successor to 5G [ITU-R (International Telecommunication Union - Radiocommunication) IMT-2020 (International Mobile Telecommunications-2020)], and is currently in the early stages of the standardization process, tracked by the ITU-R as IMT-2030 with the framework and overall objectives defined in recommendation ITU-R M.2160-0. Similar to previous generations of the cellular architecture, standardization bodies such as 3GPP (Third Generation Partnership Project) and ETSI (European Telecommunications Standards Institute), as well as industry groups such as the NGMN (Next Generation Mobile Networks (NGMN) Alliance, are expected to play a key role in its development.

6G networks will be able to use higher frequencies than 5G networks and provide substantially higher capacity and much lower latency. One of the goals of the 6G internet is to support one microsecond latency communications. This is 1,000 times faster — or 1/1000th the latency — than one millisecond throughput.

Numerous companies (Airtel, Anritsu, Apple, Ericsson, Fly, Huawei, Jio, Keysight, LG, Nokia, NTT Docomo, Samsung, Vi, Xiaomi), research institutes (Technology Innovation Institute, the Interuniversity Microelectronics Centre) and countries (United States, United Kingdom, European Union member states, Russia, China, **India**, Japan, South Korea, Singapore, Saudi Arabia, United Arab Emirates, and Israel) have shown interest in 6G networks, and are expected to contribute to this effort.

6G networks will likely be significantly faster than previous generations, thanks to further improvements in radio interface modulation and coding techniques, as well as physical-layer technologies. Proposals include a ubiquitous connectivity model which could include non-cellular access such as satellite and WiFi, precise location services, and a framework for distributed edge computing supporting more sensor networks, AR/VR and AI workloads. Other goals include network simplification and increased interoperability, lower latency, and energy efficiency. It should enable network operators to adopt flexible decentralized business models for 6G, with local spectrum licensing, spectrum sharing, infrastructure sharing, and intelligent automated management. Some have proposed that machine-learning/AI systems can be leveraged to support these functions.

The NGMN alliance have cautioned that "6G must not inherently trigger a hardware refresh of 5G RAN (Radio Access Network) infrastructure", and that it must "address demonstrable customer needs." This reflects industry sentiment about the cost of the 5G rollout, and concern that certain applications and revenue streams have not lived up to expectations. 6G is expected to begin rolling out in the early 2030s, but given such concerns it is not yet clear which features and improvements will be implemented first.

11.4 WHAT ARE THE ADVANTAGES OF 6G VS. 5G?

6G networks will operate by using signals at the higher end of the radio spectrum. It is too early to approximate 6G data rates, but it is estimated that a theoretical peak data rate of 1 terabyte per second for wireless data may be possible. That estimate applies to data transmitted in short bursts across limited distances. LG, a South Korean company, unveiled this type of technology based on adaptive beamforming in 2021.

This level of capacity and latency will extend the performance of 5G applications. It will also expand the scope of capabilities to support new and innovative applications in wireless connectivity, cognition, sensing and imaging. With 6G, access points will be able to serve multiple clients simultaneously via orthogonal frequency-division multiple access.

6G's higher frequencies will enable much faster sampling rates than with 5G. They will also provide significantly better throughput and higher data rates. The use of sub-mm waves — wavelengths less than 1 millimeter — and frequency selectivity to determine relative electromagnetic absorption rates is expected to advance the development of wireless sensing technology.

Mobile edge computing will be built into all 6G networks, whereas it must be added to existing 5G networks. Edge and core computing will be more integrated as part of a combined communications and computation infrastructure framework by the time 6G networks are deployed. This approach will provide many potential advantages as 6G technology becomes operational. These benefits include improved access to AI capabilities and support for sophisticated mobile devices and systems.

11.4.1 When will 6G internet be available?

6G internet is expected to launch commercially in 2030. The technology makes greater use of the distributed radio access network (RAN) and the terahertz (THz) spectrum to increase capacity, lower latency and improve spectrum sharing.

While some early discussions have taken place to define the technology, 6G research and development (R&D) activities started in earnest in 2020. 6G will require development of advanced mobile communications technologies, such as cognitive and highly secure data networks. It will also require the expansion of spectral bandwidth that is orders of magnitude faster than 5G.

China has launched a 6G test satellite equipped with a terahertz system. Technology giants Huawei Technologies and China Global reportedly plan similar 6G satellite launches in future. Many of the problems associated with deploying millimeter wave radio for 5G must be resolved in time for network designers to address the challenges of 6G.

11.4.2 How will 6G work?

It's expected that 6G wireless sensing solutions will selectively use different frequencies to measure absorption and adjust frequencies accordingly. This method is possible because atoms and molecules emit and absorb electromagnetic radiation at characteristic frequencies, and the emission and absorption frequencies are the same for any given substance.

6G will have big implications for many government and industry approaches to public safety and critical asset protection, such as the following:

- threat detection;
- health monitoring;
- feature and facial recognition;
- decision-making in areas like law enforcement and social credit systems;
- air quality measurements;
- gas and toxicity sensing; and
- sensory interfaces that feel like real life.

Improvements in these areas will also benefit smartphone and other mobile network technology, as well as emerging technologies such as smart cities, autonomous vehicles, virtual reality and augmented reality.

11.4.3 Do we even need 6G?

There are a number of reasons why we need 6G technology.

They include the following:

(i) **Technology convergence**. The sixth generation of cellular networks will integrate previously disparate technologies, such as deep learning and big data analytics. The introduction of 5G has paved the way for much of this convergence.

(ii) **Edge computing.** The need to deploy edge computing to ensure overall throughput and low latency for ultrareliable, low-latency communications solutions is an important driver of 6G.

(iii) **Internet of things (IoT).** Another driving force is the need to support machine-to-machine communication in IoT.

(iv) **High-performance computing (HPC).** A strong relationship has been identified between 6G and HPC. While edge computing resources will handle some of the IoT and mobile technology data, much of it will require more centralized HPC resources to do the processing.

11.4.4 Who is working on 6G technology?

The race to 6G is drawing the attention of many industry players. Test and measurement vendor Keysight Technologies has committed to its development. Major infrastructure companies, such as Huawei, Nokia and Samsung, have signaled that they have 6G R&D in the works.

The race to reach 5G may end up looking minor when compared with the competition to see which companies and countries dominate the 6G market and its related applications and services.

The **major projects underway** include the following:

(i) The University of Oulu in Finland has launched the 6Genesis research project to develop a 6G vision for 2030. The university has also signed a collaboration agreement with Japan's Beyond 5G Promotion Consortium to coordinate the work of the Finnish 6G Flagship research on 6G technologies.

(ii) South Korea's Electronics and Telecommunications Research Institute is conducting research on the terahertz frequency band for 6G. It envisions data speeds 100 times faster than 4G Long-Term Evolution (LTE) networks and five times faster than 5G networks.

(iii) China's Ministry of Industry and Information Technology is investing in and monitoring 6G R&D in the country.

(iv) The U.S. Federal Communications Commission (FCC) in 2020 opened up 6G frequency for spectrum testing for frequencies over 95 gigahertz (GHz) to 3 THz.

(v) Hexa-X is a European consortium of academic and industry leaders working to advance 6G standards research. Finnish communications company Nokia is leading that project, which also includes Ericsson, a Swedish operator, and TIM in Italy.

(vi) Osaka University in Japan and Australia's Adelaide University researchers have developed a silicon-based microchip with a special multiplex to divide data and enable more efficient management of terahertz waves. During testing, researchers claimed the device transmitted data at 11 gigabits per second compared to 5G's theoretical limit of 10 Gbps of 5G.

11.4.5 Future scope of 6G networks

About 10 years ago, the phrase "Beyond 4G" (B4G) was coined to refer to the need to advance the evolution of 4G beyond the LTE standard. It was not clear what 5G might entail, and only pre-standards R&D-level prototypes were in the works at the time. The term B4G lasted for a while. It referred to what could be possible beyond 4G. Ironically, the LTE standard is still evolving, and 5G will use some aspects of it.

Similar to B4G, Beyond 5G is seen as a path to 6G technologies that will replace fifth-generation capabilities and applications. 5G's many private wireless communications implementations involving LTE, 5G and edge computing for enterprise and industrial customers have helped lay the groundwork for 6G.

Next-generation 6G wireless networks will take this one step further. They will create a web of communications providers — many of them self-providers — much in the way that photovoltaic solar power has brought about cogeneration within the Smart grid. 6G could advance mesh networks from concept to deployment, helping to extend coverage beyond the range of older cell towers.

Data centers are already faced with big 5G-driven changes. These include virtualization, programmable networks, edge computing and issues surrounding simultaneous support of public and private networks. For example, some business customers may want to combine on-premises RAN with hybrid on-premises and hosted computing — for edge and core computing, respectively — and data center-hosted core network elements for private business networks or alternative service providers.

6G radio networks will provide the communication and data gathering necessary to accumulate information. A systems approach is required for the 6G technology market that makes use of data analytics, AI and next-generation computation capabilities using HPC and quantum computing.

In addition to profound changes within RAN technology, 6G will bring changes to the core communications network fabric as many new technologies converge. Notably, AI will take center stage with 6G.

11.4.6 Other changes 6G is likely to bring

(i) **Nano-core.** A so-called nano-core is expected to emerge as a common computing core that encompasses elements of HPC and AI. The nano-core does not need to be a physical network element. Instead, it could encompass a logical collection of computational resources, shared by many networks and systems.

(ii) **Edge and core coordination.** 6G networks will create substantially more data than 5G networks, and computing will evolve to include coordination between edge and core platforms. In response to those changes, data centers will have to evolve.

(iii) **Data management.** 6G capabilities in sensing, imaging and location determination will generate vast amounts of data that must be managed on behalf of the network owners, service providers and data owners.

6G networks are attempting to extend fast Gigabit Ethernet connectivity to commercial and consumer devices. 6G is expected to provide substantially higher throughput and data flow. As envisioned, 6G will enable the following:

- deliver a theoretical data rate of about 11 Gbps simultaneously across multiple gigahertz channels;

- deploy up to three 160-megahertz (MHz) bandwidth channels; and multiplex up to eight spatial streams.

11.5 5G, 6G, AND BEYOND: INDIAN ARMY AND MEITY FORGE CUTTING-EDGE ALLIANCE FOR DEFENCE TECH REVOLUTION

The Indian Army and the Ministry of Electronics and Information Technology (MeitY) have joined hands to lead a technological transformation in defence. This collaboration stands at the forefront of progress, introducing advancements such as indigenous 5G solutions, AI-powered tools, and heightened cybersecurity measures. The Indian Army's 5G laboratories serve as a testing ground for 6G, focusing on developing military-grade applications for both generations to meet the evolving needs of future warfare. Through pioneering projects, comprehensive training initiatives, and digital advancements, this alliance aims to fortify defence capabilities, positioning the Indian Army as a technologically advanced and strategically formidable force on the global stage.

The genesis of this strategic collaboration can be attributed to pivotal discussions between the Minister of State for Electronics and IT, and Chief of the Army Staff, as per army sources. Over the past six months, these discussions have evolved into reciprocal visits, laying the groundwork for a defence strategy heavily dependent on technological advancements. Unveiling the ongoing details of the transformative collaboration between the Indian Army and MeitY, the army sources provided clear insights into the contours of this partnership.

(i) Integration of 5G and 6G Technology: Indian Army Takes Strides in Indigenous Solutions

India has taken a significant leap in addressing the requirements of 5G technology on a national scale. The Indian Army has established 5G labs and the Military College of Telecommunication Engineering, positioning them as a groundbreaking test bed for 6G. The development of military-grade applications for 5G and 6G is underway, aligning with the demands of future warfare.

(ii) Innovative Software Development

A crucial element of this transformation involves the collaborative development of proprietary software by the Indian Army and MeitY. This joint endeavor encompasses the design of sophisticated AI-driven decision-making tools, predictive analytics, and robust cybersecurity measures. Notable accomplishments include creating software tailored for interpreting the Enemy's Electronic Order of Battle (ORBAT) and innovative platforms like the Situational Awareness Module for the Army (SAMA).

(iii) Cutting-Edge Cybersecurity Initiatives

Recognizing the paramount importance of cybersecurity, the Indian Army is undergoing integration with Security Operations Centre 2.0, representing a significant stride in handling cyber threats. The creation of state-of-the-art cybersecurity and cyber forensics tools reinforces this initiative.

(iv) Dominance in Artificial Intelligence

The Military College of Telecommunication Engineering has transformed into a centre for advanced AI research and development. Notable highlights include innovations like the Situational Awareness

Module for the Army (SAMA) and sophisticated pattern recognition software for satellite imagery analysis.

(v) Digital Infrastructure and Training Initiatives

Collaborative training programs in digital technologies are currently in progress, utilizing platforms such as the National Informatics Centre (NIC). These initiatives aim to enhance the skills of Army personnel in the latest digital and cyber practices. Specific training activities include 'Digital Video/Image and CCTV Forensics' at the National Science University (Gandhinagar) and joint training for IT staff at the National Informatics Centre (NIC).

(vi) Cutting-edge Platforms and Digital Transformation

Initiatives such as SRIJAN for transactions, *Raksha Bhoomi* for digitizing land records, and MISO for streamlined inventory management exemplify the Army's dedication to digital innovation. These software solutions enhance operational efficiency and result in cost savings for the exchequer.

Furthermore, the Indian Army has filed for intellectual property rights (IPR) for over 22 projects selected during Ideas and Innovation Competitions. An MOU has also been established with IIT Delhi to advance further the militarized versions of in-house innovations conducted by Indian Army personnel.

The collaboration between the Indian Army and MeitY and the in-house Ideas and Innovations underscores how technology can be harnessed to strengthen national defence. It boosts the Army's operational capabilities and aligns with India's broader Digital India and Make in India initiatives. As these projects advance, the Indian Army is poised to emerge as a technologically advanced and strategically formidable force on the global stage.

11.6 PENTAGON READIES FOR 6G

Since transitioning most of its 5G research and development projects to the Chief Information Office last year, the Pentagon's Future Generation Wireless Technology Office has shifted its focus to preparing the Defense Department for the next wave of network innovation.

That work is increasingly important for the U.S., which is racing against China to shape the next iteration of wireless telecommunications, known as 6G. These more advanced networks, expected to materialize in the 2030s, will pave the way for more dependable high-speed, low-latency communication and could support the Pentagon's technology interests — from robotics and autonomy to virtual reality and advanced sensing.

Staying ahead means not only fostering technology development and industry standards but making sure that policy and regulations are in place to safely use the capability, according to Thomas Rondeau, who leads the Pentagon's FutureG office. Staking a leadership role in the global competition, he said, could give DOD a level of control over what that future infrastructure looks like.

"If we can define those going into it, then as we export our technologies, we're also exporting our policies and our regulations, because they're going to be inherently part of those technology solutions," Rondeau told Defense News in a recent interview.

The Defense Department started making a concerted investment in 5G about five years ago when then Undersecretary of Research and Engineering Michael Griffin named the technology a top priority for the Pentagon.

In 2020, DOD awarded contracts totaling $600 million to 15 companies to experiment with various 5G applications at five bases around the country. The projects included augmented and virtual reality training, smart warehousing, command and control and spectrum utilization.

The department has since expanded the pilots and pursued other wireless network development projects, including a 5G Challenge series that incentivized companies to move toward more open-access networks.

The result has, so far, been a mixed bag. Most of the pilots didn't transition into formal programs within the military services. Several of the failed efforts involved commercial augmented or virtual reality technology that wasn't mature enough for DOD to justify continued funding.

Among the projects that did transfer, was that a pilot effort at Naval Air Station Whidbey Island in Washington to provide fixed wireless access to the base. The project essentially replaced hundreds of pounds of cables with radio units that broadcast the communications network to the personnel who need it. Today, the system is supporting logistics and maintenance operations at the base.

This and other transitioned pilots will likely make their way into a formal budget cycle by fiscal 2027.

DoD also saw some success from the 5G Challenges it staged in 2022 and 2023 to encourage telecommunication companies to transition to an open radio access network, or Open RAN. A RAN is the first entry point a wireless device makes into a network and accounts for about 80% of its cost. Historically, proprietary RANs managed by companies like Huawei, Ericsson, Nokia and Samsung have dominated the market.

They're driving a world where they control the entire system, the end-to-end system," Rondeau said. "That causes a lack of insight, a lack of innovation on our side, and it causes challenges with how to apply these types of systems to unique, niche military needs.

The 5G Challenge offered companies a chance to break open that proprietary model by moving to Open RAN — and according to Rondeau, it was a success. The initial challenge then expanded into a broader forum that addressed issues like energy efficiency and spectrum management. Ultimately, the effort reduced energy usage by around 30%, he said.

While much of the focus of these initiatives was on 5G, the work has informed the Pentagon's vision and strategy for 6G, which the department believes should have an open-source foundation.

That is a direct result of not only my background and push for some of these things, but also the learnings that we got from the networks we've deployed, from the 5G Challenge," he said. "All these things come into play that led us towards an open-source software model being the right model for the military and, we think, for industry.

One of the FutureG office's top priorities these days, a direct outgrowth of the 5G Challenge, is called OCUDU, which stands for open centralized unit, distributed unit. The project is focused on implementing a fully open software model for 6G that meets the needs of industry, the research community and DoD.

The office is also exploring how the military could use 6G for sensing and monitoring. Its Integrated Sensing and Communications project, dubbed ISAC, uses wireless signals to collect information about different environments. That capability could be used to monitor drone networks or gather military intelligence.

While ISAC technology could bring a major boost to DOD's ISR systems, commercialization could make it accessible to adversary nations who might weaponize it against the U.S. That challenge reflects a broader DOD concern around 6G policies and regulation – and drives urgency within Rondeau's office to ensure the U.S. is the first to shape the foundation of these next-generation networks.

They are looking at this as a real opportunity for dramatic growth and interest in new, novel technologies for both commercial industry and defense needs, but also, the threat space that it opens up for us is potentially pretty dramatic, so we need to be on top of this.

CHAPTER 12

CLOUD COMPUTING IN WARFARE

12.1 INTRODUCTION TO CLOUD COMPUTING

Cloud computing is the on-demand availability of computer system resources, especially data storage (cloud storage) and computing power, without direct active management by the user. Large clouds often have functions distributed over multiple locations, each of which is a data center. Cloud computing relies on sharing of resources to achieve coherence and typically uses a pay-as-you-go model, which can help in reducing capital expenses but may also lead to unexpected operating expenses for users.

Cloud computing consists of the following **five essential characteristics**:

(i) **On-demand self-service.** A consumer can unilaterally provision computing capabilities, such as server time and network storage, as needed automatically without requiring human interaction with each service provider.

(ii) **Broad network access.** Capabilities are available over the network and accessed through standard mechanisms that promote use by heterogeneous thin or thick client platforms (e.g., mobile phones, tablets, laptops, and workstations).

(iii) **Resource pooling.** The provider›s computing resources are pooled to serve multiple consumers using a multi-tenant model, with different physical and virtual resources dynamically assigned and reassigned according to consumer demand.

(iv) **Rapid elasticity.** Capabilities can be elastically provisioned and released, in some cases automatically, to scale rapidly outward and inward commensurate with demand. To the consumer, the capabilities available for provisioning often appear unlimited and can be appropriated in any quantity at any time.

(v) **Measured service.** Cloud systems automatically control and optimize resource use by leveraging a metering capability at some level of abstraction appropriate to the type of service (e.g., storage, processing, bandwidth, and active user accounts). Resource usage can be monitored, controlled, and reported, providing transparency for both the provider and consumer of the utilized service, although for some organizations the revenue impact of high usage may affect profitability, compared to an option of sunk capital costs.

12.2 CLOUD COMPUTING: TYPES OF AS-A-SERVICE OFFERINGS

Cloud computing services are classified in three layers in which different services are provided. A cloud computing service provider can provide infrastructure including software, application development platform or hardware such as processor, memory and disk space to users.

The cloud is here—and so are its acronyms. Since software as a service (SaaS) hit the world in 2001, the "as a service" model has been extended to just about everything you can think of. Along the way, the definition has become a little muddied.

It used to be that as-a-service offerings meant something delivered on a subscription basis via the cloud, without a physical component. Hence software is delivered to your computer continually via the internet rather than a program you purchased for a one-time fee and installed from a disc.

These days, though, as-a-service offerings can mean just about anything delivered by subscription or even just a plain old outsourced service, whether the cloud is involved or not. Some -aaS models are an industry category and some are proprietary to a single company.

Following are some of the **service models available from the Cloud**:

(i) AIaaS (Artificial Intelligence as a Service)

AI as a Service (AIaaS) refers to the provision of artificial intelligence (AI) capabilities and services through cloud-based platforms. It allows businesses to leverage AI tools and technologies without having to develop or maintain their own AI infrastructure. This includes services like machine learning, natural language processing, and data analytics, which can be accessed and utilized on-demand.

Market: AI as a Service (AIaaS) is changing the game for businesses by offering cost-effective, scalable AI capabilities through cloud-based models. This means companies can tap into advanced AI without needing a team of in-house experts. In 2023, the global AIaaS market was worth over $9.3 billion. And the growth doesn't stop there—it's expected to soar to a whopping $98.21 billion by 2030.

(ii) BaaS (Backend as a Service)

BaaS, or Backend as a Service, is a cloud service model that provides developers with the tools and infrastructure needed to build and manage the backend of applications without having to handle server management or complex backend programming themselves. It helps developers with faster app development, allowing them to focus on building the app's features instead of spending time wondering how to scale.

Market: The global backend-as-a-service (BaaS) industry made $3.1 billion in 2022 and is expected to reach $28.7 billion by 2032, growing at an impressive annual rate of 25.3% from 2023 to 2032.

(iii) CaaS (Cloud as a Service)

Also known as "cloud services," CaaS is an all-in-one cloud computing service that combines Infrastructure as a Service, Platform as a Service (PaaS), and Software as a Service(SaaS) technologies in one subscription.

Market: This CaaS market is valued at $264.8 billion in 2019 and projected to reach $927.5 billion by 2027.

(iv) CaaS (Containers as a Service)

Containers as a Service (CaaS) is a cloud computing service that lets developers deploy and manage containerized applications, a software package that bundles everything needed to run an app (code, runtime, configuration, and system libraries) so it can operate on any host system. It provides businesses of all sizes with portable and easily scalable cloud solutions.

Market: Valued at $1.3 billion in 2018, this is another CaaS market with huge growth: it reached $2.1 billion in 2023 and is projected to reach $10.77 billion by 2030.

(v) CaaS (Country as a Service)

What if you're Estonia—technically advanced with high living standards, but no one wants to live there? Offer e-residency with access to a variety of digital services in the country. Where could this go in the future? Watch for developments in banking, tax, and business.

Market: As of 2021, this tech has attracted tens of thousands of foreign entrepreneurs and businesses from across the world In four years, it produced the more unicorns—tech startups now worth over $1 billion—per capita than anywhere in the world.

(vi) DaaS (Data as a Service)

Data as a Service (DaaS) is a cloud-based service model that provides users with on-demand access to data, allowing for easy data storage, processing, and analysis. The idea is that when data is siloed off, it's not working as hard as it could. By centralizing data in the cloud, it can be accessed easily and analyzed far more deeply.

Market: The DaaS market is expected to grow from $20.74 billion in 2024 to $51.60 billion by 2029.

(vii) DaaS (Device as a Service)

Device as a Service (DaaS) is a subscription-based model where businesses lease hardware, such as laptops and other devices, bundled with lifecycle services like setup, management, maintenance, and overall support. This model provides flexibility, reduces upfront costs, and ensures devices remain updated and secure. Is this a recurring revenue opportunity for MSPs? Or a chance for hardware vendors to disinter-mediate them? HP, for one, offers device aaS as a channel program, and Lenovo is also getting in on the action. Seems most vendors will. See Surface as a service below.

Market: The global DaaS market was valued at $51.7 billion in 2021, and is projected to reach $1.8 trillion by 2031.

(viii) DRaas (Disaster Recovery as a Service)

Disaster Recovery as a Service (DRaaS) is a cloud-based solution that ensures data protection and business continuity by replicating and hosting physical or virtual servers to enable rapid recovery in case of a disaster—ideally before your users even notice a hiccup. It minimizes downtime and data loss, providing seamless access to critical applications and information.

Market: The DRaaS is a quickly expanding market that's expected to grow from $10.7 billion in 2023 to $26.5 billion by 2028.

(ix) EaaS (Environment as a Service)

EaaS (Environment as a Service) is a cloud-based solution that provides on-demand, scalable, and customizable test environments for software development and testing. Going further than virtual machines or containers, this service enabling teams to simulate real-world scenarios efficiently and cost-effectively.

Market: The EaaS market reached $13.0 billion in 2023 and is projected to reach $55.2 billion by 2032.

(x) FaaS (Function as a Service)

Function as a Service (FaaS) is a cloud computing service model that allows developers to execute code in response to events without managing the underlying infrastructure, often used for building microservices and serverless applications.

Market: The FaaS market is estimated at $17.70 billion in 2024, and is expected to reach $44.71 billion by 2029.

(xi) IaaS (Infrastructure as a Service)

One of the fundamental -aaS offerings, infrastructure as a Service (IaaS) is a cloud computing model that provides virtualized computing resources over the internet, allowing businesses to rent servers, storage, and networking infrastructure on a pay-as-you-go basis. This service makes world-class IT infrastructure available to any size of business, with setup and maintenance outsourced to third parties. A huge number of platforms, applications, and software have been built on cloud infrastructure.

Market: A booming sector that keeps growing. The IaaS market is expected to skyrocket from $51.3 billion in 2020 to $481.8 billion by 2030. Behemoths AWS, Google, IBM, Microsoft, and Oracle dominate.

(xii) IoTaaS (IoT as a Service)

IoT as a Service (IoTaaS) provides cloud-based solutions and infrastructure to manage, monitor, and analyze IoT devices and data, enabling businesses to implement IoT without significant upfront investment. It offers scalability, flexibility, and reduced complexity for IoT deployments. IoTaaS is great for scenarios where there are sudden spikes in customer need and you need to quickly set up and tear down resources and don't want to incur costs for devices you only need for a short period of time.

Market: The IoTaas market reached $330.3 billion in 2021 and is expected to grow to a staggering $650.5 billion by 2026.

(xiii) KaaS (Knowledge as a Service)

Knowledge as a Service (KaaS) is a cloud-based delivery model that provides users with access to knowledge resources, expertise, and information on-demand, often through AI and machine learning technologies. What we're talking about here is data with context, as opposed to just raw data or information.

Market: Driven by the rise in big data, the KaaS market was estimated at $0.9 billion in 2023, reaching $2.4 billion by 2028.

(xiv) LaaS (Location as a Service)

Location as a Service (LaaS) is a cloud-based service that provides location data and geolocation services to applications and devices via APIs. This service enables businesses to track, analyze, and utilize geographic information without managing the underlying infrastructure. Retail (and other) companies sit on an enormous quantity of customer location data without the tools to pull business insight from it. Location aaS lets them rent high-quality location data analysis.

Market: LaaS is an emerging sector, expected to reach $286.7 billion by 2030.

(xv) MaaS (Monitoring as a Service)

Monitoring as a Service (MaaS) is a cloud-based service model that provides tools and capabilities to monitor IT infrastructure, applications, and networks, enabling proactive management and troubleshooting. This helps organization avoid having to purchase and install a potentially costly on-premises monitoring tool.

Market: The MaaS market is expected to grow from $3.53 billion in 2024 to $7.06 billion by 2032.

(xvi) NaaS (Network as a Service)

Network as a Service (NaaS) is a cloud-based model that allows users to manage and operate their network infrastructure virtually, offering scalability, flexibility, and cost efficiency without the need for physical hardware. Rented network functionality from a third party (usually an ISP) allows companies to scale up or down on port capacity as needed. This works best for companies with highly variable demand.

Market: The NaaS market is projected to grow from $13.2 billion in 2022 to $46.6 billion by 2027.

(xvii) OaaS (Operations as a Service)

Operations as a Service (OaaS) is a managed service model where third-party providers handle an organization's operational functions, such as IT, logistics, and supply chain management, allowing the business to focus on core activities. A new name for a managed service MSPs have been offering for years.

Market: The OaaS market was valued at $62.51 billion and is projected to reach a market size of $251.46 billion by 2030.

(xviii) PaaS (Platform as a Service)

IaaS moves IT hardware to the cloud but opens new challenges for developers in configuring and operating their app deployment platforms. Platform service is the answer. Platform as a Service (PaaS) is a cloud computing model that provides a platform and environment for developers to build, deploy, and manage applications without the complexity of managing the underlying infrastructure. This service provides not just infrastructure, but operating systems, software, databases, and other useful tools.

Market: Much smaller than IaaS but growing rapidly. Estimated at $75.7 billion in 2023 and expected to reach $290.2 billion in 2030.

(xix) RaaS (Ransomware as a Service)

Ransomware as a Service (RaaS) is a cybercrime model where ransomware creators sell or lease their malware to other criminals, allowing them to deploy ransomware attacks without needing to develop the malware themselves. Aka DIY ransomware kits that would-be criminals can purchase and implement. This is one service you DON'T want to deliver. Unfortunately there's not a whole lot you can do about the sale of code kits online. What you can do to protect your business? Protect against ransomware when it attacks.

Market: Oversupply is driving the cost of kits down—that's bad. However, the growth of this hacker market is driving the growth of the ransomware protection market, which was valued $18.4 billion in 2023 and projected to reach $57.9 billion by 2030.

(xx) SaaS (Software as a Service)

The granddaddy of the entire -aaS market. Software as a Service (SaaS) is a cloud computing model where applications are delivered over the internet as a service, allowing users to access software via a web browser without needing to install or maintain it locally. Providers like SaaS (because regular monthly payments) are better than one-time deals, and clients like its flexibility.

Market: The global SaaS market was valued at $261.15 billion in 2022 and is poised to grow from $296.93 billion in 2023 to $829.34 billion by 2031.

(xxi) SECaaS (Security as a Service)

IT Security as a Service (SECaaS) is a cloud-based service that provides comprehensive security solutions, including antivirus, intrusion detection, and identity management, delivered over the internet to protect IT systems and data. This could be provided by a third-party like an MSSP (managed security service provider).

Market: The global SECaas market is expected to grow from $12.4 billion in 2021 to $23.8 billion by 2026.

(xxii) UaaS (Unified Communications as a Service)

Unified Communications as a Service (UCaaS) is a cloud-based delivery model that provides a variety of communication and collaboration applications and services, such as voice, video, messaging, and conferencing, through a unified platform accessible over the internet. VoIP, instant messaging, LinkedIn, Skype, phones, Wi-Fi, social media; new communication channels appear at ever-shorter intervals and businesses struggle to keep their networks organized, secure, and efficient. Unified communications aaS vendors take care of all the hardware and software while guaranteeing a level of quality.

Market: UCaaS has a strong, established market with steady growth. Expected to expand from $52.9 billion in 2022 to reach $311.6 billion by 2032.

(xxiii) VaaS (Video as a Service)

Video as a Service (VaaS), aka cloud-hosted video conferencing, is a cloud-based delivery model that provides video conferencing, streaming, and video content management services, allowing users to access and share video content over the internet without needing to manage the underlying infrastructure. As more companies move from phone to video conferencing, the IT headaches of keeping them running multiply. Companies moving to cloud-based video and enjoy higher-quality images and fewer dropped calls, with technicians on standby to keep things running smoothly.

Market: The VaaS market has seen huge growth during and after the pandemic due to increased remote or hybrid working models. The market is estimated at $6.62 billion in 2024 and is expected to reach $13.22 billion by 2029.

(xxiv) VaaS (Virtualization as a Service)

Virtualization as a Service (VaaS) provides virtualized computing resources such as servers, storage, and networks, allowing users to run multiple virtual machines and applications without managing physical hardware. One distant server, many accessible virtual machines. Ten years ago, turning one physical server into several virtual machines was a groundbreaking way to fully use server capacity and free up physical space. Today, it can be done via the cloud.

Market: Virtualization as a whole is considered a mature market—not many new opportunities here. Still, the market's projected to grow from $86.38 billion in 2023 and to $164.13 billion by 2030.

(xxv) WaaS (Workspace as a Service)

Workspace as a Service (WaaS) is a cloud-based service that delivers virtual workspaces or virtual desktop environments, allowing users to access their desktop environments, applications, and data from any device, anywhere, providing flexibility and enhancing remote work capabilities. Just log in to access your office desktop, exactly as you like it, with all the business data and applications you need, from any device you choose. Easy to see why this sector has taken off, especially among companies with remote workers and small businesses without resources to efficiently manage their own IT.

Market: The WaaS market is large and growing steadily: $7.43 billion in 2022 with expectations to hit $25.24 billion by 2030.

(xxvi) XaaS (Anything (or everything) as a Service)

Anything (or Everything) as a Service (XaaS) is a broad term that refers to the delivery of various services over the internet, encompassing a wide range of IT functions and resources such as software, platforms, infrastructure, and more, allowing for flexible, on-demand access.

Market: The XaaS market was valued at $409.5 billion in 2023 to $2.47 trillion by 2032.

12.3 CHALLENGES AND LIMITATIONS OF CLOUD COMPUTING

Applications hosted in the cloud are susceptible to the fallacies of distributed computing, a series of misconceptions that can lead to significant issues in software development and deployment.

One of the main challenges of cloud computing, in comparison to more traditional on-premises computing, is the data security and privacy. Cloud users entrust their sensitive data to third-party providers, who may not have adequate measures to protect it from unauthorized access, breaches, or leaks. Cloud users also face compliance risks if they have to adhere to certain regulations or standards regarding data protection, such as GDPR (General Data Protection Regulation) or HIPAA. and HIPAA (Health Insurance Portability and Accountability Act.

Another challenge of cloud computing is reduced visibility and control. Cloud users may not have full insight into how their cloud resources are managed, configured, or optimized by their providers. They may also have limited ability to customize or modify their cloud services according to their specific needs or preferences. Complete understanding of all technology may be impossible, especially given the scale, complexity, and deliberate opacity of contemporary systems; however, there is a need for understanding complex technologies and their interconnections to have power and agency within them. The metaphor of the cloud can be seen as problematic as cloud computing retains the aura of something noumenal and numinous; it is something experienced without precisely understanding what it is or how it works.

Additionally, cloud migration is a significant challenge. This process involves transferring data, applications, or workloads from one cloud environment to another, or from on-premises infrastructure to the cloud. Cloud migration can be complicated, time-consuming, and expensive, particularly when there are compatibility issues between different cloud platforms or architectures. If not carefully planned and executed, cloud migration can lead to downtime, reduced performance, or even data loss.

12.3.1 Cloud computing security

Cloud computing poses privacy concerns because the service provider can access the data that is in the cloud at any time. It could accidentally or deliberately alter or delete information. Many cloud providers can share information with third parties if necessary for purposes of law and order without a warrant. That is permitted in their privacy policies, which users must agree to before they start using cloud services. Solutions to privacy include policy and legislation as well as end-users' choices for how data is stored. Users can encrypt data that is processed or stored within the cloud to prevent unauthorized access. Identity management systems can also provide practical solutions to privacy concerns in cloud computing. These systems distinguish between authorized and unauthorized users and determine the amount of data that is accessible to each entity. The systems work by creating and describing identities, recording activities, and getting rid of unused identities.

According to the Cloud Security Alliance, the top three threats in the cloud are (i) Insecure Interfaces and APIs, (ii) Data Loss & Leakage, and (iii) Hardware Failure—which accounted for 29%, 25% and 10% of all cloud security outages, respectively. Together, these form shared technology vulnerabilities. In a cloud provider platform being shared by different users, there may be a possibility that information belonging to different customers resides on the same data server. The hackers spend substantial time and effort looking for ways to penetrate the cloud. There are some real Achilles' heels in the cloud infrastructure that are making big holes for the bad guys to get into. Because data from hundreds or thousands of companies can be stored on large cloud servers, hackers can theoretically gain control of huge stores of information through a single attack—a process he called "hyper-jacking". Some examples of this include the Dropbox security breach, and iCloud 2014 leak. Dropbox had been breached in October 2014, having over seven million of its users passwords stolen by hackers in an effort to get monetary value from it by Bitcoins (BTC). By having these passwords, they are able to read private data as well as have this data be indexed by search engines (making the information public).

There is the problem of legal ownership of the data (If a user stores some data in the cloud, can the cloud provider profit from it?). Many Terms of Service agreements are silent on the question of ownership. Physical control of the computer equipment (private cloud) is more secure than having the equipment off-site and under someone else's control (public cloud). This delivers great incentive to public cloud computing service providers to prioritize building and maintaining strong management of secure services. Some small businesses that do not have expertise in IT security could find that it is more secure for them to use a public cloud. There is the risk that end users do not understand the issues involved when signing on to a cloud service (persons sometimes do not read the many pages of the terms of service agreement, and just click "Accept" without reading). This is important now that cloud computing is common and required for some services to work, for example, for an intelligent personal assistant (Apple's Siri or Google Assistant). Fundamentally, private cloud is seen as more secure with higher levels of control for the owner, however public cloud is seen to be more flexible and requires less time and money investment from the user.

The attacks that can be made on cloud computing systems include man-in-the middle attacks, phishing attacks, authentication attacks, and malware attacks. One of the largest threats is considered to be malware attacks, such as Trojan horses. Recent research conducted in 2022 has revealed that the Trojan horse injection method is a serious problem with harmful impacts on cloud computing systems.

12.4 CLOUD COMPUTING DEPLOYMENT MODELS

In cloud computing there are four types of application models. These models are based on different scenarios and requirements.

12.4.1 Public Cloud

Public Cloud model is known as the basic cloud computing model. Individual users and/or organizations can reach the cloud infrastructure and resources through the Internet. The cloud service in this model is managed by the organization, which provides the service. Well-organized web services are available for the customers through the Internet. Public cloud services are free or in the form of pay as you go. For example, the user clients use the public cloud application while checking e-mails via e-mail clients like gmail or hotmail.

12.4.2 Community Cloud

In Community Cloud Model, common IT requirements of different organizations are met by the cloud infrastructure managed by a service provider or the organizations themselves. Unlike the public cloud, community cloud operating costs are shared between user organizations. So the costs are higher than that in the public cloud. On the other hand, a higher level of security and durability can be achieved with a little effort. There are significant efforts to establish the community cloud between the government agencies in order to get the advantages of the application. For example, in U.S. such a study has been conducted for government agencies.

12.4.3 Private Cloud

In Private Cloud Model, cloud infrastructure is established only for one organization and the overall system works under a special network managed by the organization. As a result of the limited access and high-authorization procedures, this model is a solution to reduce the risks associated with security and privacy effectively. This model is suitable for **armed forces**, **defense industry** companies and security agencies which need highly secure information. The cost of the model is much higher than that of the other models.

12.4.4 Hybrid Cloud

Hybrid Cloud Model consists of different combination of the cloud models mentioned above related to different requirements. While some of the IT requirements that have security and access concerns can be addressed by the private clouds, the community cloud can be used for the IT requirements that require low level security and the public cloud for the routine.

12.5 CLOUD COMPUTING IN DEFENCE (EUROPEAN DEFENCE AGENCY)

Cloud computing refers to the delivery of computing services— including servers, storage, databases, networking, software, analytics, and intelligence—over the internet ("the cloud") to offer faster innovation, flexible resources, and economies of scale. Cloud computing providers like can deliver storage and intelligence, offer a platform for building applications, and analyze data. It differs from the traditional approach of owning and maintaining physical data systems installed on premises, which requires significant electricity funds for power and cooling. By offloading some or all of this computing to cloud-based infrastructure through an external company or companies, civil companies and even organizations such as the Ministry of Defence can pay-by-use, and save significantly on costs.

In 2019, the European Defence Agency financed a study about cloud computing for the defence sector (European Defence Agency, 2024). The EDA's study, Cloud Intelligence for Decision-Making Support and Analysis (CLAUDIA), ended in January 2024. The study was run in collaboration with a private capital technology business group, and The Information Processing and Telecommunications Center (IP&T Center) (European Defence Agency). There are three uses of cloud in the defence sector, including (i) source analysis, (ii) edge computing. and (iii) multi-domain operations.

The European Defence Agency's CLAUDIA demonstrated that cloud computing can offer robust source analysis, significantly enhancing situational awareness and preventative capabilities in defence. Hybrid warfare, characterized by a combination of disinformation tactics, conventional warfare, non-conventional warfare and cyberwarfare, is becoming increasingly prevalent. Consequently, gathering intelligence from digital sources and discerning between fake and genuine can be vital to combating threats from non-state actors such as insurgents, terrorist organizations and hackers. The EDA developed SWAN through CLAUDIA, a Software Analysis platform tested on different cloud-based capabilities. The software was used to detect disinformation from open-source information, such as social media, media newspapers and websites. Applying of this function across European forces could result in significant security enhancements because using real-time insights leads to proactive measures against hybrid threats. This pilot project led to several new ideas to be explored in the future, including the capability of the Cloud to prevent cyber-warfare, demonstrating how artificial intelligence could exploit the cloud by analyzing open-source data and generate intelligence about early signals regarding potential hybrid warfare activity. EDA project officer for CLAUDIA, highlighted the potential expansion of this capability to supplement Common Operational Pictures, also known as COP. COP is a decision-making tool based on up-to-date information from multiple sources merged into a single display. COP functions as a go-to piece of information for decision-making in military operations. SWAN's ability to supplement COPs by generating a map that geo-locates recorded cyberattacks, recognizes their nature and calculates the possibility of a cyberattack against military operations and critical infrastructure. Attacks on critical infrastructure include energy systems, transport networks, communication systems or government services and protecting them Is essential for the security of the EU and the well-being of its citizens. As the Cloud facilitates easy and quick access to information from various locations, units have better situational awareness when making decisions regarding cyber threats. This information could also be shared with other government agencies. In an increasingly digitized world, the Cloud presents an advanced, fast and adaptable system to analyze information critical to tactical and preventative measures in defence.

(i) Edge Computing & Information Processing

CLAUDIA has demonstrated that Cloud computing can bring information processing closer to the "edge". After delivering CLAUDIA, the EDA will focus on planning projects that address edge computing planning to help reduce this data lag. The US Air Force has already tested this concept on its F-35s, using the multifunction advanced data link (MADL) and data sensors onboard each aircraft to share information amongst a squadron. This shared information creates a single comprehensive analysis of threats based on the unique vantage points of each aircraft. The concept can also be applied to tanks, drones and the "digital soldier", operating as instruments that collect and distribute information on-site. Edge computing improves information superiority, safety and operational efficiency by conducting real-time analysis on the spot.

(ii) Interoperability: Multi Domain Operations and Data Sharing

Cloud is essential for supporting multi-domain operations, as demonstrated by the NATO Allied Command Transformation. Within the NATO structure, are five distinct operational areas: Maritime, Land, Air, Space and Cyberspace, which traditionally function as separate entities. However, multi-domain interoperability becomes increasingly important as warfare and the digital landscape evolve. NATO states that integrating these military domains with instruments of power and external stakeholders is also crucial. Unlike joint operations, multi-domain operations prioritize military collaboration with non-military assets. These non-military assets refer to member states' various departments or agencies, such as the Ministry of Foreign Affairs. In NATO, a secure and scalable Cloud environment will serve as the foundation of the Digital Backbone and to enable Multi-Domain Operation. At the same time, stakeholders are external groups such as academia and private industries with whom NATO maintains business partnerships. Cloud-enabled data sharing also supports interoperability between different classification levels. In 2021, the US Government enabled Microsoft Azure Top Secret to simplify data segregation amongst different classification levels. Initiatives like these offer a seamless approach to data sharing, but outdated legacy systems still threaten the competitiveness of defence organizations such as NATO. As warfare evolves and technology advances, the demand for interoperability is escalating. Cloud computing is highly beneficial in facilitating data sharing across various defence classifications levels and with the civil sector.

12.6 CLOUD COMPUTING INITIATIVE BY THE INDIAN DEFENCE FORCES

Cloud computing has the potential to be a game changer in military warfare, and India has taken steps to implement it in its defense sector.

For the last two decades, there has been significance increase in demand of cloud computing for Indian defence forces. India has built a series of dedicated secure defence networks across tri-forces (Army, Navy and Air Force) including a robust cyber and space command agency. Cloud computing enhance the area of joint-ness, Technology led security solutions and ISR (Intelligence, Surveillance and Reconnaissance), interoperability and cyber security are one of imperative requirement of the forces.

On November 9, 2015, the then Defence Minister of India, Manohar Parrikar and Army General Dalbir singh inaugurated highly encrypted cloud system of the Indian Army similar to that of Google, which will store personnel as well as operational data.

(i) Army Cloud

The Indian Army launched a highly encrypted cloud system called Army Cloud to provide IT infrastructure, including servers, storage, and network security equipment. The system also includes a disaster recovery site to replicate critical data.

The 'Army Cloud' includes a central data centre, a near-line data centre, both in Delhi and disaster recovery site for a replication of it's critical data along with virtualized server and storage in an environmentally controlled complex. It will provide IT (Information Technology) infrastructure such as server, storage, network and other computing resources for automation of Indian Army.

(ii) Digi-Locker

On same day another 'cloud initiative' was inaugurated 'Digi-Locker' which provides a secure and exclusive dedicated network of data storage space to all the units of the Indian Army. Army officials stated that "it's crucial step towards automation and digitalize of infrastructure which will act as force multiplier and a landmark towards transforming Indian Army from platform-centric to network-centric force.

Indian Air Force has upgraded it's AFNET (Air Force Network) from AFNET 1.0 to 2.0 which will have various cloud computing aspects such as cyber security, service

12.6.1 Future of Cloud Computing in India

NASSCOMInsights has worked on the "Cloud Skills: Powering India's Digital DNA" report that highlights various factors driving cloud adoption, explores the stages of cloud maturity, its impact on the talent demand and availability and recommends course of action for India to attain its potential of becoming a cloud-first nation.

The following statistics are as per the NASSCOMInsight report that states an exponential growth in cloud investments across the globe and in India:

(i) Global Public Cloud spend, currently valued at $332 bn as of 2021, is projected to grow by 20% YoY till 2022.

(ii) SaaS remains the largest global market segment at 39% share; PaaS, IaaS are expected to grow exponentially.

(iii) Investments in Cloud Management, Storage Networks, Security & Back-up services are expected to grow by 31% YoY till 2022; reflecting demand for Cloud-enabled infrastructural upgradations.

(iv) Hybrid Cloud adoption in Indian organizations is set to grow by 49% in the next 3 years Rise of new age cloud skills.

(v) 'Digital-First' is the latest buzz word for every business that also increases the demand for cloud computing experts across the globe. Cloud Developer, Cloud Security Engineer, Front-end developer, Back-end Developer, Solutions Architect, Cloud Security Specialist, Cloud Migration Specialist and many more profiles are in demand that needs skilled cloud experts. As per the report, India alone recorded nearly 379,000 job openings for cloud roles in the past year, and this demand is only set to grow with time.

The demand for digital skills today (including cloud) is around 8-10X higher than the supply that exists in India. To make India the first cloud-nation, NASSCOMInsights has worked on the report to understand the existing cloud talent landscape in India and the demand-supply analysis by 2025. We could then realize the full potential of cloud and in the process, make India the talent nation for the digital world.

12.6.2 Adoption of cloud computing

(i) The Indian Government's increasing cloud adoption through GI Cloud (**MeghRaj**), the National and State Service Delivery Gateways, development of a Cloud-based AI platform called **AIRAWAT**, **TechSaksham** and other initiatives enable ease, provide authentic access and transparency to government benefits for the common citizens because of cloud computing.

(ii) Skilling India's tech talent for the future

With the growing interest and demand for cloud job roles, the number of job opportunities has also seen a boost and is expected to grow tremendously which also increases the need of skilling in cloud computing. From a global perspective, what sets India apart from the other leading countries, is its extensive, diverse pool of technology services talent which stands at over 4.5 mn as of FY2021, consisting of the traditional IT job roles of 2.4 to 2.6 mn, and the FutureSkills talent of 1.9 to 2 mn. Consequently, by 2025, the unmet demand is expected to grow 3X by 2025; demand for over 2.2 mn Cloud professionals versus an installed talent of 1.4 to 1.5 mn, as per the report.

(iii) Cloud is the way forward in NASSCOM

NASSCOM's Sector Skill Council (SSC) is the primary conduit that standardizes and builds IT Skills, set up under the purview of Ministry of Skills Development & Entrepreneurship (MSDE).

The launch of FutureSkills PRIME, is an expansion of the initial FutureSkills initiative, which extended the scope to skill more professionals across a wider spectrum of digital specialization areas and industry expertise, apart from the emerging FutureSkills technologies.

CHAPTER 13

RENEWABLE ENERGY TECHNOLOGIES IN WARFARE

13.1 INTRODUCTION TO RENEWABLE ENERGY

Renewable energy (or green energy) is the energy from renewable natural resources that are replenished on a human timescale. The most widely used renewable energy types are solar energy, wind power, and hydropower. Bioenergy and geothermal power are also significant in some countries. Some also consider nuclear power a renewable power source, although this is controversial. Renewable energy installations can be large or small and are suited for both urban and rural areas. Renewable energy is often deployed together with further electrification. This has several benefits: electricity can move heat and vehicles efficiently and is clean at the point of consumption. Variable renewable energy sources are those that have a fluctuating nature, such as wind power and solar power. In contrast, controllable renewable energy sources include dammed hydroelectricity, bioenergy, or geothermal power.

Renewable energy systems have rapidly become more efficient and cheaper over the past 30 years. A large majority of worldwide newly installed electricity capacity is now renewable. Renewable energy sources, such as solar and wind power, have seen significant cost reductions over the past decade, making them more competitive with traditional fossil fuels. In most countries, photovoltaic solar or onshore wind are the cheapest new-build electricity. From 2011 to 2021, renewable energy grew from 20% to 28% of global electricity supply. Power from the sun and wind accounted for most of this increase, growing from a combined 2% to 10%. Use of fossil energy shrank from 68% to 62%. In 2022, renewables accounted for 30% of global electricity generation and are projected to reach over 42% by 2028. Many countries already have renewables contributing more than 20% of their total energy supply, with some generating over half or even all their electricity from renewable sources.

The main motivation to replace fossil fuels with renewable energy sources is to slow and eventually stop climate change, which is widely agreed to be caused mostly by greenhouse gas emissions. In general, renewable energy sources cause much lower emissions than fossil fuels. The International Energy Agency estimates that to achieve net zero emissions by 2050, 90% of global electricity generation will need to be produced from renewable sources. Renewables also cause much less air pollution than fossil fuels, improving public health, and are less noisy.

The deployment of renewable energy still faces obstacles, especially fossil fuel subsidies, lobbying by incumbent power providers, and local opposition to the use of land for renewable installations. Like all mining, the extraction of minerals required for many renewable energy technologies also results in environmental damage. In addition, although most renewable energy sources are sustainable, some are not.

13.2 APPLICATION OF RENEWABLE ENERGY SOURCES IN THE US MILITARY

On September 14, 2019, the Iranian military launched a swarm of missiles and drones on the oil fields of Saudi Aramco, a Saudi Arabian state-run oil company. The calculated attack was an assault on a key U.S. ally in the region and an affront to American hegemony worldwide. Conceivably, Iran designed the attack to affect the U.S. without actually committing an act of war, which would have most likely led to retaliation from the U.S., Saudi Arabia, and Israel. While the attack did not meet the definition of an act of war, it did have global ramifications, knocking out 5% of the world's oil supply with 18 drones and three missiles, and causing a 20% spike in crude oil prices. The event highlighted the impact low-tech or ill-equipped actors can have not only on the global economy, but also on militaries who are dependent on oil for their operations.

13.2.1 Vulnerabilities

The National Resources Defense Council ranks the U.S. as the world's largest single consumer of energy. The majority of this energy consumption, especially from the U.S. military, is from fossil fuels, with the by-product, carbon dioxide, a problem because it affects the environment with greenhouse gases, creating a vulnerability in terms of initial product cost and secure delivery to the consumer, often inside hostile territory. As demonstrated by the Iranian attack on Saudi Arabian oil fields and resulting spike in global energy costs, adversaries are becoming increasingly aware of this vulnerability.

The second major vulnerability is that U.S. military bases are integrated with Industrial Control Systems (ICS) and Supervisory Control and Data Acquisition (SCADA) networks which control a variety of power, water, communication, and transportation infrastructure stateside. Currently, these bases are linked to the state and municipal ICS/SCADA grid systems, with backup power via short-term fossil fuel stockpiles running generators. This reliance on ad-hoc ICS/SCADA systems and public energy is dangerous because an attack on these systems could disrupt U.S. military operations across the globe.

To illustrate the danger in trusting these systems, Iranian hackers have already successfully infiltrated the ICS/SCADA networks of several countries worldwide, including telecommunications and energy grids in the U.S., along with the U.S. Navy's Marine Corps Intranet. An attack on any U.S. grid could cause a domino-effect breakdown to all operations in the system. In order to increase security, and protect against the associated fuel and energy vulnerabilities, the U.S. military must approach these issues with a **three-phase solution**.

13.2.2 Solutions

(i) Phase-1 Solution

The first phase in repairing the energy grid vulnerability is to remove U.S. military bases from local power grids while implementing renewable energy systems. This will limit the susceptibility to cyberattacks and reduce reliance on fossil fuels.

U.S. Department of Energy (DOE) is already working on a decentralized power solution called Autonomous Energy Grids (AEG). These AEGs are made up of Network Optimized Distributed Energy

Systems (NODES) that form a self-contained, smart energy community that could provide resources for one installation, while sending excess energy, or requesting enough to cover a deficit from other installations on the AEG. Because this new system is self-contained and largely self-reliant, an attack on one NODES would not cause a cascading effect to others.

In addition, because the system is standardized, DoD technicians can maintain and operate between different installations, despite the actual source of the energy itself, which could be tailored specifically to the environment of the installation. For example, Fort Huachuca, Arizona, situated in the Sonoran Desert, may consider a combination of nuclear and solar power as its main power source; Fort Bliss, Texas, a combination of solar power and wind; while Naval Base San Diego, California, may instead rely on wave and tidal generators.

For this new system to be secure, any outflow from bases must also be restricted. For example, Fort Bliss currently has a number of solar power panels which feed back into the local economy at subsidized rates. This practice would stop because any connection to the local grid means a potential vulnerability and weakened internal system. This shift of energy grids would not just affect the military base, but also impact the local community, many of which rely on military installations for economic reasons, especially energy costs paid to the local energy provider for day-to-day base operations.

In the short-term, there are financial hurdles and risks for all parties currently involved. Implementing major technological advancements and overhauling a previous operating system can be costly, however, the long-term benefits and financial rewards of pioneering clean and renewable energy sources while also reducing avenues of attacks for enemies is too great to ignore.

(ii) Phase-2 Solution

The second phase requires research and development of an interoperable, non-petroleum-dependent vehicle system that uses renewable resources. Commanders could tailor a number of variants and power options for specific missions and usages such as reconnaissance, logistical supply, route clearance, infantry delivery, gunnery options, and intelligence collection platforms.

While it will be costly and difficult to develop electrical vehicle systems that can survive the rigors and demands of combat, the military could partner with the civilian sector to accomplish this task. **Tesla is already developing solar banks and battery-powered trucks**, while General Motors Co. is releasing an electric "Hummer", which reportedly has a powertrain supporting up to 1,000 horsepower, which is double the Army's current Heavy Expanded Mobility Tactical Truck's 500 horsepower. This relationship could benefit both parties as the military would rapidly advance their warfighting technology, and civilian corporations would stand to profit off the product and research.

Completely removing the reliance on petroleum from military operations is a long-term goal. A first step could be the DoD mandating all government-owned non-tactical vehicles on stateside military bases convert to electric. This large-scale implementation could give investors and researchers the time and incentive to develop options for a full-scale non-petroleum program in the future.

(iii) Phase-3 Solution

After the DoD has removed military bases from the national and local power grid networks, shifted to **renewable and reliable operational energy sources**, and fielded the military with non-petroleum dependent vehicles, the final phase is to shift Army training and doctrine to account for the **pivot away from fossil fuels**. This step would involve disruption to the current structure and way of life of the Army. Not only will new equipment need to be fielded, likely brigade by brigade much like the Interim Armored Vehicle [Stryker] was rolled out in the early 2000s, but obsolete military occupation specialties (MOS) like the 92F petroleum supply specialist ("refueler") will need to retrain and reclassify into different sustainment specialties with an emphasis on the technical expertise needed in each new system. Additionally, sustainment doctrine like Army Doctrine Publication, would need a complete overhaul. Standard operating procedures, from Theater Sustainment Command down to each tactical resupply convoy, would need significant revisions to accommodate the new technology.

From 2011 to 2015 the U.S. military nearly increased its renewable power generation by 100 percent while the nation's economy added merely 2.6 percent of renewable power generation. With continued growth in renewables and further investment in the U.S. Defense Advanced Research Projects Agency (DARPA), which engages in cutting edge experimental technologies, the United States could maintain a competitive advantage with its allies in NATO and the EU as well as with rivals like China who is prepared to be the world's leading producer and consumer of alternative energy technology in the world.

Despite the United States increasing its domestic oil and gas reserves over the past decade and decreasing its dependence on foreign energy supplies, the military should continue its pursuit of alternative energy sources if the United States aims to achieve true energy security for both its warfighters and the country. he Defense Department's largest installation energy user is the U.S. Army, which is particularly at risk of experiencing disruption from various commercial power grids, including natural disasters. To mitigate those risks the "Army Net Zero Initiative" will establish twenty-five net zero installations by 2030 which will depend on the use of renewables.

In 2016, the Navy deployed the **Great Green Fleet**, a year-long event that highlighted the military's commitment to reducing fossil fuel use. The Great Green Fleet consisted of ships that used hybrid-electric propulsion technology, a 50/50 mix of biofuels and diesel, fuel cells and nuclear power to reduce greenhouse gas emissions and dependence on foreign oil.

Ships that use energy more efficiently can go farther, deliver more firepower and remain at sea longer. It makes them more flexible and brings the military closer to fuel independence. Additionally, efficient energy usage reduces the time ships are tied to oilers, increasing security for Marines involved in refueling. It also takes fuel convoys off the road.

The US Air Force produces more carbon emissions than the Army, Marines and Navy combined, with much of this pollution stemming from jet fuel use. **Some aircraft now use alternative energy sources such as biofuel blends to reduce emissions.**

U.S. military renewable energy use will mitigate climate change, improve troop safety and stabilize the military's budget. It is also necessary for complying with government orders. In 2021,

President Biden signed Executive Order 14057, "Catalyzing Clean Energy Industries and Jobs Through Federal Sustainability." It instructs the national government to use electricity that is 100% free of carbon pollution by 2035.

EO 14057 allows some exemptions for the military, stating any vehicle or aircraft used in tactical and relief operations, combat support or military training may continue to use fossil fuels as necessary. However, the DoD is not exempt regarding energy usage for heating, cooling, lighting and communications on military bases.

The U.S. Armed Forces has already started using renewable energy for transportation and military bases, and it has its sights on more comprehensive programs.

The Tactical Garbage to Energy Refinery Program (**TGER**) employs deployable, tactical biorefineries that turn garbage into ethanol, composite gas and benign ash. When soldiers add a small amount of diesel to the mix, they can use the fuel to power a generator that produces electricity. This process cuts down on waste disposal, energy and vehicle fuel costs.

Under the 2023 National Defense Authorization Act, the DoD must prepare a plan to use more sustainable aviation fuel and begin testing it. The DoD must also identify a refinery that can produce the new biofuel.

The Army is investing in tactical electric vehicles through General Motors Defense, a business unit of GM that focuses on advanced defense mobility requirements. The DoD's Defense Innovation Unit has already begun testing them. It has also commissioned General Motors to develop a heavy-duty battery pack that can power military vehicles in the field.

As the military starts incorporating more electric vehicles, it will be important that they keep the vehicles up to code with their gasoline fueled counterparts. Military-grade trucks must adhere to rigorous standards like the MIL-STD-1275. This code addresses the voltage, electrical noise and maximum current of parts for ground vehicles. Meanwhile, MIL-STD-810 ensures a military vehicle can handle extreme temperature shifts, high altitude, humidity, sand and vibrations.

Electric vehicles are more powerful than those with internal combustion engines and can accelerate rapidly. They will enable longer travel distances and better performance, and their silence makes them stealthier in the field. Although charging infrastructure must expand before the military can roll out an all-electric tactical fleet, the technology for electrifying vehicles at military bases is already established.

Hydrogen fuel cells are another option for powering military vehicles. The Army has begun experimenting with the Chevrolet Colorado ZH2 fuel cell electric truck, which runs on hydrogen instead of a battery charge. The Air Force also drives hydrogen-fueled nontactical vehicles in Hawaii, indicating hybrid powertrain designs might be viable for tactical vehicles.

Microgrids

The Army installed its first microgrid in 2013 in Fort Bliss, Texas, which includes a solar array, energy storage system and interconnection to the larger energy grid. This installation foreshadowed the solar industry's explosive growth, with the U.S. solar market on track to quadruple by 2030.

The military has become increasingly reliant on microgrids — it plans to incorporate them in 100% of its bases by 2035, the same year it intends to deploy an all-electric fleet of non-tactical vehicles. Microgrids provide renewable, reliable energy the military can use in case of a widespread power outage. The Army currently has 950 renewable energy projects that supply it with 480 megawatts of power and plans to add 25 new microgrids by 2024.

Other U.S. military renewable energy technology includes solar-powered blankets and backpacks that can recharge the batteries in communications equipment, letting soldiers power their equipment as they walk or rest.

Thermoelectric power is another hypothetical energy source the Army is considering utilizing. The technology works by generating electricity from the small temperature gradient between the skin and surrounding air. As electrons move from the hot side to the cold side of the wearable device — which could be a helmet or clothing — they generate an electric current. This technology would vastly reduce the battery weight a soldier has to carry.

Additionally, the Army has unveiled a floating solar array on Big Muddy Lake at Fort Bragg in North Carolina. It will provide backup power to the nearby training facility during blackouts, as in the case of possible climate-change-related hurricanes in the area. The cooling effect of the water allows the solar farm to generate much more electricity than traditional solar panels.

13.3 ENERGY STRATEGY FOR THE INDIAN ARMED FORCES

The Indian Armed Forces have a sterling record on ecological conservation - stabilization of sand dunes, afforestation of the Thar Desert, managing watersheds in Himachal and Uttarakhand or reclaiming the barren Mussoorie hills, mines, all of these projects have been carried out by with military-like work culture and commitment. The Army, in particular, has contributed to the conservation of nearly extinct species in wildlife - the Great Indian Bustard, The Ring -necked Crane and the snow leopard. The forces have been in the forefront in the fight to save the planet.

The Ecological Task Force battalions of the Territorial Army (TA) have been raised to execute specific ecology-related projects by enrolling ex-servicemen and have created history with their success stories.

The Indian Air Force has made a beginning with experimenting with use of biofuels for their transport aircraft. The Indian Navy has undertaken numerous projects for the cleaning of the beaches, protection of mangrove swamps, marine life and corals.

It is time for the Armed Forces to now make a mark by taking steps to contribute to the environment by reducing their carbon foot-print where they can, without compromising on their operational efficiency.

There are many areas, particularly in peace stations and bases where substantial measures can be taken to reducing carbon emissions, using cleaner energy, energy savings through better building designs, efficient transportation, heating & lighting methods, recycling waste and such other measures.

13.3.1 Aim of Energy Strategy for the Indian Armed Forces

The aim of Energy Strategy for the Armed Forces is to enable information exchange on best practices and technologies for advancing energy efficiency in the military. The thrust is to find ways and means to make the military less dependent on fossil fuel and to increase the effectiveness and improve the security of future military missions, while reducing the military's carbon footprint.

The event will bring together numerous experts from military, industry and academia and create the platform to present the expertise, discuss lessons learned. The virtual exhibition will offer an opportunity for companies to present their innovative energy efficiency technologies.

13.3.2 Scale of the Problem

The climate costs of mobilizing military assets and personnel—including movement of manpower, millions of tons of hardware and equipment, food supplies and related services - is huge.

Vast swathes of military are big carbon emitters – tanks, trucks jet planes, much of the navy except nuclear-powered submarines. The military's carbon "boot" print comes from production of military equipment, energy use at military bases (energy use, food, waste management) and vehicle use (aircraft, marine vessels and land vehicles). The impact of war on the environment is entirely different level of concern. This does not include emissions of contractors and suppliers.

Training of the Armed Forces is large in scale involving movement over long distances, using live

A long-range bomber produces 251 tonnes $CO_{2-}e$ per mission. A fighter bomber 28 tonnes $CO_{2-}e$ per mission. The average freight truck emits 161.8 grams of CO_2 per ton-mile.

13.3.3 Armed Forces and Climate Change

Today, each country is required to render reports to the UN on their emissions, but these exclude any fuels purchased and used by the military. As a result it is still difficult to calculate the exact responsibility of the world's military forces for greenhouse gas emissions.

Armed forces of countries around the world will no longer be automatically exempted from emissions-cutting obligations under the UN Paris climate deal, when enforced. Decisions will be left to nation states as to which national sectors should make emissions cuts before 2030. Exemptions can only be sought through legislative exemption.

While the atmosphere counts the carbon from the military, it is politically inconvenient to reduce military emissions.

In many countries, activities including intelligence work, law enforcement, emergency response, tactical fleets and areas classified as national security interests are also exempted from reporting obligations.

13.3.4 The Need to Discuss Solutions

The beginning point could be an audit of the carbon footprint of the Armed Forces so that scale and dimensions of the problem can be understood, a long-term plan can be worked out and progress can subsequently be measured.

Some countries have launched strategies and adopted measures to reduce the carbon footprint of their military. Many have set targets to be achieved in the next 10 and 20 and 30 years.

Solutions must be discussed so that they can be incorporated in the planning of habitat, construction of buildings, operational functioning and training, General Staff Qualitative Requirements of some of the equipment, and allocations of financial support to implement the schemes.

While it is recognized that energy security should not impede the forces from performing their primary mission, the importance of energy efficiency for the conduct of military operations has come to the fore over the past decade. The weight of batteries to power the wide range of electronic equipment used by the military adds a substantial burden to soldiers. Moreover, fuel convoys are vulnerable to attack.

We must examine whether new technologies allow us to change the way we plan our missions, procure equipment, and conduct campaigns. The possibilities are endless - Solar power, hybrid power and microgrids for bases, better insulation for soldiers' habitat in extreme cold areas, fuel cells to power the equipment of individual soldiers, biofuels for military vehicles and so on.

Atmospheric water generator, intelligent power storage and management system, tents lined with insulation material, photovoltaic solar panels, light-emitting diode (LED) lamps, hydrogen fuel cell that produces electricity could replace diesel generators.

13.3.5 Role of Indian Industry

Enhancing energy efficiency in the military focuses on reducing the energy consumption of military vehicles and camps, as well as minimizing their environmental footprint. Energy-saving logistics solutions must come from private companies that can contribute equipment and expertise for 'smart energy' production, storage, distribution and consumption. Public sector experts from ministry of defence and universities have a role to play.

CHAPTER 14

INTERNET OF THINGS (IOT) IN WARFARE

14.1 WHAT IS INTERNET OF THINGS (IOT)?

The Internet of Things (IoT) is a network of physical objects that are connected to the internet and can exchange data with other devices and systems. These objects, also called "smart objects", are embedded with sensors, software, and other technologies.

The IoT can include a wide range of devices, such as:

- Household objects like appliances, thermostats, and lighting systems;
- Wearable devices that monitor health;
- Industrial machinery;
- Transportation systems;
- Security cameras; and
- Vehicles.

The IoT can simplify and automate tasks, and devices can often communicate with each other and act on the information they receive without human intervention. For example, smart speakers can add items to shopping lists and turn lights on and off.

The IoT can use artificial intelligence and machine learning to make data collection easier and more dynamic. Data can be analyzed to identify patterns, offer recommendations, and identify potential issues.

The term "Internet of Things" was coined by someone working in supply chain optimization who wanted to attract senior management's attention to a new technology called RFID (Radio-Frequency Identification) trackers.

The term IoT, or Internet of Things, refers to the collective network of connected devices and the technology that facilitates communication between devices and the cloud, as well as between the devices themselves. Thanks to the advent of inexpensive computer chips and high bandwidth telecommunication, we now have billions of devices connected to the internet. This means everyday devices like toothbrushes, vacuums, cars, and machines can use sensors to collect data and respond intelligently to users.

The Internet of Things integrates everyday "things" with the internet. Computer Engineers have been adding sensors and processors to everyday objects since the 90s. However, progress was initially slow because the chips were big and bulky. Low-power computer chips called RFID tags were first used to track expensive equipment. As computing devices shrank in size, these chips also became smaller, faster, and smarter over time.

The cost of integrating computing power into small objects has now dropped considerably. For example, you can add connectivity with Alexa voice services capabilities to MCUs with less than 1MB embedded RAM, such as for light switches. A whole industry has sprung up with a focus on filling our homes, businesses, and offices with IoT devices. These smart objects can automatically transmit data to and from the Internet. All these "invisible computing devices" and the technology associated with them are collectively referred to as the Internet of Things.

14.1.1 How does IoT work?

A typical IoT system works through the real-time collection and exchange of data. An IoT system has three components:

(i) Smart devices

This is a device, like a television, security camera, or exercise equipment that has been given computing capabilities. It collects data from its environment, user inputs, or usage patterns and communicates data over the internet to and from its IoT application.

(ii) IoT application

An IoT application is a collection of services and software that integrates data received from various IoT devices. It uses machine learning or artificial intelligence (AI) technology to analyze this data and make informed decisions. These decisions are communicated back to the IoT device and the IoT device then responds intelligently to inputs.

(iii) A graphical user interface

The IoT device or fleet of devices can be managed through a graphical user interface. Common examples include a mobile application or website that can be used to register and control smart devices.

14.1.2 What are examples of IoT devices?

Let's look at some examples of IoT systems in use today:

(i) Connected cars

There are many ways vehicles, such as cars, can be connected to the internet. It can be through smart dashcams, infotainment systems, or even the vehicle's connected gateway. They collect data from the accelerator, brakes, speedometer, odometer, wheels, and fuel tanks to monitor both driver performance and vehicle health.

Connected cars have a range of uses:

- Monitoring rental car fleets to increase fuel efficiency and reduce costs;
- Helping parents track the driving behavior of their children;
- Notifying friends and family automatically in case of a car crash;
- Predicting and preventing vehicle maintenance needs.

(ii) Connected homes

Smart home devices are mainly focused on improving the efficiency and safety of the house, as well as improving home networking. Devices like smart outlets monitor electricity usage and smart thermostats provide better temperature control. Hydroponic systems can use IoT sensors to manage the garden while IoT smoke detectors can detect tobacco smoke. Home security systems like door locks, security cameras, and water leak detectors can detect and prevent threats, and send alerts to homeowners.

Connected devices for the home can be used for:

- Automatically turning off devices not being used;
- Rental property management and maintenance;
- Finding misplaced items like keys or wallets; and
- Automating daily tasks like vacuuming, making coffee, etc.

(iii) Smart cities

IoT applications have made urban planning and infrastructure maintenance more efficient. Governments are using IoT applications to tackle problems in infrastructure, health, and the environment.

IoT applications can be used for:

- Measuring air quality and radiation levels;
- Reducing energy bills with smart lighting systems;
- Detecting maintenance needs for critical infrastructures such as streets, bridges, and pipelines; and
- Increasing profits through efficient parking management.

(iv) Smart buildings

Buildings such as college campuses and commercial buildings use IoT applications to drive greater operational efficiencies.

IoT devices can be use in smart buildings for:

- Reducing energy consumption; and
- Lowering maintenance costs.

14.2 INTERNET OF THINGS (IOT) IN DEFENCE

Every nation continuously tries to strengthen its military and defence sectors with the latest technologies, and skills. The world's top countries with powerful militaries like the USA, Russia, **India**, and China have a solid backup of technology. And now, because of this, implementing IoT in defence is no longer a "want" but a "need."

This has given rise to a very new sector- **IoMT (Internet of Military Things)**. The use of IoT in defence and military is termed IoMT. The IoT is used in the military to boost situational awareness, improve decision-making, and remotely monitor and manage assets. In the defence sector, physical objects like helmets, weapon systems, and combat vehicles are integrated with sensors and other related technologies. This allows the physical objects to communicate with each other, making them capable of collecting and transferring data over the internet to carry out a wide range of activities.

14.2.1 Applications of IoT in Defence

(i) Smart Bases

Smart bases are military bases where a network of sensors and IoT devices are deployed to smartly monitor the real-time status of important infrastructure like water and power systems, and other security-related data monitoring to ensure safety and security.

(ii) Asset Tracking

In the field of military, weapons, tanks, equipment, food, and several more items are required. Now, the data on the availability of these items should be conveyed securely, but still, organizing all this data can be troublesome, even for our military personnel. With the use of RFID (Radio-Frequency Identification) trackers, there will be real-time visibility of the stocks. Smart sensors can also be embedded in combat equipment to give real-time data on their health and whether they need maintenance.

(iii) Shipment Tracking & Management

For all successful military operations, it is crucial to manage the military vehicles in which ammunition, men, and other important stocks are transported. IoT based shipment tracking solutions can be utilized to track the movement of goods from their origin to their dropped location. IoT devices installed on the shipment vehicles will also monitor parameters like location, fuel consumption, damage, vehicle health, and much more.

Read our shipment tracking case study to understand how the solution works.

(iv) Soldiers-Health Monitoring using IoT

IoT sensors can be integrated into the **wearables**, making them smart wearables that can be worn by defence personnel to track soldier's health vitals and movements in real-time, even remotely. This will offer real-time communication with the defence personnel and ensure their safety.

(v) Battlefield Monitoring

The Internet of Things technology is meant to track and monitor data remotely and in real-time without the involvement of humans. With IoT, the military can smoothly conduct battlefield monitoring using drones that are fitted with smart cameras and sensors. The battle team can keep a watchful eye on the battleground and make prompt decisions using the data collected from the site.

14.3 INTERNET OF MILITARY THINGS (IOMT)

The Internet of Military Things (IoMT) is a class of Internet of things for combat operations and warfare. It is a complex network of interconnected entities, or "things", in the military domain that continually communicate with each other to coordinate, learn, and interact with the physical environment to accomplish a broad range of activities in a more efficient and informed manner.

The Internet of Military Things (IoMT) is an evolving field integrating advanced military technology with the Internet of Things (IoT).

The concept of IoMT is largely driven by the idea that future military battles will be dominated by machine intelligence and cyber warfare and will likely take place in urban environments. By creating a miniature ecosystem of smart technology capable of distilling sensory information and autonomously governing multiple tasks at once, the IoMT is conceptually designed to offload much of the physical and mental burden that warfighters encounter in a combat setting.

Over time, several different terms have been introduced to describe the use of IoT technology for reconnaissance, environment surveillance, unmanned warfare and other combat purposes. These terms include the Military Internet of Things (MIoT), the Internet of Battle Things, and the Internet of Battlefield Things **(IoBT).**

The Internet of Military Things (IoMT) encompasses a large range of devices that possess intelligent physical sensing, learning, and actuation capabilities through virtual or cyber interfaces that are integrated into systems. These devices include items such as sensors, vehicles, robots, UAVs, human-wearable devices, biometrics, munitions, armor, weapons, and other smart technology.

14.3.1 Key Components of Military IoT

(i) Sensors and actuators

These are the primary devices that gather environmental data. Sensors can detect everything from troop movements to environmental conditions, while actuators can carry out actions like triggering alarms or adjusting equipment.

(ii) Communications networks

Secure and resilient networks are essential for transmitting data between IoT devices. These networks must withstand harsh environments and potential cyber threats.

(iii) Data processing and analytics

Collected data is analyzed using advanced algorithms, providing actionable insights for military operations.

(iv) Edge computing

Processing data close to the source reduces latency and ensures military personnel can make critical decisions quickly in the field.

In addition to connecting different electronic devices to a unified network, researchers have also suggested the possibility of incorporating inanimate and innocuous objects like plants and rocks into the system by fitting them with sensors that will turn them into information gathering points. Such efforts fall in line with projects related to the development of electronic plants, or e-Plants.

Proposed examples of IoMT applications include tactical reconnaissance, smart management of resources, logistics support (i.e. equipment and supply tracking), smart city monitoring, and data warfare. Several nations, as well as NATO officials, have expressed Interest in the potential military benefits of IoT technology.

14.4 APPLICATION OF IOMT

The Connected Soldier

The Connected Soldier project was a research initiative supported by the U.S. Army Natick Soldier Research, Development and Engineering Center (NSRDEC) that focused on creating intelligent body gear. The project aimed to establish an internet of things for each soldier by integrating wideband radio, biosensors, and smart wearable systems as standard equipment. These devices served not only to monitor the soldier's physiological status but also to communicate mission data, surveillance intelligence, and other important information to nearby military vehicles, aircraft, and other troops.

14.5 INTERNET OF BATTLEFIELD THINGS (IOBT)

Internet of Battlefield Things (IoBT) technology in an unstructured, chaotic urban environment.

In 2016, the U.S. Army Research Laboratory (ARL) created the Internet of Battlefield Things (IoBT) project in response to the U.S. Army's operational outline for 2020 to 2040, titled "Winning in a Complex World." In the outline, the Department of Defense announced its goals to keep up with the technological advances of potential adversaries by turning its attention away from low-tech wars and instead focusing on combat in more urban areas. Acting as a detailed blueprint for what ARL future warfare may entail, the IoBT project pushed for better integration of IoT technology in military operations in order to better prepare for techniques such as electronic warfare that may lie ahead.

In 2017, ARL established the Internet of Battlefield Things Collaborative Research Alliance (IoBT-CRA) to bring together industry, university, and government researchers to advance the theoretical foundations of IoBT systems.

According to ARL, the IoBT was primarily designed to interact with the surrounding environment by acquiring information about the environment, acting upon it, and continually learning from these interactions. As a consequence, research efforts focused on sensing, actuation, and learning challenges. In order for the IoBT to function as intended, the following prerequisite conditions must first be met in regard to technological capability, structural organization, and military implementation.

(i) Communication

All entities in the IoBT must be able to properly communicate information to one another even with differences in architectural design and makeup. While future commercial internet of things may exhibit a lack of uniform standards across different brands and manufacturers, entities in IoBT must remain compatible despite displaying extreme heterogeneity. In other words, all electronic equipment, technology, or other commercial offerings accessed by military personnel must share the same language or at least have "translators" that make the transfer and processing of different types of information possible. In addition, the IoBT must be capable of temporarily incorporating available networked devices and channels that it does not own for its own use, especially if doing so is advantageous to the system (e.g. making use of existing civilian networking infrastructure in military operations in a megacity). At the same time, the IoBT must take into consideration the varying degree of trustworthiness of all the networks it leverages.

Timing will be critical in the success of IoBT. The speed of communication, computation, machine learning, inference, and actuation between entities are vital to many mission tasks, as the system must know which type of information to prioritize. Scalability will also serve as an important factor in the operation since the network must be flexible enough to function at any size.

(ii) Learning

The success of the IoBT framework often hinges on the effectiveness of the mutual collaboration between the human agents and the electronic entities in the network. In a tactical environment, the electronic entities will be tasked with a wide range of objectives from collecting information to executing cyber actions against enemy systems. In order for these technologies to perform those functions effectively, they must be able to not only ascertain the goals of the human agents as they change but also demonstrate a significant level of autonomous self-organization to adjust to the rapidly changing environment. Unlike commercial network infrastructures, the adoption of IoT in the military domain must take into consideration the extreme likelihood that the environment may be intentionally hostile or unstable, which will require a high degree of intelligence to navigate.

As a result, the IoBT technology must be capable of incorporating predictive intelligence, machine learning, and neural network in order to understand the intent of the human users and determine how to fulfill that intent without the process of micromanaging each and every component of the system.

According to ARL, maintaining information dominance will rely on the development of autonomous systems that can operate outside its current state of total dependence on human control. A key focus of IoBT research is the advancement of machine learning algorithms to provide the network with decision-making autonomy. Rather than having one system at the core of the network functioning as the central intelligence component dictating the actions of the network, the IoBT will have intelligence distributed throughout the network. Therefore, the individual components can learn, adapt, and interact with each other locally as well as update behaviors and characteristics automatically and dynamically on a global scale to suit the operation as the landscape of warfare constantly evolves. In the context of IoT, the incorporation of artificial intelligence into the sheer volume of data and entities involved in the network will provide an almost infinite number of possibilities for behavior and technological capability in the real world.

In a tactical environment, the IoBT must be able to perform various types of learning behaviors to adapt to the rapidly changing conditions. One area that received considerable attention is the concept of **meta-learning**, which strives to determine how machines can learn how to learn. Having such a skill would allow the system to avoid fixating on pretrained absolute notions on how it should perceive and act whenever it enters a new environment. Uncertainty quantification models have also generated interest in IoBT research since the system's ability to determine its level of confidence in its own predictions based on its machine learning algorithms may provide some much-needed context whenever important tactical decisions need to be made.

The IoBT should also demonstrate a sophisticated level of situation awareness and artificial intelligence that will allow the system to autonomously perform work based on limited information. A primary goal is to teach the network how to correctly infer the complete picture of a situation while measuring relatively few variables. As a result, the system must be capable of integrating the vast amount and variety of data that it regularly collects into its collective intelligence while functioning in a continuous state of learning at multiple time scales, simultaneously learning from past actions while acting in the present and anticipating future events.

The network must also account for unforeseen circumstances, errors, or breakdowns and be able to reconfigure its resources to recover at least a limited level of functionality. However, some components must be prioritized and structured to be more resilient to failure than others. For instance, networks that carry important information such as medical data must never be at risk of shutdown.

(iii) Cognitive Accessibility

For semi-autonomous components, the human cognitive bandwidth serves as a notable constraint for the IoBT due to its limitations in processing and deciphering the flood of information generated by the other entities in the network. In order to obtain truly useful information in a tactical environment, semi-autonomous IoBT technologies must collect an unprecedented volume of data of immense complexity in levels of abstraction, trustworthiness, value, and other attributes. Due to serious limitations in human mental capacity, attention, and time, the network must be able to easily reduce and transform large flows of information produced and delivered by the IoBT into reasonably-sized packets of essential information that is significantly relevant to army personnel, such as signals or warnings that pertain to their current situation and mission.

A key risk of IoBT is the possibility that devices could communicate negligibly useful information that eats up the human's valuable time and attention or even propagate inappropriate information that misleads human individuals into performing actions that lead to adverse or unfavorable outcomes. At the same time, the system will stagnate if the human entities doubt the accuracy of the information provided by the IoBT technology. As a result, the IoBT must operate in a manner that is extremely convenient and easy to understand to the humans without compromising the quality of the information it provides them.

14.6 MANIFESTATION OF IOBT IN THE INDIAN ARMED FORCES

Considering a threat to our national security, both from the West and the North, rapid modernization of our military capabilities is the need of the hour. With overwhelming requirements of equipment modernization and other operational and logistic challenges, employment of niche technologies comprising robotics, artificial intelligence (AI) and IoBT is yet to begin in a full-fledged manner. However, limited use of niche technology has started in some critical formations using commercial off-the-shelf equipment. In the Indian context, IoBT R&D is still in the nascent stages. This needs immediate attention considering the size of the economy and also the fact that India has arrived on the global stage and many nations especially, those belonging to the Global South, see India as their leader. 5G technology is the bedrock of implementing IoBT and the all-pervasive nature of 5G will be crucial for fast flow of information in the IoBT network. The testbed of Indian 5G research at IIT Madras, has achieved complete end-to-end solutions indigenously and is customizable including a secure NB-IoT Chip. There is a need for deploying indigenized 5G solutions across varied terrain configurations while utilizing the existing radio, fiber, and satellite data links. Project Network for Spectrum (NFS). In 2012, under the Network for Spectrum (NFS) project, the Ministry of Defence (MoD) and Department of Telecommunication (DoT) conceptualized a network to connect critical defence locations. The project encompasses creating an optical-fiber-cable-based network exclusively for defence communication. OFC cables have been laid all across the nation in varied terrain profiles including intrusion-proof cable technology, network monitoring and GIS mapping, making it the most advanced and secure network for the armed forces. This can be further exploited for creating IoBT network exclusively for the armed forces. Under the *'Atmanirbhar Bharat'* slogan, several new projects are being implemented by the Department of Defence Production (MoD). As per reports of July 2022, several projects integrated with AI are being developed. Some of them are enumerated as under:iSentinel - Intelligent Automated Threat Tracking and Identification System. Designed as a surveillance system for CI/CT environment, iSentinel is a deep learning-based threat detection and tracking system. It can be effectively incorporated with Anti Infiltration Obstacle Systems (AIOS) with an aim to address cross-border terrorism.

(i) Smart Helmets

To address the need for real-time situational awareness, the 'Smart Helmet' can capture 3D information of any unknown environment in real-time and assist in various decisions. The 'Smart

Helmet' comprises of an optical sensor, mounted on the helmet of an active combat soldier, and can be effectively used for special operations in LC scenario and also in border management postures along the northern borders. This helps in tracking potential threats and also location of own troops inside an unknown environment thereby creating a comprehensive situational awareness for decision making.

(ii) Permissive Block Chain Mechanism

The solution is intended to create a Trusted Communication Platform using Blockchain. The idea is to establish a secured network for data transfers between the entities. Considering the ongoing progress in terms of developing niche technology under the *'Atmanirbhar Bharat'* Project and Indian Army's 'Year of Technical Absorption', there is a need to integrate outputs from sources like AI, Robotics, drones, deep learning, blockchain technology along with IoT, to come up with tailor-made solutions for specific problems such as operating in CI/CT grid or countering Chinese threat in High Altitude Areas (HAA) along the northern borders.

14.6.1 Implementation of IoBT in the Indian Army

Owing to the development of niche technology in warfighting, there will be substantial change in the way the Indian Army will operate in recent future. The armed forces need to focus on a 'bottom-up' approach and enable the smallest of teams to work with 'intelligent' systems. However, implementation of IoBT, where human soldiers and machines form tailormade units/teams for specific tasks, comes with numerous challenges and warrants various changes to approach, training, organization etc., some of which are as under:

(i) Approach to Warfare

There is a need to adopt a holistic approach to warfare and even maintain defensive posture that incorporates several changes to tactics, drills due to emerging technologies. It requires decisionmakers, junior leaders as also the troops on ground to get abreast with handling the technology and operating the same in fast paced and vulnerable environment. At present, the Indian Armed Forces are driven by specific arms and services, but there is a need to break specific silos and focus on more jointness in operations in a well-connected IoBT network.

(ii) Information Warfare

In today's warfare, information is the most lethal weapon. Hence, there is a need to incorporate an Information-driven culture that emphasizes specific procedures for collecting, collating and disseminating data using the IoBT network.

(iii) Man v/s Machine

With the implementation of autonomous systems like robots, swarm drones and UAVs, there is a huge risk of taking humans 'off the loop'. Human logic cannot replace machine intelligence to win wars. Hence, there is a need to reserve decision-making with commanders in the Observe, Orient,

Decide, Act (OODA) loop while creating an IoBT network infused with 'artificially intelligent' sensors and smart weapons capable of delivering immense firepower within a short span of time.

(iv) Independent 'cloud' for the Armed Forces

The Indian Army has launched a highly encrypted 'Army Cloud', as part of its technology absorption initiative, in similar lines with **'Megh Raj'** cloud service of National Informatics Centre. The **'Army Cloud'** along with Network for Spectrum (NFS) project will revolutionize strategic decision-making at the highest level once the smart sensors are integrated to provide a comprehensive picture of the battlefield in real-time.

(v) Focus on Specialization

Research and Development in emerging technologies demands officers to be specialized in the domain. Innovation, being an important aspect, requires individuals to work for a considerable time for a project to fructify.

Recently, the India Army introduced an HR policy wherein officers specializing in niche technologies can continue in the same domain on promotion to Colonel instead of getting posted out for command assignments. Such policies incentivizes R&D of niche technologies and help in speedy transformation into a technologically superior force. A similar effort can be undertaken at the tri-services level.

(vi) Training Aspects

Along with R&D, it is also important to train the tactical commanders, junior leaders and troops on ground to effectively utilize niche technology while in active operations. Manual war gaming has been a method for a long time. However, emerging technologies requires a new approach to training. Computerized war games can be introduced that cater to various possible scenarios like cyberattacks, jamming, drone warfare, enhanced operations using IoBT network, etc.

(vii) Cyber Security

As future battles will be fought with machines connected over various networks, cyber security assumes greater importance. We are already in an age where wars are fought on the field and also off the field, as we have seen in the case of the Russia-Ukraine war, the conflict in the Middle East.

14.7 MOSAIC WARFARE

Mosaic Warfare is a term coined by former DARPA Strategic Technology Office director Tom Burns and former deputy director Dan Patt to describe a "systems of systems" approach to military warfare that focuses on re-configuring defense systems and technologies so that they can be fielded rapidly in a variety of different combinations for different tasks. Designed to emulate the adaptable nature of the lego blocks and mosaic art form, Mosaic Warfare was promoted as a strategy to confuse and overwhelm adversary forces by deploying low-cost adaptable technological expendable weapon systems that can play multiple roles and coordinate actions with one another, complicating the decision-making process for the enemy. This method of warfare arose as a response to the current monolithic system in the military, which relies on a centralized command-and-control structure fraught with vulnerable single-point communications and the development of a few highly capable systems that are too important to risk losing in combat.

The concept of Mosaic Warfare existed within DARPA since 2017 and contributed to the development of various technology programs such as the System of Systems Integration Technology and Experimentation (SoSIT), which led to the development of a network system that allows previously disjointed ground stations and platforms to transmit and translate data between one another.

14.8 OCEAN OF THINGS

In 2017, DARPA announced the creation of a new program called the Ocean of Things, which planned to apply IoT technology on a grand scale in order to establish a persistent maritime situational awareness over large ocean areas. According to the announcement, the project would involve the deployment of thousands of small, commercially available floats. Each float would contain a suite of sensors that collect environmental data—like sea surface temperature and sea state—and activity data, such as the movement of commercial vessels and aircraft. All the data collected from these floats would then be transmitted periodically to a cloud network for storage and real-time analysis. Through this approach, DARPA aimed to create an extensive sensor network that can autonomously detect, track, and identify both military, commercial, and civilian vessels as well as indicators of other maritime activity.

The Ocean of Things project focused primarily on the design of the sensor floats and the analytic techniques that would be involved in organizing and interpreting the incoming data as its two main objectives. For the float design, the vessel had to be able to withstand the harsh ocean conditions for at least a year while being made out of commercially available components that cost less than $500 each in total. In addition, the floats could not pose any danger to passing vessels and had to be made out of environmentally safe materials so that it could safely dispose of itself in the ocean after completing its mission. In regards to the data analytics, the project concentrated on developing cloud-based software that could collect, process, and transmit data about the environment and their own condition using a dynamic display.

14.9 FUTURE CAPABILITIES OF IOMT

Given the nature of the IoMT, there is much scope for the development of additional capabilities in the future through the integration of sensors, robots, munitions, wearable devices, vehicles, and weapons. For example, with current and applicable technology, **weapon magazines** could easily be made "smart", allowing them to track and report ammunition states directly for the automated generation of logistical requests.

Wearable devices placed on the soldier could aid in human performance and management on the battlefield, and even allow for more accurate triage and improved medical treatment if required. These wearable devices could include sensors which augment the traditional Battle Management Systems (BMS) by collecting more accurate data on individual troop movements and the surrounding environment.

Similar sensors could be remotely deployed from an uncrewed aerial vehicle (UAV) to allow for the remote exploitation of sites – generating a detailed and **manipulable three-dimensional model** that could be explored and preserved by offsite intelligence analysts. The opportunities are only limited by the imagination and innovation of the creators.

14.10 SECURITY CONCERNS OF IOMT

One of the largest potential dangers of IoMT technology is the risk of both adversarial threats and system failures that could compromise the entire network. Since the crux of the IoMT concept is to have every component of the network—sensors, actuators, software, and other electronic devices—connected together to collect and exchange data, poorly protected IoT devices are vulnerable to attacks which may expose large amounts of confidential information. Furthermore, a compromised IoMT network is capable of causing serious, irreparable damage in the form of corrupted software, disinformation, and leaked intelligence.

According to the U.S. Department of Defense, security remains a top priority in IoT research. The IoMT must be able to foresee, avoid, and recover from attempts by adversary forces to attack, impair, hijack, manipulate, or destroy the network and the information that it holds. The use of jamming devices, electronic eavesdropping, or cyber malware may pose a serious risk to the confidentiality, integrity, and availability of the information within the network. Furthermore, the human entities may also be targeted by disinformation campaigns in order to foster distrust in certain elements of the IoMT. Since IoMT technology may be used in an adversarial setting, researchers must account for the possibility that a large number of sources may become compromised to the point where threat-assessing algorithms may use some of those compromised sources to falsely corroborate the veracity of potentially malicious entities.

Minimizing the risks associated with IoT devices will likely require a large-scale effort by the network to maintain impenetrable cybersecurity defenses as well as employ counterintelligence measures that thwart, subvert, or deter potential threats. Examples of possible strategies include the use of "disposable" security, where devices that are believed to be potentially compromised by the enemy are simply discarded or disconnected from the IoMT, and honeynets that mislead enemy eavesdroppers. Since adversary forces are expected to adapt and evolve their strategies for infiltrating the IoMT, the network must also undergo a continuous learning process that autonomously improves anomaly detection, pattern monitoring, and other defensive mechanisms.

Secure data storage serves as one of the key points of interest for IoMT research. Since the IoMT system is predicted to produce an immense volume of information, attention was directed toward new approaches to maintaining data properly and regulating protected access that don't allow for leaks or other vulnerabilities. One potential solution that was proposed by The Pentagon was 'Comply to Connect' (C2C), a network security platform that autonomously monitored device discovery and access control in order to keep pace with the exponentially-growing network of entities.

In addition to the risks of digital interference and manipulation by hackers, concerns have also been expressed regarding the availability of strong wireless signals in remote combat locations. The lack of a constant internet connection was shown to limit the utility and usability of certain military devices that depend on reliable reception.

14.11 SOME TRENDS IN INTERNET OF MILITARY THINGS (IOMT)

Here below are mentioned some trends of Internet of Military Things (IoMT) that can be critical contributors to protect human life and acting on real-time data:

(i) Advanced Sensor Networks

The advancement of military digital infrastructure goes hand-in-hand with the improvements in personnel equipment, mobile devices and vehicles on the field. These complex tools thereby call for more data points to enable sufficient monitoring and support with state-of-the-art networking solutions. With the proliferation of connected devices, IoMT enables the deployment of advanced sensor networks in military operations. These networks provide real-time data on various parameters, such as location, environmental conditions, and enemy activity. By utilizing sensor networks, military personnel can gain enhanced situational awareness and make better-informed decisions on the battlefield.

There are different kinds of sensors that are deployed to strengthen the capabilities of IoMT backbones. These include EO/IR sensors that can work exceptionally well in challenging atmospheric and climatic conditions, and phased-array radars that can multi-task and simultaneously collect intelligence across land, water, and air accurately.

(ii) Artificial Intelligence (AI) Integration

AI plays a significant role in allowing the effective analysis of the vast amounts of data flowing from several edge devices. Security or defense-related intelligence flows in the form of logistics, open-source intelligence (OSINT), maintenance, and support alongside battlefield intelligence. AI technologies such as machine learning, in conjunction with big data analytics, can complete scans on larger volume of data and reduce the associated noise at the same time to enable predictive analysis.

Recently, the United States Army Research Lab funded $25million for Internet of Battlefield Things Research on Evolving Intelligent Goal-driven Networks (IoBT REIGN) to develop new predictive battlefield analytics.

Some of the other key benefits include anomaly detection, and autonomous decision-making. This integration enhances military capabilities, enabling faster response times, better threat detection, and more effective missions.

(iii) Cybersecurity and Data Protection

There are serious concerns surrounding the rise of cyber-attacks in relation to the expansion and development of IoMT. The use of smart devices and an increasing dependence on the same within a communications network can potentially put military networks and communication systems at risk. Moreover, cyber security breaches can materialize at any level, such as applications, network, devices, data levels, and storage.

Innovations in encryption techniques, secure communication protocols, and threat detection systems are being developed to protect critical military assets and sensitive information from hostile actors.

(iv) Interoperability and Network Integration

The IoMT brings together a wide range of devices, platforms, and systems from different manufacturers and military branches. Ensuring interoperability among these diverse components is vital for seamless information sharing and collaboration, but at the same time, it is quite difficult to accomplish. One approach is standardizing data formats, communication protocols, and network architectures facilitate the integration of disparate IoMT devices, enabling better coordination and interoperability between military units.

(v) Edge Computing and Cloud Infrastructure

The IoMT generates enormous volumes of data that need to be processed, analyzed, and stored. To address this challenge, edge computing and cloud infrastructure are gaining prominence. Tactical edge computing is a decentralized approach to cloud computing that that essentially decouples technology resources from the centralized cloud. It allows for data processing and analytics at the edge of the network, thus reducing latency and enabling real-time decision-making. On the other hand, cloud infrastructure offers scalable storage and computing resources besides facilitating data aggregation, advanced analytics, and remote access to critical information.

The Internet of Military Things is set to revolutionize military operations by leveraging connectivity, advanced sensors, AI, and data analytics. That said, for defense organizations, IoMT is more than just a trend – it is developing into a necessary infrastructure for the modern battlefield that is becoming more complex. In addition, it will be possible to anticipate and covert countless future crises with quicker actions driven by real-time IoT data from the field as well.

CHAPTER 15

WEARABLE DEVICES IN WARFARE

15.1 WEARABLES FOR SOLDIERS

Wearables are revolutionary products that could impact our armed forces in the coming years. Napoleon Bonaparte quotes "To have good soldiers, a nation must always be at war." This quote is germane, but do we care for the survival of our soldiers? Well, wearables can do it for us. Many equipment and gadgets can become an integral part of a soldier's wearable or his uniform, making him free from carrying additional payloads.

Wired soldiers mounted with a series of microsensors can be integrated to a wireless captive network and to a cloud-based centralized server. Dedicated applications can be developed to monitor and control the soldiers deployed in hazardous missions by 'Digitally Twinning' them. We can create 'Net Warriors' in future with Wireless Body Area Networks (**WBANs**).

Remarkable progress has been made in the sports and medical fields with the support of wearables. Armed forces are yet to exploit this technology for the comfort of soldiers. Wearables can be tailored suiting to the specific requirements of soldiers.

The military wearables market is projected to grow from USD 4.2 billion in 2019 to USD 6.4 billion by 2025, at a CAGR of 7.2% from 2019 to 2025.

15.1.1 Network Infrastructure

Building network infrastructure along the border is a very important stride in bringing wearables to soldiers. Establishing a reliable backbone fiber connectivity is an absolute requirement for keeping soldiers connected. These fiber networks can be terminated with wireless networks similar to 4G/5G for establishing the last mile connectivity. India has given great impetus for the up-gradation of road infrastructure along the borders in the last few years. Communication networks can be the next in line. Implantable sensor technology includes technology like Wireless Body Area Networks (WBANs) coupled with technologies like the Internet of Battlefield Things (IoBT), would make battlefield management fully digital. A Multi-Level Fusion Framework (MLFF) based on Body Sensor Networks (BSNs) of soldiers would make a unique heterogeneous sensor network model.

15.1.2 Biometric Sensors

Development of miniature sensors for biometric applications has opened expansive opportunities. Sensors can even be implanted under the skin for specific needs. Powering sensors and getting appropriate voltage proportional to the varying parameters of the human body is a difficult exercise.

Sensors are charge-coupled devices or CMOS image sensors. Textile sensors can be designed to measure temperature, biopotential for cardiogram, myographs, encephalography, acoustic signals for diagnosing heart, lungs, digestion, and joints. Ultrasound can be used for blood flow, motion for respiration, pressure for BP, radiation for IR, spectroscopy as well as odor and sweat for skin parameters.

15.1.3 Applications of Wearable Devices

Many wrist-worn wearable devices are currently available in the market from Fitbits, Garmins, Apple, and Samsung as well as strap-based devices from Polar and Garmin products, besides high-end devices from Zephyr and Hidalgo Equivital. Our soldiers undertake many hazardous missions on a daily basis. They are deployed at hostile and remote areas, where their mental and health conditions are always tested for survivability. In this information era, we should explore and identify how we can wire our soldiers and put them at a comfort level during their missions.

Some applications or wearable devices are given below.

(i) Health Monitoring

'More you sweat in the peacetime, less you bleed in war'. This is very true, but it does not imply that we should keep our warriors unsafe while on a mission. The success story of warriors on missions is directly linked to the mental and physical well-being. A range of sensors integrated into the soldier's body can monitor the soldier's health in real-time. The biometric sensors integrated to clothing can sense vitals such as heart rate, respiratory rate, ECG, EEG oxygen level, body temperature etc. Army men at high altitudes, navy commandos deployed for underwater missions and fighter pilots under G load can be connected up and monitored during the mission to keep them safe. Even predictive analysis can be done on their vital data to avoid disaster using Artificial Intelligence.

(ii) Performance Monitoring

Training is an incessant activity in the armed forces. Soldiers and men in uniform are engaged in training activity for more duration during peacetime. The threshold level of every warrior is tested and ascertained during the training. Performance monitoring can be gauged during the training/simulation phase for many activities. This performance data would certainly help in selecting and nominating men for specific missions. The biometric sensors can sense and monitor muscle activity, muscle fatigue, muscle symmetry, speed, acceleration levels, and distance travelled, blood flow to the brain.

(iii) Warming Jacket

Outdoor missions remain a challenge for soldiers in winter and at high altitudes. The temperature can be automatically controlled/regulated at the optimum level through wired costumes which could keep the soldiers comfortable. Providing these jackets to soldiers would save a large amount of energy compared to bunker/room heating.

(iv) Tagging Soldiers

Radio Frequency Identification (RFID) tagging is one of the most popular applications in the military environment. Planting a microchip under the skin with an RFID detection technology may be an option to tag the soldiers. Identification chips can be made as a part of the wearable device and can be used for monitoring the movement of soldiers in a tactical area, ships and even inside submarines. This could disrupt the fingerprint biometric system.

(v) InfraRed (IR) Shielding Uniform

IR and thermal stealth uniforms can effectively block IR and thermal radiation emitted from the human body. The material used for this shield uses a polyurethane–antimony tin oxide (PU–ATO) composite fiber. By wearing this shield/ uniform, the soldiers will remain stealth from IR and thermal cameras. This will provide added security for soldiers during night operations.

(vi) Conductive Textile

Conductive textiles can be either made by depositing conductive layers onto non-conductive textiles or using conductive fibers. Textiles can be a part of electronic component circuitry that create systems capable of sensing, heating, lighting or transmitting data. Electronics flushed on textile fabrics can sense, compute, communicate, and actuate. Electronic equipment being carried for missions can become an integral part of wearables/ uniforms.

(vii) Battery Charging

Power management is a very critical factor for missions. If footwear/boots can convert the pressure of heels to electrical energy, this energy can be used to power equipment and batteries integrated with soldiers. Sole of the shoe is connected to a power pack fitted outside the shoe and collects kinetic energy using a gear mechanism that converts linear motion generated while walking into rotational motion which would spin a generator which charges the battery continuously. This will be very useful for foot soldiers under patrolling duties along the border.

(viii) Printed Antenna

Carrying a communication device antenna is always a painful activity for foot soldiers. Textile antennas are a special class of antennas that are partially or entirely made out of textile materials, unlike conventional antennas which have rigid materials. The textile antennas have electrically conductive fabrics with radiating and grounding parts and also have dielectric materials for the insulation. These smart textile systems can facilitate body-centric communication between tower and soldiers as well as a soldier to soldier.

(ix) Medical Curing Wearables

Movement of casualty during an evacuation should be made safe and there is a need to move in specially designed warming and IR radiating suits while evacuation to keep the body parts safe and healing. This is purely a medical application

(x) Helmet Mount Display System (HMDS)

HMDS which improves situational awareness with icons and graphics that are visible within the user's Field of View (FOV) like what pilots have while flying fighter aircraft. The Net Warrior (NW) can see-through a helmet-mounted display in limited visibility. Even target acquisition with bore-sighted weapon sight imagery can be integrated into the soldier's FOV. The device can be designed with voice activation using a boom microphone. Even augmented reality can be integrated into the display system. AI can be facilitated generating predictive data on display aiding soldiers for making decisions.

(xi) Helmet Mount Cameras

Helmet mount cameras can be of IR or Night Vision enabled to get a clear vision during night operations. A reverse looking camera across the face can be mounted to view the face of the networked soldier from a remote location. The miniaturization of components associated with camera technology has certainly helped mounting lightweight gadgets to helmets. A rear looking camera coupled to Helmet Mount Display could become a lifesaver for soldiers.

(xii) Communication and Navigation

Blockchain-based secure communication systems with every soldier can be deployed as an added layer of security along with crypto. The current smartwatches integrated with functions currently available in our smartphones with wireless technology similar to 4G/5G communication infrastructure can help the soldiers remain connected from their base station while navigating. Many sensors can be integrated into such wearable devices for centralized monitoring and control.

15.2 SOME TECHNOLOGIES FOR THE MILITARY WEARABLES

(i) Thin and Flexible Heaters in Military Clothing

Stay warm in any weather condition by adding a printed heater to your gloves, pants, jackets, helmets, and boots. Printed heaters use silver and carbon conductive inks and can be heat transferred or sown into almost any kind of fabric.

Safety and mobility are key benefits of printed heaters. Unlike traditional wired heaters, printed PTC heaters do not have any constricting or uncomfortable wires.

Innovative printed flexible heaters are also self-regulating, so there is no need for any outside controls. This is a great feature because they will not overheat, spark, or start any electrical fires.

By adding a printed heater to the military personnel's clothing, troops will be able to withstand the temperature extremes in harsh climates. These heaters also distribute heat evenly without any hot spots. When looking for a reliable form of heat, a printed heater is what is needed for continuous mobility.

(ii) Biometric Sensors for Health Monitoring

Biometric sensors are a perfect solution for remote health monitoring in soldiers. By using conductive inks, sensors are printed onto a film, that can stretch with the clothing. Biometric sensors, with no movement limitations, can detect EKG, ECG, EEG, and other biometric data. Like the printed heaters, the sensors can be added to just about any fabric and incorporated into military gear.

Biometric sensors make it possible to monitor critical patient information from a remote location. The same goes for wearables for military personnel. Soldiers can be monitored while being out in the field. The biometric sensors can be used to track soldiers' vitals such as heart rate or respiratory rate. They also provide remote access to information such as body temperature, which can be very important, especially when military personnel find themselves in dangerous situations.

The important part of the sensor implementation is that it must be applied to some type of compression fabric. Printed biometric sensors are dry electrodes, which means that there is no need for any messy gels, like with traditional biometric sensors. Compression is important so that the sensors are continuously touching the skin, but the sensors would also work inside of helmets that are directly touching the forehead.

However, the most common application would be via the use of compression shirts, shorts, pants, and sleeves. The information being tracked from the sensors can get sent via Bluetooth to a device that a person can easily track the health status of the military personnel while training or in the field. Another great advantage of using flexible printed biometric sensors is that they are waterproof and machine washable.

(iii) Tracking Devices

One of the technology profiles that is showcased on the DoD's website features a monitoring program for the Navy. The profile is similar to that of Radio Frequency Identification or RFID tags. Usually, those kinds of tags are popular with products inside brick-and-mortar retail stores to help with inventory management.

An application for printed antennas in the defense industry is for each tag to be attached to an inventory item, person, or product. Security is a major concern for everyone, especially for the military. The tag can be programmed to sound off an alarm if the tag leaves a restricted room, or could be programmed to sound an alarm if the tag enters a restricted area.

Antennas can be made to send signals to receptors located throughout a building or ship. This way the product/person is being tracked while it is in motion. The printed antenna tag is so small that it can be inserted onto the smallest inventory objects or seamlessly added into clothing.

(iv) Speedy Recovery Time

When it comes to healthcare in the military field, biometric sensors are not the only helpful tool. Imagine the use of a heated bandage that could help with quicker wound healing in harsh conditions. With the use of printed technology, heaters are now able to be printed into bandages to help protect the wound and provide quicker recovery times. The heat can also soothe the muscle to increase movement without pain.

Another innovative application that is ideal in a military setting is using electronic stimulation during a personnel's rehabilitation time. Electronic stimulation, also known as e-stim, is another kind of sensor printed. E-stim sensors can be applied to bandages or compression garments to speed up rehab time. The sensor sends electrical impulses into the muscles to make them contract. This loosens up the muscles after an injury, making the recovery time painless and quicker than without muscle stimulation. E-Stim can also assist with training, by helping the recruits or injured soldiers quickly gain muscle mass.

Another type of sensor is transcutaneous electrical nerve stimulation (TENS). This kind of sensor has two different benefits. When the TENS treatment is used at a lower frequency, pain is eliminated by stimulating sensory nerves to release endorphins. When used at a higher frequency, pain is eliminated by stimulating sensory nerves to block signals trying to get to the brain. Both frequencies are great for reducing pain and increasing movement. Printed sensors are very thin, this way they can easily slip into any first aid kit, backpack, and medical vehicle.

(v) Performance Monitoring

Another use of wearables for military personnel is performance monitoring. Smart clothing with the use of biometric sensors can monitor the performance levels of individuals. One great way of measuring the physical strain of an individual is to track their heart and breathing rates. By implementing biometric sensors into the military personnel's clothing, it is easy to monitor muscle activity, muscle fatigue, and muscle symmetry.

Biometric sensors can also measure speed, acceleration levels, and distance traveled. When monitoring muscle activity, biometric sensors provide real-time data. This would be critical when determining the status of a soldier in the field.

Sensors like capacitive touch and force-sensitive resistors can be added to military devices to make the product more ideal for the environment. Capacitive touch is usually attached to a device instead of being implemented into a wearable.

Force sensing resistors (FSRs) can be added to the insoles of boots to track how effectively the person can run. It can also locate the muscles in the feet that get the most strain. By monitoring the strained muscles with FSRs, injuries can be avoided with proper training to eliminate any strain.

15.3 HOW SEMICONDUCTORS ARE SHAPING WEARABLE DEVICES?

As the use of wearable electronic devices has exploded in recent years, the military sector has taken notice. Wearable devices for military use now include health sensors, printed electronics, and communications systems. Advancements in semiconductor technology have created improved sensing capabilities, allowing wearable sensors to detect subtle changes in a soldier's health and performance in real time. **As semiconductor technology continues to shrink in size, these sensors have become smaller, lighter, and easier to wear.** They can be embedded in clothing, helmets, or wristbands. The US Army is studying how to use wearable sensors to enhance troops' readiness and performance by collecting data over time and adapting training programs accordingly.

Printed electronics technology is allowing devices to shrink even more by printing electronics onto a variety of substrates, including textiles. Printing allows further enhancement of the critical SWaP (size, weight, and power) factor, making devices lighter and easier for soldiers to carry in the field. Over the last few years, the Department of Defense has invested hundreds of millions of dollars into flexible electronics research to develop better wearable devices for military applications.

15.3.1 Wearable devices in military communications

Communications are a critical part of the defense industry. Soldiers must carry many electronic devices to ensure they maintain a clear line of communication throughout the chain of command—but these devices add weight and bulk, impeding movement in the field.

Wearable devices for military use are helping address this challenge. The development of modular load-carrying systems, which can easily attach to vests and backpacks while holding many pieces of equipment, has helped streamline the amount of gear troops must carry. Advancements in semiconductor technology have made communications devices smaller and lighter, helping reduce weight. Advancements in connector design have reduced the number of wires and cables required for military communications, further streamlining each soldier's load.

The next step is integrating communications systems into clothing or backpacks. Some connectors are now being sewn directly into garments, offering features like automatic magnetic latching, auto-alignment, and self-coupling to make it easy to mate and de-mate with one hand.

15.3.2 The challenges of wearable devices

Although wearable devices offer many benefits for military applications, they also come with challenges. The defense industry requires rigorous performance and quality standards; wearable devices must meet standardized requirements while also factoring in comfort and ease of use.

Durability is always a concern in military applications, where devices can be used in tight spaces and dirty environments. Electronics are sensitive to environmental factors like heat, sunlight, dust, rain, and abrasion, which become an even bigger threat when devices are worn directly on the body. Wearable devices for military use must be designed to withstand all of the elements without snagging on clothing or abrading fabric.

15.3.3 How will this impact the global semiconductor market?

Wearable devices aren't the only area in which the military needs semiconductors. Every aspect of national defense, from transportation to cybersecurity, relies on countless semiconductors to perform at the highest standard.

With the recent advancements in wearable devices and electronic warfare, the military is expected to drastically increase its demand for semiconductors. The defense and aerospace semiconductor market is predicted to more than double between 2021 and 2031. The next generation of military equipment will need not only more advanced chips, but higher quantities of them, which could put additional strain on an already volatile market. This is one of the motives behind the CHIPS Act, which aims to strengthen the US semiconductor market by providing federal funding for semiconductor research, design, and production.

The expansion of wearable devices is poised to positively impact soldier training and performance. How the semiconductor industry will meet the rising demand for chips is something a lot of people will be watching.

15.4 ADDRESSING THE LEGAL ISSUES FOR WEARABLE TECHNOLOGY IN INDIA

Wearables serve two main purposes: To track a specific condition, and to help users maintain good health. Wearables can collect a host of data to execute these roles. Some of the collectable data include steps taken, food and water intake, calories burned, sleep movement, and breathing. Wearable technology can also collect biometric data such as heart rate (ECG and HRV), brainwave (EEG), and muscle bio-signals (EMG) from the human body to provide valuable information in the field of health care and wellness.

This acts as the starting point for the myriad of **legal issues** that arise in the usage of this technology. In this point of discussion, there are two scenarios that emerge: i) while one makes use of wearable device there is a lot of exchange of data between the user and the wearable device. This raises lots of questions regarding the ownership of such data; ii) whether such data collected in a wearable device belongs to the wearable device manufacturer or the service provider providing the said service regarding operations and maintenance of the device. Data protection is the crucial aspect that needs to be considered with regard to wearable technology and devices – this includes issues regarding the data collection and processing in a wearable device. Furthermore, there is sharing of sensitive and personal data between the device, user, and several third parties.

In view of this, deciding the jurisdiction of the data processing and storage of a wearable device is another issue to look into – situations such as i) location of data storage; ii) location of data processing and sharing; and iii) the recipient of such information shall be integral in deciding the jurisdiction of the data, which will then be subject to the laws of the same.

15.4.1 Laws and Regulations

(i) In the case of S. Puttaswamy vs. Union of India, the Supreme Court held that the Right to Privacy is a fundamental right protected under Article 14, 19 and 21 of the Constitution of India. Users of wearable devices and social media networks may not conceive of themselves as having volunteered data but their activities of use and engagement result in the generation of vast amounts of data about individual lifestyles, choices, and preferences. Data such as medical information would be a category to which a reasonable expectation of privacy attaches. This depicts that the Court believes in securing medical information about an individual and it is reasonable for an individual to expect its right to privacy over violation by anyone.

(ii) The Personal Data Protection Bill 2019, was introduced in Lok Sabha on 11th December 2019 and referred to the Standing Committee of Information Technology of Parliament on the same day itself. Its main purpose is to tackle issues regarding data privacy in this digital world. It is applicable in India as well as outside India where foreign companies are dealing with the data security of Indian individuals. The Bill enumerated the rights of an individual, grounds for processing personal data, defined steps for the data protection authorities for tackling data protection issues and penalties on violations and defined the social media intermediaries. It also ensures that the data collected by the device should be consensual and should be stored fairly and reasonably having transparency to the customers.

(iii) The Ministry of Family and Health Welfare issued the Digital Information Security in Healthcare Act 2018also known as DISHA. This legislation would cover the aspects regarding the rights of data owners and restricting the collection and processing of health data. It has also established a panel of digital health authorities consisting of a National Authority and various State authorities. It aims at the data collected at medical institutes on a consensual basis but it also deals with other ways in which medical data gets generated.

(iv) The Information Technology Act 2000 is the main legislation that regulates the issues regarding data protection in India. It is the first legislation that addressed the issues and in regard to privacy, established a notice and consent model. It also prescribed penalties to the offences related to breach of data privacy. Under the Act, it is the Central Government that decides as to what type of data comes under sensitive personal data. In addition to the legislation, the Government also issued Information Technology (Reasonable security practices and procedures and sensitive personal data or information) Rules 2011 under Section 87 of the IT Act which laid down various measures for data collectors as well as for data processors in regard to data protection. These rules play a major role because it created rules regarding medical records under the shelter of Sensitive Personal Data and also included the physical, physiological and mental health conditions. This eventually depicts that the data collected by fitness trackers are regulated under these rules.

(v) The Electronic Health Record Standards 2016, are advisory but not enforceable by law. Many companies use them in their policies regarding the data protection of their customers. By considering the recommendations of these standards, they are also applicable to fitness trackers as they are recognized as self-care health devices. These standards also imply the consent model where there should be a consent of patient whether his/her medical information can be disclosed to a third party or not. It also includes recommendations concerning ownership of medical data, data security standards, prescribe to follow ISO standards etc.

15.4.2 Assessing the Indian Legal Scenario

While wearable devices make use of exchange of data between users and them, this data is ultimately nothing but information in electronic form and can be considered as an electronic record under the Information Technology Act, 2000. Further, these wearable devices collect data and provide a service with respect to that data turning themselves to be referred to as 'intermediaries' under the IT Act (as mentioned under Section 79). It is pertinent to note that the Intermediary guidelines along with the cyber legal compliance envisaged under the IT Act and the rules and regulations made under shall be made applicable to wearable devices and technology.

In case of the Indian context, in terms of data stored and analyzed at the hands of a wearable device – currently, there are a couple of landmark judgments that highlight the individual's right to privacy – Puttaswamy and Puttaswamy II, among others. However, there is a lack of a data protection legislation, which shall uphold users' rights in the course of the introduction of new technological initiatives and schemes. At this juncture, the Data Protection Bill, 2019 is the first step into this direction. The need for an established legislation is imminent due to the fact that that there have been several concerns and controversies around the assimilation of personal data at the hands of the government (reference is being made towards issues such as the Aadhar Card controversy, the collection of data in the National Digital Health Mission, the deployment of FRTs in law enforcement agencies, the Arogya Setu application, the DISHA scheme etc.) – a similar situation has been observed towards activities taken up by private entities as well, and due to the absent adequate legislation(s), they do not fall under the radar of any defaults/punishments.

Considering this, the privacy of an individual in the usage of wearable technology becomes essential due to the amount of information that is taken up by such devices, and potential of trouble it holds during a breach. At this point, companies will have to adopt focal thrust strategies and policies in the light of wearable devices – although there exists a lacuna in the law, judicial precedents do not leave an individual/entity completely handicapped in this regard. Another pertinent point to note is as follows: while it is just not the privacy relating to wearable devices that needs to be considered, but its inter-operability with respect to the relevant mobile/computer applications also need to be investigated, since these wearable devices are connected to such mobile apps. It is pertinent to note that wearable technology could also be used for surveillance and monitoring by both state and non-state actors.

SECTION-V

DISRUPTIVE TECHNOLOGIES FOR MODERN AND FUTURE WARFARE

CHAPTER 16

GENERATIVE AI IN WARFARE

16.1 WHAT IS GENERATIVE ARTIFICIAL INTELLIGENCE (GENERATIVE AI)?

Generative artificial intelligence, also known as Gen AI, is a type of AI that can create new content and ideas, including conversations, stories, images, videos, and music. It can learn human language, programming languages, art, chemistry, biology, or any complex subject matter. It reuses what it knows to solve new problems. For example, it can learn English vocabulary and create a poem from the words it processes.

16.1.1 Generative AI Applications

(i) Financial services

Financial services companies use generative AI tools to serve their customers better while reducing costs.

- Financial institutions use chatbots to generate product recommendations and respond to customer inquiries, which improves overall customer service.
- Lending institutions speed up loan approvals for financially underserved markets, especially in developing nations.
- Banks quickly detect fraud in claims, credit cards, and loans.
- Investment firms use the power of generative AI to provide safe, personalized financial advice to their clients at a low cost.

(ii) Healthcare and life sciences

One of the most promising generative AI use cases is accelerating drug discovery and research.

- Generative AI can create novel protein sequences with specific properties for designing antibodies, enzymes, vaccines, and gene therapy.
- Healthcare and life sciences companies use generative AI tools to design synthetic gene sequences for synthetic biology and metabolic engineering applications.
- For example, Gen AI can create new biosynthetic pathways or optimize gene expression for biomanufacturing purposes.

- Generative AI tools also create synthetic patient and healthcare data. This data can be useful for training AI models, simulating clinical trials, or studying rare diseases without access to large real-world datasets.

(iii) Automotive and manufacturing

- Automotive companies use generative AI technology for many purposes, from engineering to in-vehicle experiences and customer service. For instance, they optimize the design of mechanical parts to reduce drag in vehicle designs or adapt the design of personal assistants.
- Auto companies use generative AI tools to deliver better customer service by providing quick responses to the most common customer questions. Generative AI creates new materials, chips, and part designs to optimize manufacturing processes and reduce costs.
- Another generative AI use case is synthesizing data to test applications. This is especially helpful for data not often included in testing datasets (such as defects or edge cases).

16.1.2 Generative AI benefits

(i) Content creation

Generative AI can assist in generating compelling and personalized content for marketing, advertising, and social media campaigns. It can create engaging visuals, write persuasive content, and even compose music or design graphics.

(ii) Virtual assistants

Generative AI-powered virtual assistants can provide personalized and interactive support to employees. They can answer queries, schedule appointments, automate tasks, and even engage in natural language conversations.

(iii) Creative design

Generative AI tools can aid designers in creating unique and innovative designs. They can generate variations of design elements, layouts, or color schemes, providing inspiration and saving time in the creative process.

(iv) Language translation

Generative AI models can be used for language translation, enabling businesses to communicate effectively across different languages and cultures. They can provide real-time translation services, improving global collaboration and customer support.

(v) Data analysis

Generative AI techniques can be employed for data analysis and anomaly detection. They can identify patterns, outliers, and potential risks in complex datasets, enhancing data-driven decision-making.

(vi) Accelerates research

Generative AI algorithms can explore and analyze complex data in new ways, allowing researchers to discover new trends and patterns that may not be otherwise apparent. These algorithms can summarize content, outline multiple solution paths, brainstorm ideas, and create detailed documentation from research notes. This is why generative AI drastically enhances research and innovation. For example, generative AI systems are being used in the pharma industry to generate and optimize protein sequences and significantly accelerate drug discovery.

(vii) Enhances customer experience

Generative AI can respond naturally to human conversation and serve as a tool for customer service and personalization of customer workflows. For example, you can use AI-powered chatbots, voice bots, and virtual assistants that respond more accurately to customers for first-contact resolution. They can increase customer engagement by presenting curated offers and communication in a personalized way.

(viii) Optimizes business processes

With generative AI, your business can optimize business processes utilizing machine learning (ML) and AI applications across all lines of business. You can apply the technology across all lines of business, including engineering, marketing, customer service, finance, and sales.

For example, here's what generative AI can do for optimization:

- Extract and summarize data from any source for knowledge search functions.
- Evaluate and optimize different scenarios for cost reduction in areas like marketing, advertising, finance, and logistics.
- Generate synthetic data to create labeled data for supervised learning and other ML processes.

(ix) Boosts employee productivity

Generative AI models can augment employee workflows and act as efficient assistants for everyone in your organization. They can do everything from searching to creation in a human-like way.

Generative AI can boost productivity for different kinds of workers:

- Support creative tasks by generating multiple prototypes based on certain inputs and constraints. It can also optimize existing designs based on human feedback and specified constraints.
- Generate new software code suggestions for application development tasks.
- Support management by generating reports, summaries, and projections.
- Generate new sales scripts, email content, and blogs for marketing teams.
- You can save time, reduce costs, and enhance efficiency across your organization.

16.1.3 How did generative AI technology evolve?

Primitive generative models have been used for decades in statistics to aid in numerical data analysis. Neural networks and deep learning were recent precursors for modern generative AI. Variational autoencoders, developed in 2013, were the first deep generative models that could generate realistic images and speech.

(i) VAE

VAEs (variational autoencoders) introduced the capability to create novel variations of multiple data types. This led to the rapid emergence of other generative AI models like generative adversarial networks and diffusion models. These innovations were focused on generating data that increasingly resembled real data despite being artificially created.

(ii) Transformers

In 2017, a further shift in AI research occurred with the introduction of transformers. Transformers seamlessly integrated the encoder-and-decoder architecture with an attention mechanism. They streamlined the training process of language models with exceptional efficiency and versatility. Notable models like GPT emerged as foundational models capable of pretraining on extensive corpora of raw text and fine-tuning for diverse tasks.

Transformers changed what was possible for natural language processing. They empowered generative capabilities for tasks ranging from translation and summarization to answering questions.

16.1.4 How does generative AI work?

Like all artificial intelligence, generative AI works by using machine learning models—very large models that are pre-trained on vast amounts of data.

(i) Foundation models

Foundation models (FMs) are ML models trained on a broad spectrum of generalized and unlabeled data. They are capable of performing a wide variety of general tasks. FMs. are the result of the latest advancements in a technology that has been evolving for decades. In general, an FM uses learned patterns and relationships to predict the next item in a sequence. For example, with image generation, the model analyzes the image and creates a sharper, more clearly defined version of the image. Similarly, with text, the model predicts the next word in a string of text based on the previous words and their context. It then selects the next word using probability distribution techniques.

(ii) Large language models

Large language models (LLMs) are one class of FMs. For example, OpenAI's generative pre-trained transformer (GPT) models are LLMs. LLMs. are specifically focused on language-based tasks such as such as summarization, text generation, classification, open-ended conversation, and information extraction.

What makes LLMs. special is their ability to perform multiple tasks. They can do this because they contain many parameters that make them capable of learning advanced concepts.

An LLM like GPT-3 can consider billions of parameters and has the ability to generate content from very little input. Through their pretraining exposure to internet-scale data in all its various forms and myriad patterns, LLMs. learn to apply their knowledge in a wide range of contexts.

16.2 MILITARY APPLICATIONS OF GENERATIVE AI

(i) Generative AI can improve military training and educational programs

AI-powered language models can read training manuals and other sources and use them to create new training materials, including notes, quizzes, and study guides. AI can also help evaluate students' current abilities and tailor training to their specific needs. With NLP, generative AI can answer students' questions and explain concepts similar to the way that a human instructor would. By analyzing large amounts of intelligence data, records of previous combat experiences, and more, AI can craft more comprehensive training, including detailed military simulations. Conversational AI can also provide customized feedback to help students build their skills and help commanding officers know where a particular student may be struggling.

While AI holds a lot of potential for military training applications, it should never entirely replace human instructors. To avoid issues like bias or misinformation, leadership should always review AI-generated materials and be in charge of the ultimate analysis of students' skills. Human instructors should determine the overall syllabus, while AI can craft individualized lessons that human instructors can then review for accuracy and other issues. However, with AI assistance, instructors can create and administer more effective training programs, because of individualized attention to students that humans may not be able to always provide, and do so more quickly due to AI's processing speed.

(ii) Leveraging LLM in Military Simulations

A recently developed naval wargaming simulation, Fleet Emergence, leverages state-of-the-art LLM together with ACI architecture. The sophistication of this simulation lies in the intricate scenarios the LLM can create, as well as its ability to generate realistic communications and to respond in a manner similar to the way real-life adversaries would.

One of the most important details about combat simulation is that it is far safer than reality. Many casualties can occur from training with real weapons and situations. This allows soldiers to experience the best simulation of the realities of warfare, without being endangered. These virtual realities can aid soldiers in understanding how to handle clones of weapons just like their real-life counterparts, make decisions in stressful situations, as well as work with their teammates. The training software can prepare soldiers for just about anything, and can save them in the long run. Not only can AI-based simulation train soldiers, but it can personalize training programs, as well as make fair assessments in order to make future adjustments to the programs. Combat simulation can also save time and money due to being more efficient at certain tasks than humans are. Check out our innovative AI model, Strat Agent, which acts as a modern-day battlefield commander that can be used in combat simulation.

(iii) Target Recognition

Artificial intelligence can aid in making target recognition more accurate in combat environments. AI can improve the ability for systems like this to identify the position of their targets. It can also allow defense forces to acquire a detailed understanding of an operation area by examining reports, documents, news, and other forms of information, aggregating and analyzing these sources much more quickly than humans would be able to do so. With generative AI's conversational abilities, there can be a two-way discussion about this information, so military decision makers can ask questions to make sure the most relevant information comes to the surface. AI systems have the ability to predict enemy behavior, anticipate vulnerabilities, weather and environmental conditions, assess mission strategies, and suggest alleviation plans. This can save time and human resources, putting soldiers a step ahead of their targets, but as always requires humans to make the ultimate decision.

(iv) Threat Monitoring

Threat monitoring, as well as situation awareness uses operations that gain and analyze information to aid in many different military activities. There are unmanned systems that can be remotely controlled or sent on a pre-calculated route. These systems use AI in order to aid defense personnel in monitoring threats, and thus leveraging their situational awareness. Drones with AI can also be used in these situations. They can monitor border areas, recognize threats, and alert response teams. Additionally, they can strengthen the security of military bases, as well as increase the safety of soldiers in combat.

(v) Cybersecurity

Even highly secure military systems can be vulnerable to cyber-attacks, which is where AI can be of great help. Attacks can put classified information at risk, as well as damage a system altogether, which can endanger military personnel and jeopardize the mission. AI has the ability to protect programs, data, networks, and computers from persons not authorized to access them. AI also has the skills to study patterns of cyber-attacks and form protective strategies in order to fight against them. These systems can recognize the smallest behaviors of malware attacks far before they enter a network.

Generative AI's analysis, scenario generation, and communication capabilities can also improve cybersecurity in military settings. With the ability to analyze large quantities of data and find patterns, generative AI can detect potential threats and use predictive analytics to help predict future attacks. At the same time, generative AI also poses its own threats in the wrong hands, such as the concern that attackers can leverage the power of generative models for social engineering. The military will also have to take care to counteract this possibility as well, with consistently updated training and mitigation plans. When applied with care and close supervision, generative AI can enhance cyber defense, even for critically important military applications.

As it does in many other areas, advanced AI has a mixed impact on cybersecurity. Functions such as the ability to write malware may make AI dangerous in the hands of bad actors, but AI can also help to detect and mitigate these threats. In essence, the military applies AI to counter adversaries who may also have access to AI. This means that it is critical for the military to have access to the most advanced and tailored AI cybersecurity solutions in order to stay safe amid a constantly evolving landscape of AI-driven cybersecurity risks.

(vi) Transportation

AI is able to play a role in the transportation of ammunition, goods, armaments, and troops. The logistics and transportation of these things is obviously vital to the success of military operations. AI can lower transportation costs and reduce the need for human input by, for example, plotting the most efficient route to travel under current conditions. It can also pre-identify problems for military fleets in order to increase efficiency of their performance. As the combination of innovation in computer vision and autonomous decision making over time also continues to bring self-driving vehicles closer to common use in the commercial space, this technology may also prove useful in the military context.

(vii) Casualty Care and Evacuation

Because soldiers and medics have to make decisions in high-stress situations, AI is able to aid them when a fellow service member may need help. The dangerous environment of the battlefield presents a lot of obstacles to providing medical treatment to the wounded. The assistance of AI in a highly emotional environment can help with analyzing the situation and suggesting the best course of action. This advice can help humans to base their split-second decisions on information provided by an analytical rather than emotional mind. However, obviously, AI cannot make these decisions for humans, precisely because AI does not have an understanding of the emotional and contextual factors involved in a life-or-death situation.

This type of AI uses an algorithm and large medical database that is able to access data containing medical trauma cases, which include diagnoses, vital sign sets, medications given, treatments, and outcomes. It then takes this data, combined with manually entered information in order to provide indications, warnings, and suggestions for treatment. This is another situation in which AI needs human guidance in order to operate effectively; while the AI will make recommendations without emotional considerations as a hindrance, humans must use their emotional abilities to make appropriate decisions that take into account these recommendations. AI is not qualified to make medical decisions but it can provide rapid analysis to give humans more information on which to base their decisions.

16.3 HOW AI WILL REVOLUTIONIZE WARFARE?

The latest AI—known as generative pre-trained transformers (GPT)—promises to utterly transform the geopolitics of war and deterrence. It will do so in ways that are not necessarily comforting, and which may even turn existential.

On the one hand, this technology could make war less lethal and possibly strengthen deterrence. By dramatically expanding the role of AI-directed drones in air forces, navies and armies, human lives could be spared. Already, the U.S. Defense Department is experimenting with AI bots that can fly a modified F-16 fighter jet, and Russia has been testing autonomous tank-like vehicles. China is rushing to roll out its own AI-run systems, and the effectiveness of armed drones will also take off in coming years. One of the largest, although still nascent, efforts to advance AI is a secretive U.S. Air Force program, Next Generation Air Dominance, under which some 1,000 drone "wingmen," called collaborative combat aircraft, operate alongside 200 piloted planes.

On the other hand, AI-driven software could lead the major powers to cut down their decision-making window to minutes instead of hours or days. They could come to depend far too much on AI strategic and tactical assessments, even when it comes to nuclear war. The danger is that decision-makers could gradually rely on the new AI as part of command and control of weaponry, since it operates at vastly greater speeds than people can.

The real problem is how little it takes to convince people that something is sentient, when all GPT amounts to is a sophisticated auto-complete. Given AI's propensity to hyperbole when people start to believe that machines are thinking, they're more likely to do crazy things.

In a report published in early February, the Arms Control Association said AI and other new technologies, such as hypersonic missiles, could result in blurring the distinction between a conventional and nuclear attack. The report said that the scramble to "exploit emerging technologies for military use has accelerated at a much faster pace than efforts to assess the dangers they pose and to establish limits on their use. It is essential, then, to slow the pace of weaponizing these technologies, to carefully weigh the risks in doing so, and to adopt meaningful restraints on their military use."

U.S. officials have said they are doing so, but they may be navigating a slippery slope. This January, the Defense Department updated its directive on weapons systems involving the use of artificial intelligence, saying that at least some human judgment must be used in developing and deploying autonomous weapon systems. At the same time, however, the Pentagon is experimenting with AI to integrate decision-making from all service branches and multiple combatant commands. And with the Biden administration cracking down on high-tech exports to China, especially advanced semiconductors, in order to maintain the current U.S. lead in AI, the Pentagon is likely to accelerate those effort

While the Defense Department was eagerly pursuing integration of AI capabilities, this would definitely not include nuclear command and control. There should be a role for AI in determining how to use lethal force—once a human decision is made

Better Tech Regulations Can Save Democracy

Another concern is that advanced AI technology could allow rogue actors such as terrorists to gain knowledge in building dirty bombs or other lethal devices. And AI is now shared by far more actors than during the Cold War, meaning that it could be used to detect nuclear arms sites, reducing the deterrent effect of keeping their locations secret. AI will change the dynamic of hiding and finding things that much of the data today is held by private companies that might be vulnerable to AI-driven espionage and probing of weapons systems.

The idea that governments are going to sanction a delay for safety's sake is unlikely in the extreme. This is not only because the world's biggest tech companies are engaged in vicious competition, especially in Silicon Valley, but also because the new technology is being rolled out in an international environment in which the U.S., China, and Russia are now embroiled in a grim struggle for dominance.

As far back as 2017, though, Russian President Vladimir Putin declared that "the one who becomes the leader in this sphere [AI] will be the ruler of the world," and that future wars will be decided "when one party's drones are destroyed by drones of another."

DARPA—the Pentagon's Defense Advanced Research Projects Agency—is conducting a wide-ranging research program called AI-Forward. DARPA spokesman Matt Turek said recently that generative AI systems such as ChatGPT have raised serious new issues about "the level of reliability that we probably need for life-or-death decision-making."

16.4 GENERATIVE AI FOR INDIAN MILITARY AND AEROSPACE

Generative AI has opened up a whole lot of applications for defence and aerospace in India

(i) Image and Video Generation

Deep learning models like Generative Adversarial Networks (GANs) and Variational Autoencoders (VAEs) can be used for military-specific images and videos. These generative models can create synthetic visuals, which can be valuable for data augmentation, simulation and training purposes. AI-generated images and videos can be used for deception operations or disinformation campaigns to mislead adversaries and protect their own sensitive information. These models can help generate virtual training environments, target and threat simulation and even create synthetic satellite images, aerial footage or surveillance data to simulate various intelligence and reconnaissance scenarios. It can generate synthetic video streams for autonomous navigation, target tracking and even for situational awareness. It can even develop camouflage patterns and techniques to improve the concealment of military assets as well as create synthetic visual deception strategies to confuse adversaries.

(ii) Data Augmentation

Generative AI can be used to augment datasets for training machine learning models used in various military applications such as object recognition, target identification and autonomous systems. This process will enhance the diversity and quantity of training data available for machine learning models in various military applications which would improve the performance and robustness of machine learning algorithms by artificially expanding the dataset. This is inevitable where the collection of large amounts of real-world data is challenging or risky. The data sets can be images, audio clips, videos or even texts.

(iii) Strategic and Operational Planning

Generative AI can aid in generating diverse scenarios and predicting potential outcomes during strategic planning, war gaming, and decision-making processes. Models can be employed to simulate various military scenarios, taking into account factors like terrain, weather conditions, enemy positions and friendly forces. Simulations can help military planners explore potential outcomes and identify optimal strategies. AI can also generate alternative courses of action for military operations and air combat based on available data and historical patterns. This can provide military planners with a broader range of options for developing strategic plans. Models can even help to assess risks and allocation of optimal resources for operational tasks.

By analyzing historical data and current intelligence, generative AI can offer predictive insights into potential enemy actions and responses, helping military planners anticipate and prepare for various contingencies. Gen AI, combined with natural language processing, can assist in analysing large volumes of intelligence data like reports, satellite imagery and intercepted communications to identify patterns and trends relevant to strategic planning. It would facilitate integration across different military domains, such as land, sea, air, space, and cyber, allowing military planners to develop comprehensive and coordinated strategies.

(iv) Simulation and Training

Generative AI can create highly realistic and dynamic training environments for military missions. This includes generating virtual battlefields, adversaries and scenarios to train soldiers/pilots to improve their tactical decision-making skills. These models can be used to create realistic and diverse training scenarios like terrain conditions and even enemy behavior. It can create environments that can simulate complex, dynamic scenarios which allow military personnel/pilots to practice decision-making and tactical skills in a safe and controlled setting. Synthetic data can be created for training machine learning models that aid military applications like target recognition and intelligence analysis. It can even simulate enemy tactics and strategies based on historical data or own expert knowledge which would facilitate developing counter-strategies.

(v) Autonomous Systems

Generative AI can be employed to enhance the capabilities of autonomous systems like UAVs, autonomous underwater vehicles and ground vehicles, enabling them to adapt to changing environments and scenarios. Generative AI is used to create realistic virtual environments, where autonomous systems can undergo training and testing. These simulations enable the platforms to practice navigating complex airspace and terrains, encountering diverse scenarios and learning from virtual challenges without any risks associated with real-world training. These models can enhance the algorithms' ability to recognize and interpret real-world visuals. It can help autonomous systems predict and adapt to various environmental conditions and obstacles in real time, improving their ability to navigate complex terrains and unknown airspace. Any deviations from these learned patterns can also be detected. By learning from a wide range of simulated scenarios, these models can generate adaptive strategies for dynamic environments and mission objectives. Models can simulate realistic adversary behavior and tactics for autonomous systems, which can be taught to anticipate and respond to potential threats in real time. These models enable adaptive decision-making and the ability to respond to changing environments and threats.

(vi) Natural Language Generation (NLG)

NLG can be utilized to create automated reports, summaries and analyses from vast amounts of raw intelligence data, assisting military analysts in decision-making and situational awareness. NLG models can be used to develop machine translation systems for real-time communications between different languages. This is particularly useful for military operations and intelligence analysis during joint exercises. Sentiment Analysis helps military personnel understand public perception, assess the impact of operations, and monitor sentiment within the ranks. Named Entity Recognition (NER) assists in intelligence analysis, tracking key individuals and understanding the affiliations of various entities. NLG helps military commanders generate intelligent reports for quick eyes. Chatbots for assisting soldiers with expert systems, Speech-to-Text and Text-to-Speech conversion wherever necessary including multilingual communications, threat analysis and fake news detection are other possible NLG applications.

(vii) Deception and Stealth Technology

AI-generated models can assist in designing and optimizing stealth technology, helping military aircraft and vessels evade detection by radar and other sensors. The AI models could generate and evaluate numerous design variations, seeking shapes, materials, and configurations that minimize radar cross-sections, infrared signatures and underwater sonar signatures while maintaining performance requirements. Generative AI could be utilized to create camouflage patterns that blend seamlessly with various environmental backgrounds, making it difficult for adversaries to detect and identify military assets. Generative AI might be employed to generate decoy signals or signatures that mimic real military assets. These decoys could divert enemy attention and confuse their targeting systems. Models could be used to optimize communication protocols and strategies for stealthy communications, making it harder for adversaries to intercept or detect military communications. Generative AI might assist in designing and optimizing electronic warfare systems that jam or disrupt enemy sensors and communications.

(viii) Electronic Warfare

By generating signals and emulating various communication and radar signals, generative AI can be used to test and improve electronic warfare systems. Generative AI can be used to optimize electronic warfare systems that perform jamming of enemy communication and radar signals. They can analyze signal characteristics and adapt jamming strategies in real time to counter evolving enemy countermeasures. Models can process and analyze large amounts of electronic signals, classifying them based on the type of emitter, frequency, modulation, and other characteristics. This helps in identifying potential threats and determining the appropriate response. It can assist in managing the electromagnetic spectrum efficiently by predicting and preventing congestion, interference, and overlap of friendly signals. It can even create realistic decoy signals to mimic friendly or enemy electronic emissions, confusing adversaries and making it harder for them to identify real targets.

Gen AI can be used for Cyber Electronic Warfare (CEW) by creating synthetic cyber-attacks and responses, enhancing the understanding of cyber threats and defense. A large amount of ELINT data comprising intercepted radar signals and communication intelligence can be analyzed to extract meaningful information and identify patterns relevant to electronic warfare operations. Generative AI can enable electronic warfare systems to adapt to changing airspace scenarios/battlefield conditions, learn from past engagements and adjust their strategies accordingly. It can also be used to develop Electronic Counter-Countermeasures (ECCM) to counter the adversaries to neutralize the ECM efforts.

Even signal denial and deception can be done for adversaries to detect friendly signals or identify the location of enemies' electronic warfare assets.

(ix) Predictive Maintenance

AI can generate predictive models to forecast equipment failures and maintenance requirements, ensuring optimal military asset readiness and availability. It can analyze historical sensor data from equipment and systems to learn normal behavior patterns with the new data. It can also detect anomalies or deviations from expected behavior, indicating potential equipment malfunctions or impending failures in advance. Gen AI can process real-time sensor data from military assets and platforms such as temperature, pressure, vibration and other parameters. Models can predict potential failure modes and patterns based on historical failure data. By understanding the common failure scenarios, maintenance strategies can be tailored to address these specific issues proactively. AI can even estimate the remaining useful life/residual life of critical components and equipment. It can simulate different operating conditions and scenarios to forecast future equipment performance and predict maintenance requirements accurately. Models can even augment sensor data to mimic real-world conditions.

Optimization of Maintenance Schedules can be done by Generative AI by analyzing historical maintenance data and equipment performance to optimize maintenance schedules. It helps determine the most cost-effective time to perform maintenance activities while minimizing disruptions to operations. It can assist in Root Cause Analysis(RCA) as well as Prognostics and Health Management (PHM) for aircraft and military assets.

(x) Cybersecurity

Generative AI can be also used to simulate cyber-attacks and assess the resilience of military networks, systems and software against various threat scenarios. Generative Adversarial Networks (GANs) can be used to generate adversarial examples that can help in evaluating and testing the robustness of machine learning models. Can be trained for intrusion detection, anomaly detection, and malware detection. We can even generate realistic honey-pots and deception techniques to divert attackers' attention and gather information about their tactics, techniques, and procedures. Models can analyze, detect and predict phishing emails or messages and alert users and military personnel about potential social engineering attacks.

CHAPTER 17

EDGE COMPUTING IN WARFARE

17.1 INTRODUCTION TO EDGE COMPUTING

Edge computing is a distributed computing model that brings computation and data storage closer to the sources of data. More broadly, it refers to any design that pushes computation physically closer to a user, so as to reduce the latency compared to when an application runs on a centralized data centre. Edge computing is a distributed information technology (IT) architecture in which client data is processed at the periphery of the network, as close to the originating source as possible.

The term began being used in the 1990s to describe content delivery networks—these were used to deliver website and video content from servers located near users. In the early 2000s, these systems expanded their scope to hosting other applications, leading to early edge computing services. These services could do things like find dealers, manage shopping carts, gather real-time data, and place ads.

Edge computing involves running computer programs that deliver quick responses close to where requests are made. Edge computing is computing that occurs outside the cloud, at the network's edge, particularly for applications needing immediate data processing.

In simplest terms, edge computing moves some portion of storage and compute resources out of the central data center and closer to the source of the data itself. Rather than transmitting raw data to a central data center for processing and analysis, that work is instead performed where the data is actually generated — whether that's a retail store, a factory floor, a sprawling utility or across a smart city. Only the result of that computing work at the edge, such as real-time business insights, equipment maintenance predictions or other actionable answers, is sent back to the main data center for review and other human interactions.

Thus, edge computing is reshaping IT and business computing. Take a comprehensive look at what edge computing is, how it works, the influence of the cloud, edge use cases, tradeoffs and implementation considerations.

The world's data is expected to grow by 61 percent to 175 zettabytes by 2025.

[A zettabyte (ZB) is a unit of measurement for the storage capacity of devices and computers, and is equal to: 1,000 exabytes, 1 billion terabytes, 1 trillion gigabytes, 2 to the 70th power, and 1 sextillion bytes. To put this into perspective, a zettabyte is equivalent to the amount of data on 250 billion DVDs. It would take about a million supercomputers or a billion powerful home computers to store a zettabyte of data.]

Around 10 percent of enterprise-generated data is created and processed outside a traditional centralized data center or cloud. By 2025, it is predicted that this figure will reach 75 percent. The increase in IoT devices at the edge of the network is producing a massive amount of data — storing and using all that data in cloud data centers pushes network bandwidth requirements to the limit. Despite the improvements in network technology, data centers cannot guarantee acceptable transfer rates and response times, which often is a critical requirement for many applications. Furthermore, devices at the edge constantly consume data coming from the cloud, forcing companies to decentralize data storage and service provisioning, leveraging physical proximity to the end user.

In a similar way, the aim of edge computing is to move the computation away from data centers towards the edge of the network, exploiting smart objects, mobile phones, or network gateways to perform tasks and provide services on behalf of the cloud. By moving services to the edge, it is possible to provide content caching, service delivery, persistent data storage, and IoT management resulting in better response times and transfer rates.

The distributed nature of this paradigm introduces a shift in security schemes used in cloud computing. In edge computing, data may travel between different distributed nodes connected through the Internet and thus requires special encryption mechanisms independent of the cloud. Edge nodes may also be resource-constrained devices, limiting the choice in terms of security methods. Moreover, a shift from centralized top-down infrastructure to a decentralized trust model is required. On the other hand, by keeping and processing data at the edge, it is possible to increase privacy by minimizing the transmission of sensitive information to the cloud. Furthermore, the ownership of collected data shifts from service providers to end-users.

Edge computing is all a matter of location. In traditional enterprise computing, data is produced at a client endpoint, such as a user's computer. That data is moved across a WAN such as the internet, through the corporate LAN, where the data is stored and worked upon by an enterprise application. Results of that work are then conveyed back to the client endpoint. This remains a proven and time-tested approach to client-server computing for most typical business applications.

But the number of devices connected to the internet, and the volume of data being produced by those devices and used by businesses, is growing far too quickly for traditional data center infrastructures to accommodate. As stated earlier that by 2025, 75% of enterprise-generated data will be created outside of centralized data centers. The prospect of moving so much data in situations that can often be time- or disruption-sensitive puts incredible strain on the global internet, which itself is often subject to congestion and disruption.

So IT architects have shifted focus from the central data center to the logical edge of the infrastructure — taking storage and computing resources from the data center and moving those resources to the point where the data is generated. The principle is straightforward: If you can't get the data closer to the data center, get the data center closer to the data. The concept of edge computing isn't new, and it is rooted in decades-old ideas of remote computing — such as remote offices and branch offices — where it was more reliable and efficient to place computing resources at the desired location rather than rely on a single central location.

17.2 EDGE COMPUTING FOR NATIONAL SECURITY AND MILITARY

While the government and the military are innovative in many ways, they don't like to live on the edge when it comes to information technology. Stable, reliable, and secure are the name of the game. But in order to support the most advanced technology, such as next-gen warfare and highly complex supply chains, agencies need the power of edge computing with more choices that go beyond proprietary vertical stacks. Edge computing eliminates the dependencies of on-premise infrastructure and decouples devices in any form from the data center, providing powerful, always available computing wherever the user is—from our communities to our borders to the battlefield.

Imagine the real-time connectivity that we're used to in our social and personal lives and applying it to federal and military use cases. Instead of hauling around a 60-pound server and battery pack, which is not feasible, today's low-powered devices such as an assault vehicle, border patrol agents' watch, soldier's backpack, or first responder's phone, are lightweight and just as powerful, able to run applications where your mission is taking place.

Today's sensor-based devices collect and analyze the data, then return it to you at the "edge," rather than sending it back to the data center or your cloud environment for processing.

Edge computing enables faster decision making because processing is done where the data is actually consumed, with real-world applications such as:

- Secure communications on combat vehicles and for command and control;
- Data collection and analysis at our borders or in communities for infectious diseases;
- On-site claims processing in natural disasters;
- Detection of supply chain disruptions;
- High-volume transportation optimization;
- Situational awareness in fighting forest fires, in law enforcement, and in coastal waters protection; and
- Comprehensive soil, weather, pesticide, and other data for advanced agricultural strategies.

By decoupling computing from traditional or virtual infrastructure, we can extend compute power and processing to the front lines or "far edge" and allow for customization that is not available in the cookie-cutter virtual cloud experience. When necessary, we can also operate at the "near edge" with a base camp of sorts—multiple processing components grouped between the consumer and the data center.

By centralizing on open-source software and focusing on the application level, IT leaders can stop doing one-off installations and development and have the flexibility to use any cloud, whether it's running on a computer or on a spy plane. A mission-critical application developed to run in a data center in a stateless manner (with no session or status data needed) can just as easily be pushed to the tactical edge. You can build once and deploy anywhere with embedded security and compliance, and quickly scale up or down as needed.

17.3 RISE OF EDGE COMPUTING IN THE MILITARY

Edge computing advances life-saving possibilities for warfighters and the defense community. Thanks to edge computing, troops have access to insights in remote locations with little connectivity. Weather conditions, machine performance data and other sensitive information can now be turned into actionable decision-making. As possibilities at the edge advance, these applications continue to expand.

At the same time, with millions of remote workers and strained networks, there's a greater need for computer power, capacity, and storage closer to another new network edge — home offices. The result is a boom in edge-related hardware, software and applications.

To fully take advantage of possibilities at the edge, the first step is determining how best to deploy such solutions for each unique situation. As an emerging technology, there may not be a prior example or tested solution, and each branch of the department has different circumstances and needs. It is important for senior leaders to evaluate where edge computing is most needed and how to utilize it most efficiently.

For example, how can warfighters in theater — operating under the most extreme of circumstances — have the ability to utilize actionable intelligence where asynchronous operations and connectivity are to be expected?

Once the mission is clear, it's critical to think about data protection at the edge. As the Defense Department explores new applications, data protection needs to advance along with the possibilities. Considering cyber basics, a smart backup strategy, connectivity and unique requirements for the technology's footprint at the edge can ensure sensitive data information is reliable and secure.

To ensure security at the edge, strong governance programs are key — beginning with an understanding of what data is being generated as well as how it is processed and transferred. All edge devices must be properly secured despite their less-central location and data should be encrypted at rest and in flight.

The implementation of edge can be an opportunity for the department to place strong cybersecurity practices at the onset. It can use this opportunity to assess its own risk appetites and where it can manage those risks accordingly. For example, if the department continues to move toward a zero-trust model, the approach should be integrated into edge computing applications versus implementing edge and then trying to change it to fit zero trust afterwards.

Additionally, IT security teams can choose to keep certain data at the endpoints, limiting the amount of information that gets sent back to the network and potentially keeping threats away from the data center. Edge computing may provide more endpoints for attack, but it can also prohibit bad actors from reaching the data center and mission-critical resources.

The department also needs to continue to evolve effective data backup and management strategy.

In choosing a solution, reliability, ease of use and versatile restore options are crucial features for backup — the moment data is lost in a remote location isn't the right time to discover a backup solution is overly complex. The right solution will include all recovery mechanisms including backup, replication, storage snapshots and continuous data protection.

No matter where they are in the world, defense forces need correct and up-to-date information. Mission success depends on it. If a warfighter becomes disconnected from crucial information, there could be a lag in decision-making or lack of vital information while government workers try to reconnect. Every moment disconnected is critical.

This is another place where backup comes in. If the necessary information is available reliably and separately from the network at the edge, defense forces won't need to depend on connectivity to be productive and complete missions. When warfighters become disconnected, they can operate offline and batch changes at the edge, then connect back to the network when possible. Depending on the need, edge-based deployment can asynchronously or sporadically back up at the edge. This flexibility cuts down disconnect times and increases agility in situations where network connectivity isn't reliable.

For special operations or forward operating bases, data backup can't add hardware — more equipment and additional bulk limits room for other mission-critical essentials. In some cases, cloud computing can cut down on additional hardware, but the Defense Department must ensure that the agency has a clear understanding of what the cloud provider is responsible for in terms of backup and protection.

At the edge, software-based solutions can make backup accessible in situations where there is no space to spare. Ideally, this is a complete software platform that provides benefits like scalability and the flexibility to change components whenever needed.

17.3.1 Considerations to start workloads on edge devices.

(i) Ease of deployment

Most service staff members aren't technology experts and deploying the apps they need in such environments — as well running and managing the clusters after setup — is always complex.

It allows someone right out of basic training to be able to operate a system and also manage the swapping in and out of hardware too as time goes on.

(ii) Security

Security is a big challenge when running Kubernetes clusters in edge environments. You may not be able to physically protect the edge device from theft or physical tampering. You can't guarantee network access for authentication and authorization — or patching new vulnerabilities. And that's just for starters.

That's why edge security for these clusters needs to be holistic. It must be founded on immutable, tamper-proof software stacks, use encryption and have secure software supply chains.

(iii) Scalability

Depending on the workload and use case, military teams may need to deploy Kubernetes clusters to many different environments, even ones with intermittent or degraded connectivity, or where a cluster is air-gapped — having no network connections by design.

The software and applications developed and should be created on a variety of platforms and typically housed in hybrid, multi-cloud infrastructures.

17.4 BENEFITS AND DRAWBACKS OF EDGE COMPUTING

Agencies are increasingly deploying cloud computing solutions, but now must deploy and manage devices and network assets all the way out to the network edge.

It is very straightforward to manage assets if agencies have them in an enterprise data center, via a singular or cohesive pool of technology resources. As agencies move out to the edge, those technology resources begin to diversify and fragment. However, with edge computing, agencies can manage devices at the network edge as one common virtual enterprise because as long as I can touch the device, I can manage it.

That adds in complexity in terms of cybersecurity and tampering. But you can unlock greater power of your data and your information and manage it in a more cohesive, virtual manner if you choose common platforms, both hardware and software, to build your ultimate solution from.

It allows you to take the power of edge computing and tactical cloud into areas where you haven't been able to take them before, so that you can produce great results from a mission perspective.

The remote sites require robust software platforms that can support specialized hardware and, more importantly, quickly and securely process data on-site, eliminating the challenges of transmitting vast amounts data back to a centralized cloud over unreliable network connections.

Edge computing is being used in a variety of agencies. For example, Dell partners with Microsoft to deliver cloud capabilities to tactical environments for the Air Force's air operations centers around the globe. These are essentially military versions of the Federal Aviation Administration that allow the Air Force to manage, control and protect all of its air assets. With Microsoft Azure Stack, the Air Force can build next-generation apps in the tactical cloud.

The Air Force has been able to save almost $1 million a week in tanker refueling costs via edge and cloud computing. Using the Dell Pivotal Cloud Foundry platform, the company helped train Air Force developers to build and deploy a tanker refueling application, he says. The software provides better predictive logistics, especially for refueling planes when they are in the air.

The Marines and Special Forces are also using deployable, tactical clouds on missions. That gives them improved command and control when they are out at the tactical edge. Soldiers are able to run more complex and sophisticated applications at the tactical edge that create situational awareness, such as 3D-rendering applications or geospatial apps. Soldiers can dynamically change the applications based on the changing mission on the ground.

For example, soldiers may be on a monitoring mission, but when they get to a specific area, something may have changed that turns the mission into an intelligence, surveillance and reconnaissance mission. Edge computing allows them to reconfigure their edge devices and bring up apps that allow them to collect more information and perform better analysis on site, all while connected to or disconnected from the network.

Microsoft and Dell announced the release of the Dell EMC Tactical Microsoft Azure Stack and the Azure Data Box family of products, which will help agencies with remote operations have access to the full range of cloud data analytics.

The richer the data services are, the more consumers use data services. Such has been the case with the advent of smartphones and mobile apps like video streaming. That requires more compute, memory and storage in smartphones.

As devices become smarter, there will be a requirement that those enhanced compute and storage capabilities at the edge in order to deliver that richer human-to-machine experience that is the era we're going into.

SECTION-VI

CRITICAL TECHNOLOGIES FOR MODERN AND FUTURE WARFARE

CHAPTER 18

ADVANCED COMPOSITE MATERIALS IN WARFARE

18.1 INTRODUCTION TO COMPOSITE MATERIALS

A composite material is a material which is produced from two or more constituent materials. These constituent materials have notably dissimilar chemical or physical properties and are merged to create a material with properties unlike the individual elements. Within the finished structure, the individual elements remain separate and distinct, distinguishing composites from mixtures and solid solutions. Composite materials with more than one distinct layer are called composite laminates.

Composites are formed by combining materials together to form an overall structure with properties that differ from that of the individual components.

Typical engineered composite materials include:

- Reinforced concrete and masonry;
- Composite wood such as plywood;
- Reinforced plastics, such as fiber-reinforced polymer or fiberglass;
- Ceramic matrix composites (composite ceramic and metal matrices);
- Metal matrix composites; and
- other advanced composite materials.

There are various reasons where new material can be favored. Typical examples include materials which are less expensive, lighter, stronger or more durable when compared with common materials, as well as composite materials inspired from animals and natural sources with low carbon footprint. More recently researchers have also begun to actively include sensing, actuation, computation, and communication into composites, which are known as robotic materials.

Composite materials are generally used for buildings, bridges, and structures such as boat hulls, swimming pool panels, racing car bodies, shower stalls, bathtubs, storage tanks, imitation granite, and cultured marble sinks and countertops. They are also being increasingly used in general automotive applications.

One of the most common and familiar composite is fiberglass, in which small glass fibers are embedded within a polymeric material (normally an epoxy or polyester). The glass fiber is relatively strong and stiff (but also brittle), whereas the polymer is ductile (but also weak and flexible). Thus the resulting fiberglass is relatively stiff, strong, flexible, and ductile.

Composites can also use metal fibers reinforcing other metals, as in metal matrix composites (MMC) or ceramic matrix composites (CMC), which includes bone (hydroxyapatite reinforced with collagen fibers), cermet (ceramic and metal), and concrete. Ceramic matrix composites are built primarily for fracture toughness, not for strength. Another class of composite materials involve woven fabric composite consisting of longitudinal and transverse laced yarns. Woven fabric composites are flexible as they are in the form of fabric.

Organic matrix/ceramic aggregate composites include asphalt concrete, polymer concrete, mastic asphalt, mastic roller hybrid, dental composite, syntactic foam, and mother of pearl. **Chobham armor is a special type of composite armor used in military applications.**

The most advanced examples perform routinely on spacecraft and aircraft in demanding environments.

18.2 ADVANTAGES AND DISADVANTAGES OF USING COMPOSITES

18.2.1 Advantages

(i) Strength and stiffness

One of the main advantages of composites is that they can offer high strength and stiffness, while being lightweight and flexible. This means that composites can withstand high loads and stresses, and reduce the weight and fuel consumption of mechanical systems. For example, carbon fiber reinforced polymer (CFRP) is a composite that has a higher strength-to-weight ratio than steel or aluminum, and is widely used in aircraft and car structures. Composites can also be tailored to have specific mechanical properties, such as modulus, toughness, and fatigue resistance, by varying the type, orientation, and arrangement of the fibers and matrices.

(ii) Corrosion and wear resistance

Another advantage of composites is that they can resist corrosion and wear better than some metals and alloys. This means that composites can last longer and require less maintenance and repair, especially in harsh environments. For example, glass fiber reinforced polymer (GFRP) is a composite that has good corrosion resistance to salt water, chemicals, and acids, and is widely used in marine and offshore applications. Composites can also be coated or treated with additives to enhance their resistance to abrasion, erosion, impact, and thermal shock.

Nanomaterial-based composites are revolutionizing the field of materials science and engineering, offering unprecedented properties and performance. These composites incorporate nanoscale materials, such as nanoparticles, nanotubes, or nanofibers, into a matrix material to form a composite with enhanced properties. For example, carbon nanotube (CNT) reinforced composites exhibit a much higher strength-to-weight ratio than traditional composites. This allows for the development of lightweight structures with superior mechanical performance. Additionally, they can also be used in applications that require high thermal resistance such as spacecraft re-entry systems.

18.2.2 Disadvantages

(i) Cost and availability

One of the main **disadvantages** of composites is that they can be more expensive and less available than some metals and alloys. This means that composites can increase the initial and operational costs of mechanical systems, and limit the design and manufacturing options. For example, CFRP is a composite that has a high cost of raw materials, processing, and quality control, and is not readily available in large quantities or shapes. Composites can also be difficult to recycle or dispose of, due to their complex and heterogeneous structure.

(ii) Damage and failure

Another **disadvantage** of composites is that they can be more prone to damage and failure than some metals and alloys. This means that composites can reduce the reliability and safety of mechanical systems, and require more inspection and monitoring. For example, composites can suffer from delamination, cracking, buckling, and debonding, due to the weak interface between the fibers and matrices, or the external loads and environmental factors. Composites can also be difficult to repair or join, due to their anisotropic and non-homogeneous nature.

(iii) Thermal and electrical properties

A final factor to consider when using composites in mechanical engineering is their thermal and electrical properties. Composites can have advantages or disadvantages in this aspect, depending on the application and the desired performance. For example, composites can have low thermal conductivity and expansion, which can prevent heat transfer and distortion, but also limit heat dissipation and cooling. Composites can also have high or low electrical conductivity and resistance, which can enable or hinder electrical functions and signals.

Carbon nanotubes (CNT) based matrix materials are excellent candidates for use in high temperature applications such as laser power measurement test panels and spacecraft re-entry cones.

18.3 DETECTION OF DAMAGE IN ADVANCED COMPOSITE MATERIALS USED IN AEROSPACE INDUSTRY

Advanced composite materials are being increasingly used in the production of aerospace components, wind turbines, and marine equipment owing to their high stiffness-to-weight ratio, excellent strength and corrosion resistance, low weight, and improved mechanical properties. Significant progress has been made in improving the product efficiency and cost-effectiveness of advanced composites, as well as in enhancing specific properties, such as strength and modulus. Composite materials are gradually becoming alternatives to metallic materials. Advanced composites are widely used in various applications, including the manufacturing of chemical containers, pressure vessels, machine spindles, power transmission shafts, and robot arms.

Advanced composite materials are prone to various types of damage during - manufacturing and long-time service. **Ultrasonic detection methods have been widely employed to detect damage** because of their high accuracy and reliability.

Owing to the complicated curing and drilling processes during composite manufacturing, damages, such as delamination, fiber frying, spalling, chipping, and fuzzing, can occur. In addition, during the service life of aircraft structures, the extent of the damage in the composite structure may gradually increase owing to various environmental factors, such as load, vibration, and hygrothermal conditions. For example, delamination damage occurs in the interior of composites during service, the generation and propagation of which are not visible. This damage adversely affects the mechanical properties of aircraft structures. If the occurrence and accumulation of damage cannot be detected in time, catastrophic failure of aircraft structures may occur, thus endangering human life. Over the past 30 years, research on the damage detection of composite materials in specific fields has focused on crack damage detection, determination of bonding quality in laminated plates, large-scale thin- or thick-walled structures, intelligent structures, and sandwich structures.

Damage must be detected during the manufacturing and processing stages, as well as during the service period, for quality control and safety performance evaluation of aircraft structures. Numerous researchers have investigated the internal damage of composite structures using methods such as X-ray, optical testing, magnetic particles, infrared thermography, and ultrasonic testing. Among these, ultrasonic inspection technology is highly reliable for assessing structural safety performance. It is advantageous in detecting damage in advanced composite materials owing to its ability to scan the cross-sectional area of a structure; further, it utilizes low-cost and portable inspection equipment, and boasts improved scanning speed and resolution.

However, detecting and evaluating the structural integrity of composite materials via ultrasonic inspection is particularly challenging because of the multilayered structures with anisotropic velocity distributions. The complexity of ultrasonic wave propagation in composites renders difficulty in measuring material parameters and accurately detecting any damage. Furthermore, damage may occur at multiple locations in different components of advanced composite materials, thus leading to complex acoustic wave propagation mechanisms, which render difficulty in tracking all the damaged parts. Owing to these complicated characteristics, the difficulty in damage identification through ultrasonic inspection increases with the level of target execution, including the (i) qualitative indication of damage occurrence and type, (ii) accurate location of damage based on precise material measurement parameters, and (iii) structural safety prediction.

Various non-destructive testing techniques for composites have been developed. Their effectiveness in the lifecycle of composite products has been proven, with focus on various aspects such as process design and optimization, manufacturing inspection, in-service testing, and structural health monitoring. A combination of several technologies has been proposed to improve the diagnosis of damaged composite structures and their evolution during their lifespan. In addition, it is indicated that the necessity to promote the mature application of ultrasonic-guided Lamb waves for cost-effective non-destructive testing and structural health monitoring technologies. A roadmap for future research has been developed to take advantage of artificial intelligence and big data to develop digital twins in the digital era. However, with the emergence of new composites and inspection methods, a comprehensive description of the application and the latest progress in ultrasonic methods is required for the performance evaluation and damage diagnosis of composites.

Standard practices for the measurement and evaluation of ultrasonic detection methods are provided by the American Society for Testing and Materials (ASTM) to ensure the structural integrity of advanced composite materials, and guide researchers and engineers.

18.3.1 Typical damage types and causes in composite materials

The main parameters of advanced composite materials include elastic constant, acoustic velocity, and density. Because the accuracy of damage detection is closely related to the accuracy of composite parameter measurements, the actual composite parameters must be accurately determined by direct measurement or inversion methods. In particular, ultrasonic measurement methods for characterizing the mechanical properties of anisotropic composite materials are summarized.

(i) Damage detection based on bulk waves

Ultrasonic bulk waves, that is, waves treated as acting in an infinite medium, were employed to detect damage to composite materials in the thickness direction. These methods include pulse reflection and penetration. Based on the acoustic model, an image of the damage to the composite material was generated using imaging algorithms that processed the data obtained by ultrasonic testing.

(ii) Damage detection based on the guided waves

Compared with the detection range of the bulk-wave-based detection technology, those using ultrasonic-guided waves have a wide detection range and high detection efficiency and are widely used for damage detection in large composite plates in the aerospace field. After clarifying the propagation characteristics of guided waves, they can be combined with phased array, air-coupled, and laser ultrasound detection technologies to excite guided waves and detect damage in composite laminates.

(iii) Damage assessment and service life prediction for composites

Damage evaluation comprises damage detection/identification, damage location, damage quantification, and prediction of the remaining service life of the structure modules. Non-destructive testing/structural health monitoring technologies currently focus on the detection and location of damage; that is, damage is detected during manufacturing and service. Inputting the test results into a structural residual life model and thereby developing a digital twin for structural state.

Ultrasonic testing technology for parameter estimation and damage detection of composite materials has substantially progressed in recent decades. New technologies and methods have brought new development momentum and direction to this field. However, with the continuous development of new composite materials with complex structures, certain challenges regarding the efficient, reliable, and accurate detection and evaluation of various damage types to these materials

18.4 APPLICATIONS OF COMPOSITES IN THE MILITARY

Composites have become pivotal in advancing military technology and are used in lightweight armored vehicles, stealth aircraft, naval ships, submarines, missiles and rockets. Known for their unparalleled strength-to-weight ratio, composites offer unique advantages that traditional materials cannot match.

The defence sector employs various types of composites to enhance the capabilities of military equipment. These materials include carbon fiber and glass fiber composites, which are selected for their unique properties; contributing significantly to the performance, durability, and safety of defence applications.

(i) Carbon Fiber Composites

Carbon Fiber composites are renowned for their exceptional strength-to-weight ratio. Predominantly used in aerospace and aviation, these composites offer unparalleled structural integrity while maintaining lightness. This makes them ideal for aircraft, drones, and missiles where weight reduction is crucial for performance and fuel efficiency. Additionally, their resistance to fatigue and corrosion makes them a preferred choice in challenging environments.

(ii) Fiberglass Composites

Fiberglass composites, known for their robustness and cost-effectiveness, are a staple in naval defence applications. Their resistance to water and corrosion makes them suitable for ship and submarine hulls, ensuring longevity and reducing maintenance needs. Additionally, these composites provide good tensile strength, making them useful in a variety of other military applications, including vehicle body panels and protective gear.

18.5 ADVANCEMENTS IN COMPOSITE TECHNOLOGY FOR MILITARY APPLICATIONS

The significant advancements in composite technology have revolutionized how defence equipment is designed and manufactured, enhancing performance and opening up new avenues for innovative applications.

18.5.1 Innovations in Composite Manufacturing

Innovations in composite manufacturing have been pivotal in developing more efficient and robust military equipment. The introduction of **advanced fabrication techniques**, such as **3D printing** and automated layup processes, has improved the precision and strength of composite components. These methods allow for complex shapes and designs, essential for aerospace and naval applications, and have significantly reduced production time and costs, making high-performance composites more accessible for various military uses.

18.5.2 Future Trends in Composite Materials in Military

The future of composite materials in military applications looks promising, with research focusing on developing **smarter**, more adaptive defence composites. These materials are expected to have enhanced functionalities, such as self-healing properties, improved ballistic resistance, and the ability to adapt to environmental changes. Shape memory alloys may also enable components to return to their original shape after deformation. Plus, the integration of **nanotechnology** is anticipated to bring forth lighter, stronger, and more versatile composites.

18.5.3 Applications of Composites in Military Equipment

Composites are used in a wide range of military equipment including lightweight armored vehicles, stealth aircraft, naval ships, submarines, missiles and rockets. Here are examples of past and present uses of composites in military defence vehicles.

(i) Advanced composite armor protection systems for military vehicles

In the present era, vehicle armor primarily comprises steel. However, due to the growing size of projectiles (>7.62 mm) and the necessity to reduce overall armor weight, steel has been deemed unsuitable for military vehicles. Therefore, alternative composite materials have been introduced to enhance the ballistic resistance of armored vehicles. These materials aim to offer the maximum possible protection against projectiles while minimizing the armor's mass and production expenses.

Because lives are at stake, the quality expectations are very high. To achieve consistently high quality sustainably and efficiently, advanced quality assurance methods are necessary, such as Zero-Defect Manufacturing (ZDM). ZDM has a clear advantage over the traditional quality improvement methods (e.g., Six Sigma), taking full advantage of Industry 4.0 technologies; more importantly, it implies 100% product inspection. For components critical to the defense of vehicles or humans, particularly in their armor, assuring 100% quality is vitally important. On the other hand, the materials designed to be used for such purposes should be defined such that manufacturing defects are improbable. Inspecting 100% of the production process is difficult. However, leveraging modern advanced scheduling tools and decision-making methods, this process can be performed efficiently.

Here the focus is on the development and production of new laminated composite lightweight protection systems with three different backplates composed of steel, aluminum alloy and metal matrix composite materials (i.e., metal matrix composites [MMCs]) reinforced with carbon nanoparticles (NPs), to be used for vehicles with protection Level 4+ according to STANAG 4569. NATO has a strict standard that can be applied and accepted for vehicles in NATO countries. Moreover, an innovative and lightweight multi-layered armor concept is introduced, designed specifically for the passive protection of contemporary light, medium, and heavy armored vehicles. The three protection proposals advanced in this work can adapt to any ground combat vehicle application (e.g., 4 × 4 or 8 × 8 vehicles) where Level 3 and Level 4 (according to STANAG 4569) protection are needed, as well as for main battle tanks and, in general, heavy armored vehicles, where Level 5 and Level 5+ are needed. The ballistic testing can also be performed on armor composite solutions, offering comparisons thereof. These comparisons show that Al-CNT could soon be a highly competitive backplate solution.

(ii) Aerospace and Aircraft Applications

In the aerospace and aircraft sectors, composites play a crucial role. They are used in the construction of fighter jets, helicopters, and unmanned aerial vehicles (UAVs) due to their lightweight yet strong nature. This results in enhanced maneuverability and fuel efficiency, which is crucial in combat and reconnaissance missions. Composites also allow for stealth features in military aircraft, such as reducing radar signatures and enhancing mission success.

Military helicopters have come a long way since the Sea King composite main rotor blade development program was set up between Westland and the British Ministry of Defence in 1977. This program developed main rotor blades using a mixture of carbon and glass fibers to create a blade with improved aerodynamic properties. However, composites are now being used extensively for enhanced properties, such as ballistic protection in the ultra-high molecular weight polyethylene composite panels in the Merlin Helicopter MK3.

Upcoming advancements in aerospace and aircraft applications include 'The Tempest aircraft', which is set to join the RAF fleet in 2035. Currently under development, this aircraft will integrate advanced composite materials to produce lightweight, power-dense configurations that can operate at high temperatures.

(iii) Naval Vessels and Submarine Construction

Composites have significantly impacted naval vessel and submarine construction. Their resistance to corrosive marine environments makes them ideal for hulls and superstructures of ships and submarines. This results in lighter, faster, and more durable naval vessels, capable of enduring prolonged exposure to harsh sea conditions. Additionally, the non-magnetic properties of certain composites are advantageous in mine countermeasure vessels, reducing the risk of triggering mines.

An example is the Sandown Class of minehunter vessels, first commissioned on the 9th of June, 1989, built out of fiberglass. Similarly, The Royal Navy's Astute Class of Submarines use composite materials, as does the HMS Glasgow, which belongs to the fleet of the global Navy combat ships known as Type 26 Frigates. The HMS Glasgow is the first Royal Navy vessel to have a mainmast made from composites and also hosts an active/passive bow-mounted sonar array in a fiberglass dome.

18.5.4 Impact of Composites in Defence Applications & Strategies

Composite materials have not only enhanced the capabilities of military equipment; they have also influenced the tactics and approaches used in defence scenarios, redefining the landscape of military technology. So, let's look at how they can enhance performance and safety, as well as outline the operational advantages and challenges.

(i) Enhancing Performance and Safety

Composites have significantly enhanced the performance and safety of defence equipment. By offering superior strength and durability while being lighter than traditional materials, they have enabled the development of faster, more agile, and more efficient military hardware. This includes aircraft capable of higher performance and armored vehicles offering better protection. The inherent properties of composites, such as resistance to heat and corrosion, further ensure the safety and longevity of military equipment in various environments.

(ii) Operational Advantages and Challenges

The advantages of composites include improved stealth capabilities, enhanced performance, improved safety, corrosion resistance and increased longevity. However, they also present unique challenges such as high development costs, specialized maintenance requirements and complex manufacturing processes. While some composites have increased longevity and corrosion resistance, others may be sensitive to environmental factors such as moisture and UV radiation. Therefore, the advantages must be carefully considered against the cost of materials, manufacturing and maintenance, as well as ensuring the correct composite is chosen for the required application.

18.6 US ARMY INVESTS IN COMPOSITE MATERIALS TO LIGHTEN GROUND VEHICLES

The U.S. Army is actively pursuing new technologies to reduce the weight of its combat and transport vehicles by incorporating lighter composite materials. This initiative is part of a broader effort to enhance the mobility and payload capacity of its ground fleet. For the fiscal year 2025, the Army has requested nearly $1 billion to support these efforts, reflecting a significant investment in modernizing its ground platforms.

This focus on weight reduction is not limited to new vehicles but also extends to key in-service platforms, including Abrams tanks, Bradley Fighting Vehicles, and the Family of Heavy Tactical Vehicles (FHTV). By reducing the weight of these vehicles, the Army aims to improve their maneuverability on the battlefield and increase their ability to carry additional payloads, which is crucial for operational effectiveness.

The US Army's investment in composite materials is part of a broader trend within the defense sector to leverage advanced materials for performance improvement. As the service continues to modernize its fleet, the integration of these technologies will likely play a crucial role in maintaining the Army's readiness and effectiveness in future conflicts.

In the military context, particularly for vehicles like Abrams tanks and Bradley Fighting Vehicles, composites are essential for several key reasons:

(i) They allow for a significant reduction in vehicle weight. By replacing heavier materials such as steel with composites, the Army can make its vehicles lighter, which improves mobility, speed, and maneuverability on the battlefield.

(ii) Composite materials offer excellent mechanical strength and increased durability, meaning they can withstand heavy loads and extreme conditions without deforming or deteriorating. This is especially important for armored vehicles that must remain operational under enemy fire or in harsh environments.

(iii) Composites can be designed to provide better protection against ballistic threats, as they can absorb and dissipate the energy from impacts more effectively than traditional materials. This enhances the protection of crews inside the vehicles.

(iv) Composites are also more resistant to corrosion, which extends the lifespan of vehicles in environments where they are exposed to moisture, salt, and other corrosive agents. These characteristics make composite materials a strategic choice for improving the overall performance of military platforms while meeting modern demands for flexibility and efficiency.

18.7 ADVANCED COMPOSITE MATERIAL AND DRDO, INDIA

The Defence Metallurgical Research Laboratory (DMRL), DRDO, India, has signed a Transfer of Technology (ToT) to Carborundum Universal Limited (CUMI) to manufacture add-on composite armor for Wheeled Armored Platform (WhAP) vehicles. The lightweight composite armor will reinforce combat vehicles providing all-round ballistic protection at STANAG Level II and III.

This ToT is a key milestone in India's quest for indigenization of advanced materials to build world-class armoring capabilities for the highest protection of vehicles. The WhAP 8×8, developed by the Defence Research & Development Organisation (DRDO), has been built with a global design philosophy. It can be adapted to fulfil diverse roles for the military since its design is modular, scalable, and can be easily reconfigured. As India's first indigenously designed and built amphibious wheeled infantry combat vehicle, the 8X8 presents a significant opportunity for Indian manufacturers to contribute to India's defence modernization while augmenting export potential of military hardware.

CUMI's indigenously manufactured elastomeric add-on armor composite panels with DRDO – DMRL technology are a pioneering innovation. The customized ceramic segments are sandwiched between rubber and backed up by a metallic mother plate which is impact-, UV- and weather-resistant. Each wheeled armored platform will require a maximum of 300 unique ceramic armor panels, engineered to precisely contour the vehicle for highest levels of safety and ballistic protection. The segments are designed using high-end technical ceramic grade of CUMI's Zirconia Toughened Alumina which has been thoroughly tested and qualified by DRDO – DMRL.

With 70 years of expertise in materials science engineering, CUMI is well-positioned to deliver advanced armor materials for defence applications. Over the years, CUMI, as a long-standing partner of DMRL, has been involved in cutting-edge innovations in materials that have helped design robust armoring solutions for military vehicles. Key priorities addressed include lightweighting, easy maneuverability regardless of terrain, improved ballistic protection through precision engineering, while being resistant to impact and extreme weather conditions. For instance, a key feature of the WhAP is superior mobility, to navigate through the toughest terrains on land and water with ease. This can be attributed to CUMI's lightweight composite armor panels, crafted with advanced ceramics for exceptional protection with minimal increase in weight.

The global market size for armored vehicles is forecast to expand from $23.73 billion in 2024 to $37.85 billion by 2032. The global market for advanced materials is growing in parallel to meet evolving security threats. It is projected to touch $10.10 billion by 2032, from $9.3 billion by 2029 and $6.7 billion in 2024.

By making significant investments in R&D and capacity expansion, materials science-led companies like CUMI have been gearing up to ably fulfill domestic requirements while also contributing to realise India's global ambitions. They are also helping to solve inherent sectoral challenges, especially in the area of critical materials – by bridging supply gaps, reducing import dependence, and making cost-effective, globally-benchmarked materials available for OEMs. to deploy the best in tech.

All these factors are not only enabling India's Armed Forces march faster towards their goal of modernization but also helping domestic players gain a wider market for their products and solutions worldwide. In the long-term, this will help India to build strategic alliances across nations, consolidate its position as a global defence manufacturing hub and realise its defence exports ambition of INR 50,000 crores by 2028-29, ultimately leading to a stronger and more self-reliant nation.

CHAPTER 19

EXPLOSIVES AND ENERGETIC MATERIALS IN WARFARE

19.1 INTRODUCTION TO ENERGETIC MATERIALS

Energetic materials, also known as explosives or propellants, are substances that release a large amount of energy when activated. They are essential to weapon systems and are critical for the performance and reliability of military forces.

An energetic material is a compound that can undergo rapid, self-sustaining, exothermic, reduction-oxidation reactions. Energetic materials may be categorized according to their intended uses: (a) **explosives**, (b) **propellants**, and (c) **pyrotechnics**. Explosives and propellants evolve large volumes of hot gas when burned; they differ primarily in their rates of reaction. Pyrotechnics (i.e., a powder or ammunition used for igniting a rocket or producing an explosion; the term is also used in the military to designate flares and signals evolve large amounts of heat but much less gas than explosives or propellants. Energetic materials may also be grouped according to their rate of reaction. Both propellants and pyrotechnics are considered to be low explosives, and the velocity at which the combustion proceeds through these materials is usually 400 m/sec or slower. In comparison, high explosives are detonated, a process in which the very rapid rate of the combustion reaction itself produces a shock wave, capable of shattering objects, in the surrounding medium. The shock wave moving through the explosive material causes further explosive decomposition of that material, and the reaction rate is determined by the speed of the shock wave. The range of velocities of the shock wave is 1,000 to 9,000 m/sec. In addition to being used as explosive charges, many high explosives are also used in propellant formulations. For purposes of this discussion, the term explosive is used generically to indicate any energetic material.

(i) Explosives

Modern explosive devices employ an explosive train that takes advantage of the specific explosive properties of its components: the initiator, the detonator, the booster charge, and the main charge. The initiator, or primary explosive, consists of a small quantity of material that is very sensitive to heat, spark, impact, or friction. Primary explosives may intensify the energy up to 10 million times that of the initiating stimulus. Geometric arrangement of the explosive device directs either the flame or the detonation wave of the initiator toward the detonator charge. The detonator, a larger amount of less sensitive but more powerful explosive material, then detonates either the booster charge or the main charge. The booster charge is an optional component that further magnifies the explosive impulse. The main explosive (or bursting) charge contains the largest amount of an insensitive, but powerful, explosive. Explosives used as booster and main charges are usually not capable of being initiated by impact, friction, or the brief application of heat, and are known as secondary explosives. The secondary explosives used currently in most military explosive devices are physical mixtures of one or more high explosives with various additives. (The use of mixtures, rather than single compounds, provides for greater flexibility in explosive design, and additives extend the range of performance even further.

Melt-loading, commonly used with TNT mixtures, is a process in which a molten explosive mixture is introduced into an empty shell casing and allowed to cool and harden. Secondary explosive mixtures are used to facilitate the melt-loading process to optimize (a) the oxygen balance of the explosive, (b) explosive characteristics such as blast and fragmentation, and (c) engineering criteria such as malleability and strength.

(ii) Propellants

Propellants are explosive materials formulated and engineered to react at carefully controlled rates, producing a sustained pressure effect over a longer period of time than high explosives. In contrast to the detonation of high explosives, the process of propellant burning is referred to as deflagration, wherein the rate of heat transfer determines the rate of the reaction, which proceeds at subsonic speeds.

Solid propellants may be classified by their chemical composition. Each class has unique properties that render it useful in certain applications. All solid propellants may contain additives similar to those used in explosive mixtures. The additives can be more toxic than the principal components of the propellant and must be considered in occupational-hazard analysis. Regardless of the composition class, the chief advantages of solid propellants include their compactness, safety, ease of storage, tolerance of temperature extremes, and ease of handling. In comparison, liquid propellant systems permit greater thrust control and deliver higher specific impulses. Liquid propellants have been limited to use in high-performance missile systems until recently, when research has focused on using liquid gun propellants for howitzers.

(iii) Pyrotechnics

Pyrotechnic materials are relatively slow-burning, nonexplosive powders such as metals, alloys, and hydrocarbon mixtures. Pyrotechnics are also widely used in the military as flares, signals, relays, delays, and fuses.

19.2 MILITARY APPLICATIONS OF ENERGETIC MATERIALS (EM)

The ordnance systems must be adaptable in size to fit a family of delivery systems, contain sufficient energy to defeat the target, have the capability to fly further and faster, while being insensitive munition (IM) compliant and affordable. The Energetic Materials (EM) program explores materials/synthetic chemistry, advanced dynamic diagnostics and theoretical/computational/predictive approaches to provide novel energetic material concepts (explosives, propellants, reactive materials) that maximize molecular and formulation energy densities, synthesis efficiencies and predicted properties to achieve performance goals. These goals include delivering maximum energy in compact volumes, significantly extending weapon range, and improving resistance to unintended catastrophic failure in stressful environments.

EM is the pillar which establishes future advanced warhead and solid rocket motor performance and characteristics, and considered with associated weapon systems can be "game changers" by increasing warfighters' lethality and area of dominance. Advanced warhead development provides catastrophic damage, improving battlefield damage assessment and reducing sorties while equally powerful, but smaller weapons optimize internal carry and facilitate higher weapon load outs. Similarly, improved propellant ingredients and design concepts provide reduced time to target and extended ranges needed in volume limited ordnance systems.

19.2.1 Research Concentration Areas in EM

The broad thrust areas within the EM program include high performance (rocket) propulsion where current needs include improved control over energy release and the ability to use energy more efficiently, and explosives development where the desire is to provide greater lethality in smaller form factors and couple energy output to targets more effectively. These thrusts require fundamental understanding and combined efforts in the current EM program research concentrations areas:

- New energetic ingredients/materials;
- Advanced dynamic diagnostics;
- Modeling/simulation for properties prediction and energetic material performance.

19.2.2 EM Research Challenges and Opportunities

Develop new classes of ingredients to enable substantially higher performance ordnance with acceptable insensitivity characteristics (i.e., high-energy density oxidizers that can surpass the oxygen content and overall performance of ammonium perchlorate; novel fuels and reactive materials; and new binder systems that will improve formulation properties (i.e., higher solids loading, higher energy output and IM enhancements) and facilitate emerging formulation/manufacturing techniques;

- Development of macroscopic mechanical and chemical models and advanced diagnostic techniques to establish the connectivity between molecular structure, crystal morphology prediction and synthesis chemistry; establish methodologies to model, measure and predict molecular and crystal energetic material response to external shock and thermal loading; and provide an understanding of molecular and reaction dynamics and material strength/reactivity correlations related to energetic ingredients and composites;
- Consistent processing and performance results; process research and development (including "scale-up" and new formulation technologies); areas of concern are critical thermal management/safety, batch to batch reproducibility, standardized process for the chemistry, and conditions and product quality and purity assurance;
- Insensitive Munitions: The Navy has concerns over conventional munitions and propellant systems, since all munitions are stored on maritime platforms. It is critical that conventional munitions display maximum insensitivity when stowed, handled, carried or otherwise

exposed to friendly forces and environments, but have sufficient energy/lethality to perform mission expectations reliably.

19.3 NEW ENERGETIC MATERIAL TECHNOLOGIES

There are several new technologies that are being developed in the field of energetic materials, also known as explosives or propellants. These technologies are focused on improving the performance, safety, and environmental impact of these materials.

Some examples of new energetic material technologies include:

(i) Green Propellants

Researchers are working on developing new propellants that are safer and more environmentally friendly than traditional propellants. These "green" propellants typically use hydrogen peroxide or hydroxyl ammonium nitrate as the oxidizer, which are less toxic and less flammable than traditional oxidizers.

(ii) Nanoenergetics

Researchers are working on developing new energetic materials at the nanoscale, which can have unique properties and performance characteristics. For example, nanoscale aluminum particles can be used to create more powerful and efficient propellants.

(iii) Insensitive Munitions

The development of new energetic materials that are less sensitive to shock, friction, and heat, making them safer to handle and transport. This can reduce the risk of accidental detonation and improve the safety of military and industrial operations.

(iv) Bio-based Propellants

Researchers are working on developing new propellants that are made from renewable, biobased materials such as plant sugars, starches, and cellulose. These biobased propellants are expected to be more sustainable and less toxic than traditional propellants.

(v) Self-healing Energetic Materials

Researchers are working on developing new energetic materials that can repair themselves after being damaged. This can improve the safety and performance of the materials and reduce the need for maintenance and repair.

(vi) Smart Energetic Materials

Researchers are working on developing new energetic materials that can be programmed to respond to specific triggers, such as temperature or pressure. This can improve the performance of the materials and allow for more precise control over the timing and intensity of the explosive reactions.

These are just a few examples of the new technologies being developed in the field of energetic materials. The energetic materials market is constantly evolving, and companies are investing in R&D to develop more advanced technologies to meet the evolving needs of customers.

Novel Energetic Materials consists of fundamental research programs to expand and validate physics-based models and experimental techniques to devise chemical formulations that will enable the design of novel insensitive high-energy propellants and explosives with tailored energy release for revolutionary Future Force lethality and survivability. This program supports demonstration of advanced energetic materials with ability to tune energy release for precision munition & counter-munition applications (e.g., propellants, explosives, thermobarics, multi-purpose warhead, APS).

These energetic materials may have the potential of providing factors of 3 to 4 in increased energy release rate compared with conventional formulations. The Army's Novel Energetic Materials for the Objective Force effort seeks to mature advanced energetic materials to provide a 40% increase in deliverable energy from advanced gun propellant systems and a 20-50% increase in warhead effectiveness (munitions, active protection).

Like advanced initiation, improved energetic materials are enabling technology for the next generation of weapon systems that will be safer, smaller and more lethal. Under this program a combination of evolutionary and novel technologies are under development. Conventional chemistry has been used to develop more powerful, less sensitive explosives.

Nano-structured and engineered materials are being explored to increase energy density and energy on target by factors of three or more. In general, initiation and detonation properties of energetic materials are dramatically affected by their microstructural properties. It is generally known in material science that the mechanical, acoustic, electronic, and optical properties are significantly and favorably altered in materials called "nanostructures," which are made from nanometer-scale building blocks. Modern technology, through sol-gel chemistry, provides an approach to control structures at the nanometer scale, thus enabling the formation of new energetic materials, generally having improved, exceptional, or entirely new properties.

Higher risk efforts are also underway to explore the possibility of metastable High Energy Density Materials (HEDM). Using conventional chemistry, a number of new candidate molecules have been synthesized, characterized and formulated. The development of new materials is based on theoretical molecular design. The structure, performance and sensitivity of new molecules are predicted computationally, then synthesis is attempted. The focus is in two areas: molecules with significantly increased energy over current materials and very insensitive materials with reasonable energies.

Another emerging technology that holds promise in the energetic materials field is the use of **biological processes** for advanced manufacturing. In effect, Synthetic biology has turned the bioscience into the future manufacturing paradigm where Companies can engineer and manufacture an infinite quantity of things, cell by cell, from scratch. These bioengineered microorganisms, plants, and animals can produce pharmaceuticals, repair defective genes, develop new generations of vaccines, destroy cancer cells, detect toxic chemicals, break down pollutants, and generate hydrogen for the post-petroleum economy. Synthetic biology will also enable the production of complex molecules for use in next-generation energetic systems. It also offers the potential of environmentally benign manufacturing solutions.

19.3.1 Boron-powered missiles

In Sep 2022 it was reported that China is developing a supersonic anti-ship missile that will be able to travel further and faster than any traditional torpedo. The 5 metres (16.4 feet) missile will be able to cruise at 2.5 times the speed of sound at about 10,000 metres (32,800 feet) – the same altitude as a commercial airliner – for 200km (124 miles) before diving and skimming across the waves for up to 20km.

One of the biggest challenges for the developers is the power system, because of the need to produce considerable thrust while breathing in either air or water. But Li's team said the problem could be solved by using boron – a light element that reacts violently when exposed to both, releasing a huge amount of heat. The team, from the college of aerospace science and engineering in the National University of Defence Technology in Changsha, Hunan province, unveiled a blueprint for the missile's power system in the September 8 issue of the peer-reviewed Journal of Solid Rocket Technology, published by the Chinese Society of Astronautics.

Boron was briefly added to jet fuel by the US Air Force in the 1950s to increase the power of supersonic bombers. But the project was abandoned because the ignited boron particles were hard to control and formed a layer of debris that gradually reduced engine performance.

A Nasa study funded by the US Navy last year found that nanotubes made using boron nitride, a combination of boron and nitrogen, could potentially be used to power hypersonic weapons travelling at speeds above 6,400km/h (4,000 miles per hour).

But most boron-powered engines are designed to work only in the air. Researchers usually choose aluminum or magnesium as fuel to drive supercavitating torpedoes as they react more easily with water. Li's team said they had designed a boron-powered ramjet engine that could work both in the air and underwater.

There are some unique components, such as adjustable inlets and exhaust nozzles to maintain the boron's burn efficiency in different environments, but the biggest change is in the fuel rods, according to their paper.

Boron usually accounts for about 30 per cent of the total fuel weight in an air-breathing missile because of the many other chemicals required to control and prolong the strong combustion.

Li's team has doubled the share of boron in the fuel and estimates the result could produce a thrust greater than that of aluminum in water.

"The cross-media ramjet uses a fuel-rich solid propellant, which burns with the external air or seawater entering into the ram to generate high-temperature gas and generates thrust through the nozzle," the paper said. "It has the high specific impulse and simple structure as an ideal power source for a cross-media anti-ship missile."

5E Advanced Materials, Estes Energetics, announced in Dec 2022 to produce boron-based materials for rockets. In a press release, the companies said that the ultimate goal of their collaboration is supporting the US space and military industries and the specific focus will be on solid rocket motor ignitors.

19.3.2 Machine learning-based Energetic Materials

Researchers have developed a machine-learning-assisted method for accelerating the discovery of new energetic materials via efficient prediction and quick screening. Suitable neural networks are established for accurately predicting the detonation properties of various N-containing molecules based on their structures, including density (ρ), detonation velocity (D), and detonation pressure (P).

Purdue, AAE play key research role in MURI grant focused on energetic materials and machine learning.

In collaboration with six universities, Purdue was awarded a Multidisciplinary University Research Initiative (MURI) award from the Department of Defense (DoD) to build a capability to allow scientists to predict the behavior of energetic materials using advanced machine learning tools, which cover a wide range of applications including military munitions, propellants, pyrotechnics, and industrial explosives. The University of Missouri is the principal investigator on the project that includes researchers from Purdue, the University of Illinois, the University of Iowa, the University of Illinois-Chicago, Columbia University, and the Rensselaer Polytechnic Institute in New York.

AAE Professor Vikas Tomar says the project, "Integrating Multiscale Modeling and Experiments to Develop a Meso-Informed Predictive Capability for Explosives Safety and Performance," will address a long-standing problem that never has been addressed before. "So far, a significant focus of energetic materials research has been on material processing. However, microstructures for such materials have random character making them unsuitable for systematic microstructural characterization tools available for other materials such as metals or ceramics," Tomar says. "This characteristic makes these materials also suitable for data science tools incorporating machine learning, a focus of the proposed work. Purdue's Interfacial Multiphysics Lab has unique capabilities to perform high throughput experimental measurements of thermal and mechanical properties in energetic materials at nanometers to micrometers length scales with picosecond time resolution, a key requirement for models needed for the proposed work."

Tommy Sewell, a professor at Missouri who is the PI on the project, says the team will use artificial intelligence or machine learning to sift through mountains of experimental and simulated data and to identify correlations in the data that scientists might miss. But he also says machine learning will only take his team so far. "If all that you seek is knowledge, maybe that is good enough for some purposes, but it's not good enough for us," Sewell says. "What we seek is understanding, and that comes from 'carbon-based' computing (human thought and physical models), not silicon-based. A long-term goal is to minimize the amount of experimentation and the assorted costs and do most of the work in computer simulations, and then use experimentation to validate the results. If we're successful, we will end up with a framework that can be adapted to treating the initiation phenomenon for a wide variety of explosives."

Another goal is to reduce accidents involving energetic materials, Sewell says. If rocket propellant transitions from a stable burn over to a detonation, it can result in catastrophic consequences. "What we are trying to do ... is to develop a theoretical framework that will allow us to derive the next generation of reactive burn models that are far more predictive than models currently in use," Sewell says. "The goal is to reduce accidents, to improve safety, and to be able to design energetic formulations that would have much more tightly tailored performance."

19.3.3 3D Printing of Energetic Materials and Components

Additive manufacturing of energetic materials is an enabling technology in that it affords unique geometries and unprecedented control over dynamic behavior. This for example could be a complex energetic materials structure, or a component of an energetic system.

3D printing involves the layer-by-layer deposition of one, or more, materials. The spatial placement of the material, if carefully controlled, can influence a desired static or dynamic property. The use of 3D printing to build complex and unique energetic components is at the center of LLNL's architected energetic materials and structures effort. LLNL has developed several different methods for using 3D printing to create articles of energetic materials applicable to high explosives, propellants, and pyrotechnics. Methods being explored include direct printing of energetic materials as well as creating unique scaffold structures for integration with energetics. Scaffolds can be fabricated with metal, polymer, plastic, ceramic, and reactive composites, and some applications are charge liners, fragmentation packs, plane-wave generators, and reactive casings.

Major benefits of 3D printing of energetics range from reducing the cost and footprint of manufacturing, to rapid prototyping of energetic components. Additionally, performance and safety properties can be controlled in new ways through the spatial control offered by additive manufacturing. This, for example, could include an assembly of two or more energetic materials spatially arranged in such a way that the collective effect yields a new or optimized behavior. Another benefit of 3D printing is that, if the performance can be optimized, it allows for the use of less mass of material. This attribute is beneficial from both a cost and safety standpoint. The versatility of 3D printing of energetic components allows much more flexibility in the design of explosive, propellant, and pyrotechnic systems.

19.4 NANOENERGETIC MATERIALS FOR THE MILITARY

Military missions require small energy-dense formulations to power future generations of miniature autonomous systems and satellites, and to provide sufficient destructive energy yields in small explosive payloads. The weapons effects will need to be tailorable; that is, tuned for explosive yield or impact resistance, and the materials sufficiently robust to operate within the unique constraints of hypersonic missiles that are subject to extreme aerodynamic forces.

In addition to their formulation as explosives and propellants, nanoenergetic materials can also be employed as pyrotechnics for breaching fortifications by destroying advanced materials such as ceramics, composites, and metal alloys, which are used for collective protection, and in antitamper devices and systems. In addition, their superior energy density relative to traditional mono-molecular materials such as TNT make them ideal candidates for future energy generation and storage devices. The figure illustrates the range of applications of nanoenergetic materials.

A key reason to explore nanoenergetic materials for fuels and propellants hinges on potential improvements resulting from altered chemical kinetics rather than thermodynamics. For example, 1-nm aluminum particles (nanoscale) reacting with oxygen release only 1.04 times as much energy as does ultrafine aluminum (traditional micron particle-size range), but the rate of energy release (i.e., the kinetics) of the former is potentially faster because the balance of the rate-controlling factors shifts as the particle size is reduced. This is because the rate of combustion, or the balance of the generally slower mass transport rates and faster chemical reaction rates, controls the explosion process.

This fact frequently makes mass transport a controlling energy release process in conventional munitions and propellants in which micron and larger particle sizes of energetic materials are used. Mass transport simply means that with larger-scale traditional explosives, the longer distance between a molecule of fuel and a molecule of oxidizer is a more important factor than the speed or kinetics of reaction. By contrast, the opposite is true of nanomaterials: the high surface area of small particles and the short diffusion length between particles are expected to enhance the role of chemical kinetics, an important consideration in designing energetic materials because particles that are closer together are going to react more quickly. In addition, an unprecedented degree of control of the energy release rate may be possible by varying the composition on the nano-dimensional scale. The burning rate might be accelerated, the delivered specific impulse, or shock wave, could be increased by improved combustion efficiency, and the detonation might achieve greater results while not increasing the size of the fuel package.

Nanoenergetic materials such as nanothermites are generally formulated as an elemental metal such as aluminum combined with a metal oxide (i.e., a metal with an oxygen bond; for example, rust); the former is the fuel and the latter is the oxidizer. These have superior reaction rates and energy yields relative to their "meso-scale" traditional formulations and to conventional explosives but pose problems unique to reactions at these small scales. Recent advances in understanding the physical and chemical properties of nanomaterials have begun to address these problems, and formulations with superior energy yields now show promise for applications in miniature military systems and to be the next generation of explosives and propellants. This is due to their decreased sensitivity to impact, friction, and shock waves, and increased energy release and burning rate. These characteristics make them much safer to handle than current munition fills.

In addition to traditional military mission applications, there is an increased emphasis on the use of autonomous systems such as unmanned aerial, ground, and underwater vehicles, and other autonomous systems to perform surveillance, reconnaissance, search and recovery, and search and destroy missions. There are advantages to miniaturizing these systems such as signature reduction, ability to penetrate small, enclosed spaces, and reduced logistics. In addition, the potential for such micro-robotic systems to be deployed as distributed, coordinated networks or swarms both increases mission flexibility and complicates an adversary's countermeasures.

Space assets are now at risk from numerous potential adversaries, especially at the onset of a conflict, and will have to be rapidly reconstituted, most likely with **swarms of "nanosatellites,"** each about the size of a shoebox. Compact, energy-dense nanoenergetic materials will power the tiny thrusters needed to maneuver such satellites in orbit.

19.4.1 Expected Operational Developments in Nanoenergetic Materials by 2040

The next twenty years will see the development of small, highly energetic explosive payloads for **microdrones**, which could be deployed in virtually undetectable swarms to disable power grids and communications networks, penetrate life support systems of underground facilities, and disable electronics and life support systems. In addition, smaller payloads for hypersonic missiles and conventional artillery will result in reduced logistics and smaller radar, heat, and optical signatures. Ultimately, the energetic yields of such materials could be dialed up or down to restrict collateral damage or maximize destructive effects.

The military will also need propulsion systems to enable maneuvering in orbit of critical space assets for communication and targeting. Nanoenergetic materials will be critical for solid fuel propulsion systems with much lighter weight and greater energetic yields than current solid fuel systems with minimal environmental impact.

Breaching and antimateriel applications rely on much the same chemistries as propellants and explosives but have their own unique requirements. These include stable and low-temperature ignition but extremely high-temperature combustion for breaching fortifications such as underground bunkers or destroying advanced materials such as ceramic composites and metal alloys, as well as for welding during forward repair or industrial manufacturing.

Finally, energy storage and power generation in forward-deployed units with limited logistics reach-back will require new energy dense materials for batteries with greater yields per mass than current materials.

19.4.2 General Technology Limits of Nanoenergetic Materials

Controlling particle size of nanomaterials and their tendency to stick together in larger clusters during synthesis is a major challenge that has yet to be resolved. The goal is to prepare monodisperse particles—that is, particles of identical size and shape, in the optimal range of 10–100 nm, since metal particles smaller than 10 nm tend to be pyrophoric and spontaneously combust in the presence of an oxidizer such as air. New surface particle coating materials are needed to enhance energy content, control ignition temperatures, tailor the rate of combustion, and mitigate processing issues.

Ignition behavior of nanoparticles is not well understood, and most research has focused on aluminum. Therefore, a systematic analysis of other elements and chemical compounds that may be superior to aluminum is needed. A better understanding of the thermomechanical properties of the oxide layer, a layer that spontaneously forms and interferes with combustion, on nanoparticles is required to develop a complete model for particle ignition. Combustion temperatures are dependent on several properties, and the relative importance of vapor-phase and surface reactions is not currently sufficiently worked out to allow for the predictive modeling needed to tailor the materials.

Nanofluid fuels hold great promise, but some basic mechanisms need to be examined. These include the relative catalytic and thermal effects of the addition of nanoparticles to fuel, the agglomeration of the nanoparticles in the fuel that can lead to uneven burn, and the multiphase flow dynamics in the exhaust nozzle of rocket engines. In short, will adding nanoparticles enhance propellant burn or have unpredictable negative effects on the combustion process?

19.5 HIGH ENERGY MATERIAL RESEARCH IN INDIA

High Energy Materials Research Laboratory (HEMRL) is one of the premier laboratories of the Defence Research and Development Organisation (DRDO) located in Pune Maharashtra, India. Main area of works of the lab include research and development of high energy materials and related technologies. HEMRL is organized under the Armaments Directorate of DRDO.

HEMRL is the main DRDO laboratory and one of the few labs in India that is involved in basic and applied research in all areas of high energy materials. Under this mandate, it conducts R&D in formulation, design and development of propellants, high explosives, pyrotechnics, polymeric materials, liners/insulators, and other materials. These include studies on the physiochemical and combustion characteristics of materials, the study of detonation phenomena and the development of new systems. HEMRL also carries out the synthesis of new explosive materials and compounds of importance in the field of HEMs. Design and development of pilot plants for synthesis of high energy materials and related non-explosive chemicals is also carried out. After the development of technologies and products, HEMRL is involved in the successful production and transfer of technology of these products to appropriate agencies.

In addition to a number of laboratory analytical instruments, HEMRL has state-of-the-art research and production facilities for the study of explosive, propellant and pyrotechnic phenomena. These include facilities for Measurement of flame intensity and temperature, study of detonation phenomenon, determination of mechanical properties of propellants & polymers and facilities for static Rocket testing and Non-destructive testing of materials. HEMRL also has labs and pilot production plants for synthesizing high energy materials and their allied chemicals.

HEMRL handles different projects for the Indian Ordnance Factories, Indian Armed Forces and other organizations like BARC and the Indian Space Research Organisation.

In 2024, HEMRL successfully designed and implemented an indigenous, cost-efficient manufacturing process for large-scale production of amorphous boron powder (BP) using metallothermic reduction of boric anhydride.

Air Regenerating Composition in HEMRL is used for maintaining breathable air within a closed space. It regenerates air inside confined space by liberating oxygen and simultaneously absorbing carbon dioxide. It can be used in civil applications like rescue work in mines, fire/ toxic environments, toxic gas/chemical plant operation, mountaineering expeditions, complete air regeneration inside a spacecraft, under sea exploration and mining etc.

Chemical Kit for Detection of Explosives (CKDE) in HEMRL is a compact, low-cost and handy explosive detection kit has been designed and perfected for field detection of traces of explosives.

The kit yields a colour reaction, based on which explosives can be detected in minutes. It is used for identification of all common military, civil and home-made explosive compositions, and is being used by Police and BSF for the detection of explosives.

Indian CL-20 in HEMRL is a new high explosive is in the making at a DRDO lab here that could replace other standard explosives of the armed forces such as RDX, HMX, FOX-7 and Amorphous Boron. Scientists at the Pune-based High Energy Materials Research Laboratory (HEMRL) have already synthesized adequate quantity of CL-20, the new explosive, in their laboratory.

The powerful explosive can substantially reduce the weight and size of the warhead while packing much more punch. The compound, 'Indian CL-20' or 'ICL-20', was indigenously developed in HEMRL using inverse technology. CL-20, so named after the China Lake facility of the Naval Air Weapons Station in California, US, was first synthesized by Dr. Arnold Nielson in 1987.

CL-20 or Hexanitrohexaazaisowurtzitane is a Nitramine class of explosive which is more powerful than HMX which in turn is more powerful than RDX

CL-20-based shaped charges significantly improve the penetration over armors and could potentially be used in the bomb for the 120-mm main gun mounted on the MBT Arjun tanks. The CL-20 in its reduced sensitivity enables easy handling and transportation and reduces the chances of mishap and loss to men, money, materials and machines.

CHAPTER 20

ADVANCED MAGNETS AND SUPERCONDUCTORS IN WARFARE

20.1 INTRODUCTION TO SUPERCONDUCTIVITY

Superconductivity is a set of physical properties observed in superconductors: materials where electrical resistance vanishes and magnetic fields are expelled from the material (Meisner effect). Unlike an ordinary metallic conductor, whose resistance decreases gradually as its temperature is lowered, even down to near absolute zero, a superconductor has a characteristic critical temperature below which the resistance drops abruptly to zero. An electric current through a loop of superconducting wire can persist indefinitely with no power source.

The superconductivity phenomenon was discovered in 1911 by Dutch physicist Heike Kamerlingh Onnes. Like ferromagnetism and atomic spectral lines, superconductivity is a phenomenon which can only be explained by quantum mechanics. It is characterized by the **Meissner effect**, the complete cancelation of the magnetic field in the interior of the superconductor during its transitions into the superconducting state. The occurrence of the Meissner effect indicates that superconductivity cannot be understood simply as the idealization of perfect conductivity in classical physics.

In 1986, it was discovered that some cuprate-perovskite ceramic materials have a critical temperature above 90 K (−183 °C). Such a high transition temperature is theoretically impossible for a conventional superconductor, leading the materials to be termed **high-temperature superconductors.** The cheaply available coolant liquid nitrogen boils at 77 K (−196 °C) and thus the existence of superconductivity at higher temperatures than this facilitates many experiments and applications that are less practical at lower temperatures.

A superconductor can be Type I, meaning it has a single critical field, above which all superconductivity is lost and below which the magnetic field is completely expelled from the superconductor; or Type II, meaning it has two critical fields, between which it allows partial penetration of the magnetic field through isolated points. These points are called vortices. Furthermore, in multicomponent superconductors it is possible to have a combination of the two behaviors. In that case the superconductor is of Type-1.5.

20.2 SUPERCONDUCTING MAGNET

A superconducting magnet is an electromagnet made from coils of superconducting wire. They must be cooled to cryogenic temperatures during operation. In its superconducting state the wire has no electrical resistance and therefore can conduct much larger electric currents than ordinary wire, creating intense magnetic fields. Superconducting magnets can produce stronger magnetic fields than all but the strongest non-superconducting electromagnets, and large superconducting magnets can be cheaper to operate because no energy is dissipated as heat in the windings. They are used in MRI instruments in hospitals, and in scientific equipment such as NMR spectrometers, mass spectrometers, fusion reactors and particle accelerators. They are also used for levitation, guidance and propulsion in a magnetic levitation (maglev) railway system being constructed in Japan.

20.3 TECHNOLOGICAL APPLICATIONS OF SUPERCONDUCTIVITY

Technological applications of superconductivity include:

- the production of sensitive magnetometers based on SQUIDs (superconducting quantum interference devices);
- fast digital circuits (including those based on Josephson junctions and rapid single flux quantum technology);
- powerful superconducting electromagnets used in maglev trains, magnetic resonance imaging (MRI) and nuclear magnetic resonance (NMR) machines, magnetic confinement fusion reactors (e.g. tokamaks), and the beam-steering and focusing magnets used in particle accelerators;
- low-loss power cables;
- RF and microwave filters (e.g., for mobile phone base stations, as well as **military ultra-sensitive/selective receivers)**
- fast fault current limiters;
- high sensitivity particle detectors, including the transition edge sensor, the superconducting bolometer, the superconducting tunnel junction detector, the kinetic inductance detector, and the superconducting nanowire single-photon detector;
- railgun and coil gun magnets; and
- electric motors and generators.

20.4 WORLD'S STRONGEST IRON-BASED SUPERCONDUCTING MAGNET MADE USING AI

In a possible breakthrough for affordable magnetic resonance imaging (MRI) machines and the electrified transport systems of the future, scientists have developed the world's strongest iron-based superconducting magnet using artificial intelligence.

The major benefit of superconducting magnets is that they can produce very strong and stable magnetic fields without extensive power needs.

This enables their use in several technologies such as MRIs which use the magnetic field to produce clear and 3-dimensional photos of soft tissue.

The researchers from King's College London and Japan have fabricated the cheap and powerful iron-based superconducting magnet using machine learning (ML), which can pave the way for widespread and affordable use of the technology.

Dr. Mark Ainslie from the King's Department of Engineering collaborated on the project with researchers from Tokyo University of Agriculture and Technology, the Japan Science and Technology Agency, the National Institute for Materials Science and Kyushu University.

The project resulted in the creation of a superconducting magnet which has a "magnetic field that is 2.7 times stronger than that previously reported."

Using a new machine learning system called BOXVIA; the scientists developed a framework that could optimize superconductor creation in the lab faster than before.

BOXVIA spotted patterns that improve performance and fine-tuned parameter changes to come up with the most optimal design for the magnet. The process would have otherwise taken months to create and then test its properties.

The researchers also discovered that the superconducting magnet they had made using the machine learning system had larger iron-based crystals within the magnet structure. This is remarkably different from the structure of magnets made without BOXVIA.

The sample produced by AI was remarkably different as it had wide range of sizes of iron-based crystals, as opposed to the uniform structure that is favored by researchers.

"Superconducting magnets are the backbone of the future. Not only are they used to image cancers with MRI machines, but they will be vital for electric aircraft and nuclear fusion," said Ainslie.

"However, the materials and technology required to create traditional copper-based wire superconductors are typically expensive, which has resulted in limited market penetration," he added.

Ainslie went on to add that using them in bulk form, as a magnet that doesn't lose its magnetism once magnetized, "can result in a smaller footprint in comparison to heavier coils of wire, but copper-based bulk superconductors can take weeks to fabricate."

According to Ainslie, the research lays down the groundwork for making superconducting magnets that are powerful enough for industrial applications at speed.

"Using artificial intelligence (AI), we've produced a cost-effective and scalable alternative using iron, which is a lot easier to work with and opens the door for smaller and lighter weight devices," he said.

"The first iron-based superconductors were made over ten years ago, but the magnetic fields they produced were nowhere near strong or stable enough for widespread use."

According to him, this will make the MRI machines cheaper and could also lead to the creation of a "new generation of smaller (MRI) units that could be deployed at a GP's office, rather than requiring large rooms in hospitals, widening accessibility."

20.5 APPLICATIONS OF ADVANCED MAGNETS AND SUPERCONDUCTORS IN WARFARE

Advanced magnets and superconductors have potential applications in warfare, primarily in areas like mine detection and sweeping, enhanced radar systems, high-power directed energy weapons, improved propulsion systems for ships, and advanced stealth technologies, all leveraging the unique properties of superconductors like zero electrical resistance and strong magnetic fields to create highly efficient and powerful military capabilities.

20.5.1 Key applications

(i) Mine Countermeasures

High-temperature superconducting magnets can be used to detect and detonate magnetic influence mines underwater by creating a strong magnetic field that interacts with the mines, allowing for more effective mine sweeping operations.

(ii) Enhanced Radar Systems

Superconducting magnets could power extremely powerful radar systems with high resolution and detection capabilities, enabling better surveillance and target acquisition.

(iii) Directed Energy Weapons

Sensors and Weapons are the most prominent differentiating feature between a commercial and naval platform. Traditional weapons of the naval platform e.g. guns and missiles are being replaced with novel directed energy weapons. These new weapons use high power pulses of up to 200 MJ electrical energy. A very recent development is the order placed by US Navy for the EM rail gun, which can propel projectiles or explosives at distance of 100 nautical miles travelling at Mach 6 or 7.

(iv) Ship Propulsion

Superconducting magnets could be used to create highly efficient electric propulsion systems for naval vessels, leading to increased speed and maneuverability.

(v) Stealth Technology

"Magnetic cloaking" concepts, where materials with tailored magnetic properties are used to manipulate magnetic fields around a vessel, could potentially reduce its detectability by magnetic sensors.

20.5.2 Advanced magnets and superconductors in warfare

(i) High Magnetic Field Strength

Superconducting magnets can generate significantly stronger magnetic fields compared to traditional electromagnets, enabling more powerful applications.

(ii) Zero Electrical Resistance

Superconductors have zero electrical resistance, allowing for highly efficient energy transfer and reduced heat dissipation.

(iii) High-Temperature Superconductors

Recent advancements in high-temperature superconductors allow for operation at less extreme cooling conditions, making them more practical for military applications.

20.5.3 Degaussing / Mine Protection

Degaussing is the process of neutralizing the magnetic field generated by the steel hull of the ship to avoid mines. Large current carrying copper wires around the ship are used for degaussing. Superconducting degaussing coils save 80% of the copper system weight resulting in smaller footprints and improved efficiency. Smaller footprint of HTS degaussing system (of US Navy) is particularly suited for submarines. US Navy retrofitted USS Higgins with HTS degaussing coils in 2008. Trials of the system were conducted from June 2009 through February 2010 with over 9,000 hours of operation and 37,000 nautical miles underway at times encountering 8-10 ft seas. This trial successfully demonstrated the use of HTS for degaussing.

20.5.4 Challenges and Considerations

(i) Cooling Requirements

Many superconducting materials still require very low temperatures to function, which can be challenging to maintain in field operations.

(ii) Cost and Complexity

Developing and implementing superconducting technologies can be expensive and complex.

(iii) Ethical Concerns

The potential for highly destructive weapons based on advanced magnetic technologies raises ethical concerns regarding their development and deployment.

20.6 APPLICATION OF HIGH-TEMPERATURE SUPERCONDUCTING MAGNETS TO COUNTER-MINE SYSTEM FOR NAVY

U.S. Navy ocean mine warfare experts are asking Textron Systems Corp. in Hunt Valley, Md. to build a counter-mine system for surface warships and uncrewed surface vessels that capitalizes on enabling technologies in high-temperature superconducting magnets.

Officials of the Naval Sea Systems Command in Washington announced an $18 million order to Textron in October for production of the Mine Sweep Payload Delivery Systems (MSPDS). Options could increase this order's value to $106.2 million.

The MSPDS surface counter-mine system locates and detonates acoustic and magnetic influence mines, and capitalizes on experimental Magnetic and Acoustic Generation Next Unmanned Superconducting Sweep (MAGNUSS) technologies, developed by Textron under supervision of the Office of Naval Research in Arlington, Va. MAGNUSS uses a high-temperature superconducting magnet with an advanced acoustic generator.

(i) Uncrewed systems

Textron's MSPDS mine warfare system reached initial operating capability in 2022. As of October 2023, the company has delivered two pilot systems and four low-rate production crafts and payloads. The Navy use these existing units to support its Littoral Combat Ship operations.

MSPDS is for the Navy's Mine Countermeasures Unmanned Surface Vessel (MCM USV), which will use several payload delivery systems, including the MSPDS, Mine-hunting Payload Delivery System, and Payload Delivery Systems For Mine Neutralization.

The Navy has been moving the mine countermeasures mission to the Littoral Combat Ship (LCS) as a suite of mission modules like the MCM-USV as a deployable system on the LCS to complete the minesweeping mission.

The advantage of using the MAGNUSS and its high-temperature superconducting magnet aboard an unmanned surface vessel (USV) is the ability to run at very high electrical currents with near-zero resistance, which can sweep magnetic influence mines when coupled to an acoustic generator.

In addition to the recent advancement of high-temperature superconductor magnets to provide a non-towed magnetic source, Navy researchers have been developing a non-towed, underwater acoustic source with low-drag as an alternative to a legacy acoustic generator that still enables additional benefits to the non-towed magnetic source.

(ii) Modular systems

The modularity of the high-temperature superconductor magnet and acoustic generator potentially could be deployed on any craft of opportunity — not just aboard the MCM-USV.

Textron engineers are building a high-temperature superconductor magnet and an acoustic generator; integrating the mechanical, electrical, and command and controls (C2) systems of the two systems with each other; and integrating the complete payload with the MCM-USV and its hull.

CHAPTER 21

PROTECTIVE CLOTHING AND EQUIPMENT IN WARFARE

21.1 MILITARY PROTECTIVE CLOTHING AND EQUIPMENT

Military protective clothing defends armed service personnel from a host of threats. Such threats include blunt force trauma, ballistic and blast-related effects, as well as chemical, biological, radiological, nuclear, and high-yield explosive (CBRNE) hazards. Armed forces require high-performance protective clothing and CBRNE equipment to ensure soldier safety and preserve combat effectiveness.

The Personal Protective Equipment (PPE) includes protective suits, helmets, goggles, gloves, boots, and decontamination and first-aid kits, as well as collective protection systems such as tents, containers, special-purpose vehicles and CBRNE detection equipment for military applications.

Defence and homeland security forces use different special-purpose PPE and CBRNE protection systems designed to help avoid harm, and to contain hazardous materials.

Some of the most effective types of protective equipment and systems used in military applications include, but are not limited to:

- Ballistic-resistant and blast-protective body armor;
- High-impact protective suits, helmets, gloves and boots;
- Military goggles, eye shields and sunglasses;
- Chemical protective, firefighter and explosive ordinance disposal suits;
- CBRN protective gas masks and hoods;
- Bomb suppression blankets;
- Waterproof and fire-retardant tents and mosquito nets;
- Armor plates and bullet-resistant glass for military vehicles;
- Chemical detectors and dosimeters;
- CBRN detection systems for defence vehicles;
- Decontamination systems for personnel and casualty operations, and
- Rapidly deployable laboratories for detecting and identifying chemical and biological warfare agents

21.1.1 What is an Armor?

An armor is a protective clothing with the ability to deflect or absorb the impact of projectiles or other weapons that may be used against its wearer. Until modern times, armor worn by combatants in warfare was laboriously fashioned and frequently elaborately wrought, reflecting the personal importance placed by the vulnerable soldier on its protection and also frequently the social importance of its wearer within the group. Modern technology has brought about the development of lighter protective materials that are fashioned into a variety of apparel suited to the hazards of modern warfare. With the rise of terrorism and the use of powerful personal weapons by criminals, armor is now frequently worn by police, by private nonmilitary security forces, and even by noncombatants who might be targets of attack.

Types of armor generally fall into one of three main categories: (1) armor made of leather, fabric, or mixed layers of both, sometimes reinforced by quilting or felt, (2) mail, made of interwoven rings of iron or steel, and (3) rigid armor made of metal, horn, wood, plastic, or some other similar tough and resistant material. The third category includes the plate armor that protected the knights of Europe in the Middle Ages. That armor was composed of large steel or iron plates that were linked by loosely closed rivets and by internal leathers to allow the wearer maximum freedom of movement.

Presumably, the use of armor extends back beyond historical records, when primitive warriors protected themselves with leather hides and helmets. In the 11th century BCE, Chinese warriors wore armor made of five to seven layers of rhinoceros skin, and ox hides were similarly used by the Mongols in the 13th century CE. Fabric armor too has a long history, with thick, multilayered linen cuirasses (armor covering the body from neck to waist) worn by the Greek heavy infantry of the 5th century BCE and quilted linen coats worn in northern India until the 19th century.

The advantage of chain mail is that it is quite flexible yet relatively impervious to slashing strokes (though a thrusting weapon can force the rings apart in spite of their riveted closure). In the form of a simple shirt, mail was worn throughout the Roman Empire and beyond most of its frontiers, and mail formed the main armor of western Europe until the 14th century. In Europe strips of mail were also worn underneath plate armor to close any gaps left between the rigid plates. Mail shirts were worn in **India** and Persia until the 19th century, and the Japanese used mail to a limited extent from the 14th century, though the rings in Japanese mail were arranged in a variety of ways, producing a more open construction than that found in Europe. Mail sleeves, leg harnesses, and hoods have also been worn.

Ancient Greek infantry soldiers wore plate armor consisting of a cuirass, long greaves (armor for the leg below the knee), and a deep helmet—all of bronze. The Roman legionary wore a cylindrical cuirass made of four to seven horizontal hoops of steel with openings at the front and back, where they were laced together. The cuirass was buckled to a throat piece that was in turn flanked by several vertical hoops protecting each shoulder.

Apart from helmets, armor made of large plates was probably unknown in western Europe during the Middle Ages. Mail was the main defense of the body and limbs during the 12th and 13th centuries. Mail hoods covered the head and neck, and mail leggings covered the legs. Mail, however, did not possess the rigid glancing surface of plate armor, and, as soon as the latter could be made responsive to the movements of the body by ingenious construction, it replaced mail. Thus, plate armor of steel superseded mail during the 14th century, at first by local additions to knees, elbows, and shins, until eventually the complete covering of articulated plate was evolved. A complete suite of German armor from about 1510 shows a metal suit with flexible joints covering its wearer literally from head to toe, with only a slit for the eyes and small holes for breathing in a helmet of forged metal. The armor suits of royalty and aristocrats were often elaborately gilded, etched, and embossed with fine decoration.

Japanese armor (yoroi) consisting of lacquer, leather, and silk and overlapping iron plates held together by laces, from the 18th century, imitating 12th–13th century yoroi; in the Metropolitan Museum of Art, New York City.

In the 16th and 17th centuries, improvements in hand firearms forced armorers to increase the thickness and, therefore, the weight of their products, until finally plate armor was largely abandoned in favour of increased mobility. Armor cuirasses and helmets were still used in the 17th century, but plate armor largely disappeared from infantry use in the 18th century because of its cost, its lowered effectiveness against contemporary weapons, and its weight.

21.1.2 Military protective clothing industry market trends

With an increasing number of terror attacks around the world, the need for a growing preparedness is of paramount importance.

Military protective clothing is being designed to tackle chemical, biological and nuclear warfare. With this, the increasing defence expenditures by countries for developing and procuring advanced CBRN protection systems are fueling the demand for CBRN defence equipment.

As one would imagine, Armed forces around the world are seeking to protect their soldiers against any kind of CBRN agents.

They are also looking to ensure their ability to conduct operations even in the case of pandemics, industrial disasters, or the use of weapons of mass destruction (WMD) by state or non-state adversaries.

21.1.3 Innovations in Modern CBRNE protective equipment

Materials like Kevlar, ultra-high-molecular-weight polyethylene (UHMWPE), and advanced ceramics are used in body armor, helmets, and vehicle armor. These materials provide excellent protection against bullets, shrapnel, and explosions. They are also lightweight, ensuring mobility.

Recent advancements in materials science and technology have led to the development of more lightweight, durable, and flexible CBRNE protective suits.

Innovations such as **nanotechnology-based materials**, **self-decontaminating fabrics**, and **breathable textiles** allow for greater mobility and comfort while maintaining high levels of protection.

Additionally, **smart sensors** embedded in suits can now monitor the environment and alert the wearer to the presence of dangerous substances, making these suits more effective in real-time threat detection.

21.1.4 How are CBRNE protective suits tested for effectiveness?

CBRNE protective suits undergo rigorous testing to ensure they meet the necessary safety standards. These tests evaluate the suit's resistance to various agents, including toxic chemicals and biological pathogens, as well as its durability under different environmental conditions.

Testing also includes assessing the suit's ability to maintain a seal and protect against contamination while allowing for mobility and comfort during prolonged use in hazardous environments.

21.1.5 What role do CBRNE suits play in military and emergency response scenarios?

In military operations, CBRNE suits are essential for protecting personnel from chemical and biological warfare agents, as well as radiological threats from nuclear devices.

They are also critical in peacekeeping and humanitarian missions where soldiers may be exposed to hazardous materials. In civilian emergency response, these suits are used by hazmat teams, firefighters, and medical responders during industrial accidents, terrorist attacks, or natural disasters involving toxic substances.

Their deployment ensures that responders can operate safely in contaminated environments while preventing the spread of hazardous materials.

21.2 TRENDS IN PERSONAL PROTECTIVE EQUIPMENT (PPE)

With the rapid development of technology, the traditional combat unit will take a qualitative leap, and the safety of soldiers will be seriously threatened, thus having an essential impact on the overall war situation. The safety of soldiers on the battlefield plays an important role in maintaining the overall combat power and the evolution of the war situation. As the life barrier of soldiers, Personal Protective Equipment (PPE) has significant value in protecting their own safety and improving combat effectiveness.

PPE mainly includes protective clothing and other protective systems and accessories, which protect the safety and health of soldiers, without affecting the basic requirements of normal operation, and enable soldiers to reduce the harm caused by various ordnance. It is the last line of defense to ensure the safety of soldiers. At present, according to the protection function, a single set of PPEs is mainly divided into **hard protection components and soft protection components**. The hard protection components include helmet, bulletproof clothing, stab-resistant clothing, explosion-proof suit and other protective equipment against mechanical injury. Soft protection components include gas masks, anti-toxic clothing, noise-proof ear-wear, anti-glare and anti-laser glasses, high temperature fireproof insulation clothing, nuclear radiation and nuclear pollution protection clothing, etc., which are mainly used to prevent various physical, chemical and biological violations. However, with the development of scientific and technological ordnance, the threats faced by soldiers are complex and diverse. The combat soldiers may be injured by traditional ordnance such as gunfire, explosive fragments, nuclear ordnance, biological ordnance and chemical ordnance, high-temperature flame, blade stabbing, and by new concepts such as radiation, high-speed pulse, laser and acoustics. Therefore, the use of various high-power and high-energy new ordnance will lead to the loss of some functions of traditional defense equipment. At present, the soldier's protective equipment still needs to adapt to the combat tasks, provide effective protection for soldiers facing various threats, the improvement in the performance of PPE will become more and more vital.

With the development of science and technology, **IoT technology** provides new ideas for PPE development and promotes the revolution of intelligent PPE. Combining the IoT technology with clothing design, to improve the communication between individual soldiers and the digital world, enhancing the life protection capabilities. Through the exploration in a long period, the integrated protection system for soldiers has gradually entered the vision of national armies. The design concept is based on arms, combining various equipment through integrated and systematic design, and use the soldier's body as a carrying platform to form a protection unit to greatly enhance the soldier's integrated protection capability, battlefield situational awareness capability and information processing capability.

In 1996, the French Army proposed the future individual integrated protection system. Since then, more than 20 countries such as Australia, Britain, Germany, Russia, South Korea, and **India** have also put forward their soldier system development plans, and the protection system technology has become increasingly mature and will play an important role in the future information battlefield. The individual integrated protection system has become an important element in the construction of 2lth century armies in western countries.

With the progress of **artificial intelligence**, big data and cloud computing, modern warfare form has transformed from information technology to intelligent, networked, data. The development of PPE for preserving military forces and win the future technological warfare has great significance in the new era.

Informatization is a key part of connecting soldiers to the digital battlefield and providing them with an information advantage. With the intensification of the digital battlefield, the requirements for the informatization of protective equipment are increasing. At present, the U.S., UK and other countries have incorporated the SIPE into the digital battlefield. The British Army will increase the digital test content of the individual system after launching the digital master plan. The U.S. Land Warrior system, based on the subsystem of computer helmets and sensors, connects soldiers to the digital battlefield, enabling them to rapidly discover targets and understand the battlefield situation, greatly improving their decision-making ability, reaction time, and overall combat effectiveness. Therefore, the development of **information technology** will significantly improve the active protection capability of the individual system.

In 2011, the French 1st Infantry Regiment made public for the first time the FELIN Future Soldier System, an integrated infantry system developed by Sagem in conjunction with 25 contractors. The core function of FELIN is the integration capability to integrate soldiers into the broader digital networked battlefield. Each soldier is equipped with a Bone-guide transmitter, allowing clear calls in a noisy battlefield environment. The FELIN system is also equipped with JIMMR target detectors and surveillance scopes with a maximum detection range of 3500 m against enemy forces. However, the FELIN system has a mass specification of 25 kg, which is a large load for soldiers to wear. In future, the French Army plans to integrate the FELIN system into the "Scorpion" modernization program and conduct research on new goggles using augmented reality technology. The goggles display a range of data such as friendly and enemy positions, tactical layouts, and munitions information, further improving the soldier's situational awareness.

In 2019, the U.S. Army introduced a new lightweight soldier system, the TEP (Torso and Extremities Protection), which is lighter, more flexible, more comfortable, and more functional. The TEP system integrates more advanced technology to ensure protection while improving man-machine ergonomics. The special leader in U.S. Soldier Program Executive Office stated that the mass of the vest in the TEP is 10.4 kg, which is 26% less than the Army's current Improved Outer Tactical Vest (IOTV), which protects against fragmentation kills and 9 mm gunshot wounds.

At present, the integrated protection system has become increasingly perfect, but also faces many challenges, of which the load has become the primary problem. Currently, a soldier in Iraq needs to carry a load of 45.4 kg, and if the soldier is equipped with electronic systems, the load will be even greater, thus affecting the soldier's mobility and sustained combat capability, so **the weight reduction is the biggest problem facing the application of the integrated protection system.** In addition, the development of information technology follows the trend of the world's new military revolution, and will occupy an important position in the future of individual combat. In summary, the development of information technology, lightweight, modularity and miniaturization of equipment will become the inevitable trend of the future integrated protection systems.

The addition of flexible rigid materials such as wire and metal rings to stab-resistant clothing has resulted in a semi-rigid suit with a lower surface density, greater flexibility and greater dexterity. At present, the representative semi-hard stab-resistant clothing on the market include the TurtleSkin MFA stab-resistant clothing from Warwick Mills and the semi-hard stab-resistant vest from Chongqing Shield King Company; The TurtleSkin MFA stab-resistant clothing combines high-performance fibers with stab-proof metal panels that are thin, less than 3.5 mm thick, with soft textiles for added flexibility and metal fasteners for stab-resistant protection; The semi-hard stab-resistant vest developed by Shield King Co. is made of a variety of new materials according to a special composite new process, which not only has the function of stab-resistant, but also can effectively block the general explosives and fragments, and has the functions of waterproof, acid and alkali resistance, UV protection, etc.

21.3 PPE FOR NUCLEAR, BIOLOGICAL AND CHEMICAL PROTECTION

Nuclear, Biological and Chemical (NBC) ordnances occupy a special place in security deterrence and military-strategic competition, therefore, countries around the world attach great importance to the protection of personnel from nuclear, chemical and biological ordnances. Since the First World War chemical warfare, military battlefields accordingly appeared to protect soldiers from skin poisoning anti-gas suits, then national armies began to use impermeable tarpaulin, rubber and plastic film, etc. to manufacture isolated anti-services suits, with good anti-gas properties, but not breathable, impermeable to moisture, in the temperature conditions above 30°C only allowed to wear time 20–30 min. In the 1960s, the British Army successfully developed and equipped its troops with breathable gas-proof clothing containing activated carbon for the first time, and various countries and armies began to develop their own distinctive gas-proof clothing containing activated carbon. After more than 40 years of research and development, forces in various countries have developed and equipped many different types of NBC protective clothing containing activated carbon, the more typical of which are micro spherical activated carbon, fibrous activated carbon, microporous PTFE, membrane layer fabrics and activated carbon cloth combinations.

21.4 INTELLIGENT PPE

As a product of the integrated application of materials, physics, biology disciplines, intelligent individual protective equipment connects isolated combat individuals with the digital world and continues to develop in the direction of intelligence and refinement, playing an important role in the harsh battlefield environment. Through literature research and summary, the research progress of intelligent PPE in four aspects of situational awareness, performance enhancement, physiological monitoring and environmental adaptation.

Russia's next-generation "Warrior" future soldier system will integrate Ratnik medical sensors to monitor combat soldiers' physiological and psychological information, including pulse, blood pressure, and emotional state, and to transmit this data to medical units and emergency commanders. Medical sensors can connect wounded soldiers on the battlefield with medical personnel, so that medical personnel can provide effective treatment to the wounded according to their type of injury and overall condition; Australia's "Future Soldier" System, disclosed in October 2018, is equipped with a new combat suit that combines camouflage, ballistic protection and ambulance functions, and can use special shaped memory fibers with millisecond response time, which can apply pressure to the wounded part of the body to stop bleeding and effectively reduce the amount of blood loss.

To perform a real-time assessment of the mission-specific physiological limitations of combatants, the **Spanish Ministry of Defense launched the ATREC project**, which aims to achieve the goal of determining the suitability of a soldier to perform a certain military mission based on the mental state he or she is in and whether there is a risk of endangering the life of the individual or others. At present, researchers have developed wearable testing systems consisting of sensing garments and fabric-based non-invasive devices. Among them, the sensing clothing is mainly composed of functional components such as gloves with built-in sensors, upper armbands and chest-bands with fabric electrodes. The non-invasive device includes a skin galvanometer, a respiratory recorder with a single-channel ECG amplifier, and two sensors used to measure skin and ambient temperature, respectively. The data of heart rate, respiratory rate, skin electrical response and peripheral temperature can be collected in real time, and used for the establishment and analysis of subsequent evaluation models.

In the future development of the 21st century, **biosensor technology** will certainly be a new growth point between information and biotechnology, and has a wide range of application prospects in military medicine, equipment development and other fields. At present, wearable soldier physiological monitoring technology still has problems of power endurance, wearing comfort, information security, etc. There is an urgent need for high sensitivity, high precision, high mechanical flexibility and low-cost sensors as well as high strength algorithm processing system. However, with the progressive research on new materials represented by **carbon nanotubes** and graphene, the continuous emergence of high-performance and differentiated functional fabrics, the continuous progress of electronic components flexibility, miniaturization, **3D printing technology**, and the increasing development of artificial intelligence and Internet of Things technology based on cloud computing and big data. Integrated electronic fabrics that combine functionality, comfort and safety will provide more options and implementations for the development of future soldier physiological monitoring equipment.

With the development and advancement of intelligent fabric materials, environmental adaptation equipment technology has been widely used in military individual protection, covering various aspects from camouflage color change and enemy identification to detection and simple treatment of injuries and assessment of combatants' mental state, thus significantly improving soldiers' protection ability to cope with dangerous environments.

A **smart fabric** is a material or structure that can sense and respond (mechanically, thermally, chemically, electromagnetically or otherwise) to stimuli from the environment, and is a further integration of traditional fabrics with electronics, sensor technology, and IoT technology.

Active protection is usually completed by active or highly intelligent fabrics that contain functional modules such as sensing, data transmission and processing, and execution, which can actively make judgments and reactions according to the external environment, and protect the wearing object in advance or issue a timely distress signal to facilitate subsequent rescue. Active smart fabrics are stimuli that can be sensed from the environment and can produce a response with sensing and actuation functions. For example, microclimate regulation system does not require external power supply, has a simple structure, and relies on phase change material (PCM) to absorb heat. Phase change material is an intelligent material that regulates environmental temperature by absorbing or releasing heat, which is reversible and can regulate the balance of body temperature through a "dynamic" process of heat storage and heat release.

Highly intelligent fabrics can sense, respond and adapt to a given environment. The U.S. Army is applying photosensitive and thermosensitive dye technology, electrochemical dye technology and dynamic biological camouflage technology to develop new camouflage training uniforms with a **"chameleon" effect**. Thermal and photosensitive dyes have the property of changing color with the back, and the camouflage color changes when the environmental conditions change. Electrochemical dyes use electronic simulation technology, which enables camouflage combat uniforms to change color by obtaining signals from the colors of the surrounding environment. Dynamic biological camouflage technology, which obtains electrical signals through visible light detectors connected to conductive polymers, performs color-light conversion so that soldiers' **combat uniforms constantly change color to match the surrounding environment.** The U.S. Army is applying **"invisible" technology** to the development of future combat uniforms, they believe that the development of combat uniforms to prevent the detection of mid- and far-infrared spectrum thermal imager has two key technologies, one is to ensure that the soldier's body temperature matches the background in which he is located, even if the temperature difference is as small as one tenth of a degree, it can also be detected by the thermal imager; the second is to make the soldier's face, head heat dissipation with the body parts wearing combat uniforms, the solution is to develop special skin coatings. Another important technology that is urgently needed in the current transformation of soldier systems is passive protection technology capable of countering a variety of advanced detectors to provide a robust counter-surveillance means for soldier systems. For example, all exposed parts of a soldier can automatically change color according to the surrounding environment, and if coupled with an anti-infrared detection coating, a good stealth can be achieved. At present, graphene-based infrared stealth coating applied to standardized clothing can be a good solution to the above problems.

21.5 FUTURE TRENDS IN PPE

Today, with the support of technological support, one-dimensional combat operations will no longer exist, replaced by a comprehensive strike from the ground, air, sea and other tangible space and electromagnetic, network, cognitive and other intangible space, the battlefield situation is increasingly brutal and complex, and put forward higher requirements for the development of PPE.

The future direction of PPE should include the following:

(i) The widespread use of new concept ordnance in the future battlefield will pose a great threat to individual soldier protection, and is a new issue facing individual soldier protection equipment. Future PPE should focus on the development of full protection against multiple coupled injuries, which requires effective response to a variety of same-space, multi-temporal domain, full-scale coverage of the damage elements;

(ii) The future development of PPE should take full consideration of ergonomics to ensure the wearing comfort and work efficiency. Focus on the development of flexible wearable equipment to make the equipment fit the human body more closely; in the process of equipment design should fully consider new technologies such as EMG, EEG, etc., so that the human-computer interaction of equipment becomes more diversified and friendlier, and the equipment system operating interface develops in the direction of multi-channel and adaptive to adapt to the complex and changing battlefield environment;

(iii) The future PPE will develop in the direction of integration, informationization, modularization, miniaturization and intelligence. Integrated design is an important strategy for the lightweighting of equipment systems, and systems and components should be considered comprehensively and integrated with multiple functions in the early stages of design; informatization is a key step in the soldier's journey to the digital battlefield, thereby greatly improving the soldier's situational awareness; modular equipment design should make use of open thinking, so that when the application scenario changes only the corresponding module needs to be modified to ensure the integrity of the function; miniaturization technology uses a system-level packaging approach to simplify equipment design by integrating components into a single, extremely small package, ultimately reducing system mass and thereby safeguarding soldier mobility; future intelligent development will comprehensively improve the degree of automation of the individual soldier system, and the man-machine electronic countermeasures will play a greater role in the future battlefield;

(iv) Strengthen the deep integration of high precision technology and PPE to build an intelligent development system of PPE in the future, as shown in Fig. 16. For materials technology, with the rise of AI technology, nanotechnology, biotechnology, more advanced brain-controlled exoskeleton, intelligent combat clothing began to appear in the field of individual combat; for sensors, integrated electronics and the introduction of multi-sensor data fusion technology to make the sensor to more miniature, lightweight, high-precision direction, the construction of real-time monitoring of the state of the soldier sensor network, to provide support for the development of intelligent PPE; for energy supply, the development of advanced battery energy technology and low-power devices can provide effective support for soldiers in sustained combat; for communication, flexible and reliable communication technology is used to improve the interactivity of intelligent equipment.

21.6 RECOMMENDATIONS FOR MILITARY PERSONAL PROTECTIVE EQUIPMENT (MPPE)

(i) Cutting-Edge Body Armor and Ballistic Protection

Advances in material science have led to the creation of lightweight yet highly effective body armor systems as part of military personal protective equipment. Modern military personnel now benefit from body armor plates made of advanced composite materials that can withstand high-velocity projectiles and provide enhanced protection against various ballistic threats. These plates are meticulously designed to disperse impact energy, reducing the risk of injury significantly. Moreover, flexible armor solutions have been developed to ensure soldiers maintain their mobility without compromising protection, allowing them to operate effectively in diverse combat scenarios.

(ii) Helmets: The Vanguard of Safety

Helmets are a cornerstone of military PPE, safeguarding soldiers from head injuries and providing ballistic protection. Recent innovations in helmet technology have resulted in lighter and more ergonomic designs, catering to the dynamic needs of the modern battlefield. Advanced materials like ballistic fibers and composites offer improved impact resistance and shock absorption, ensuring soldiers are shielded from a variety of threats. Integration of communication systems, night vision devices, and other sensors directly into the helmet design further enhances situational awareness and mission effectiveness, making helmets more than just protective gear.

(iii) Communication Integration for Enhanced Coordination

Effective communication is vital on the battlefield, and modern military PPE incorporates advanced communication systems directly into the gear, ensuring a holistic approach to personal protective equipment for military personnel. Helmet-mounted displays provide real-time data, maps, and critical information to soldiers, empowering them with the tools they need for informed decision-making. These integrated communication systems also allow for secure and seamless information exchange between team members, contributing to coordinated efforts and the accomplishment of successful missions.

(iv) Thriving in Challenging Environments: Night Vision and Thermal Capabilities

Soldiers often operate in challenging environments that demand versatile personal protective equipment. Modern advancements in thermal and night vision technologies have been seamlessly integrated into military PPE, allowing soldiers to navigate and engage effectively in adverse conditions. These innovations provide a significant edge by assisting in identifying hidden threats and enhancing target identification, while the incorporation of night vision capabilities extends soldiers' operational effectiveness during darkness.

(v) Safeguarding Against Chemical and Biological Hazards

Safeguarding against chemical and biological hazards stands as an essential element within the realm of military personal protective equipment. Cutting-edge PPE encompasses ensembles, respirators, and filtration systems meticulously crafted to offer a robust shield against an extensive array of perilous substances. These defensive protocols guarantee the capability of soldiers to function in scenarios featuring chemical or biological dangers, thus diminishing the likelihood of exposure and associated health hazards. This stands as an indispensable component of military personal protective gear.

(vi) Prioritizing Comfort and Mobility

In addition to protection, comfort, and mobility are paramount in the design of modern military PPE. Ergonomically designed gear, as a key element of military personal protective equipment, allows soldiers to move freely and perform their duties without hindrance. Adjustable straps, strategically placed padding, and efficient ventilation systems contribute to overall comfort during extended periods of wear, ensuring that soldiers can focus on their missions without being encumbered by their protective gear.

(vii) Targeted Ergonomics for Maximum Performance

Recognizing that soldiers must operate in a wide range of conditions, modern Military Personal Protective Equipment (MPPE) prioritizes targeted ergonomics. Adjustable straps, strategically placed padding, and ventilation systems ensure that gear fits securely and comfortably, regardless of the soldier's body type. This attention to detail allows soldiers to move freely and perform their duties without restrictions, boosting overall performance and mission success.

(viii) Mobility and Flexibility: Unleashing Soldier Agility

Incorporating mobility and flexibility into MPPE has become a cornerstone of modern design. Innovations in materials and construction techniques have led to gear that adapts to the soldier's movements, allowing for agile response and quick adjustments. From body armor that doesn't hinder mobility to gloves that provide dexterity for intricate tasks, every aspect of MPPE contributes to maintaining soldiers' fluidity on the battlefield.

(ix) Sustainability and Durability: Gear Built to Last

Sustainability and durability have become integral considerations in the development of MPPE. With a focus on reducing waste and minimizing environmental impact, military gear is now designed to withstand harsh conditions and extended use. Utilizing advanced materials and engineering, MPPE is built to endure the rigors of combat, ensuring that soldiers are equipped with gear that remains reliable and effective throughout their missions.

(x) Training and Familiarization: Enhancing Soldier Preparedness

Incorporating new technology and design elements into Military Personal Protective Equipment (MPPE) requires proper training and familiarization. Military personnel receive specialized training to ensure they understand the features and functions of their gear. This training not only boosts their confidence in using the equipment effectively but also maximizes the advantages offered by the latest advancements. By being well-versed in the operation of their MPPE, soldiers are better prepared to face diverse challenges on the battlefield.

(xi) Adaptive Modular Systems: Tailored Protection for the Mission

Adaptive modular systems have revolutionized the customization of MPPE. These systems allow soldiers to tailor their gear to the specific requirements of their mission. Components such as additional armor plates, specialized pouches, and equipment attachments can be added or removed based on the needs of the operation. This flexibility ensures that soldiers are neither burdened by unnecessary weight nor lacking essential tools, optimizing their ability to perform effectively and respond dynamically to changing scenarios.

CHAPTER 22

APPLICATION OF COATINGS IN WARFARE

22.1 PROTECTIVE COATINGS FOR MILITARY APPLICATIONS

Some coatings are engineered to be self-decontaminating to protect military firearms, tactical vehicles, equipment, and uniforms by neutralizing organo-phosphorous chemical agents such as sarin. Such government-specified Chemical Agent Resistant Coatings (CARC) have been developed for military personnel and first responders, creating a safer environment for soldiers and first responders by mitigating the immediate impact of contact with nerve agents.

In 2014, the U.S. Department of Defense (DoD) honored the team that developed a new, lightweight, longer-lasting aircraft coating system, which DoD plans to use on a new generation of **stealth jets**. The new coatings system being tested for the F-35 Lightning II aircraft not only reduces aircraft weight while maintaining the required coating corrosion resistance properties, but also cuts more than 50 percent of the paint/paint-removal waste stream, according to DoD. The development team has said it could cut the manufacturing process for an F-35 by four days, trimming production costs by $435 million and saving $1.07 billion in operations and sustainment costs over the life cycle of the Joint Strike Fighter Program.

A military-specified coating creates a hardening effect that offers blast mitigation in buildings, including embassies and defense installations, and is designed to provide a fast, cost-effective alternative to traditional structural steel or concrete systems. This coating was specifically designed to meet criteria requirements of the U.S. Department of Defense Anti-Terrorism Force Protection.

The U.S. coatings industry has been recognized as a key player in the fight against corrosion of our military ships, planes and tanks. Corrosion of steel, aluminum, and other structural metals erodes the safety and financial stability of industries and countries alike. Fighting corrosion in ships, tanks, planes and equipment costs the Pentagon $22.9 billion a year. Corrosion costs advanced industrialized nations about 3.5% of GDP to replace damaged material and components, plus a similar amount due to lost production, environmental impact, disrupted transportation, injuries, and fatalities.

Saltwater is the ever-present, costly enemy of Navy ships — a major cause of corrosion. New anti-corrosion marine coatings with military performance specification battle the costs and offer longer service life (2-3 times the longevity). They require less maintenance and can be cleaned versus repainted, which is more environmentally prudent because it requires fewer coats.

22.2 THERMAL PROTECTION COATING FOR ARTILLERY PROJECTILES

At high Mach speeds the projectiles may be exposed to high temperatures and heat fluxes up to 3500°C and 1000 W/cm^2, respectively. These are new environments to which conventional gun launched ammunition has not been subjected to. Along with qualifying artillery for new weapons platforms such as Extended Range Cannon Artillery (ERCA), they also have to survive the extended range environment.

The US Army is currently looking for novel **thermal protection coatings** for artillery shells in an effort to extend the capability of conventional ammunitions and enable integration of other aero-structural materials such as polymer matrix composites and high strength alloys.

The proposed solution must be able to protect the underlying base material against high heat flux and high temperature damage. The technology should be capable of surviving typical artillery gun launch loads and should conform to the geometry of an artillery projectile.

22.2.1 Phase I

During the Phase I contract, successful proposers shall conduct a proof-of-concept study that focuses on thermal protection coating technologies that can withstand and operate within varying thermal loads ranging from 5 W/cm^2 to 700 W/cm^2 and temperatures ranging from ambient to 2000°F (objective) for up to 5 minutes (objective).

Coating thickness should not exceed 5 mm (objective) and can be ablative in nature so long as sufficient thermal protection is sustained to meet the objectives. Investigations should include analysis of material performance under transient thermal loading and thermos-structural performance of a coated Inconel steel substrate.

A final proposed concept design, including a detailed description and analysis of potential candidate coating technology is expected at the completion of the Phase I effort.

22.2.2 Phase II

Using the data derived from Phase I, in Phase II the proposer shall fabricate and integrate a prototype of the technology into a nominal projectile form-factor. The proposer shall further their proof-of-concept design and determine the applicability of the coating for different surface materials.

Upon evaluation of the design through a critical design review, the prototype hardware's survivability shall be demonstrated via high G testing (35,000 G objective) in an air launched munition and aerothermal ground testing. Information and data collected from these tests will be used to validate operational performance.

22.2.3 Phase III

Phase III selections shall identify large scale production alternatives and fabricate 20 prototypes that can be integrated into a nominal projectile form-factor to be identified by the 'SBIR: Army 20 Topics and Concepts Government'. Live fire tests will be conducted, and the prototype integrated with projectile form-factor will have to withstand shock loads approaching 35,000g's.

Phase III selections will develop of a cost model of expected large scale production to provide estimates of non-recurring and recurring unit production costs. Production concept for commercial application will be developed addressing commercial cost and quality targets.

Phase III selections might have adequate support from an Army prime or industry transition partner identified during earlier phases of the program. The proposer shall work with this partner to fully develop, integrate, and test the performance and survivability characteristics of the design for integration onto the vendor's target platform.

22.3 HOW COATING SYSTEMS HELPS MILITARY PERFORMANCE?

When it comes to military applications, industrial coatings perform many vital functions that the average layman may not be aware of. Because of the chemical compositions of certain coatings, called Chemical Agent Resistant Coatings or CARCs, coated military vehicles and combat gear can better resist dangerous substances. This is just one of the ways that coatings help protect troops and reduce casualties in combat situations.

(i) Coating for Weapons

The performance of firearms on the battlefield is crucial. Military personnel may find themselves in combat in any weather conditions or in a variety of landscapes. So whether they are fighting in a dry, dusty desert with gusting winds or a humid, swampy forest, they need their weapons to be performing optimally. Coatings help firearms hold up to harsh conditions by discouraging the accumulation of rust that may compromise their functionality. This increased durability allows for a longer and better life for the weapon.

(ii) Armored Vest Protection

When applied to armor plates, an extra layer of protection can be created that slows and traps fragments from small arms fire and explosive shrapnel. Small arms fire and explosives are responsible for a large number of casualties in military actions, so any extra protection that it can be important for military personnel. The paint is sprayed onto ballistic plates, and it's ready to be back out in the field after a short curing period. The low weight of the material is great for service members who are already carrying a lot of gear, often in parts of the world that are known for their high temperatures.

(iii) Coating for Vehicles

Military vehicles are expected to work in the most extreme conditions. From boiling heat to freezing cold, and under constant threat of attack, our military men and women need vehicles that can protect them from all of it. To prevent scuffing, scratches, and corrosion, while also mitigating blast damage, a protective coating is applied. It also creates a slip-proof surface for safe exits, entrances, and standing, all while reducing cabin temperatures. The low weight also prevents adding extra strain to the body or engine of military vehicles.

A very important function of coating for military vehicles is reducing its signature. By "signature" we mean its signature as it may appear on a radar, an infrared scanner, or other detection technology.

Coatings can help reduce a vehicle's signature by blocking or interfering with the detection method, which greatly reduces the likelihood that the personnel inside will be harmed. In this way, vehicles become "camouflaged," allowing them to better accomplish their tasks efficiently and safely.

(iv) Coating for Sea Vessels

The U.S. Marine Corps relies upon protection on the torsion bars of its Assault Amphibious Vehicle (AAV). The 7-foot, 130-pound torsion bars are mounted underneath the tracked vehicle and are a critical component to the suspension system that must endure harsh conditions as it travels in the ocean and over rough terrain beach landings.

For large shipping companies, the reliability of their sea vessels is of the utmost importance. Because cargo ships can be so massive, they have a large number of parts that need to be functioning optimally and a large amount of surface area that is exposed to the elements. And at sea, the weather conditions can range from ice cold to high winds, heavy rains, choppy seas, and more. Needless to say, proper maintenance of one of these ships is a very involved process requiring the hands of many engineers and technicians.

The continuing development of industrial coatings allows for longer lifespans and greater resistance to the hazards of sea conditions for a variety of parts on a cargo vessel. Here,

What Are The Dangers of Rust?

Rust is the first issue that comes to mind when considering the issues a vessel may face after repeated trips at sea. This should come as no surprise, seeing as ships are most often comprised of metal components, and saltwater is highly corrosive to metal.

Rust on the hull of a ship will compromise its integrity. If the rust does not itself eat through the hull, it will make the hull more susceptible to punctures, scrapes, or other contacts. While it's not likely that a cargo ship will be making an impact with other objects and surfaces, a compromised hull is still a significant issue as it can put the lives of the crew and the safety of the cargo at risk. Rust on smaller parts of the ship, such as nuts & bolts, hinges, piping, etc. will do just the same thing – compromise its integrity and allow for a greater likelihood of malfunction and the need for replacement parts.

How Can Coatings Help?

For a hull, a coating places a layer of chemically resistant substance between the seawater and the metal. This layer prevents the saltwater from chemically interacting with the metal, allowing for smoother sailing overall. The same principle applies to various parts of the ship. With industrial coatings, cargo ships can enjoy fewer repairs, longer lifespans for their parts, and an overall improvement in performance.

22.4 INDIA DEVELOPS ALGORITHM-BASED TECH THAT MAKES WEAPONS APPEAR AS DECOYS

India's Defence Research and Development Organisation (DRDO) has developed an algorithm-based camouflage technology that reportedly makes military assets appear as decoys.

The tech utilizes special paint and stickers to effectively conceal otherwise easily recognizable equipment, such as artillery, tanks, and missile systems.

It allows these assets to blend into their surroundings, making it difficult for enemy satellites, drones, and reconnaissance aircraft to identify them as real targets.

By leveraging advanced algorithms, the technology can apply up to five multi-spectral stickers in military color shades that mimic common operational terrains like sand and grass.

Recent Testing

The DRDO recently tested the new tech on a T-90 main battle tank at the Pokhran Field Firing Range in northwestern India.

The camouflage coating reportedly deceived high-resolution cameras and infrared sensors, making the tank appear as something else.

According to local reports, the trial confirmed the technology's exceptional effectiveness in reducing the detection probability of large military equipment.

Once fielded, the camouflage system is expected to aid covert military operations, such as surprise attacks or strategic repositioning.

'Make in India'

The camouflage tech is among the latest innovations unveiled under the "Make in India" initiative, which aims to promote self-reliance in defense.

Last month, the Indian Army tested domestically-built drones capable of flying at altitudes exceeding 15,000 feet (4,572 meters).

The DRDO also presented a lightweight bulletproof vest that reportedly offers 360-degree protection for soldiers.

With these state-of-the-art military assets, India is positioning itself as a global manufacturing hub for strategic defense products.

CHAPTER 23

SMART MATERIALS IN WARFARE

23.1 INTRODUCTION TO SMART MATERIALS

Smart materials, also called intelligent or responsive materials, are designed materials that have one or more properties that can be significantly changed in a controlled fashion by external stimuli, such as stress, moisture, electric or magnetic fields, light, temperature, pH, or chemical compounds. Smart materials are the basis of many applications, including sensors and actuators, or artificial muscles, particularly as electroactive polymers (EAPs).

There are a number of **types of smart material**, of which are already common.

Some examples are as following:

- **Piezoelectric materials** are materials that produce a voltage when stress is applied. Since this effect also applies in a reverse manner, a voltage across the sample will produce stress within sample. Suitably designed structures made from these materials can, therefore, be made that bend, expand or contract when a voltage is applied.
- **Shape-memory alloys and shape-memory polymers** are materials in which large deformation can be induced and recovered through temperature changes or stress changes (pseudo-elasticity). The shape memory effect results due to respectively martensitic phase change and induced elasticity at higher temperatures.
- **Photovoltaic materials** or optoelectronics convert light to electrical current.
- **Electroactive polymers (EAPs)** change their volume by voltage or electric fields.
- **Magnetostrictive materials** exhibit a change in shape under the influence of magnetic field and also exhibit a change in their magnetization under the influence of mechanical stress.
- **Magnetic shape memory alloys** are materials that change their shape in response to a significant change in the magnetic field.
- **Smart inorganic polymers** showing tunable and responsive properties.
- **pH-sensitive polymers** are materials that change in volume when the pH of the surrounding medium changes.
- **Temperature-responsive polymers** are materials which undergo changes upon change in temperature.

- **Halochromic materials** are commonly used materials that change their color as a result of changing acidity. One suggested application is for paints that can change color to indicate corrosion in the metal underneath them.
- **Chromogenic systems** change color in response to electrical, optical or thermal changes. These include electrochromic materials, which change their color or opacity on the application of a voltage (e.g., liquid crystal displays), thermochromic materials change in color depending on their temperature, and photochromic materials, which change color in response to light—for example, **light-sensitive sunglasses** that darken when exposed to bright sunlight.
- **Ferrofluids are magnetic fluids** are affected by magnets and magnetic fields.
- **Photomechanical materials** change shape under exposure to light.
- **Polycaprolactone (polymorph)** can be molded by immersion in hot water.
- **Self-healing materials** have the intrinsic ability to repair damage due to normal usage, thus expanding the material's lifetime.
- **Dielectric elastomers (DEs)** are smart material systems which produce large strains (up to 500%) under the influence of an external electric field.
- **Magnetocaloric materials** are compounds that undergo a reversible change in temperature upon exposure to a changing magnetic field.
- **Thermoelectric materials** are used to build devices that convert temperature differences into electricity and vice versa.
- **Chemoresponsive materials** change size or volume under the influence of external chemical or biological compound.

23.2 IMPORTANCE OF SMART MATERIALS

Smart materials are a prime example of how science and engineering are revolutionizing our lives. These materials possess the extraordinary ability to change their properties in response to external stimuli, offering a vast array of potential applications across various industries.

Smart materials, also known as responsive materials or intelligent materials, are substances that can alter their properties in response to external factors such as temperature, pressure, light, magnetic fields, or electrical currents. This responsiveness allows them to perform tasks or adapt to their environment, making them invaluable in a wide range of applications.

Smart materials have the potential to significantly enhance the efficiency of various systems and processes. For instance, in the construction industry, self-healing concrete can repair small cracks without human intervention, prolonging the lifespan of structures and reducing maintenance costs.

In the field of healthcare, smart materials have given rise to a new era of diagnosis and treatment. Shape-memory alloys, for example, are used in minimally invasive surgery to create self-expanding stents. This reduces the need for open surgery and speeds up recovery times for patients.

Self-cleaning and anti-fouling coatings can be applied to surfaces to prevent the build-up of dirt and pollutants, reducing the need for harsh chemical cleaning agents and conserving water resources. Additionally, materials that can harvest energy from their surroundings are being used to power sensors and small devices, decreasing the reliance on traditional batteries.

Shape-changing materials can be used to design aircraft wings that adapt to different flight conditions, making aviation more efficient and environmentally friendly. Moreover, smart materials are employed in stealth technology, enabling the creation of adaptive camouflage systems for military purposes.

While the importance of smart materials is undeniable, there are challenges to overcome. These include high production costs, limited scalability, and the need for further research and development. As technology advances, these challenges are being addressed, and the potential applications for smart materials continue to grow.

23.3 SMART MATERIALS FOR THE MILITARY

Smart materials are used in the military for a variety of applications, including adaptive camouflage, self-healing coatings, and responsive armor systems. These materials are able to respond to changes in temperature, pressure, or electric fields.

Some **examples** of smart materials include:

(i) Shape memory alloys (SMAs)

These materials can return to their original shape after being deformed, usually when heated. SMAs can improve the performance and durability of aircraft and missile components.

(ii) Magnetostrictive materials

These materials are used for vibration control and can be used for active, passive, or semi-active vibration control.

(iii) Carbon fiber reinforced polymers (CFRPs)

These materials are lightweight, strong, and resistant to corrosion. They are often used in the design of aircraft structures.

(iv) Conductive polymers

These materials exhibit electrical conductivity, similar to metals, but are made up of long-chain macromolecules.

(v) Magnetorheological fluids (MRFs)

These fluids contain ferromagnetic particles that attract each other when a magnetic field is applied. This increases the fluid's viscosity and makes it behave more like a solid.

(vi) Smart textiles

These textiles can integrate multiple functionalities into a single piece of clothing or equipment, which can save time and resources, and increase soldier mobility.

23.4 SMART MATERIALS ARE CHANGING SOLDIERS' PERSONAL PROTECTION

Modern warfare has changed irrevocably. The operations in Afghanistan and Iraq have been characterized by highly agile insurgents who have promoted the urban warfare and whose weapon of first choice is often the Improvised Explosive Device (IED) very unpredictable weapons causing severe and injuries through violent impacts.

Moreover, with the focus of armies changing – from operational to contingent forces – and the great diversity of warfare environments in which they serve, the physical demands on armed forces personnel and the challenges on their equipment have become significant issues for global government defence departments.

One of the biggest equipment performance challenges is how to address the high incidence of lower limb injuries caused by load carriage. In the UK alone, over the past five years, the most common principal cause of medical discharge by armed forces personnel remains musculoskeletal disorders and injuries. The number of army personnel medically discharged with principal cause of musculoskeletal disorders and injuries has significantly increased in this time. Knee and back pain remain the two biggest causes of the musculoskeletal disorders which lead to servicemen and women being medically discharged from all three UK armed forces each.

Technology can play a significant role in helping to protect soldiers, and providing solutions to adapt personal protective equipment to meet the changing demands of modern warfare.

In meeting these demands with industry-leading solutions, technology SME are best placed to develop innovative products rapidly, enabling Government defence procurement agencies to acquire the latest advanced technology with which to equip troops.

D3O's Trauma Reduction Unrivalled Shock Technology:

D3O, a British technology innovation SME, is at the forefront of applying unique patented smart materials to engineer real-world impact protection solutions for the defence sector. D3O's technology is based on non-Newtonian principles. Its team of chemists, materials scientists and engineers has developed a range of patented smart materials which contain a rate-sensitive dilatant fluid, which means its viscosity increases with the rate of shear strain. In its raw state the material's molecules flow freely allowing the material to be soft and flexible, but under impact, the molecules lock together absorbing and dissipating the impact energy before instantly returning to a soft and flexible state. D3O material scientists incorporate the dilatant with different polymers to create a range of stable materials that retain these soft, flexible and shock absorbing properties.

D3O has applied this branded ingredient technology to develop the Trauma Reduction Unrivalled Shock Technology (TRUST) range of products which has been designed specifically for the defence and law enforcement sectors to offer a head-to-toe personal protection system that is lightweight, comfortable and flexible. These ready-made solutions can be sold within the defence industry via approved distributors or can be integrated into existing apparel and PPE produced by Government prime contractors, to be then purchased by the Government procurement agencies.

High Performance Integrated Knee Pad System

D3O has built on its long-standing partnership with outdoor apparel company Patagonia to co-develop a patented technology for the defence and workwear market. This patented technology enables a novel integration of the TRUST High Performance Internal Knee Pad into Active Combat Pants (ACP). The protective padding system will revolutionize how D3O currently sees protection for military and workwear personnel.

The system, that has both an inner pad and outer clip-on shell, uses D3O impact protection material and is secured and detached using clips through tailor-made process.

23.5 SMART MATERIALS GETTING SMARTER FOR THE ARMED FORCES

Army and Naval scientists have demonstrated a new class of solid-state phase change material for thermal energy storage, with applications and immediate impact ranging from large-scale power generation to the suppression of thermal transients associated with **high-energy laser directed energy applications.**

"In recent years, researchers and companies have attempted to develop smarter, more efficient and environmentally conscientious materials and energy systems," said Dr. Darin Sharar, principal investigator for the U.S. Army Combat Capabilities Development Command's Army Research Laboratory, the Army's corporate research laboratory also known as ARL.

"With directed energy, next-generation systems provide unique thermal management challenges due to their high heat flux and short pulse duration, requiring thermal design with transient loads in mind", Sharar said. Architectures using phase change materials are quickly becoming the preferred thermal management solution because of their ability to absorb thermal energy with minimal temperature increase, ensuring the directed energy systems can operate within spec. However, standard designs require engineering measures in the form of metallic fin structures to provide mechanical support, prevent liquid phase change materials leakage and enhance poor phase change materials thermal conductivity. Because of this, metallic fins account for ~1/2 to ~2/3 of the mass of state-of-the-art directed energy phase change materials heat exchangers, leaving as little as 1/3 for latent energy storage. This imposes significant design restrictions, hindering transition from the laboratory setting to real-world systems.

The discovery of thermal energy storage via shape memory alloys provides an unprecedented two order of magnitude improvement in the cooling figure of merit, defined by the product of the material latent heat and thermal conductivity. This opens a new paradigm of phase change material design, through which scientists can eliminate the need for heavy/large volume fin structures and fabricate thermal energy storage and heat transfer structures entirely out of metallic shape memory alloys.

This promises increases duty cycle for directed energy applications, along with significant size and weight savings, enabling more-capable, compact directed energy assets on smaller platforms.

23.6 SMART MATERIALS FOR ARMED FORCES IN INDIA

Smart materials have really changed the style of joint warfighting. They have enhanced both performance and adaptability of equipment. Shape-memory alloys and self-healing polymers are the biggest examples of smart materials. If India incorporates self-healing materials into its armored vehicles, it can increase their durability and cut down maintenance time. India can also enhance its operational readiness. These smart materials facilitate more stronger yet light weight equipment. We can get highly mobile strong protective gear for our forces in all kind of terrain and operational environment.

In the context of the Indian military, "smart materials" are primarily being researched and developed by the Defence Research and Development Organisation (DRDO) for use across all branches - Army, Navy, and Air Force - with applications including advanced armor, self-healing materials, adaptive camouflage, and noise-reducing components for quieter underwater operations, all designed to enhance the performance and adaptability of military equipment based on external stimuli like temperature or stress.

23.6.1 Key points about smart materials in the Indian Defence sector:

(i) Applications

Smart materials like shape-memory alloys and self-healing polymers can be used in protective gear, armored vehicles, naval vessels, and aircraft to improve durability, reduce weight, and adapt to changing environments.

(ii) DRDO focus

DRDO's Young Scientist Laboratory (DYSL-SM) is keen on developing indigenous smart materials for novel applications. To this end, the laboratory is working on two-fold activities wherein efforts are put to develop new smart materials with computational aided approach and to use existing smart materials in defence applications.

(iii) Benefits of Smart Materials

- **Enhanced protection**: Smart materials can provide superior ballistic resistance and protection against various threats.
- **Lightweight design:** By incorporating smart materials, military equipment can be lighter while maintaining high performance.
- **Adaptive capabilities:** Smart materials can react to environmental changes, adjusting their properties as needed.

23.6.2 Examples of smart material applications in the Indian military

(i) **Army:** Smart textiles for advanced combat uniforms with temperature regulation and enhanced protection.

(ii) **Navy:** Shape-memory alloys to reduce vibrations and noise on naval vessels, enhancing stealth capabilities.

(iii) **Air Force:** Smart materials for aircraft components that can adjust to changing aerodynamic conditions.

CHAPTER 24

SMALL SATELLITES IN WARFARE

24.1 IMPORTANCE OF SMALL SATELLITES IN WARFARE

The very first satellites launched into orbit were small, such as "Sputnik-1" (USSR), which was 58 cm in diameter and weighed 83 kg., and "Vanguard-1" (US) was 16 cm in diameter and weighed only 1.6 kg. The size of the first satellites was determined by the technical capacity of the available rocket carriers and the desire to receive a radio signal from space, so they were not very complicated. Over time, systems improved, and user needs and expectations changed, so satellites grew and reached unprecedented sizes. Most of the large satellites were launched during the Cold War e.g. reconnaissance satellite "Hexagon" (US) whose length was 16.2 m and mass was over 13 t.

Satellites are classified according to many parameters, the most popular of which is the classification according to the mass they have, i.e. femto (1-100 g), pico (0.1-1 kg), nano (1-10 kg), micro (10-100 kg), mini (100-1000 kg), and large satellites whose mass starts from 1000 kg. **Small satellites are that whose mass does not exceed 500 kg**.

The beginning of the development of small satellites was greatly stimulated, starting in 1980, by the international community of radio satellite amateurs and universities, who were looking for ways to develop cost-effective space systems performing various functions with minimal financial and technical resources. As a pioneer in the field of small satellites, the University of Surrey can be mentioned for its "UoSat program", which later influenced the creation of "Surrey Satellite Technology Ltd", the first small satellite manufacturer. This significant breakthrough facilitated an increase of small satellites available in orbit, e.g. from 2012 to 2017 increased from 70 to 380. Of the 300 small satellites put into orbit in 2017 **about 80% were satellites of the micro-nano class**. Meanwhile, even 94% of the number of satellites launched in 2020 consisted of small satellites.

Small satellites seem to be conquering space at an extraordinary rate today. After examining the companies' plans and issuing permits, it can be seen that at least 1,800 small satellites will be launched annually in the future. Materials and technologies freely available on the market are used for producing these spacecraft, allowing the discovery of new and cheaper ways of performing significant observation missions. Thanks to technological innovations, continuous miniaturization of key components, and advances in mechanical systems, batteries, and sensors, small satellites can provide high-quality data and services in areas such as remote sensing; data transmission, positioning, and timing, navigation, approach operations. The standardized implementation of hardware and software components has also greatly boosted the optimization of the production, integration, and testing of satellites and their subsystems, which is especially important for the launch of satellite clusters.

24.1.1 Advantages of small satellites

It would be naive to expect that a single small satellite could be superior to a large satellite, but small satellites can still perform many relevant functions:

(i) Small satellites are focused on performing a single function than large satellites, which may contain many complex sensors or diverse functionalities. **Small satellites are launched in clusters**, so the area covered by them can be equal to that of a large satellite. The performance of a single function leads to a smaller size of the satellite while at the same time influencing its complexity, which allows for faster production. A large satellite takes an average of 4 to 5 years to design and build, while small satellites can be built in a significantly shorter time frame, sometimes even months. The shorter development time of small satellites also allows for **the renewal of low earth orbit (LEO) satellites**, whose on-orbit life cycle is significantly shorter (1-5 years) than that of large satellites. Scientists are currently still experimenting with new solutions to increase the duration of small satellites in space by up to 7 years, such as using microwave electrothermal launchers. Engine technologies have also been developed for small satellites to change their orbits and take the correct position in the cluster.

(ii) The constant updating of small satellites allows us to expect that the newly launched satellites will have the most recent and maximally improved technologies installed.

(iii) Small satellites' relative simplicity and short development time generally result in lower costs. While traditional large satellites typically cost hundreds of millions or even billions of dollars, small satellites can be built for tens of millions or even less. For example, the creation and launch of the "Maxar" company's "WorldView-4" satellite weighting 2500 kg would cost USD 850 million, while the price of the "OneWeb" company's small satellite together with its launch services is estimated at over USD 1 million. Experts estimate that small satellites' average production and launch costs are up to 90% lower than those of large satellites.

(iv) It would be possible to create **clusters of small satellites** or a group of linked satellites that, e.g. have different sensors fly over the same point on the ground simultaneously, allowing different types of data to be combined. In this way, several small satellites would provide identical or even more detailed information than those large satellites.

(v) The small satellites that are launched frequently are part of a cluster/system that consists of over a hundred small satellites in total, which allows for an increase in the revisit frequency, e.g. the company "Planet" with over 150 satellites in LEO can collect information from any place on earth at least five times a day. Meanwhile, the **"SpaceX"** company, which plans to launch a total of up to 12,000 and later a total of up to 42,000 **"Starlink" satellites**, will be able to ensure constant and uninterrupted internet access and data transmission anywhere on Earth.

Currently, the prices of launching satellites into orbit have significantly decreased, so it can be expected that the interest in small satellites will not decrease in the future. LEO launches prices per 1 kg. ranging from USD 5-18 thousand, depending on orbit and waiting time. The "Falcon 9" launch vehicle of the "SpaceX" company can launch the cargo cheapest. Also, "SpaceX" created the

"Falcon Heavy" carrier, which, when fully loaded (up to 60 tons) would cost USD 2,500 per kilogram. Next to "SpaceX", several other companies are also trying to claim a share of the mentioned market. "Blue Origin" is developing a new heavy rocket, the "New Glenn", with two- and three-stage variants. "United Launch Alliance" is developing the "Vulcan" launch vehicle. Individual solutions are also offered on the market for launching small satellites: the US Defense Advanced Research Projects Agency (DARPA) project spacecraft "XS-1", which can achieve 10 flights in a 10-day period, companies "Virgin Orbit" and "Stratolaunch" air launch systems, and the "Rocket Lab Electron" and "Firefly Aerospace Alpha 2.0" test rockets.

Next SpaceX launch to deploy fewer Starlink satellites into higher orbit – Spaceflight Now. 60 "Starlink" satellites are placed on the front of the "Falcon 9" rocket "SpinLaunch" is currently testing an alternative method of launching up to 200 kg mass satellites – using a suborbital mass accelerator. The accelerator spins the payload up to 10000 g and launches it through the launch tube into space. When the rocket reaches the required height, its outer casing falls, and then the engines of the first and second stage are activated, which deliver the cargo to the desired orbit. The 33 m diameter accelerator is currently being tested, with all tests being completed by 2026. Following the tests, the 100 m diameter accelerator will be built to ensure the launch of small satellites into the required orbits. This method is expected to reduce the cost of launching cargo into space up to 10 times.

It is likely that the current possibilities of electronic miniaturization, the launch of small satellites into space, the constant decrease in costs, and, at the same time, the possibility for small states to have independent global monitoring systems that cover the whole world allow us to expect that small satellites will be even more popular.

24.1.2 Trends of Small Satellites in the Civil Sector

In many countries, commercial companies offering optical observation services from space have been founded, which are precisely based on the currently popular small satellites. Perhaps the most famous is the "Planet" company. "Planet" has developed and launched more than 450 satellites. Currently, two clusters of this company are functioning in orbit: 150 satellites "Dove" and 21 satellites "SkySat" (resolution 1 px. – 50 cm.). Both types of satellites have the ability to maneuver in space, so they can be directed in the desired direction. The company "Satellogic" currently has 26 satellites in orbit (1 px. – 70 cm.), and by 2025 is planning over 300 satellites intending to ensure a weekly update of the global map. For the sake of interest, it should be mentioned that the Albanian government has signed a three-year EUR 6.2 million contract with "Satellogic" to launch and maintain two satellites. "Albania-1" and "Albania-2" will ensure the transmission and processing of images, which will then be evaluated by environmental protection, border, and security services.

A $6 million contract is signed for the Albania 1 and 2 satellites/ Rama: They will monitor constructions without permission, soon with armed drones - E-TJERA. Small satellites "Albania-1" and "Albania-2

Another US company "BlackSky Global", which provides surveillance services enabled by optical and synthetic aperture radar technology, has launched 12 small satellites (1 px. – 30 cm.) into orbit and has signed a contract with the US National Reconnaissance Office for USD 1 billion for a period of

10 years. The Chinese company "Chang Guang Satellite Technology Co. Ltd." currently operates the "Jilin-1" cluster of 46 satellites (1 px. – 75 cm.), which it plans to expand to 138 satellites.

Meanwhile, other companies manufacture and operate remote sensing satellites with different types of sensors. For example, e.g. the company "Spire" manages 110 different types of satellites that collect not only radio frequency data but also allow monitoring of weather, sea traffic, and aviation activities. Another company, "Capella Space", uses synthetic aperture radar technology to monitor the earth in any weather, day and night. The company launched the first 7 satellites from a planned cluster of 36 satellites.

The company "HawkEye 360" has three types of clusters of small satellites deployed, that scan radio frequencies and asses information about the spectral environment and radiation locations. The company plans to launch up to 80 satellites of this type in total.

With the significant increase in the amount of data exchanged in space, the need for the **emergence of platforms** that enable this also increases. There are companies that have set themselves the goal of providing Internet connection in those parts of the world where terrestrial networks are not developed. "SpaceX's" "Starlink" satellite cluster is the most developed in the world, with nearly 3,000 satellites launched so far, which can provide data transmission speeds of 50-200 Mb/s in different parts of the world. The "Starlink" cluster is planned to have 12,000 satellites in its final phase. Another international company, "OneWeb", which is partly owned by the UK government, has almost 400 satellites in orbit and plans to increase the number to 648 by 2023. The "Kuiper project", overseen by "Amazon", plans to launch as many as 3,236 satellites by 2022. which would provide high-speed, low-latency broadband services at an affordable price anywhere in the world.

China-based companies "Guodian Gaoke" and "Commsat Tech Dev Co" are also active in this market. The first company will launch 38 satellites into space by the end of this year, and the second will deliver 8 satellites.

Over the past decade, many new satellite companies have emerged, but almost all of them are in the early stages of investment and development. Notably, no commercial small satellite cluster company has yet proven to be commercially viable without government support. Many experts are hesitant to assess when "SpaceX", the company that makes the "Starlink" communications satellites, will establish a market share that will allow it to maintain its manufacturing, launch, space and ground infrastructure, and support operations. Regardless of these circumstances, investors believe that the capacity of small satellites will increase, the technology will continue to develop, and companies will be able to take a sufficient share of the market. Companies manufacturing small satellites for remote sensing, communications, on-orbit servicing, and debris reduction are expected to dominate the space ecosystem in the coming decades.

24.1.3 Potential for small satellites to be used by the military and projects under development

There are noticeable trends in the US defence sector that, even though the US defence sector organizations have over 150 satellites that are capable of conducting optical and electronic reconnaissance, ensuring communication and navigation, weather forecasting, and early warning of ballistic missile launches, they still cover only 20% of their needs. The necessary remaining services are procured from the commercial sector, which is usually provided by small satellites. Noticing this trend, the US Department of Defense and the US intelligence community are investing more and more in small commercial satellites and their clusters.

Clusters of small satellites provide the required revisit frequency, which gives military decision-makers the advantage of knowing the near-real situation on the battlefield. The US military is currently testing this conceptual idea through the "Convergence" experimental project, where the capabilities of commercial and government-operated satellites have been used to photograph the battlefield and rapidly exchange information with the help of communications satellites. In addition, the latest data synthesis and artificial intelligence capabilities were used for data processing, which made it possible to shorten the decision-making speed from 20 minutes to 20 seconds (i.e., from target identification to firing a shot).

Clusters of small satellites provide another important advantage – resistance to kinetic attacks (e.g. against the Russian-made "Nudol" missile). Large satellites are seen as easy targets, while the destruction of one or more small satellites would not significantly affect the operation of the entire cluster or its functionality, so the adversary may decide not to launch this type of attack at all. Of course, clusters, like large individual satellites, can be vulnerable to ongoing cyber-attacks or could be destroyed by nuclear explosions in space. According to the latest research, small satellites are more resistant to cyber-attacks than large ones.

Due to the very short production time and cost of small satellites, in the event of a conflict, it is possible to quickly replace satellites in orbit, launch new ones for specific tasks, restore damaged ones, or increase the number of them in a certain orbit to increase the observation time.

DARPA runs the "BlackJack" program, which aims to encourage the development of small satellites in LEO related to US national security. One of the more exciting projects financed by the program – the production and launch into LEO of small satellites designed to demonstrate data exchange capabilities using lasers – the project took only 9 months from approval to test in space. Laser communication systems allow the safe transmission of large amounts of information (1 Gb/s) that cannot be affected by external interference or data interception, as with radio communication.

In line with plans to integrate small satellites into a common surveillance system, the US Department of Defense plans to expand data processing capabilities to include advanced analytics, artificial intelligence, and machine learning. For that purpose, contracts were concluded with "Ball Aerospace" and "Microsoft" companies in order to create solutions that allow the processing of large amounts of data generated by various types of sensors in clusters of small satellites.

As governments in the future will increasingly use small satellites operated by private companies to provide services to other commercial or government actors around the world, this will make it more difficult for adversary forces to make the decision to destroy several small satellites in a crisis situation knowing that this action can be met with a direct or indirect reaction from other states.

Small satellites can also be used as non-standard weapons in space. Satellites must reach a minimum speed of 25,000 km/h to maintain LEO orbit, so even a small object flying at that speed can damage or even destroy the satellite if they "accidentally" collide.

24.1.4 Potential of small states in the market for small satellites: a Lithuanian case study

Of course, many small states have their own research and production potential, some of which specialize in this particular market. In this article, we will review only the case of Lithuania because the author is from Lithuania.

It should be mentioned that Lithuania has set an ambitious strategic goal – to aim for the Lithuanian space sector to create about 1 per cent of the country's GDP. Currently, the support concept of the Lithuanian space sector is being developed, and the cooperation in the sector is being matured – a recently created space technology cluster, in which, in addition to the "Visoriai Information Technology Park", the companies registered in Lithuania "NanoAvionics", "Elsis Pro", "Geomatrix" and "Blackswan Space" participate.

24.1.5 Nanosatellites & CubeSats | NanoAvionics

"NanoAvionics", which produces small satellites (up to 220 kg) and their subsystems, manages satellites and ground stations, and organizes services for launching satellites into the required orbits, is probably the most widely known in the world. The company, which started operations in Vilnius, currently has branches in Kaunas, the US, and the UK and is one of the world's largest manufacturers of small satellites. Starting this year, "NanoAvionics" is owned by "Kongsberg Defense & Aerospace", a Norwegian technology company. Since its inception, "NanoAvionics" has completed more than 120 small satellite missions and other related projects. The Vilnius-based division is the company's main strategic, production, research, and satellite management centre, providing small satellites and mission management services to NASA, the European Space Agency, Thales Alenia Space, and other organizations around the world, including those related to in the creation of clusters of satellites of various configurations.

The "Elsis Pro" company develops data exchange solutions, and according to the European Space Agency contract, it will develop a data exchange platform on the "Galaxy" nanosatellite cluster. The "Geomatrix" company provides solutions for satellite data processing. Startup "Blackswan Space" specializes in developing autonomous technologies for satellites, while startup "Astrolight" develops wireless laser communication technologies for satellites and drones.

24.1.6 Updates – Astrolight

In terms of the potential of scientific institutions, the most outstanding could be "Vilnius Tech", whose scientists specialize in the development and research of precision mechanical drives, miniature piezo drives required for the control of satellite mechanisms and optical instruments, and in the development of adaptive antenna technologies.

24.1.7 Options for small states while using small satellites to increase national security

The war in Ukraine has clearly shown the importance of surveillance tools in space is evident and having said tools gives a very big advantage. Data on the deployment of troops, equipment, and weapons, as well as approach routes, were obtained from both state-run and privately owned satellites.

According to current trends and the level of technology development, it is clear that small satellites will become popular, the cost of their production and launching into space will decrease, the technical characteristics of satellites will improve, and they will be able to stay in orbit for a longer period of time. At the same time, there will be an increase in the number of companies that will invest in this niche by creating clusters of satellites designed to perform certain functions. Many companies will also offer users data processing and other services.

For a small state, it is important to properly assess the amount of information required for national security and ensure that the requested and received information is available to a wider range of users. Centralization of information retrieval and order would be required to avoid duplication. As the security situation changes, so should the periodicity of obtaining and updating information.

It is necessary to perform an analysis in order to assess what type of sensors would be the most effective, what would be the optimal number of small satellites, their orbits, and other important parameters. It can be assumed that in the case of a small state, the solution would be: a) to have several state-owned satellites (e.g. the case of Albania) and, b) in case of need, to be ready to order the data of the required areas from foreign companies or other states. In this way, the state institutions would constantly and periodically receive data on the areas of interest to the security services in the state and outside of it. They could ensure almost continuous monitoring and evaluation of the said areas if necessary. Cost reduction in implementing the first step would be possible by implementing this idea with neighboring countries.

Also, it would be appropriate to evaluate other possible alternatives to ensure the monitoring of the areas of interest from the space. It could be joining the already running programs of neighboring countries, but in this case, it should be ensured that the necessary resources are always available.

24.2 SMALL SATELLITE, MIGHTY POTENTIAL (INPUTS FOR INDIA)

Small satellites have a transformative impact on modern military strategies. India needs to harness the capabilities of these diminutive marvels, which range from reshaping warfare strategy and enhancing military resilience to improving communication, intelligence, and bolstering space warfare preparedness.

In the past, military satellites were expensive and heavy, limiting their availability to a select few nations. However, advancements in electronic, electrical, and optical technologies have made it possible to miniaturize components, resulting in the development of **small satellites**. This technological progress, coupled with innovation, has created an ecosystem of small satellites. The number of small satellites in orbit has increased rapidly over the last decade, and the industry is experiencing a significant amount of investment, leading to an increase in the launch, development, and operation of small satellites. The infographic illustrates the growth of small satellites in the last decade.

Small satellites, owing to the miniaturization of key components and technological innovation, can now provide high-quality data and services across various space-based capabilities. While the technical capabilities of a small satellite may be less than those of a larger satellite of the same type, several attributes make their military use operationally advantageous.

24.2.1 Key attributes of small satellites

(i) Lower Cost

The reduced development costs and technological innovations, coupled with large-scale manufacturing, further drive down the costs of smaller satellites. In comparison to traditional large satellites that cost hundreds of millions of dollars to develop, small satellites can be built for one-tenth or even less, making them commercially feasible and enabling new capabilities and applications.

The number of small satellites in orbit has increased rapidly over the last decade, and the industry is experiencing a significant amount of investment, leading to an increase in the launch, development, and operation of small satellites

(ii) Manufacturing at Scale

Small satellites are often deployed in large constellations with relatively short lifespans, requiring fast replacements. This necessity leads to production at mass scales using efficient assembly-line processes, further decreasing costs and development time.

(iii) Frequent Technological Upgrades

Due to the quicker replacement of satellites in orbit and shorter development cycles, small satellites facilitate frequent technological upgrades.

(iv) On-orbit Technology Testing

With a lifespan of 1-3 years and lower costs, small satellites enable the testing of technologies in orbit.

(v) Short Development and Deployment Cycle

Small satellites have a relatively short development and deployment cycle, making it easier to improve the manufacturing process and train the space domain workforce.

(vi) Huge Constellations

Small satellites enable the creation of large constellations, offering advantages such as increased temporal resolution or revisit rate. This is possible due to short development and manufacturing times and lower costs.

(vii) Formation Flying

Given the cost advantage, it is feasible to develop multiple small satellites designed to work in tandem. This opens up numerous opportunities, such as having multiple satellites with different sensors flying over the same area simultaneously, allowing for tip and cue, data fusion, and other innovative uses.

24.2.2 Military Significance of Small Satellites

Small satellites have become increasingly crucial in military operations due to the benefits they offer. Primarily utilized in Low Earth Orbit, launching these satellites is cost-effective and more accessible. Simultaneously, the detailed images/data received and faster communication provide crucial advantages in military operations. The emerging ecosystem that facilitates deploying small satellites in large numbers has the potential to impact military operations at both tactical and operational levels.

The faster information and improved communication offered by small satellites for land forces can enhance the effectiveness of various offensive systems at tactical, operational, and strategic levels. Additionally, they strengthen defensive systems by providing early warnings.

The Ukrainian forces have demonstrated innovative use of **Starlink** satellite systems, showcasing the potential achievements using small satellites during conflicts. According to Chinese researchers, the recent US Defence Department's contract with SpaceX has the potential to increase data transmission for drones and fighter aircraft by 100 times.

The potential for utilizing small satellites in the military is immense.

Militarily significant aspects of small satellites are enumerated below:

(i) High Revisit Rate & Tactical Use of Space-Based Imagery

Small satellite constellations equipped with EO (Electro-Optic) /IR (Infra-Red) /SAR (Synthetic Aperture Radar) offer an exceptionally high revisit rate, providing military decision-makers with close-to-real-time operational and intelligence inputs. The integration of the latest artificial intelligence capabilities in data processing shortens the decision-making process. This capability to generate massive amounts of data almost in real time creates an opportunity to use space-based imagery at the tactical level.

The faster information and improved communication offered by small satellites for land forces can enhance the effectiveness of various offensive systems at tactical, operational, and strategic levels. Additionally, they strengthen defensive systems by providing early warnings

(ii) Targeting

Networking satellites, along with drones and ground attack munitions/missiles significantly alters the military's targeting paradigm. Project Convergence, the US Army's testing of new space capabilities to improve targeting and networks, reported shortening the sensor-to-shooter timeline **from 20 minutes to 20 seconds.** Space-based communications also facilitate the creation of networked missiles, allowing mid-flight re-programmability to react to changes in battlefield target requirements.

(iii) Communications

Mega constellations using small satellites in low Earth orbit offer consistent global coverage with very low latency. Such communication/data services provide a force multiplier effect for special operations and out-of-area contingencies, as well as effective communication means in conflict zones.

(iv) Internet of Military Things (IoMT)

IoMT is a network of interconnected sensors or "things" in the military domain. These sensors communicate, coordinate, and interact with the physical environment, increasing intelligence and allowing for more informed decision-making. IoMT presents an opportunity for enhancing capabilities in various areas of the military.

Some use cases for the IoMT include:

- **Border Management:** Indian security forces, having thousands of kilometres of border in difficult terrain to monitor against threats, can deploy satellite-connected sensors detecting movement or activity along the borders, alerting troops to possible intrusions and enabling a better and more effective response.
- **Garrison/Base Security:** A small satellite-based network can also be used for large garrison/base security by employing appropriate sensors, making security more responsive and effective.
- **Managing Battle Casualties:** Using sensors, critical medical parameters of an injured soldier in the combat zone can be monitored. The collected data can then be shared with doctors/specialists in real-time, enabling remote casualty management, which is essential for critical intervention.
- **Formation Flying:** The concept of formation flying satellites involves multiple satellites with onboard control to maintain relative positions and preserve an appropriate topology. This entails several satellites arranged to function as a single, large, and virtual instrument. For instance, the "SAR Mirror" design envisions a large synthetic aperture radar (SAR) satellite surrounded by a **swarm of small satellites** equipped with receivers, each capable of detecting radar echoes and relaying them to the primary satellite. Similarly, a formation of Earth observation (EO) satellites could provide an image of the ground target from different angles simultaneously, enhancing swath width. The US Space Force recently announced the use of new satellite swarms equipped with infrared detectors in low Earth orbit (LEO) as part of their missile defence strategy to improve detection sensitivity and the timeliness of missile warnings.

India needs to swiftly enhance its small satellite capabilities to keep pace with military communication, ISR, and PNT capabilities and to counter China's offensive capabilities in space warfare. Small satellites can significantly enhance the operations of the Indian Armed Forces at both operational and tactical levels

(v) Operationally Responsive Space

Due to their very short development times and low costs, small satellites can be rapidly replaced in orbit. This allows for the addition of satellites with new technologies or increased revisit capabilities. During conflicts, a constellation that has been attacked by an adversary can be quickly reconstituted, enhancing resilience. Low costs also facilitate the stockpiling of satellite reserves, achieving a launch-on-demand capability.

(vi) Resilience

Small satellite constellations offer a crucial advantage — resistance to kinetic attacks, such as anti-satellite weapons. While large satellites are considered easy targets, the destruction of one or more small satellites would not significantly impact space-based capabilities. This deters adversaries from launching this type of attack.

(vii) Commercial Capabilities

Much of the recent development in small satellite systems has occurred in the commercial sector. The primary benefit is the ability of military users to leverage existing commercial systems for military purposes, avoiding the need to develop new systems and resulting in significant cost savings. It is essential to note that commercial assets need to be considered national assets for assured availability during times of conflict.

(viii) Satellite Tracking, Safety of Flight, and Space Traffic Management

While the minimal size of small satellites provides several advantages, one major challenge is the difficulty of detecting and tracking them in orbit. This limitation can hinder Space Domain Awareness and increase the risk of collisions in orbit. The large number of small satellites, especially mega constellations comprising thousands of satellites, amplifies the challenges of ensuring orbital safety.

24.2.3 Small Satellites – What are the US and China doing?

US: Small satellites are increasingly vital for military use, with both the US and China heavily investing in their development. The US Space Force is shifting away from relying on large, expensive satellites in favour of smaller, more resilient constellations. The space acquisition tenets have listed building smaller satellites as the first tenet, clearly indicating the importance of small satellites to the US Space Force. This shift is expected to improve missile warning capabilities and enhance military communication networks.

China: While detailed information on China's programs for developing small satellites for military use is not available, recent developments highlight the country's achievements and ongoing progress in the small satellite space. China is advancing its small satellite capacity and plans to launch a 13,000-satellite constellation to rival Western projects. Open sources indicate that the Chinese government has established two manufacturing plants capable of producing over 400 small satellites per year to support this ambitious effort.

24.2.4 Small Satellites for the Indian Armed Forces

India needs to swiftly enhance its small satellite capabilities to keep pace with military communication, ISR, and PNT capabilities and to counter China's offensive capabilities in space warfare. Small satellites can significantly enhance the operations of the Indian Armed Forces at both operational and tactical levels. Over the past five years, various policies and initiatives have been implemented to develop the Military-Civil ecosystem in the country for satellite technology and manufacture. This, combined with the technical prowess of **ISRO**, has led to the establishment of a vibrant small satellite ecosystem in the country. However, certain recommendations for fast-tracking capability achievement include:

(i) Space Roadmap

An industry version of the roadmap for enhancing the space and small satellite capabilities of the Armed Forces will help the industry prepare and expand to meet the demands of the Armed Forces.

Small satellites are poised to be a game-changer for military operations. However, two major technological challenges for small satellites are propulsion systems affecting maneuverability and communication means limited by power availability. Various efforts are underway to miniaturize and develop alternatives to address these challenges

(ii) Procurement Policies

Special acquisition procedures tailored to the nuances of the satellite manufacturing ecosystem are needed to facilitate the procurement of satellites and related capabilities. Simplified procurement procedures, startup-friendly funding mechanisms, and export promotion are recommended policy changes.

(iii) Technology Development Initiatives

While initiatives like IDEX and TDF provide a valuable platform for firms to participate in developing niche technologies for the space sector, there is a need to refine procedures, improve funding amounts, and establish linkages to guaranteed orders.

(iv) Market Creation

The government should lead in placing orders to create a market for space-related products. Risk-sharing vehicles for technological development and innovation should also be established.

Planning and implementation are crucial at this juncture to leverage the opportunities provided by this technological shift. The small satellite revolution, primarily led by the commercial sector, is already underway. Military forces stand to benefit by utilizing commercial offerings and exploring military-specific opportunities. Small satellites offer cost savings, the latest technologies, near-constant surveillance, increased transparency, and tactically relevant satellite imagery. However, careful consideration is required to address challenges related to space tracking and space traffic management. By leveraging small satellites thoughtfully and proactively, militaries can gain significant strategic advantages.

24.3 INDIAN DEFENCE STARTUPS BREAK INTO HYPERSPECTRAL IMAGING SMALL SATELLITES IN SPACE

A breakthrough in small and advanced multi-sensor satellites opens further ground for critical hyperspectral imaging in Indian space. Small satellites have the potential to radically improve and contribute to Telecommunication by providing mobile services and Internet of Things (IoT) connectivity and purpose built for specific task (ISR) in military and civilian space.

Indian defence startup breaks into coveted areas of hyperspectral imaging. SpacePixxel Technologies and GalaxEye will design and develop a 'Miniaturized satellite capable of carrying Electro-Optical, Infrared, Synthetic Aperture Radar, and Hyperspectral payloads up to 150 kg. The development is part of a contract under the aegis of Innovations for Defence Excellence (iDEX) in the area of hyperspectral imaging satellites in space.

The breakthrough is a milestone as the miniature satellites will be the world's first advanced multi-sensors (EO+SAR) satellite, designed, built, and launched from India by domestic private players. While there will be many elements of key systems and components from outside, the capability is certainly a substantial step in broadening the space clusters to domestic private players, working on existing and futuristic space tech for the military and commercial arena.

SpacePixxel plans to launch hyperspectral imaging satellites, enabling high-resolution images which is critical for advanced mapping, Intelligence, surveillance, and reconnaissance (ISR).

The breakthrough is a milestone as the miniature satellites will be the world's first advanced multi-sensors (EO+SAR) satellite, designed, built, and launched from India by domestic private players.

The small satellites will unfold such possibilities through the efficient constellations of the small satellites which can focus with much greater details and accuracy to provide detailed earth observation data. Moreover, the small satellites has the potential to radically improve and contribute to Telecommunication by providing mobile services and Internet of Things (IoT) connectivity in military and civilian space.

In communication, small satellites facilitate global internet coverage, especially in remote and underserved regions. These satellites are instrumental in establishing IoT networks for real-time data collection from sensors on the ground, which is crucial for industries like logistics, utilities, and smart grids. In space technologies, the miniaturization of the satellites— small satellites — is no small breakthrough as it further opens great new possibilities in space-based tech capabilities.

Defence applications of multi-sensor small satellites are equally diverse and critical. They enhance situational awareness on the battlefield through constant surveillance and reconnaissance. These satellites can provide critical threat assessment data to the armed forces in wartime and conflict scenarios. The versatility of multi-sensor small satellites allows for rapid reconfiguration and adaptation to new missions, making them a strategic asset in modern warfare.

GalaxEye has formally signed an agreement with iDEX to build systems in a Defspace Challenge of designing & develop a miniaturised satellite capable of carrying multiple payloads up to 150 kg for the Indian Air Force

The development of mini, nano, and microsatellites brings new dimensions and advantages, making space exploration more accessible to companies and organizations of all sizes.

Component-wise it is radically easier to customize and fabricate such as batteries, antenna material, and radio equipment. Additionally, it can be modelled by producing custom components, for example on the 3D printer. This benefits the construction of the satellite and its payload for various technological or scientific experiments. Further advancement is taking place in limiting a number of pre-tests (testbed) and time.

(i) LEO advantage

What adds to the perspective is the use of low LEO orbit (polar orbit type) which flies over the Earth's poles, up to 1500 km on the Earth which enables quicker deployment and experiments. Further, the LEO allows global coverage of radio communication between 98 ° north and 98 ° south latitude. In that case, a smallsat can perform radio communications in a much lower cost-effective bracket and transmissions with less power increase rechargeable batteries and the performance of photovoltaic cells with less expensive antenna systems with less weight.

(ii) AI and the future of small satellites

Now the Indian space program is also geared to embrace artificial intelligence (AI) in its various elements which show the potential of small satellites and their applications.

Recent developments show the immense possibilities in areas like data processing and automatic diagnoses based on telemetry. The very concept of fault-finding and autonomous corrections in spacecraft is itself a revolutionary process. Another groundbreaking AI application here is about the administration of constellations. With smallsat the areas of further research and gauging the potential of such satellite-provided services and spacecraft control will establish new frontier in space tech.

ISRO with six other countries — Canada, Finland, France, South Korea, the UK, and the US —led the co-passenger satellites of one Microsatellite and one Nanosatellite from India and 3 Microsatellites and 25 Nanosatellites from six countries. The successful mission involved the combined total weight of all the 31 satellites carried onboard PSLV-C40 turns out to be 1,323 kg.

Innovations for Defence Excellence (iDEX) is fostering a burgeoning community of start-ups within the defence sector, currently engaging with over 400 start-ups and MSMEs.

Demonstrating the applications, space tech expert, Ankit Bhateja, Founder and Director, Xovian Aerospace, further explains: "Multi-sensor small satellites are a game-changer, offering cost-effective, scalable, and responsive solutions to various needs. Their potential use cases span from enhancing agricultural productivity to ensuring national security, demonstrating their indispensable role in the advancement of both civilian and defence capabilities. As technology continues to evolve, the applications of these agile spacecraft will only expand, further integrating space assets into our daily lives and strategic operations."

Importantly, as a space startup, he echoes that such Initiatives will not only help in capacity building for the domestic space sector but will also help the Indian Space startups to go global with these Indigenous developments.

(iii) iDEX's initiative

Innovations for Defence Excellence (iDEX) is the flagship initiative of the Ministry of Defence (MoD). The contract was inked with SpacePixxel Technologies. While the concept was selected last year, iDEX contract was signed and exchanged in June 2024 between Additional Secretary (Defence Production) & CEO, Defence Innovation Organisation (DIO) Anurag Bajpai and SpacePixxel and GalaxEye.

This iDEX contract enables innovation in space electronics, wherein many payloads earlier deployed on dedicated large satellites are now being miniaturized. The MoD said the modular small satellite will integrate multiple miniaturized payloads as per requirement, providing advantages like faster and economical deployment, ease of manufacturing, scalability, adaptability, and less environmental impact.

iDEX has launched 11 editions of the Defence India Start-up Challenge (DISC), and recently unveiled the Acing Development of Innovative Technologies with iDEX (ADITI) scheme to promote innovations in critical and strategic defence technologies. iDEX has successfully gained momentum, fostering a burgeoning community of start-ups within the defence sector. It is currently engaged with over 400 start-ups and MSMEs. According to the MoD, so far, procurement of 35 items, worth over ₹2,000 crore, has been cleared.

24.4 NANOSATELLITES COULD PLAY PIVOTAL ROLE IN DEFENSE AGAINST ENEMY MISSILES

Two Missile Defense Agency nanosatellites — known as CubeSats — that launched into low-earth orbit from the Mojave Air and Space Port in California could play a large role in the future of U.S. missile defense.

The CubeSat Networked Communications Experiment Block 1 — part of MDA's Nanosat Testbed Initiative — uses small, low-cost satellites to demonstrate networked radio communications between nanosatellites while in orbit. MDA will conduct a 90-day demonstration, with a mission extension of up to one year, to ensure the two CubeSats can navigate properly, receive and send signals to radios and networks and operate as intended.

"These satellites will test key technologies that mitigate risk for systems, such as the Hypersonic and Ballistic Tracking Space Sensor," Walt Chai, MDA director for space sensors, said. "The CNCE Block 1 mission will demonstrate the viability of advanced communications technologies using reduced size, weight and power in support of missile defense communications architectures."

MDA is developing the Hypersonic and Ballistic Tracking Space Sensor payload. When eventually deployed on satellites in low earth orbit, it will detect and track hypersonic and ballistic missile threats and provide critical data to the Missile Defense System and the warfighter.

"The missile defense architecture will require communications between interceptors, sensors and command and control systems to quickly identify, track and destroy incoming enemy missiles before they reach their targets. The CubeSats will allow the agency to demonstrate the capabilities quickly and affordably," Chai said.

CubeSats are a subset of the small satellite family of satellite systems known as nanosatellites. A small satellite is generally considered to be any satellite that weighs less than 300 kilograms (660 pounds). Within the small satellite family, CubeSats are defined by standardized characteristics such as shape, size and weight.

By conforming to very specific CubeSat standards, reduced mission costs are realized — including costs associated with transporting CubeSats to, and deploying them into, space, Feth said.

Most CubeSats are produced as commercial off-the-shelf products. This is due, in part, to the standardization inherent to CubeSats, making mass-produced components and off-the-shelf parts attractive for commercial vendor production.

The cost per satellite is about $1.3 million versus hundreds of millions required for traditional satellite construction. The hardware was built, and the bus and payloads were integrated, under a Rapid Innovation Fund contract.

"The ability to leverage the rapid advances in commercial CubeSat technology, as well as the growing base of commercial small launch providers, enables a unique testing capability never before available," Eric Cole, NTI project lead for MDA, said. "The ability to test in the relevant environment of space enables testing to achieve higher technology readiness levels, making the technology transition path into operational systems much more viable."

According to Jeff Keller, chief engineer for technology maturation at MDA, a primary advantage to maturing technology through NTI is the ability to divide complex challenges into discrete parts. "This allows us to effectively balance risk and cost by utilizing a series of phased demonstrations. Each mission leverages lessons learned from the previous mission," he said.

The overall result is that engineering and development of CubeSats is less costly than more highly customized small satellites. CubeSat payloads also enjoy the cost benefits from commercial CubeSat technology, but they tend to be more specialized for the missions selected by the CubeSat user.

The CubeSats went to space aboard a VOX Space LLC, a subsidiary of Virgin Orbit, LauncherOne rocket as part of a payload-sharing arrangement with the DOD Space Test Program.

Other agencies involved in CNCE Block 1's CubeSat development and experimentation are: defense department-led Mobile CubeSat Command and Control, or MC3, ground station network, Space Dynamics laboratory (Mission Integrator), Space Micro Inc. (Payload) and Blue Canyon Technologies (Spacecraft Bus).

CHAPTER 25

QUANTUM COMPUTING IN WARFARE

25.1 INTRODUCTION TO QUANTUM COMPUTING

A quantum computer is a computer that exploits the phenomena of quantum physics. On small scales, physical matter exhibits properties of both particles and waves, and quantum computing leverages this behavior using specialized hardware. Classical physics cannot explain the operation of these quantum devices, and a scalable quantum computer could perform some calculations exponentially faster than any modern "classical" computer. Theoretically, a large-scale quantum computer could break some widely used encryption schemes and aid physicists in performing physical simulations; however, the current state of the art is largely experimental and impractical, with several obstacles to useful applications.

The basic unit of information in quantum computing, the **qubit (or "quantum bit"**), serves the same function as the bit in classical computing. However, unlike a classical bit, which can be in one of two states (a binary), **a qubit can exist in a superposition of its two "basis" states, which loosely means that it is in both states simultaneously.** When measuring a qubit, the result is a probabilistic output of a classical bit. If a quantum computer manipulates the qubit in a particular way, wave interference effects can amplify the desired measurement results. The design of quantum algorithms involves creating procedures that allow a quantum computer to perform calculations efficiently and quickly.

Quantum computers are not yet practical for real work universally. Physically, engineering high-quality qubits has proven challenging. If a physical qubit is not sufficiently isolated from its environment, it suffers from quantum decoherence, introducing noise into calculations. National governments have invested heavily in experimental research that aims to develop scalable qubits with longer coherence times and lower error rates. Example implementations include superconductors (which isolate an electrical current by eliminating electrical resistance) and ion traps (which confine a single atomic particle using electromagnetic fields).

In principle, a classical computer can solve the same computational problems as a quantum computer, given enough time. Quantum advantage comes in the form of time complexity rather than computability, and quantum complexity theory shows that some quantum algorithms are exponentially more efficient than the best-known classical algorithms. A large-scale quantum computer could in theory solve computational problems unsolvable by a classical computer in any reasonable amount of time. This concept of extra ability has been called "quantum supremacy". While such claims have drawn significant attention to the discipline, near-term practical use cases have remained limited.

Computer engineers typically describe a modern computer's operation in terms of classical electrodynamics. Within these "classical" computers, some components (such as semiconductors and random number generators) may rely on quantum behavior, but these components are not isolated from their environment, so any quantum information quickly decoheres. While programmers may depend on probability theory when designing a randomized algorithm, quantum mechanical notions like superposition and interference are largely irrelevant for program analysis.

Quantum programs, in contrast, rely on precise control of coherent quantum systems. Physicists describe these systems mathematically using linear algebra. Complex numbers model probability amplitudes, vectors model quantum states, and matrices model the operations that can be performed on these states. Programming a quantum computer is then a matter of composing operations in such a way that the resulting program computes a useful result in theory and is implementable in practice.

Quantum computing is the process by which computers use principles of quantum physics, quantum mechanics, and information science for increased computational power and speed. Quantum physics, mechanics, and information science, allow computers to perform simultaneous calculations by measuring the physical photons, electrons, or atom nuclei of data. This is possible due to the continuous physical nature of photons that can contain values that exists in more than one state at once, making contributions in both states for computational power and speed. Their existence in multiple states, and contributions to different but simultaneous calculations to accomplish a collective goal, are real examples of the utility of science-fiction theories that attempt to prove parallel existence.

Quantum computing accomplishes this feat through the use of information known as qubits. Qubits encapsulate the existence of data in more than one state through information science methodologies. In classical computing, information exists in individual electronic signals or voltages known a bits; these bits can only carry values of zero or one (0, 1), but not both. This forces classical computers to use algorithms for computing in the bits limited binary state. Because qubits are physical in nature, its existence and properties are continuous. It can be harnessed in a two-dimensional form known as superposition, where information exists in multiple states at once through a process in which these superimposed qubits are intimately correlated and connected (even over a certain distance), known as entanglement; or, measured by process of noise known as squeezing. The physical nature of each photon and the harnessing and measurement of its continuous state reveals qubits as probabilities which distribute value at the time it is measured. Thus, in theory, these qubits carry an infinite number of values of these "zeros and ones" or combinations thereof at all times until each measurement is conducted (0, 1, [00], [01], [10]...etc.). As this methodology and process of harnessing and measuring information is perfected, encoding of information in those qubits occurs; and the serious implications of surpassing and replacing classical computing which uses binary states is evident.

Quantum computing's increased power due to qubits contains the ability to dramatically reduce the time needed to factor the integers of current encryption, thus rendering moot conventional security protocols for protecting information. Quantum computing would be able to potentially find the solution to the integer by simultaneously generating answers for every possible scenario, as well as their alternatives at increasing speed. From the description of Quantum computing, it is easy to see how current computing power and encryption schemes which rely on classical computing will be impacted.

Some key differences between quantum and classical computing systems are as follows:

(i) **Computational Power:** While classical computers perform calculations sequentially, quantum computers leverage qubit superposition to perform many calculations simultaneously, leading to potentially exponential increases in computational power.

(ii) **Data Handling:** Classical systems process data in binary form, managing one binary decision at a time. Quantum systems manipulate qubits through quantum gates that operate on superposed states, thereby enabling the processing of complex datasets more efficiently.

(iii) **Problem-Solving:** Quantum computers excel at solving specific problems currently intractable for classical systems, such as simulating molecular interactions at the quantum level.

(iv) **Error Correction:** The fragile nature of qubit states requires sophisticated error correction techniques in quantum computing, whereas classical systems benefit from well-established digital error correction methods.

25.2 QUANTUM COMPUTING: A GAME CHANGER IN WARFARE

The potential applications of quantum computing span various fields, including healthcare and finance. This could enable advancements such as accelerated drug discovery and more accurate financial modelling in these areas. However, one domain in which its impact could be particularly transformative is the **military**. As countries realise the strategic edge that quantum computing can provide, there is a growing interest in how it can be applied to defence.

25.2.1 Applications of Quantum Computing in Military Operations:

(i) Space Warfare

The advent of quantum computing has heralded transformative opportunities for military applications, particularly space warfare. The ability to send and receive information **securely** over astronomical distances is critical for strategic communication and operational success.

(ii) Secure Communication and Navigation

Quantum computing has facilitated the development of communication systems impervious to conventional hacking methods. This is achieved through quantum key distribution (QKD), a technique in which cryptographic keys are transmitted via qubits. Any attempt to intercept these qubits invariably alters their state, thereby revealing the presence of an intruder and preserving the integrity of communication.

(iii) **Quantum-resistant Satellite Communications**: The vulnerability of satellite communications to interception and jamming poses a significant risk to modern warfare. Quantum technology mitigates these risks by enabling quantum-resistant encryption and ensuring secure data transmission between command centres and deployed units or satellites.

(iv) **Unbreakable Codes for Interstellar Probes:** Interstellar probes dispatched for reconnaissance or communication purposes require robust encryption to protect the data from adversaries. Quantum computing provides a platform for creating codes that, under the principles of quantum mechanics, cannot be broken using traditional code-breaking techniques.

(v) **Enhanced Situational Awareness:** The application of quantum computing extends beyond secure communication, offering heightened situational awareness through advanced navigation capabilities.

(vi) **Quantum-enhanced Detection Systems:** Spacecraft equipped with quantum sensors can detect signals or objects at vast distances with unprecedented precision, enhancing detection capabilities against stealth technologies.

(vii) **Navigation in GPS-compromised Scenarios:** In instances where GPS signals are denied or spoofed, quantum-enhanced inertial navigation systems offer a viable alternative, using ultrasensitive quantum accelerometers and gyroscopes to maintain accurate positioning.

25.2.2 Strategic Advantages of Quantum Computing in Warfare

(i) **Real-time Strategy Adaptation:** Quantum computers can process complex simulations at unattainable speeds by classical computers, allowing military strategists to adapt tactics in real-time based on evolving scenarios.

(ii) **Resource Optimization:** Quantum algorithms optimize logistics and resource allocation across various platforms in space, ensuring efficient use of energy, fuel, and other critical supplies.

These use cases exemplify the potential for quantum computing to revolutionize warfare operations. The security and accuracy provided by quantum-based systems enhance the survivability and efficacy of military assets.

(iii) Data Security in Military Systems

In the digital age, data integrity forms the backbone of military operations. The importance of securing sensitive information against potential threats cannot be overstated. With its unique capabilities, Quantum technology presents a promising solution for enhancing data security in military applications.

(iv) Data Security

Quantum computing brings forth a paradigm shift in encryption, a critical aspect of data security. Traditional encryption techniques hinge on the difficulty of factoring large numbers—a problem easily tackled by quantum computers. Thus, existing cryptographic systems face the threat of becoming obsolete due to advancing quantum technology.

(v) Quantum Key Distribution (QKD)

Against this backdrop, quantum key distribution (QKD) emerges as a formidable countermeasure. QKD allows two communicating parties to generate a shared secret key that can be used to encrypt and decrypt messages. Any attempt at interception irrevocably alters the quantum state of the key, thereby alerting legitimate users about the breach.

(vi) Enhanced Security for Military Networks

The application of QKD in military networks offers unprecedented levels of security. For instance, consider a scenario where defence networks employ QKD for the secure transmission of classified information. In this case, any attempt by adversaries to intercept or tamper with the data would disrupt the quantum state of the key and trigger an instant alert. This ensures that sensitive military information remains shielded from unauthorized access.

(vii) In GPS-Denied Navigation in Battlefield Environments

An unspoken law of conflict zones is that securing a competitive edge correlates with winning the battle. One such advantage emerges from superior situational awareness, a facet profoundly impacted by advances in Quantum Technology. While instrumental in warfare scenarios, traditional Global Positioning Systems (GPS) can fall prey to jamming or complete unavailability. This presents an enormous challenge for maintaining situational awareness on the battlefield. Pinpointing one's location and navigating accurately is crucial, particularly in hostile terrains and inclement weather conditions where visibility is compromised. Quantum-enhanced Inertial Navigation Systems (INS) could solve this problem. INS are independent of external signals, relying instead on measuring velocity changes to estimate shifts in position over time. However, these systems are not without their faults – they suffer from 'drift', which means minor measurement errors can accumulate over time, leading to significant inaccuracies. Leveraging principles such as superposition and entanglement, quantum-enhanced INS can dramatically reduce this drift. One key innovation is the development of quantum accelerometers that can measure changes in velocity with unprecedented precision.

(viii) Precision-Guided Munition (PGM)

PGMs. are weapons designed to hit specific targets with high accuracy, thereby reducing collateral damage and improving operational efficiency. However, their effectiveness hinges on accurate navigation data – a necessity for compromised GPS signals. By integrating quantum-enhanced INS into PGMs, militaries could maintain elevated levels of accuracy even under GPS-denied conditions. This is due to quantum sensors' exceptional sensitivity and ability to detect minute acceleration or rotation rate changes – data crucial for precise navigation.

(ix) Intelligence Gathering and Surveillance with Advanced Sensors

In modern warfare, having advanced tools for measuring and sensing is crucial. Technology has come a long way, allowing the development of sensors to find even the most silent enemy submarines.

Submarines have become increasingly stealthy with advanced technology, making them difficult to detect using traditional sonar systems. **Quantum-enabled sensors**, based on the principles of quantum mechanics, offer a potential solution to this problem. A superconducting quantum interference device (SQUID) is a sensitive magnetometer that can measure extremely weak magnetic fields. It is already used in Magnetoencephalography (MEG), magnetic field imaging (MFI), and magneto-gastrography. A military application under development is for using SQUIDS in anti-submarine warfare by mounting a SQUID array on maritime patrol aircraft to detect submarines from the air on the magnetic anomaly detector (MAD) lines.

(x) Quantum radar systems

The quantum-enabled sensors rely on quantum entanglement, where two particles become connected so that their states are instantly linked, no matter how far apart they are. Using the above concept, quantum radars can create pairs of entangled photons (particles of light). One photon is sent towards a target, while the other is kept for comparison. If any changes are observed in the saved photon, it means that its pair interacted with an object, indicating the presence of a stealthy object. Unlike classical radars that emit detectable signals, quantum radars are hard to detect because they do not emit a signal that can be easily picked up. This makes them nearly impossible to evade for stealthy targets. With their ability to detect stealthy objects without being detected themselves, quantum-enabled sensors are becoming crucial for gathering intelligence in naval warfare scenarios.

(xi) **Advanced measurement capabilities:** Quantum technology enhances the ability to measure things more precisely than ever.

(xii) **Reshaping warfare strategies**: As quantum technology advances, it will lead to new ways of thinking about and conducting military operations.

(xiii) **Bolstering armaments:** Quantum technology has the potential to improve existing military equipment and create new ones with enhanced capabilities.

25.2.3 Future of Quantum-Enhanced Warfare

Quantum arms race and military dominance are two phrases that accurately represent the future of warfare. The rapid advancements in quantum technologies are transforming military strategies, leading to an arms race on a global scale. Nations worldwide are investing heavily in quantum research to achieve quantum supremacy.

A significant aspect of future warfare lies at the **intersection of quantum computing with other emerging technologies:**

(i) **Artificial Intelligence (AI):** Quantum computing could significantly enhance AI's performance by enabling faster processing speeds and more complex computations. This advancement could lead to more sophisticated automated systems in warfare, including advanced drones and autonomous vehicles.

(ii) **Biotechnology:** Quantum computing may accelerate advancements in biotechnology, potentially leading to improved medical treatments for soldiers or the development of biological weapons.

(iii) **Nanotechnology:** Nano-sized devices powered by quantum computing could be used for various military applications, including surveillance, reconnaissance, or targeted attacks.

The potential integration of these technologies implies a battlefield revolutionized by speed, precision, and automation. Quantum-enhanced capabilities will undeniably shape the future landscape of warfare. As militaries venture into this new era, balancing technological progress with ethical considerations and risk mitigation strategies is crucial.

25.3 HOW QUANTUM COMPUTING WILL CHANGE THE FUTURE OF WARFARE?

Adding more qubits is not the only strategy being made to gain quantum supremacy. Many innovations from academia and industry are being made by advancements in entanglement. Quantum entanglement, which Albert Einstein referred to as a "spooky action at a distance", at the time being considered a "bedrock assumption" in the laws of physics. It is when two systems are strongly in tune with one another in gaining information about one system, where one will give immediate information about the other no matter how far apart the distance is between them.

It is too early to tell who is successfully going to achieve quantum supremacy. However, the prospects are turning towards China and the US. A report by the RAND Corporation stated, "China has high research output in every application domain of quantum technology." And unlike the US, "Chinese quantum technology R&D is concentrated in government-funded laboratories, which have demonstrated rapid technical progress.".

Under the Biden Administration, the US has engaged in a full-on trading war with China. It had focused on the exports of technology to China, which includes quantum technology, however the same way Russia cut access to supply of natural gas when they were engaged in a war with Ukraine. Cutting off exports might backfire on the US as China could still purchase advanced technology from other nations like Japan.

Quantum computing is still an emerging tech that is achieving breakthroughs. There is a lot of innovation occurring at this very moment. We will only have to wait a short while until it performs military exercises and is considered officially in warfare.

25.4 IMPACT OF QUANTUM COMPUTING ON CYBER WARFARE STRATEGIES

In the landscape of 2024, quantum computing has emerged as a transformative force in the realm of cyber warfare. Its profound impact on cybersecurity strategies is reshaping how nations approach digital defense and offensive cyber capabilities.

Quantum computing's ability to perform complex calculations at unprecedented speeds presents both an opportunity and a threat in cyber warfare. Its potential to crack existing cryptographic defenses can render traditional security measures obsolete, prompting a reevaluation of cyber warfare tactics.

The threat posed by quantum computing has accelerated the development of quantum-resistant cryptographic algorithms. Nations and military organizations are investing heavily in these technologies to protect sensitive data and secure communication channels.

(i) Shift in Offensive Cyber Warfare Strategies

On the offensive front, quantum computing enables the execution of complex cyber-attacks that can bypass conventional security systems. This shift necessitates new strategies in cyber defense, emphasizing the need for quantum-proof security systems.

(ii) The Race for Quantum Supremacy in Cybersecurity

Quantum supremacy, the point where quantum computers outperform classical computers in specific tasks, has significant implications for cyber warfare. Nations are in a race to achieve quantum supremacy to gain an upper hand in cybersecurity and cyber warfare capabilities.

(iii) Collaborative Efforts in Quantum-Resistant Technologies

The development of quantum-resistant technologies is a collaborative effort, involving governments, academia, and the private sector. International alliances and partnerships are key to addressing the challenges posed by quantum computing in cyber warfare.

(iv) Training and Preparedness for Quantum-Era Warfare

Training military and cybersecurity professionals in quantum computing and its implications is crucial. Developing a workforce skilled in quantum technologies and strategies is essential for effective defense and offense in the quantum era.

(v) Ethical and Legal Considerations

The integration of quantum computing into cyber warfare raises ethical and legal questions. Nations must navigate these considerations carefully, ensuring that the use of quantum technologies in warfare complies with international laws and norms.

As we move through 2024, the impact of quantum computing on cyber warfare strategies is undeniable. It requires a fundamental shift in how nations approach cybersecurity, from developing quantum-resistant defenses to rethinking offensive capabilities. The era of quantum computing in cyber warfare demands innovation, collaboration, and a deep understanding of the new quantum landscape to ensure national security in an increasingly digital world.

CHAPTER 26

QUANTUM CRYPTOGRAPHY IN WARFARE

26.1 INTRODUCTION TO QUANTUM CRYPTOGRAPHY

Quantum cryptography is the science of exploiting quantum mechanical properties to perform cryptographic tasks. The best-known example of quantum cryptography is quantum key distribution (QKD), which offers an information-theoretically secure solution to the key exchange problem. The advantage of quantum cryptography lies in the fact that it allows the completion of various cryptographic tasks that are proven or conjectured to be impossible using only classical (i.e. non-quantum) communication. For example, it is impossible to copy data encoded in a quantum state. If one attempts to read the encoded data, the quantum state will be changed due to wave function collapse (no-cloning theorem). This could be used to detect eavesdropping in quantum key distribution (QKD).

Cryptography is the strongest link in the chain of data security. However, interested parties cannot assume that cryptographic keys will remain secure indefinitely. Quantum cryptography has the potential to encrypt data for longer periods than classical cryptography. Using classical cryptography, scientists cannot guarantee encryption beyond approximately 30 years, but some stakeholders could use longer periods of protection. Take, for example, the healthcare industry. As of 2017, 85.9% of office-based physicians are using electronic medical record systems to store and transmit patient data. Under the Health Insurance Portability and Accountability Act, medical records must be kept secret. Quantum key distribution can protect electronic records for periods of up to 100 years. Also, quantum cryptography has useful applications for governments and militaries as, historically, governments have kept military data secret for periods of over 60 years. There also has been proof that quantum key distribution can travel through a noisy channel over a long distance and be secure. It can be reduced from a noisy quantum scheme to a classical noiseless scheme. The process of having consistent protection over a noisy channel can be possible through the implementation of quantum repeaters. Quantum repeaters have the ability to resolve quantum communication errors in an efficient way. Quantum repeaters, which are quantum computers, can be stationed as segments over the noisy channel to ensure the security of communication. Quantum repeaters do this by purifying the segments of the channel before connecting them creating a secure line of communication. Sub-par quantum repeaters can provide an efficient amount of security through the noisy channel over a long distance.

26.1.1 Position-based quantum cryptography

The goal of position-based quantum cryptography is to use the geographical location of a player as its (only) credential. For example, one wants to send a message to a player at a specified position with the guarantee that it can only be read if the receiving party is located at that particular position. In the basic task of position-verification, a player, Alice, wants to convince the (honest) verifiers that she is located at a particular point. It has been shown that position-verification using classical protocols is impossible against colluding adversaries (who control all positions except the prover's claimed position). Under various restrictions on the adversaries, schemes are possible.

Under the name of 'quantum tagging', the first position-based quantum schemes have been investigated in 2002 by Kent. A US-patent was granted in 2006. The notion of using quantum effects for location verification first appeared in the scientific literature in 2010. After several other quantum protocols for position verification have been suggested in 2010, it has been claimed that a general impossibility result: using an enormous amount of quantum entanglement (by using a doubly exponential number of EPR pairs, in the number of qubits the honest player operates on), colluding adversaries are always able to make it look to the verifiers as if they were at the claimed position. However, this result does not exclude the possibility of practical schemes in the bounded- or noisy-quantum-storage model. Later a couple of scientists improved the amount of EPR pairs needed in the general attack against position-verification protocols to exponential. They also showed that a particular protocol remains secure against adversaries who controls only a linear amount of EPR pairs. It is argued in that due to time-energy coupling the possibility of formal unconditional location verification via quantum effects remains an open problem. The study of position-based quantum cryptography also has connections with the protocol of port-based quantum teleportation, which is a more advanced version of quantum teleportation, where many EPR pairs are simultaneously used as ports.

26.2 MILITARY APPLICATIONS OF QUANTUM CRYPTOGRAPHY

"Quantum cryptography" is considered a key technology for the "future" of "**military**" and "warfare" as it has the potential to revolutionize secure communication within the military by providing unbreakable encryption, making it a critical element in modern and future battlefields; however, it is still under development and not widely implemented yet.

Key points about **quantum cryptography and its military implications**:

(i) Unbreakable encryption:

Quantum cryptography leverages the principles of quantum mechanics to create encryption methods that are theoretically impossible to crack with current computing power, offering a significant advantage in secure military communications.

(ii) Future warfare:

While not yet fully operational, major military powers are actively researching and developing quantum cryptography technology, anticipating its potential impact on future battlefields.

Despite its promise, quantum cryptography faces technical challenges related to implementation and maintaining the necessary quantum states, limiting its widespread use currently.

(iii) Quantum sensing:

Beyond encryption, quantum sensing technology is also being explored for military applications like detecting enemy submarines, navigating in GPS-denied environments, and enhancing intelligence gathering.

26.2.1 Post-quantum cryptography provides the next generation in data protection

Asymmetric encryption is in almost all software, billions of devices worldwide and most of the communications over the internet. Yet by 2029, advances in quantum computing will make asymmetric cryptography unsafe and by 2034 fully breakable. "Harvest-now, decrypt-later" attacks may already exist.

To resist attacks from both classical and quantum computers, organizations must transition to post-quantum cryptography (PQC). But that's hardly a simple switch. It will require more work than preparing for Y2K, and failure could have dangerous consequences. Further, many organizations haven't yet planned or budgeted for this shift.

26.2.2 Future of Quantum Cryptography

The ability to encrypt information is an essential part of military command and control, just as breaking military codes has been a decisive factor in modern warfare. With that in mind, the nations should take steps now to prepare for a day when adversaries could have quantum computing-enabled decryption capabilities.

Ensuring that data is successfully encrypted and thus inaccessible to attackers is key to maintaining a strong cyber defense posture. To that end, cryptographic technologies are widely employed to authenticate sources, protect stored information, and share data in a confidential and secure manner. Algorithms currently in use are so advanced and have revolutionized data security to such an extent that even the fastest classical computers could take years, in some cases decades, to unlock encrypted files. As a result, rather than attempt brute force decryption, hackers have instead preferred to steal encryption keys or find weak links in a security network to bypass secure channels and steal decrypted data.

For example, in the recent Colonial Pipeline incident, attackers obtained access to the IT system through a legacy Virtual Private Network, or VPN, profile that had not been used or monitored for years. Better enforcement of cyber hygiene is a short-term solution, but in the long-term, security networks must also be overhauled to implement cryptographic algorithms designed to fend off future attacks made possible with emerging technologies such as quantum computing.

With the potential that quantum could have on the international economy, it is no surprise that billions of dollars are being invested to fund research in this emerging technology area. In the United States, efforts are being led by academia, government labs and technology companies across the industrial base. However, China is investing heavily and is close behind. President Xi

Jinping's government has spent more than $10 billion to set up the National Laboratory for Quantum Information Sciences, and at the current rate will spend more on quantum research than any other nation by 2030.

Larger computations would also require bigger quantum chips with many millions of connections. Even if that were possible, we do not currently have the capability to control multiple qubits on the time scales required for useful operations, on the order of tens of nanoseconds.

Notably, only a limited set of problems have been identified that can currently be solved more effectively on a quantum computer than a traditional one.

However, given the pace of advancement and magnitude of investments by peer competitors, we should not wait to implement quantum-resistant algorithms on our security networks. There are steps we can take now to guard against future quantum computational capabilities, including the implementation of post-quantum cryptography algorithms that are secure against both classical and quantum computers. Of course, systems protected by even the most robust quantum-resilient algorithms would still be vulnerable to attack via weak links in a network, so these are necessary but not sufficient steps.

It would be difficult to predict when, or even if, quantum computing will provide our adversaries, or even bad actors, with the ability to creak previously unbreakable codes. But regardless of the timeline for that threat, we can take steps today that will significantly reduce the potential risks posed by that future capability.

26.3 MILITARY-GRADE ENCRYPTION BROKEN BY QUANTUM COMPUTER

Chinese researchers have said they have made significant inroads in using a current quantum computer to break encryption algorithms used in banking and defense. A team led by Wang Chao from Shanghai University used a D-Wave Advantage quantum computer to attack encryption algorithms such as Present, Gift-64 and Rectangle.

These algorithms are part of the Substitution-Permutation Network (SPN) structure, which is integral to widely used encryption methods like AES-256, which is considered one of the strongest standards and is used in fields including banking and cryptocurrency. It is also used in military systems to secure satellite communication, radio communication, network and email security, voice over internet protocol and secure voice communication.

The researchers published their findings in a peer-reviewed paper titled "Quantum Annealing Public Key Cryptographic Attack Algorithm Based on D-Wave Advantage," published in the Chinese Journal of Computers (in 2024).

D-Wave's quantum computing technology uses a technique known as quantum annealing to search for solutions to a problem. This allows quantum computers to bypass the limitations that traditional computing methods face when solving highly complex encryption problems.

Wang described quantum annealing in his paper as like an AI algorithm with the ability to globally optimize solutions. His team used it alongside traditional thematic methods to devise a new computational architecture.

He also noted that while quantum computing has potential, current systems are limited by environmental interference, hardware constraints and the inability of a single attack algorithm to target multiple cryptographic systems.

While this breakthrough does not yet mean that quantum computers can easily break encryption on a mass scale, it highlights the growing threat that quantum technology poses to current cryptographic systems.

The development emphasizes the need for widespread implementation of post-quantum cryptography, which aims to design encryption methods resilient to quantum attacks.

CHAPTER 27

QUANTUM COMMUNICATION IN WARFARE

27.1 INTRODUCTION TO QUANTUM COMMUNICATION

Quantum communication is a field of physics that uses quantum mechanics to transmit and process information. It uses quantum bits, or qubits, instead of classical bits.

Quantum communication is secure because the information cannot be accessed or cloned without tampering the quantum state. This is due to quantum principles like quantum superposition, quantum entanglement, and the no-cloning theorem.

Over the past few decades, wireless communication has witnessed tremendous advancements. Yet, the forthcoming wireless network demands even faster, more efficient, and secure information exchange, where quantum communication appears to be a promising solution. The quantum world, with its counter-intuitive properties, has gained significant attention from physicists for decades. Quantum mechanics, the fundamental theory governing the behavior of particles, creates a pathway to a wide range of technological advancements. Indeed, the most promising implementation of quantum mechanics lies in its potential to revolutionize the way we communicate with each other. Quantum communication precisely makes use of two fundamental theories namely quantum entanglement and quantum superposition.

The quantum entanglement is a phenomenon which interconnects qubits in such a way that the state of one is linked to the state of another, irrespective of the travelled distance. Whereas, superposition is another phenomenon, in which quantum can exist in multiple states simultaneously. This duality conservation of spinning quantum pairs is mainly utilized for information encoding and is known as Quantum Shift Keying (QSK). Unlike classical communication which uses classical bits, quantum communication uses quantum bits, popularly known as qubits. Apart from new definitions of information exchange, quantum-enabled transmission provides several other complementary features, with a particular research focus on enhancing network security. The phenomenon of entanglement is utilized for producing encryption keys, known as Quantum Key Distribution (QKD). QKD is a secure communication protocol that ensures the confidentiality of cryptographic keys by leveraging the principles of quantum mechanics.

Quantum communications is similar to quantum information processing and quantum teleportation, but focuses specifically on encryption. One type of encryption is known as Quantum Key Distribution (QKD). QKD describes the use of quantum mechanical effects to perform cryptographic tasks or to break cryptographic systems. The principle of operation of a QKD system is quite straightforward: two parties (Alice and Bob) use single photons that are randomly polarized to states representing ones and zeroes to transmit a series of random number sequences that are used as keys in cryptographic communications. Both stations are linked together with a quantum channel and a classical channel. Alice generates a random stream of qubits that are sent over the quantum channel. Upon reception of the stream Bob and Alice — using the classical channel — perform classical operations to check if an eavesdropper has tried to extract information on the qubits stream. The presence of an eavesdropper is revealed by the imperfect correlation between the two lists of bits obtained after the transmission of qubits between the emitter and the receiver. One important component of virtually all proper encryption schemes is true randomness which can elegantly be generated by means of quantum optics.

Quantum communication has already made its debut in 2021 when China launched the world's first quantum satellite. "**Micius**," the satellite, has been able to use QKD to encrypt sensitive information, including a video conference meeting between Beijing and Vienna. The US military will continue to use quantum technology to enhance its advantage over other countries. As long as quantum physics keeps advancing, the military will be right behind it, figuring out different ways to apply this new phenomenon.

In January 2024, scientists from Russia and China successfully established quantum communication over a distance of 3,800 kilometers. This suggests that a quantum communication network among the BRICS nations could be feasible.

27.2 INTRODUCTION TO QUANTUM COMMUNICATION IN WARFARE

In warfare, quantum communication refers to the use of quantum mechanics principles to transmit information securely, essentially creating unhackable communication channels between military units by leveraging the unique properties of quantum particles, making it a potential game-changer for secure data transmission on the battlefield, especially for highly sensitive information. This technology is considered a key aspect of "quantum warfare" as it could significantly enhance military operations by preventing eavesdropping on critical communications.

27.2.1 Applications of quantum communication in warfare

This technology could be used to secure communication between command centers, troops on the ground, drones, and other military assets, allowing for the transmission of critical battlefield intelligence without risk of interception.

27.2.2 Challenges of quantum communication in warfare

While promising, quantum communication is still in its early stages of development and faces challenges like maintaining the fragile quantum states over long distances and integrating it with existing military communication systems.

27.2.3 Potential impact of quantum communication in warfare

(i) **Enhanced situational awareness:** Securely sharing real-time intelligence could provide military commanders with a significant advantage in decision-making.

(ii) **Improved operational security:** By preventing enemy access to sensitive communications, quantum technology could significantly enhance military operations.

(iii) **Deterrence factor:** The development and deployment of quantum communication capabilities could act as a deterrent against potential adversaries.

27.3 QUANTUM COMMUNICATION FOR THE MILITARY

Four different military use cases will be considered.

(i) Secure communication

The future quantum computers will be able to break current widely used asymmetric cryptographic protocols using Shor's algorithm. This will require new, quantum-safe approaches for encryption. Likewise, symmetric keys will be targeted with Grover's algorithm, though doubling the key length will probably ensure a quantum-safe situation for symmetric cryptography. This is very relevant in the military domain. Information dominance will be lost and missions will be endangered if communications are not reliable, robust and secure anymore. To achieve secure communication, usually symmetric encryption is used. This symmetric encryption in turn requires keys to both encode and decode data. It is however not always possible to distribute these keys between all participating parties before a mission starts, or to change keys very frequently during a mission. In order to still safely communicate, a shared secret key must be exchanged. Using quantum key distribution protocols, this shared key can be generated in a theoretically secure manner. Consider the situation, where multiple possible scenarios for quantum key distribution in a military setting are shown. Most of the communication links use free-space quantum key distribution, including (laser) communication between ships, satellites, underwater vessels, planes and/or drones, and ground-based stations and vehicles. Furthermore, a fiber optic communication channel between two ground-based stations can be used. Fiber optic communication also allows for quantum key distribution between land-based strategic assets, such as headquarters, command posts, naval stations, bases and airfields, which usually have a fixed or semi-static location. Each of the links is a possible application of quantum key distribution between two parties. However, each of the situations also comes with its own specific challenges. For example, for free-space quantum key distribution a direct line-of-sight is a necessity.

Different experiments have been performed for potential military applications of quantum key distribution, even though the experiments themselves showcase QKD without a particular user Quantum key distribution is just one of the options for quantum-safe communication. Instead, also trusted couriers can be used and/or post-quantum cryptography. Post-quantum cryptography does not make use of quantum technologies, but instead is based on different hard mathematical problems that are expected to remain safe against quantum computers and hence burden adversaries with increased computational power requirements. When using post-quantum cryptography, the already available IT infrastructure, used by current encryption mechanisms, can be used, which eases the transition from one to the other encryption protocol. Note that both the trusted courier solution and post-quantum cryptography come with their own challenges and limitations.

(ii) Exact position determination

Exact position determination is of the utmost importance when it comes to military missions, especially accurately determining the location of friendly units, guiding precision munition, or operating autonomous or unmanned vehicles (UxVs). This position can accurately be determined using GNSS. By perfectly synchronized clocks, the precision of GNSS can be made arbitrarily accurate. Currently, most of the navigation and positioning determination is based on GNSS-satellites. This is in turn can however be spoofed, though very hard in practice, or access can be denied by jamming. For these scenarios it is not known if quantum communication offers solutions, however, the field of quantum sensing certainly does.

(iii) Position-based access to data

The key idea of position-based cryptography is that data can only be decrypted by being on a specific location. As this application is new, it currently is not used in practice. It would also require a different approach to data access management than currently used, so not just role-based, but conditionally- or risk-based access. One potential application can be that data is only available if you are at the compound. This protects the data from malicious parties that have no access to the compound. Another use case would be the other way around: specific information should only become available at a certain place and a certain time during a sensitive mission.

(iv) Improved sensing capabilities

For military application many different sensors are used, for instance to detect the use of certain transmitters. With larger detectors and telescopes, more accurate signals can be detected and noise can be suppressed more efficiently. The technique currently used is a phased array technology, where detectors are distributed over different locations. Increasing the area over which the detectors are scattered, increases the detection capabilities and the detection precision. Furthermore, this scattering potentially allows for a better protection against failure or blockade of specific detectors in specific regions. When scattering detectors over a larger area, the complexity of combining the incoming signals increases and the impact of noise will become too large for practical use. By using quantum communication, detectors over longer distances can coherently be combined.

CHAPTER 28

QUANTUM SENSORS IN WARFARE

28.1 INTRODUCTION TO QUANTUM SENSORS

Within quantum technology, a quantum sensor utilizes properties of quantum mechanics, such as quantum entanglement, quantum interference, and quantum state squeezing, which have optimized precision and beat current limits in sensor technology. The field of quantum sensing deals with the design and engineering of quantum sources (e.g., entangled) and quantum measurements that are able to beat the performance of any classical strategy in a number of technological applications. This can be done with photonic systems or solid-state systems.

In photonics and quantum optics, photonic quantum sensing leverages entanglement, single photons and squeezed states to perform extremely precise measurements. Optical sensing makes use of continuously variable quantum systems such as different degrees of freedom of the electromagnetic field, vibrational modes of solids, and Bose–Einstein condensates. These quantum systems can be probed to characterize an unknown transformation between two quantum states. Several methods are in place to improve photonic sensors' quantum illumination of targets, which have been used to improve detection of weak signals by the use of quantum correlation.

Quantum sensors are often built on continuously variable systems, i.e., quantum systems characterized by continuous degrees of freedom such as position and momentum quadratures. The basic working mechanism typically relies on optical states of light, often involving quantum mechanical properties such as squeezing or two-mode entanglement. These states are sensitive to physical transformations that are detected by interferometric measurements.

Quantum sensing can also be utilized in non-photonic areas such as spin qubits, trapped ions, flux qubits, and nanoparticles. These systems can be compared by physical characteristics to which they respond, for example, trapped ions respond to electrical fields while spin systems will respond to magnetic fields. Trapped Ions are useful in their quantized motional levels which are strongly coupled to the electric field. They have been proposed to study electric field noise above surfaces, and more recently, rotation sensors.

In solid-state physics, a quantum sensor is a quantum device that responds to a stimulus. Usually this refers to a sensor that, which has quantized energy levels, uses quantum coherence to measure a physical quantity, or uses entanglement to improve measurements beyond what can be done with classical sensors.

There are 4 criteria for solid-state quantum sensors:

- The system has to have discrete, resolvable energy levels.
- You can initialize the sensor and you can perform readout (turn on and get answer).
- You can coherently manipulate the sensor.
- The sensor interacts with a physical quantity and has some response to that quantity.

A good example of an early quantum sensor is an avalanche photodiode (APD). APDs have been used to detect entangled photons. With additional cooling and sensor improvements can be used where photomultiplier tubes (PMT) in fields such as medical imaging. APDs, in the form of 2-D and even 3-D stacked arrays, can be used as a direct replacement for conventional sensors based on silicon diodes.

The Defense Advanced Research Projects Agency (DARPA) launched a research program in optical quantum sensors that seeks to exploit ideas from quantum metrology and quantum imaging, such as quantum lithography and the NOON state *(NOON state is a quantum-mechanical many-body entangled state, which represents a superposition of N particles in mode a with zero particles in a mode and vice versa. Usually, the particles are photons, but in principle any bosonic field can support NOON states),* in order to achieve these goals with optical sensor systems such as lidar. The United States judges quantum sensing to be the most mature of quantum technologies for military use, theoretically replacing GPS in areas without coverage or possibly acting with ISR capabilities or detecting submarine or subterranean structures or vehicles, as well as nuclear material.

28.2 QUANTUM SENSING ENTERS THE DOD LANDSCAPE

The Defense Science Board identified three areas of quantum technology that will be most relevant to the Department of Defense's (DoD) technological superiority: sensing, encryption, and communications. Of these, quantum sensing is the most mature and promising for use in military technologies reliant on accurate positioning, navigation, and timing (PNT). PNT is critical to the function of key capabilities such as imaging, earth observation, and communication.

Recognizing these imperatives, Defense Innovation Unit (DIU) - in collaboration with the Office of the Undersecretary of Defense for Research & Engineering's Quantum directorate, kicked-off a quantum sensing initiative in late 2020. The program focuses on moving atomic sensors out of the lab and to the real world. Today, the program is bearing fruit with delivery of a quantum sensor that can meet the rigors of space.

For this project, DIU awarded a contract to Vector Atomic, a 5-year-old California start-up focused on fielding and commercializing atomic instruments. Recently, Vector Atomic, in partnership with Honeywell Aerospace, delivered a fully integrated, high-performance atomic gyroscope. This is the **first atomic gyroscope** to undergo space qualification and is expected to be the first atomic inertial sensor to operate in space.

Gyroscopes measure rotational motion, a critical function for drones, ships, aircraft, satellites, and even consumer electronics, and make up a multi-billion-dollar per year market. An atomic gyroscope uses atoms and precise laser interactions to act as rulers **to discern angular rates**, compared to current state-of-the-art approaches that rely on photons (i.e., light). Atoms, in principle, are massive and slow in comparison, and thus, the effects on them are more apparent when experiencing rotation. The added sensitivity and precision offered by the quantum inertial sensor means reducing positional error, and most importantly, the reliance on external PNT signals provided from systems like GPS. This project represents decades of investment by the Department of Defense, particularly DARPA, into quantum based inertial sensors.

DIU's quantum inertial sensor project doesn't stop with the demonstration of the atomic gyroscope in space. The next phase of the project will include the demonstration of a fully integrated atomic inertial measurement unit (IMU), composed of independent accelerometers and gyroscopes to sense motion along all degrees of freedom. The IMU is a building block of inertial navigation solutions to platforms, regardless of domain.

28.3 THE US ARMY SCIENTISTS CREATE INNOVATIVE QUANTUM SENSOR

A quantum sensor could give Soldiers a way to detect communication signals over the entire radio frequency spectrum, from 0 to 100 GHz, said researchers from the US Army.

Such wide spectral coverage by a single antenna is impossible with a traditional receiver system, and would require multiple systems of individual antennas, amplifiers and other components.

In 2018, Army scientists were the first in the world to create a quantum receiver that uses highly excited, super-sensitive atoms—known as Rydberg atoms—to detect communications signals, The researchers calculated the receiver's channel capacity, or rate of data transmission, based on fundamental principles, and then achieved that performance experimentally in their lab improving on other groups' results by orders of magnitude.

These new sensors can be very small and virtually undetectable, giving Soldiers a disruptive advantage. Rydberg-atom based sensors have only recently been considered for general electric field sensing applications, including as a communications receiver. While Rydberg atoms are known to be broadly sensitive, a quantitative description of the sensitivity over the entire operational range has never been done.

To assess potential applications, Army scientists conducted an analysis of the Rydberg sensor's sensitivity to oscillating electric fields over an enormous range of frequencies—from 0 to 10^{12} Hertz. The results show that the Rydberg sensor can reliably detect signals over the entire spectrum and compare favorably with other established electric field sensor technologies, such as electro-optic crystals and dipole antenna-coupled passive electronics.

Quantum mechanics allows to know the sensor calibration and ultimate performance to a very high degree, and it's identical for every sensor. This result is an important step in determining how this system could be used in the field. This work supports the Army's modernization priorities in next-generation computer networks and assured position, navigation and timing, as it could potentially influence novel communications concepts or approaches to detection of RF signals for geolocation.

28.4 DEFENSE EXPERTS PREDICT QUANTUM SENSING POISED TO REPLACE VULNERABLE GPS SYSTEMS

In the face of growing concerns about technologically sophisticated adversaries, including China and Russia, potentially corrupting or disabling GPS signals, quantum sensing has captured the attention of the Pentagon. The ability to spot minute changes in atomic properties presents an opportunity for highly accurate and resilient navigation systems.

The leading researchers in government and industry laboratories across the globe are in a race to refine technology and methods for detecting changes in motion and electric and magnetic fields at the atomic level. This effort has led to the **emergence of quantum sensing**, a technique that can yield extremely precise and accurate measurements.

Naval Research Laboratory research physicist Roger Easton's work was foundational for GPS, leading to the launch of NTS-2, the first satellite to transmit GPS signals in 1977, according to the magazine. Today, it is among those adapting quantum sensors for an alternative navigation technique that actually predates GPS — inertial navigation.

Inertial navigation, which has been used for military aircraft and weapons guidance since the 1960s, relies on accelerometers, gyroscopes and a computer (known as an inertial measurement unit or IMU) to calculate continuously the position, orientation, and velocity of a moving object without external references, according to the National Defense Magazine. While GPS took over by the early 1990s, inertial navigation using quantum sensors is now being explored as a less vulnerable alternative.

One of the biggest hurdles to quantum sensing devices suitable for dynamic environments like military ships, submarines, or aircraft is making them small enough and energy efficient enough. Shrinking the quantum sensors, however, presents a challenge as it diminishes their accuracy and precision.

The scientists tell the National Defense Magazine that size matters — but not in the way that phrase is commonly implemented. Smaller is better.

Operating as part of new inertial navigation systems, quantum IMUs would perform the same functions as classical IMUs but with enhanced precision and accuracy. This improvement would be especially valuable during GPS disruptions.

People are now taking the more advanced optical atomic clocks and engineering them to have field-able package sizes.

Naval Research Laboratory used digital engineering to put quantum gravimeters on ships. They developed an atomic physics level model that forecasts how the gravimeters behave without the need for large, heavy stabilization gimbals.

28.5 QUANTUM SENSING'S POTENTIAL IMPACTS ON STRATEGIC DETERRENCE AND MODERN WARFARE

Quantum sensors measure the same thing as other sensors, physical phenomena such as magnetic fields or acceleration. However, they are unique in that they make these measurements at the highest levels of sensitivity that are physically possible with quantum mechanics and often feature greatly enhanced performance as a result. The specific advantages vary, but they often include higher sensitivity, better long-term stability, or small sensor size compared with alternatives.

In a renewed period of great power competition, the U.S. military's failure to stay ahead of competitors like China in the race to field and integrate new and/or improved quantum sensors could result in disadvantageous technological asymmetries for the United States. Whether quantum sensing capabilities have the potential to disrupt long-held strategic deterrence architectures, as well as transform the dynamics of modern warfare, is worth exploring.

Quantum sensors may also play a disruptive role in modern warfighting and will require time, investment, and research to develop relevant and fieldable devices. If quantum sensors could provide PNT functionality, it would enable or bolster anti-access/area denial (A2/AD) capabilities designed to prevent and counter U.S. power projection. Conversely, U.S. military integration of new and improved quantum sensing technology may enable and improve upon access and entry operations.

CHAPTER 29

FREE-SPACE OPTICAL AND FIBER-OPTIC COMMUNICATION IN WARFARE

29.1 FREE-SPACE OPTICAL COMMUNICATION (FSO)

Free-space optical communication (FSO) is an optical communication technology that uses light propagating in free space to wirelessly transmit data for telecommunications or computer networking. "Free space" means air, outer space, vacuum, or something similar. This contrasts with using solids such as optical fiber cable.

The technology is useful where the physical connections are impractical due to high costs or other considerations.

The invention of lasers in the 1960s revolutionized free-space optics. Military organizations were particularly interested and boosted their development. In 1973, while prototyping the first laser printers at PARC, Gary Starkweather and others made a duplex 30 Mbit/s CAN optical link using astronomical telescopes and HeNe lasers to send data between offices; they chose the method due partly to less strict regulations (at the time) on free-space optical communication by the FCC. However, laser-based free-space optics lost market momentum when the installation of optical fiber networks for civilian uses was at its peak.

Many simple and inexpensive consumer remote controls use low-speed communication using infrared (IR) light. This is known as consumer IR technologies.

(i) Usage and technologies for FSO

Free-space point-to-point optical links can be implemented using infrared laser light, although low-data-rate communication over short distances is possible using LEDs. Infrared Data Association (IrDA) technology is a very simple form of free-space optical communications. On the communications side the FSO technology is considered as a part of the optical wireless communications applications. Free-space optics can be used for communications between spacecraft.

(ii) Useful distances

The reliability of FSO units has always been a problem for commercial telecommunications. Consistently, studies find too many dropped packets and signal errors over small ranges (400 to 500 meters (1,300 to 1,600 ft)). This is from both independent studies, such as in the Czech Republic, as well as internal studies, such as one conducted by MRV FSO staff.

Military-based studies consistently produce longer estimates for reliability, projecting the maximum range for terrestrial links is of the order of 2 to 3 km (1.2 to 1.9 mi). All studies agree the stability and quality of the link is highly dependent on atmospheric factors such as rain, fog, dust and heat. Relays may be employed to extend the range for FSO communications.

TMEX USA ran two eight-mile links between Laredo, Texas and Nuevo Laredo, Mexico from 1998 to 2002. The links operated at 155 Mbit/s and reliably carried phone calls and internet service.

(iii) Extending the useful distance

The main reason terrestrial communications have been limited to non-commercial telecommunications functions is fog. Fog often prevents FSO laser links over 500 meters (1,600 ft) from achieving a year-round availability sufficient for commercial services. Several entities are continually attempting to overcome these key disadvantages to FSO communications and field a system with a better quality of service. DARPA has sponsored over US $130 million in research toward this effort.

Other non-government groups are fielding tests to evaluate different technologies that some claim have the ability to address key FSO adoption challenges. As of October 2014, none had fielded a working system that addresses the most common atmospheric events.

FSO research from 1998 to 2006 in the private sector totaled $407.1 million, divided primarily among four start-up companies. All four failed to deliver products that would meet telecommunications quality and distance standards:

Terabeam received approximately $575 million in funding from investors such as Softbank, Mobius Venture Capital and Oakhill Venture Partners. AT&T and Lucent backed this attempt. The work ultimately failed, and the company was purchased in 2004 for $52 million (excluding warrants and options) by Falls Church, Va.-based YDI, effective June 22, 2004, and used the name Terabeam for the new entity. On September 4, 2007, Terabeam (then headquartered in San Jose, California) announced it would change its name to Proxim Wireless Corporation, and change its NASDAQ stock symbol from TRBM to PRXM.

AirFiber received $96.1 million in funding, and never solved the weather issue. They sold out to MRV communications in 2003, and MRV sold their FSO units until 2012 when the end-of-life was abruptly announced for the Terescope series.[21]

LightPointe Communications received $76 million in start-up funds, and eventually reorganized to sell hybrid FSO-RF units to overcome the weather-based challenges.

The Maxima Corporation published its operating theory in Science, and received $9 million in funding before permanently shutting down. No known spin-off or purchase followed this effort.

Wireless Excellence developed and launched CableFree UNITY solutions that combine FSO with millimeter wave and radio technologies to extend distance, capacity and availability, with a goal of making FSO a more useful and practical technology.

One private company published a paper on November 20, 2014, claiming they had achieved commercial reliability (99.999% availability) in extreme fog. There is no indication this product is currently commercially available.

29.2 FIBER-OPTIC COMMUNICATION IN WARFARE

In warfare, fiber optic communication plays a crucial role by providing highly secure, high-bandwidth data transmission that is immune to electromagnetic interference, making it ideal for transmitting sensitive military information without risk of eavesdropping, particularly in environments with heavy electronic activity on the battlefield; this includes command and control systems, battlefield intelligence, and real-time communication between units.

29.2.1 Key advantages of fiber optic communication in warfare:

(i) Security

Unlike traditional copper wires, fiber optic cables do not emit any detectable signals, making it extremely difficult for adversaries to intercept communications.

(ii) High Bandwidth

Fiber optic cables can transmit large volumes of data at very high speeds, enabling the transmission of complex battlefield information like real-time video feeds and sensor data.

(iii) Immunity to EMI/EMC

Fiber optic cables are not affected by electromagnetic interference and electromagnetic compatibility, which is prevalent in combat environments with heavy electronic activity.

(iv) Lightweight and Compact

Compared to copper cables, fiber optic cables are lighter and easier to deploy in challenging terrains.

29.2.2 Military applications of fiber optic communication

(i) **Command and Control Systems:** Secure communication between military headquarters and deployed units.

(ii) **Tactical Communications:** Real-time data exchange between frontline troops and command centers.

(iii) **Sensor Networks:** Connecting remote sensors on the battlefield to gather intelligence on enemy movements.

(iv) **Aircraft and Vehicle Systems:** Data transmission within military aircraft and vehicles.

(iv) **Remote Surveillance:** High-quality video transmission from surveillance cameras in sensitive areas.

29.2.3 Challenges in using fiber optic communication in warfare:

(i) Deployment Difficulty

Installing fiber optic cables in harsh environments like battlefields can be challenging.

(ii) Physical Damage

Cables can be vulnerable to physical damage from explosions or enemy actions.

(iii) Joint Interoperability

Integrating fiber optic systems with older communication technologies used by different military branches.

29.3 USE OF FIBER OPTIC NETWORKS ON NAVY SHIPS

Fiber optic networks are the main data communications medium on the Navy's ships and submarines. They provide the high bandwidth that today's data-intensive systems require, as well as immunity to electromagnetic interference (EMI) and significant weight and space advantages over copper cable. Virtually every mission critical system on modern ships now uses fiber optic networks, from weapons systems to communication, navigation, radar and sonar systems, infrastructure management, and real time monitoring and control to enable a state-of-the-art navy ship to fulfill its mission.

The demand for bandwidth continues to grow as shipboard operations become increasingly automated. For example, the new Gerald R. Ford class of supercarriers contain more than four million feet of fiber optic cable, supporting numerous advanced technologies that to improve combat readiness, ship performance and crew efficiency.

29.4 LASER COMMUNICATION SYSTEMS

With the upcoming trend of electronic warfare, military operations demand broadband capacity with the highest level of security. Nowadays, tactical operations are enabled with large volumes of ISR imagery and video data that are being transferred from sensing locations to battlefield grounds. Also, timely access to critical information delivered to soldiers in the battlefield can change the war game. For this reason, laser communication, also known as free space optics (FSO), is a good choice owing to its high carrier frequency, ultra-low latency and immunity towards EM radiation. These links allow line-of-sight communication between two parties to have a very low probability of detection, interception or exploitation (LPD/LPI/LPE). LPD means preventing an enemy from detecting the transmission whereas LPI is preventing an enemy from tapping on to the information. LPE is concerned with the prevention of exploitation of signals caused by spoofing, sniffing, decoding or position monitoring. Exploitation involves using the transmitted data for intelligence or counter-intelligence purposes. The covert nature of this technology makes the laser beam resilient to jamming or spoofing, which is essential for military operations. The probability of laser beam being detected depends upon the beam divergence and spectrum of frequencies emitted by the laser. Covert military operations demand to work in near-IR band, with narrow beam divergence and minimal spillover, or spurious emissions, like side lobes.

Laser beams can still be detected using appropriate tools like IR goggles. Since the spectral sensitivity of these goggles is from 0.4 μm to 1.3 μm, it enables the soldiers to see both visible (0.4 μm to 0.7 μm) and near-IR light (0.7 μm to 1.5 μm) through these goggles. Further, environmental conditions like smoky wartime scenario, fog, haze and dust particles, scatter the light and make the laser beam detectable. In such cases, in order to minimize the probability of detection, transmitters should not use excessive power; it would minimize the scattered light and reduce detection probability. Intercepting a laser beam requires tapping the information signal by using some sensing device in the path of the transmitted signal. It is almost difficult to intercept a laser transmission without disrupting the system, owing to the narrow divergence of the optical beam. As most of the signal falls within the detector surface area, intercepting the signal blocks the transmission path resulting in a significant drop in the received power level and therefore, raising an alarm for intrusion detection. Consequently, for security reasons, a beam with narrow beam divergence is preferred, although it causes difficulty in pointing and aligning the beam with a distant receiver. To resolve this, blockage shields helps to minimize the probability of interception.

Besides LOS communication, NLOS EO laser communication utilizing UV radiation is also studied for military applications. With the development of UV light emitting diodes (LEDs) and APDs, short-range NLOS UV communications offer significant advantages over LOS links, by relaxing, pointing and tracking requirements of IR links. Therefore, laser technology is a good alternative to traditional RF links as it is capable of providing secure, high capacity and rapid information transfer for dynamic mission planning. Laser technology is very beneficial for space applications. Laser communication links can be applied to both static and mobile platforms for ground, air or underwater environments. Despite the many benefits of laser communications, this technology has considerable limitations, that prevents it from being a direct replacement for conventional RF communication. The performance of laser links is very susceptible to varying weather conditions and it deteriorates during heavy fog, smog or high temperature circumstances. For this reason, military bodies around the world are looking at the laser communication as a technology to augment the existing RF-based system or keep it handy to provide assistance in case of jamming. Laser communication systems are generally designed for short-range point-to-point or multi-point configurations, where other communication networks are practically impossible to be installed. Various developments in laser communication have been observed by many defense organizations over the past few decades. In order to support both commercial and DoD requirements, various experimental investigations have been carried out for terrestrial laser-com links, space laser links, air-to-ground/ground-to-air laser links, air-to-ocean/ocean-to-air laser links. These developments have demonstrated increased throughput capabilities with low probability of interception for future electronic warfare scenarios.

29.5 EARTH RECEIVES A LASER MESSAGE FROM 16 MILLION KILOMETERS AWAY

In a groundbreaking experiment, NASA has successfully transmitted a laser message over an astonishing distance of 16 million kilometers. This achievement, part of the Psyche spacecraft mission, marks a pivotal step in revolutionizing deep-space communication. What does this mean for the future of interplanetary exploration?

In an ambitious leap forward for interplanetary communication, NASA has successfully transmitted a laser-based message from the Psyche spacecraft to Earth, covering a staggering distance of 16 million kilometers. This technological breakthrough, part of the Deep Space Optical Communications Experiment (DSOC), promises to transform the way humans exchange data across the cosmos, paving the way for more sophisticated and efficient space exploration.

The Psyche spacecraft, launched in October 2023 aboard a SpaceX Falcon Heavy rocket, carries multiple scientific instruments, including the DSOC laser communication system. The primary aim of DSOC is to test high-speed optical communication technology, which offers substantial advantages over traditional radio-frequency systems. By utilizing near-infrared lasers, this technology can transmit data 10 to 100 times faster than current methods.

On November 14, 2023, the system successfully transmitted a coded laser signal from the spacecraft to the Hale Telescope at the Palomar Observatory in California. This marked the first successful test of optical communication at such a vast distance, far exceeding previous experiments limited to lunar orbits.

The achievement is part of a larger mission to explore the asteroid Psyche, a metallic body believed to hold clues about the early formation of planetary cores. While the spacecraft's primary scientific mission will not commence until it reaches the asteroid in 2028, the laser communication test demonstrates a crucial step toward advanced interplanetary communication.

29.6 BATTLEFIELD POTENTIAL OF LASER COMMUNICATIONS

Free-Space Optical (FSO) Wireless, or laser communications, has been in use in the commercial sector for years. It's potential on the battlefield, though yet to be fully exploited, is now being taken evermore seriously.

Imagine a means of line-of-sight communications that is totally secure, cannot be intercepted, eavesdropped, or disrupted by EW and jamming. Well, the commercial sector has been developing and using FSO technology over the past 30 years or more for a wide range of applications and this article takes a look at the technology, its development, and its potential in defence.

FSO technology uses laser beams via a line-of-sight optical bandwidth connection to transfer data, video or voice communications across areas ranging typically from 100m to a few kilometres at throughput bandwidths up to 1.5Gbps at frequencies above 300GHz of wavelengths, typically, 785nm to 1550nm. Using FSO wireless networks eliminates the need to secure licensing found with RF signal solutions and also the expensive costs of laying fiber-optic cable. The practical side of FSO communication requires a pair of transceiver units housing an optical receiver and transmitter, allowing the sending and receiving of data simultaneously. The unit at one location transmits a beam of focused light carrying/delivering information directly to the receiving location where the light beam is transferred to an optical fiber from a high-sensitivity receiver. FSO systems offer various advantages over normal RF wireless networks; they don't suffer from RF interference or band saturation, their operation is license free, no software is needed on client devices, indoor installation is possible and unaffected by operation through glass and, importantly for the defence sector, FSO provides a very high level of security – the technology is already certified in governmental and defence applications for transferring confidential and classified information.

Due to the signal strength being affected by atmospheric conditions over increasing distance, the most efficient terrestrial FSO wireless network ranges are typically between 100m and 2km. However, when combined with LAN or WLAN networks they can provide very effective solutions to many scenarios, for example, by providing a bridge between WLAN-to-WLAN connections on campuses at Fast Ethernet or Gigabit Ethernet speeds to cater for many subscribers simultaneously. Alternatively, FSO can provide a bridge between LAN-to-LAN connections and they can also create a wireless link across an area where no physical access is possible. They can also be used to quickly re-establish high-speed network connections after incidents, as in disaster-recovery scenarios.

The speeds delivered by FSO are, today, comparable to those of optical-fiber connections, but with the flexibility and practicality of being part of a wireless network providing bandwidth speeds up to 10Mbps, 100Mbps, 155Mbps and 1.5Gbps; speeds of up to 10Gbps are likely in the future. Currently, the only other wireless technology capable of such speeds is mm-Wave RF Wireless Networking, which, in comparison, requires licensing and can be affected severely by rain.

That is not to say FSO doesn't have some atmospheric drawbacks. Unlike rain and snow, which generally have little effect on FSO communication, it is fog and water vapour droplets that hinder FSO operating performance; small water droplets can prevent light beams being received due absorption, refraction, scattering, or even complete reflection, which can significantly lower data rates. Line-of-Sight obstructions also interfere with FSO wireless transmission. As light can't travel through opaque mediums, objects such as birds, planes and people can momentarily interrupt the service by blocking the beam, with service resuming instantly when the light path is cleared. Multi-beam technology can counter this problem to an extent. On occasions, building sway due to wind can also be a problem; it disrupts alignment between two transceiver units causing signal loss. On the safety side of FSO, the technology is strictly controlled, standards are followed and dangers limited, with equipment typically meeting Class 1M, eye-safe certification.

29.7 SPACE-BASED LASER COMMUNICATION SYSTEMS

Laser technology is emerging as a game-changer in satellite communications (SATCOM), enabling the creation of ultra-secure networks capable of transmitting vast amounts of data at unprecedented speeds via satellite networks and constellations. With ongoing advancements, the industry is poised for growth and collaboration, seizing the untapped potential of unconnected populations. The ability to handle the surging volume of data is a key advantage offered by laser communications, providing a significant value proposition.

While it has its challenges, laser communications is promising for defense and space applications as U.S. government agencies continue to embrace the forward prospects for laser communications as they invest in further developments to enhance its use.

(i) A revolution in satellite communications

The global space-based laser communication market — mostly for commercial applications - was worth about $1.13 billion in 2022, and is expected to quadruple by 2031, growing at nearly 26 percent per year, according to research conducted by Straits Research in Maharashtra, India, titled Space-Based Laser Communications Market.

Space-based laser communications has emerged as a ground-breaking solution for data transfer in remote and challenging locations. This technology offers a

The NASA Lunar Laser Communication Demonstration (LLCD) will demonstrate laser communications from lunar orbit to Earth at six times the rate of the best modern-day advanced radio communication systems.

The NASA Lunar Laser Communication Demonstration (LLCD) will demonstrate laser communications from lunar orbit to Earth at six times the rate of the best modern-day advanced radio communication systems.

wide array of applications like inter-satellite communications, and satellite-to-ground connectivity. As the number of orbiting satellites continues to grow, space-based laser communications is becoming a prominent player in satellite communications systems.

Investors, while cautious, are monitoring the industry's progress closely to compete with major operators commercial like Starlink and OneWeb. The expansive customer base awaiting connectivity presents a lucrative opportunity for growth and collaboration. With laser communications offering faster speeds, improved security, and greater efficiency, it is poised to reshape the landscape of satellite communications.

Laser communications are poised to redefine secure and efficient data transmission not only in commercial markets, but in defense and government markets as well. Through technological advancements, increasing affordability, and a rapidly expanding customer base, laser technology is heralding a new era in satellite communications. As researchers continue to enhance laser diode efficiency, power output, and wavelength stability, the performance and reliability of laser communication systems are set to reach new heights.

In the international community, standardization efforts also are in progress, spearheaded by organizations such as the International Telecommunication Union (ITU) and other industry consortia. The establishment of common standards ensures interoperability and widespread adoption of laser communication systems. This enables seamless integration with existing RF infrastructure, creating hybrid systems that provide enhanced reliability and performance.

Despite atmospheric challenges affecting laser communications, innovative techniques such as adaptive optics, beam steering, and error correction algorithms are being explored to mitigate these effects. Overcoming these hurdles is crucial for the practical implementation of laser communications.

Suffice it to say that laser technology is revolutionizing the satellite communications industry. With its unparalleled capabilities, laser communications offer faster speeds, enhanced security, and increased efficiency than are available today. As the sector continues to mature, collaborations and partnerships will be instrumental in driving innovation and shaping the future of satellite communications.

(ii) Bright future for Laser Communication in Space

Those advancing the technology are focused on exploiting the advantages of laser optical communications over RF communications such as increased bandwidth, small beam size, hard to intercept/detect or jam, and the ability to utilize a largely unregulated portion of the spectrum. These benefits provide an outlook for laser communications that has a market expected to grow from hundreds of millions of dollars to multiple billions of dollars by 2030 according to multiple market analysis studies.

The typical use of laser communications is for space. Traditionally, laser communications have been developed and employed for space-based applications such as satellite to satellite and ground to satellite data transfer, this is due to the impacts of the atmosphere on laser communications for terrestrial, maritime, and air domains, to include bandwidth, pointing, acquisition, and tracking, and resiliency of communication through multiple atmospheric conditions.

Technology developers are now concentrating investments on increasing the practical use of laser communications within these domains such as bandwidth, distance, and resiliency through a variety of atmospheric conditions," Dare says. Bandwidth nominally is in the 1 to 10 gigabits per second range today going to hundreds of gigabits per second in the future. Distance today is about 1 kilometer to up to tens of kilometers going to more than 50 kilometers in the future for terrestrial-terrestrial applications.

The Orion Artemis II Optical Communications System (O2O) will bring laser communications to the Moon aboard NASA's Orion spacecraft during the Artemis II mission to transmit high-resolution images and video.

The Orion Artemis II Optical Communications System (O2O) will bring laser communications to the Moon aboard NASA's Orion spacecraft during the Artemis II mission to transmit high-resolution images and video.

The future of laser communications for satellite communications is very bright. Secure intersatellite links as well as space to ground communications will both leverage this technology in the future. While there will still be a need for spectrum management as this capability proliferates, this management should be much easier due to the narrowness of the beams.

The future of laser communications to rests much on reliable technologies that can operate through a wide array of atmospheric conditions.

Implementing reliable technologies to increase bandwidth and operate in a variety of atmospheric conditions will eventually explode the terrestrial, maritime, and air domain laser communications markets within the next 3-5 years.

The importance of creating solutions in these domains allows the use of laser communications to be leveraged in situations where radio frequency (RF) is contested, not available, or not affordable, such as in contested or anti-access/area denial (A2AD) environments, austere environments, on-the-move capabilities, and having to lay fiber in difficult terrain or urban areas.

Each of these capabilities can be utilized in multiple markets, such as defense, civil, commercial, and agricultural. It is anticipated that the markets will expand its laser communication usage in the next few years to leverage the benefits mentioned as the capabilities and practicality of laser communications through the atmosphere become widely available.

(iii) Laser communication is not without challenges

Laser communications are influenced by the atmosphere they penetrate and transmit through; if not ideal, communications may have problems. Namely, the beam that is used for directional communications can be fidgety against different atmospheric conditions. According to Booz Allen's Dare, these and other considerations mean that laser communication has its challenges.

The Integrated LCRD Low-Earth Orbit User Modem and Amplifier Terminal (ILLUMA-T) will bring laser communications to the Space Station and empower astronauts living and working there with enhanced data capabilities.

The Integrated LCRD Low-Earth Orbit User Modem and Amplifier Terminal (ILLUMA-T) will bring laser communications to the Space Station and empower astronauts living and working there with enhanced data capabilities.

One of the advantages of laser communications, the small beam size, also presents one of the most difficult challenges in implementation," explains Dare. "The small beam requires very precise control to establish and maintain the link. This is referred to as pointing, acquisition and tracking (PAT).

(iv) Pointing the laser in less-than-ideal conditions can be a challenge

The line-of-sight nature of laser communications necessitates an understanding of where to transmit (point), allow for link establishment (acquisition), and an ability to adjust to maintain the link (tracking). This PAT equation becomes more challenging when transmitting through a domain that can impact the beam-the atmosphere or water (for subsurface communications).

Converting satellite laser communications links for in-atmosphere links can be an option. "PAT systems are in use today for some laser communications such as satellite to satellite and terrestrial to satellite. These systems can be upgraded to provide the capabilities for the more complex systems that are purely within the atmospheric domain to allow for laser communications on the move (terrestrial, maritime, air, or subsurface) through a PAT system.

Establishing a laser communications link is one thing, but maintaining it can be quite another. A challenge that must be overcome includes the ability to establish and maintain an on-the-move link in potentially challenging 3-axis conditions, countering the effects of sea state, rugged terrain, air turbulence, or subsurface currents. Another basic challenge that must be overcome is the condition of the atmosphere (or subsurface). "Several conditions can cause interference with a high-quality laser communications link such as dust, heat, turbulence, rain, spray, clouds, etc. Any solution that will be resilient enough to work in a true operational environment must have the ability to counter the impacts of these conditions to maintain a reliable link."

Lockheed Martin's Huttenhoff says he also sees challenges. "Several challenges are present as we continue to mature this technology for the future," he says. "First, if systems are going to be interoperable with each other, then some form of standard will need to be generated by industry to be an enabler. Second, as we look at the increasing data-rates that are possible with laser communications, the need for processors that can perform data switching and processing in space at these rates will need to be developed."

Booz Allen's Dare says that until these challenges are overcome, space and defense leaders should utilize laser communications as an additional tool to address their communication requirements. "It is a complement to RF communications as part of an overall plan. Additionally, it is important to note that laser communication is not a one size fits all solution; laser communications can provide a highly tailorable solution based on mission and domain requirements," he says.

(v) Many use cases for laser communications

Laser communications crosslinks have been employed in the space environment for applications like satellite-to-satellite links, ground to satellite links, and "last mile" connections terrestrially, which can bypass the need for fiber optics.

These applications will continue to expand to increase distance, setup time, and dynamic situations to allow reliable on-the-move communications. The current state of laser communications for terrestrial applications are limited in throughput and distance due to the precision required for establishing and maintaining a link as well as varying atmospheric conditions. At this point laser communications is applicable to a niche market due to imitations; however systems can be provided to meet specific customer needs that utilize stable platforms — these systems can be miles apart and provide gigabits per second speeds, however they will still be impacted by atmospheric conditions.

As technology advances, laser communications increasingly will shift to cross-domain opportunities — particularly in terrestrial and ground to space applications. For RF communications this will provide an ability to expand data flows that the spectrum cannot accommodate."

There are advantages and scenarios where laser communications are best used. The government customers are exploiting the low probability of interference and low-probability-of-detection attributes and employing laser communications in covert scenarios or in an RF denied environment. Specific areas of growth for terrestrial applications will depend on the ability to acquire and maintain link, and more importantly to increase bandwidth with high resiliency to be able to communicate in a variety of atmospheric conditions. It is clear the benefits laser communications bring to contested and denied RF environments, however the ability of significantly increasing reliance in multiple atmospheric conditions and also reducing the Size, Weight, Power and Cost (SWaP-C) will be key to providing end users a capability that will meet needs across all markets.

(vi) Investments in laser communications

There are many initiatives across government agencies. The U.S. National Aeronautics and space Administration (NASA) in Washington, for example, seeks to advance laser communications. Several other research initiatives are in progress that stand to serve defense needs. They are taking steps to enhance the capabilities of laser communications and are collaborating with industry leaders to get down to the nitty gritty of optical materials that could be useful and disruptive for laser communications in important defense applications.

Optical communications terminals that use lasers to beam data across space will be tested in upcoming experiments by the Space Development Agency and the Defense Advanced Research Projects Agency.

Optical communications terminals that use lasers to beam data across space will be tested in upcoming experiments by the Space Development Agency and the Defense Advanced Research Projects Agency.

The U.S. Defense Advanced Research Projects Agency (DARPA) in Arlington, Va., issued a disruption opportunity in May for Accelerating discovery of Tunable Optical Materials (ATOM) project seeking breakthroughs in tunable optics for free-space optics and integrated photonics intended for use in laser communications.

The ATOM program's aim is to discover new tunable optical materials that possess large refractive index contrast across the entire spectrum while maintaining low loss and fast switching times. The program also seeks to demonstrate a device with a minimum area of 250 square microns, capable of achieving repeatable and stable multi-state switching, all while preserving bandwidth and performance.

Late last year, DARPA also moved forward with industry-contracted research to develop miniature optical beam steering for applications like free-space laser optical communications and light detection and ranging (lidar) for use in laser communications. Late last year they launched a microsystems exploration topic for the Steerable Optical Aperture Receivers (SOAR) project. SOAR will explore new approaches to optical beam steering in miniature form factors, and experimentally demonstrate their operation in receive mode with small aperture sizes.

Currently, the optical beam steering primarily relies on mechanical systems such as gimbals or motors to adjust the direction of optical lenses. However, these gimbal-based beam steering systems are often too large and heavy for small autonomous vehicles that require onboard laser communications and lidar capabilities.

Integrated photonics has emerged as a plausible solution to assist microscopic devices on chips replicate the functions of discrete optics. According to the announcement, this not only leads to a significant reduction in size but opens possibilities for new and intricate optical system architectures that were previously impractical at the macroscopic scale. The impetus behind the SOAR project is to address critical questions regarding optical receiver performance, scalability, and integration.

(vii) Intricate optical systems

SOAR aims to develop optical interfaces capable of receiving light from any direction without precise knowledge of the incoming angle. This is achieved by steering the angle of acceptance to capture and couple the input beam into a common output mode or detect the optical signal within the receiver interface.

SOAR is technology-agnostic; It aligns with the program's goals. Two-dimensional optical parametric amplification (OPA), non-planar integrated photonics.

The U.S. Space Development Agency (SDA) Is Investing in Low-Earth Orbit Satellite Development

In parallel to DARPA, the U.S. the Space Development Agency (SDA) is also aggressively investing in and collaborating with industry to pursue advanced research to assist laser communications. The agency announced in March of 2022, that Northrop Grumman Corporation was selected to create an innovative constellation of 42 low-Earth orbit (pLEO) satellites and establish the Tranche 1 Transport Layer (T1TL) mesh satellite communications network. The Space Development Agency, a new agency, plans to boost the delivery of space-based capabilities to support terrestrial missions.

The announcement described the T1TL network as a low-latency, high-volume data transport network used to support critical U.S. military missions. Its primary objective is to connect various components of an integrated sensing architecture, ensuring persistent and secure connectivity. The network is expected to play a vital role in the Joint All Domain Command Control initiative.

Specifically, the T1TL network will integrate advanced technologies and infrastructure to enable future space missions, prioritizing key aspects such as battle management, missile tracking, and target custody. According to the announcement of the initiative, laser communication terminals will be employed to connect the global constellation, while providing persistent, networked Link-16 and high-rate Ka-band connectivity to air, maritime, and ground users.

Northrop Grumman is collaborating with commercial marketplace partners. Their intent is to assist SDA's efforts to boost speed and expedite the delivery of an innovative system.

Northrop Grumman has a good history in this type of work and is credited with the assembly, integration, and testing of the sophisticated Iridium NEXT satellite constellation, consisting of 81 satellites that were successfully launched into low-Earth orbit in 2019

(viii) The road ahead for laser communications

The research sponsored by DARPA and SDA is among the many efforts to improve secure high data-rate laser communications via satellite. Tthe future of secure, high data-rate satellite communications along with the benefits of optical communications are becoming very compelling. It can be envisioned that a data-centric satellite communications architecture in space that is enabled utilizing laser communications. This would include not only satellites in low earth orbit, but higher data-rate terminals in medium earth orbit providing data backhaul. By leveraging such an architecture, they can envision data reaching the user with much less latency than today.

As with any new technology, there will be multiple industry and government initiatives that will try to get to market first to exploit the benefits of laser communications. The problem set in implementing a reasonable cost, reliable, high-bandwidth solution is not small, and care must be taken when thinking through the myriad of potential solutions. Developers of the technology must focus on the end user and the environment in which they will be using laser communications. It will be a non-stable, dirty environment with most likely poor atmospheric conditions. The internal mechanisms for laser communications demand a clean environment with precision in the transmit and receive modules for optical alignment. The mounts must be rugged enough to withstand the environment and handling of the system without damaging or causing internal alignment or contamination issues — these are not easy problems to solve.

The early adopters of the technology exist and there will be success and failures in being able to use the new technology. It is important that solution providers and the early adopters work together closely to allow the technology issues to be resolved. The benefits for laser communications are great and one day will become a standard part of the communications toolkit; as such, the ability to find partners to allow the technology solutions to be refined in a true operational environment will be key to achieving those positive outcomes.

CHAPTER 30

RF & MM-WAVE COMMUNICATION IN WARFARE

30.1 INTRODUCTION TO RADIO FREQUENCY (RF) COMMUNICATION

Radio frequency (RF) is the oscillation rate of electromagnetic field frequency ranging from around 20 kHz to around 300 GHz. This is roughly between the upper limit of audio frequencies and the lower limit of infrared frequencies, and also encompasses the microwave range. RF energy is ubiquitous in many electronic devices such as cell phones, radios, and televisions., etc. Different sources specify different upper and lower bounds for the frequency range.

Today, wireless technology is ubiquitous in the military and civil society; it connects people and systems and is the universally accepted way to transfer large amounts of data across (almost) unlimited space. We have become entirely dependent on wireless devices sending radio signals across the Electromagnetic Spectrum: mobile telephony, TV, Radar, broadcast radio, Bluetooth, Wi-Fi, and the IoT are all examples of technologies completely reliant on radio frequency (RF) transmission.

(i) What is spectrum management?

Spectrum management is the process of regulating the use of the radio frequency spectrum. The goal is to ensure that all wireless communication technologies can coexist, minimizing interference, maximizing efficiency, and eliminating unauthorized resource use.

There are **three key areas of spectrum management**.

- First, the process involves allocating specific frequencies and assigning them to particular uses, such as broadcasting, cellular networks, satellite communication, and other wireless services.
- Second, it entails establishing technical standards for wireless devices and ensuring they comply with them to avoid interference with other devices.
- Third, it includes monitoring and enforcing compliance with regulations to ensure that wireless services operate within their allocated frequency bands.

Effective spectrum management is essential to ensure that wireless communication networks operate efficiently and reliably. This should promote innovation and investment in new technologies and support the growing demand for wireless services.

Users sharing a segment of spectrum frequency is a crucial part of spectrum management. Efficient sharing means more spectrum is available for more users and is facilitated by national bodies working together and international organizations coordinating near border regions.

However, managing the spectrum closer to international borders is more challenging because different countries generally have different rules and regulations regarding the allocation and use of radio frequencies. As a result, frequency usage must be coordinated in border areas, where the same frequency may be allocated to different services, as there may be interference when multiple signals from different countries overlap.

As spectrum allocation is a political decision, when political tensions exist between neighboring countries, there may be a lack of cooperation or disagreements over the use of specific frequencies. This can lead to delays or difficulties in negotiating agreements on spectrum usage.

There is also a military aspect of spectrum allocation. When a unit is mobilized to a foreign region, they must ensure they follow local allocation rules. In this scenario, military spectrum managers are essential in planning and executing field operations.

(ii) Historic RF milestones

1887 – Heinrich Hertz demonstrates the existence of radio waves, proving that electromagnetic waves can be transmitted through the air. Although Hertz was not the first to produce radio waves, he was the first to understand they were electromagnetic.

1895 – Guglielmo Marconi built (and then marketed) the first wireless telegraphy system. Marconi's system used RF waves to transmit signals wirelessly over long distances. He developed a transmitter that could convert electrical signals into radio waves and a receiver that could detect those radio waves and convert them back into electrical signals.

1901 – Marconi sends the first wireless message across the Atlantic Ocean, ushering in a new era of global communication.

1906 – Reginald Fessenden made the first long-range radio transmission of voice and music from Brant Rock to ships sailing along the Atlantic coast.

Fessenden's radio broadcast used an electrical signal from a microphone to modulate the carrier wave, which was then transmitted as an RF wave. The RF wave was picked up by receivers tuned to the same frequency and demodulated to recover the audio signal.

1930 – **Pre-World War II**, several countries started developing radar technologies: radio waves sent from an antenna that bounce off objects they encounter.

The reflected signal can be used to determine the location and speed of the object. Work on radar undertaken in the 1930s by Robert Watson-Watt in the UK and Christian Hülsmeyer in Germany played a critical role in detecting enemy aircraft and ships.

1957 – Sputnik-1 was the first artificial satellite sent to orbit around the Earth (by the Soviet Union).

Satellite communications use RF to transmit signals carrying information between the satellite and the ground station or another satellite.

1964 – The Syncom-3 satellite was launched by NASA. It was an experimental communications satellite designed to demonstrate the feasibility of using satellites in geostationary orbit for global telecommunications.

1969 – Apollo 11 made the first moon landing. The spacecraft and the space station used RF to communicate, even when the spacecraft was on the far side of the moon out of line-of-sight control. The spacecraft used an S-band radio frequency to transmit data back to Earth, while the ground stations used an X-band frequency to send commands to the spacecraft.

1973 – Martin Cooper, an engineer from Motorola, invented the technology upon which the first mobile phone was built. However, it wasn't until 1984 that Motorola released their DynaTAC—affectionately known as 'the brick'.

The RF technology in mobile phones uses radio waves to communicate between devices. Mobile telephones send RF signals to nearby cell towers, which transmit the signal to the recipient's cell phone. The signal is converted into sound waves, allowing the receiver to hear the caller's voice.

1977 – The first live human was given an **MRI scan**. Magnetic Resonance Imaging was invented in the early 1970s by Paul Lauterbur and Peter Mansfield; the technology uses RF waves to generate images of the body's internal structures.

MRI machines generate a strong magnetic field, aligning hydrogen atoms in the patient's body. It then emits RF waves causing the hydrogen atoms to emit energy, which is detected by the MRI machine and used to create images of the body.

1993 – The full **Global Positioning System** constellation of 24 satellites became operational in 1993. GPS uses a network of satellites in orbit around the Earth to determine a user's location. Each GPS satellite transmits a signal with a unique code and timing information. By receiving signals from multiple GPS satellites, a GPS receiver can calculate its position by measuring the time delay between the transmission of a signal and its reception.

The signals transmitted by GPS satellites use RF as the waves can travel long distances through the atmosphere and are able to penetrate clouds, foliage, and buildings.

2003 – Internet of Things (IoT) technology became prevalent thanks to the development of low-cost sensors and the widespread availability of wireless networks. RF is crucial for IoT technology as it enables wireless communication between these devices.

2019 – The first 5G commercial network was launched in South Korea. 5G technology uses RF waves to transmit data wirelessly between devices and cellular towers, but it uses higher frequencies than previous generations of wireless technology, allowing faster data transfer rates, lower latency, and greater network capacity, making it possible to support more connected devices and applications.

30.2 ROLE OF RADIO FREQUENCY OVER FIBER (RFOF) IN AN AUTONOMOUS MILITARY FUTURE

The presence of autonomous vehicles and weaponry in the military is set to grow at a significant rate, spurred by funding from the Department of Defense (DOD) in recent years. In 2021, the Pentagon received $7.5 billion to fund unmanned systems across the Air Force, Army and Navy. Generally speaking, these are positive innovations that help the U.S. keep pace with the military advancement of other world powers, such as Russia and China, but they are not without significant risks to national security if the communication infrastructure supporting them is insufficient.

Unmanned vehicles and weapons rely heavily on continuous real-time communication, in addition to radar/LiDAR, as well as other passive sensors that are deployed without human involvement. Operational integrity is at risk if there is any communication failure or adversarial intervention based on emission profiles. Even autonomous mission-oriented apparatus that does not emit information, for the sole purpose of avoiding detection, relies on receiving signals. While always important, communication has never been on the "hot seat" quite like it is now when it determines whether our autonomous defenses work reliably. Figure 1 shows a portion of the U.S. Naval fleet, a military domain containing some of the biggest sensor platforms in existence.

Exacerbating the issue are the high frequency RF signals that provide the low latency, high bandwidth communication necessary for automation. High frequencies tend to be less resilient and more easily disrupted by outside elements, of which there are many in military environments. It is examined here the state of military communications and why radio frequency over fiber (RFoF) is important to safeguard autonomous military systems.

30.2.1 Military Communication at a Glance

Military communication relies on a wide spectrum of RF bands, each serving specific purposes tailored to the demands of modern warfare. The choice of RF bands varies depending on factors such as geographic location, technology availability and operational requirements.

What follows is a general overview of key RF bands used in military communication and their typical applications:

(i) Ultra-High and Very High Frequency (UHF/VHF) Bands

The UHF (300 MHz to 3 GHz) and VHF (30 to 300 MHz) bands, often associated with "walkie-talkie" style communication, are essential for ground troops, vehicles and aircraft. VHF radios excel in line-of-sight communication over open terrain, providing reliable voice and data transmission. In contrast, UHF radios are preferable in congested or urban environments due to their shorter wavelengths and better signal penetration capabilities. UHF satellite communication facilitates secure and encrypted data transmission among military units, including ground stations, ships and aircraft.

(ii) L-Band (1 to 2 GHz)

L-Band is used for satellite communications, GPS and certain radar systems, offering a versatile spectrum for military applications. The ability of L-Band networks to penetrate atmospheric conditions and foliage enhances utility in scenarios where communication resilience is crucial. This makes it suitable for military command and control, intelligence gathering and communication of strategic information.

(iii) S-Band (2 to 4 GHz)

S-Band frequencies support long-range surveillance and tracking radars, as well as select satellite communication systems. Some overlap exists with Wi-Fi and cellular bands, as well as certain GPS applications.

(iv) C-Band (4 to 8 GHz)

C-Band is a versatile frequency range employed in radar systems, satellite communications and various weather radar applications. It offers a balanced compromise between the need for high-resolution and atmospheric penetration.

(v) X-Band (8 to 12 GHz)

Military radar systems rely on X-Band for tasks like target identification, tracking and missile guidance. This band provides exceptional resolution and accuracy to improve situational awareness.

(vi) Ku-Band (12 to 18 GHz) and Ka-Band (26.5 to 40 GHz)

These higher frequency bands are crucial for military satellite communications, offering enhanced data transfer rates compared to lower frequencies. Ku-Band is widely used for communication between different U.S. military units.

While these RF bands are arguably the most common, frequency allocations and standards in military communication can vary globally and evolve with technological advancements. Additionally, classified or encrypted communication further complicates the disclosure of precise frequency bands and their applications.

30.2.2 Balancing Act of Digital and Analog Communications

The military has relied on communications since its inception. Even with network evolution, at the heart of contemporary military communications lies the fact that the nature of transportation over RF bands is analog. The receiver of the signal must have a wide dynamic range, so even systems deploying down-converters must have an analog transport section.

Digital communication systems are celebrated for their precision and reliability, boasting error correction capabilities, adaptability to various data types and better spectral efficiency. However, converting RF signals into digital formats introduces certain complexities. These conversions can lead to latency and jitter, along with creating vulnerabilities such as unauthorized interception, commonly known as "listening," by nefarious actors.

In contrast, analog technology is the primary medium for propagating signals through the air. Autonomous non-tethered military platforms, such as drones and unmanned aircraft, rely on analog components for receiving and transmitting signals. Since there is no need for conversion between endpoints and the transport medium, there is less latency and it drastically reduces the chance of detection. While cabled communication via fiber or CAT cables is principally digital, coaxial and waveguide transport is almost always analog. Each technology has its strengths and weaknesses, making their combination essential for effective and secure communication.

A persistent concern in military communication is the potential detection of electromagnetic radiation by adversaries. To enhance survivability, military strategists often opt for remote antennas, distancing them from sensitive command centers or main platforms. For instance, an unmanned ship may house its processing and command center below deck while needing to communicate with remote antennas on its exterior. This practice creates a foundation for resiliency, as it reduces the risk of mission failure due to antenna damage or compromise. It is also the reason that RFoF is so critical for autonomous military systems.

The presence of autonomous vehicles and weaponry in the military is set to grow at a significant rate, spurred by funding from the Department of Defense (DOD) in recent years. In 2021, the Pentagon received $7.5 billion to fund unmanned systems across the Air Force, Army and Navy.1 Generally speaking, these are positive innovations that help the U.S. keep pace with the military advancement of other world powers, such as Russia and China, but they are not without significant risks to national security if the communication infrastructure supporting them is insufficient.

Unmanned vehicles and weapons rely heavily on continuous real-time communication, in addition to radar/LiDAR, as well as other passive sensors that are deployed without human involvement. Operational integrity is at risk if there is any communication failure or adversarial intervention based on emission profiles. Even autonomous mission-oriented apparatus that does not emit information, for the sole purpose of avoiding detection, relies on receiving signals. While always important, communication has never been on the "hot seat" quite like it is now when it determines whether our autonomous defenses work reliably. Figure 1 shows a portion of the U.S. Naval fleet, a military domain containing some of the biggest sensor platforms in existence.

Exacerbating the issue are the high frequency RF signals that provide the low latency, high bandwidth communication necessary for automation. High frequencies tend to be less resilient and more easily disrupted by outside elements, of which there are many in military environments. This article examines the state of military communications and why radio frequency over fiber (RFoF) is important to safeguard autonomous military systems.

30.3 HOW SOFTWARE-DEFINED RADIO (SDR) FITS INTO VARIOUS DEFENSE-RELATED ROLES?

(i) Importance of SDR to electronic-warfare applications

The military and defense industry relies heavily on advanced radio technology to carry out its operations effectively and securely. From communication systems to radar equipment, understanding and interacting with radio frequencies plays a crucial role in maintaining national security.

In recent years, software-defined radios (SDRs) have emerged as the next generation of wireless systems, revolutionizing the way military and defense organizations operate. With their versatility and adaptability, SDRs. are having a significant impact on key markets such as radar, spectrum monitoring and recording, and electronic warfare.

(ii) How RF Devices Fit into Defense Roles

Since the early days of analog radio, RF transceivers have been an integral part of warfare and defense operations, especially considering the tremendous strategic advantage introduced by wireless communications in tactical missions.

With the advancement of radio to modern software-based systems, SDRs. became a crucial device for uses beyond telecommunication systems and into highly specialized and complex RF applications. Such apps require not only exchange of data through the air, but also detailed analysis of the electromagnetic signals, high levels of reconfigurability/adaptability, and high-throughput data interfaces for integration with host computers to allow for remote control and monitoring.

In military radar systems, RF devices represent the heart of the operation. A radar system transmits a custom waveform that's broadcast to all directions of the region of interest. It then receives/decodes the reflected signals to obtain position and velocity information about a certain target. Therefore, it must implement at least one transmitter chain, one receiver chain, a duplex switch, and a digital back end for waveform generation and signal processing.

At the transmitter side, several parameters can limit performance, including the resolution of digital-to-analog conversion of the waveform, the specific frequencies that are broadcast through the transmitter, and the antenna's output power distribution. At the receiver side, the radio chain must have excellent phase stability and coherency to accurately receive and mix the signal. In addition, it must provide a high dynamic range and low noise figure to dynamically attenuate/amplify the received reflection to eliminate clutter/jams and reduce noise.

Furthermore, beam-steering/beamforming techniques have been widely implemented to reduce the effects of jamming and interference, diminishing the received power at certain points. In this case, multiple-input, multiple-output (MIMO) systems with powerful parallel computing and high phase stability are crucial to leverage performance and precision.

In terms of software, the DSP capabilities of the digital back-end must provide very-low-latency computation and a high-throughput data interface with the host network to prevent information lost while tracking fast targets. For anti-jamming techniques, such as frequency hopping, the digital back-end must be able to quickly detect power thresholds and adapt the waveform and modulation scheme accordingly.

(iii) Spectrum Monitoring

Another application of RF devices that's crucial to modern warfare is spectrum monitoring, especially considering that most tactical endeavors rely heavily on wireless communication and electromagnetic attacks. Spectrum monitoring is the process of analyzing and monitoring the usage of the electromagnetic spectrum for the purpose of detecting RF interference, monitoring adversarial communication and radio activity, and obtaining situation awareness in the battlefield.

RF devices must provide several key features to ensure spectrum coverage and high-performance detection. First, the receiver of the radio front-end (RFE) has to provide a wide tuning range and high instantaneous bandwidth, so that the same device can be used to cover a wide range of signals.

Moreover, MIMO devices with independent channels can significantly expand the spectrum coverage by assigning a bandwidth portion to each channel, which increases the effective spectrum coverage and maximizes the probability of detecting signals. Spurious-free dynamic range (SFDR) of the radio device significantly impacts the accuracy of measurements; thus, selecting SDRs. with high SFDR will certainly improve performance.

On the software side, a powerful digital back end is fundamental in handling the huge amounts of data involved, especially if implementing multiple channels. In this case, FPGAs with on-board DSP capabilities are the way to go, as they provide very-low-latency computation and can handle parallel computing without major hardware specialization. Furthermore, the FPGA must offer high-throughput interfaces with host computers and storage solutions to ensure the integrity of the measured data.

(iv) What's an SDR?

An SDR is basically an RF transceiver with two main parts: a RFE and a digital back end. The RFE contains all of the receive (Rx) and transmit (Tx) RF chains, including amplification, filtering, and antenna coupling interfaces. The RFE must be able to receive signals over a wide frequency range (typically 0 to 18 GHz, upgradable to 40 GHz), while also providing a high instantaneous bandwidth. Some of the highest-bandwidth SDRs. offer up to 3 GHz/channel.

Each channel of the RFE is connected to the digital back end via high-performance analog-to-digital and digital-to analog converters (ADCs/DACs) that are independent for each Rx/Tx chain, enabling parallel operation that makes MIMO SDRs. a "several-in-one" type of equipment.

(v) Digital back-end

The digital back-end of an SDR contains an FPGA with onboard DSP capabilities for tasks such as modulation, demodulation, upconverting, down-converting, and so on. It's also highly configurable and upgradable, allowing for the latest radio protocols and DSP algorithms to be incorporated into the system. Thus, SDRs. are ideal for military and defense applications in which adaptability, performance, and reliability are critical.

Furthermore, the digital back end offers embedded data packaging and high-speed optical interfaces for high data-throughput communication with the host and storage solutions, making them ideal for spectrum-monitoring systems. The figure shows a high-level diagram of the SDR, specifying the main functions of each board, including the RFE, the digital backend, the power distribution, and the clock generation.

(vi) Radio front-end

The RFE of an SDR is a critical component that plays a vital role in the device's performance. The Rx chain includes multi-stage chains (such as low/baseband and high band), LNAs and power amplifiers, attenuators for dynamic adjustment of signals, IQ downconverters, anti-aliasing filters, and ADCs that connect to the FPGA via JESD204B.

At the Tx side, the signal path also goes through multi-stage chains, beginning at the FPGA, passing through the DAC, anti-imaging filters, frequency synths and local oscillators, IQ upconverters, and RF gain blocks. High-performance SDRs. typically include completely independent Tx and Rx chains for optimal signal processing and parallel configuration.

(vii) Timing and power boards

To support these functions, SDRs. also rely on other boards for timing and power. Timing boards provide clocks for ADCs, DACs, local oscillators, mixers, and the FPGA, while power boards ensure stable power supply to all other components. These elements are essential to achieving high performance and reliable operation in these transceivers.

(viii) FPGA

The digital back-end of a high-end SDR includes an FPGA equipped with onboard DSP capabilities that can be optimized for a range of functions including modulation, demodulation, and up/down-converting. These capabilities are used to perform tasks like CORDIC mixing, data packetization, and FIFO buffers, as well as any application-specific requirements such as channelization, security schemes, and artificial-intelligence/machine-learning algorithms.

The FPGA also allows for ultra-low-latency communication, and is responsible for communicating with the host system, network, or storage solution. This communication, done through Ethernet (SFP/qSFP+) ports and MGMT ports, enables the host system to remotely configure and control the SDR. Through these interfaces, the host also can send, receive, capture, and monitor raw IQ data, using proprietary, open-source (GNU-Radio), and custom software. With achievable data rates of 10 to 400 Gb/s, the FPGA plays a crucial role in optimizing the SDR's performance.

In the current RF landscape, defense systems require a variety of RF devices and systems to meet their needs for spectrum monitoring, EW systems, and radars. SDRs. provide a flexible solution for these needs by enabling the RFE and digital back end to be adapted and upgraded on-the-fly and remotely.

The radio front-end of an SDR contains one or more receive and transmit channels, capable of working with signals in a wide tuning range. The digital back end implements an FPGA with DSP capabilities for modulation, demodulation, up/down-converting, and more.

In modular SDRs, all components can be customized to comply with a variety of SWaP requirements, from powerful ground station racks to onboard radars for light aircraft. The FPGA also communicates with the host system and the Ethernet ports to transport and capture data, enabling storage solutions and fast, reliable communication with the operator. The combination of flexibility, modularity, and RF performance makes SDRs. the ideal choice for any military and defense situation.

30.4 INDIAN ARMY SEEKING MODERN, SATELLITE-BASED COMMUNICATION CAPABILITY

The Indian Army is acquiring various advanced satellite communication (SATCOM) systems to assist its troops on missions. It has contracted Bharat Electronics to supply more than 160 mobile secure satellite terminals. This is on top of the defense deals inked earlier, which include 150 man-portable Ku-band satellite terminals, 400 S-band hand-held terminals, and 300 S-band manpack terminals.

The army also ordered over 80 light vehicle-based Ku-band satellite communications terminals, according to a report by The Times of India.

The Indian Army has SATCOM sets that assist troops on long-range patrols along the line of actual control (LAC) bordering China. However, such systems have become obsolete over the years, with operators complaining about not being able to communicate effectively with their operating bases. Such forward deployed troops have to depend on terrestrial radio and satellite phones, which are not secure and have poor connectivity.

The SATCOM systems will reportedly "plug this void" because they "cannot be intercepted" by enemies.

The source also described the communication systems as a "significant force multiplier" that offers "resilient military communication support" to soldiers deployed in remote, far-flung areas.

Apart from those patrolling the LAC, the terminals will be used to assist soldiers who are performing surgical operations.

30.5 THE MILLIMETER-WAVE SPECTRUM: AN ASSET FOR DEFENSE

Extremely high frequency (EHF) is the International Telecommunication Union designation specifically included in the electromagnetic spectrum classification group with 8 other principal dedicated channel allocation. EHF is a large broadband that span a radius of about (30 GHz to 300 GHz) for the molecular spectra of radio frequencies. It lies between the super high frequency (3 GHz to 30 GHz) band and the far infrared band (300 GHz to 1015), for which the lower part is the terahertz band. Radio waves in this band have **wavelengths from ten to one millimeter**, so it is also called the **millimeter band** and radiation in this band is called **millimeter waves**, Millimeter-length electromagnetic waves were first investigated by **Jagadish Chandra Bose**, who generated waves of frequency up to 60 GHz during experiments in 1894–1896.

For many years, the millimeter-wave spectrum remained an untamed frontier for defense systems because its high atmospheric attenuation, high costs, and lack of suitable semiconductor technologies rendered it a massive technological challenge. While missile seekers and a few specialized systems tapped into its unique properties, electronic warfare, radar, and communications remained in the more accessible microwave frequencies. The only viable technology that could deliver high RF power levels was vacuum electronics, such as traveling wave tubes and klystrons, which remains true today.

That said, semiconductor technology has come a long way. As the 5G standards added these frequencies into the mix, there is finally a financial incentive to focus more intensely on technologies and fabrication techniques that will propel millimeter-wave and terahertz frequencies forward for defense systems in the coming years. Consequently, the military today is embracing millimeter-wave frequencies more intensely, leading to advances in high-resolution radar, high-data-rate communication systems, electronic warfare, drone detection, and other applications essential to ensuring future battlefield success.

30.5.1 What Millimeter Wavelengths Bring to the Table?

Each portion of the microwave spectrum has unique characteristics that make it better suited than others for specific applications. For example, sky-wave and ground-wave propagation are the most significant contributors to operation at frequencies up to about 30 MHz. As frequencies increase into the VHF and UHF regions, propagation increasingly becomes line-of-sight. Once they pass through about 1 GHz, atmospheric attenuation becomes a limiting factor along with free-space path loss.

At frequencies considered "millimeter-wave," above about 30 GHz, attenuation from precipitation, obstacles, and the characteristics of the atmosphere become more pronounced. These impediments and a lack of affordable components and enhanced technologies are the main reasons why the millimeter-wave region, especially at higher frequencies, has remained unused for so long.

Even among some veterans, attenuation at millimeter-wave frequencies has been considered to progress linearly upward as frequency increases, all the way to the infrared region. This is not the case because the resonant frequencies of oxygen, hydrogen, and other atmospheric gases absorb more electromagnetic energy at some frequencies than others. Frequencies with lower attenuation are called atmospheric windows, where long-range propagation is possible.

It's these "windows" that make millimeter-wave propagation unique. For example, at about 18 GHz, attenuation is about 0.1 dB per kilometer, and it's not that high at 40 GHz. At around 60 GHz, attenuation increases dramatically, rising to at least 12 dB per kilometer and rapidly falling to about 0.8 dB per kilometer. Surprisingly, attenuation at 80 GHz is still only about 0.6 dB per kilometer, and even at 95 GHz, attenuation is just 0.8 dB or so. When the high gain achievable with small directional antennas is considered, substantial effective isotropic radiated power (EIRP) is feasible and can virtually eliminate the additional losses due to frequency incurred from atmospheric absorption. In fact, it's been demonstrated that for the same RF output power and when antennas at the end of the path are the same, the signal strength at 140 GHz in free space is actually 5.7 dB higher than at 73 GHz and 14 dB greater than at 28 GHz.

However, millimeter-wave frequencies do have limitations. Propagation is inherently line-of-sight, and virtually any obstruction significantly reduces system performance. In addition, propagation changes with altitude, precipitation, and height above sea level. Nevertheless, the assumption that millimeter-wave frequencies, even those near-infrared, are not worth pursuing is simply incorrect.

The takeaway is that even though attenuation is higher at some millimeter-wave frequencies than others, these differences offer exciting possibilities for defense (and other applications). For example, for secure communications, operating at frequencies where attenuation is high is desirable to reduce the likelihood of inception and jamming. At frequencies where a more extended range is desired, designers have a choice of frequencies both lower and higher than this for applications such as high-resolution radar and remote sensing. A system that combines transceivers that can operate at both frequencies that can be changed in milliseconds presents formidable challenges for adversaries.

Millimeter wave technology is revolutionizing military operations through diverse applications. **This high-frequency range enables precise control over swarms of unmanned aerial vehicles,** enhancing tactical flexibility. It also powers immersive augmented and virtual reality systems, elevating the quality of simulation-based training and mission preparation. For intelligence gathering, millimeter wavelengths enable real-time surveillance and reconnaissance, and command and control structures benefit from the technology's ability to support distributed networks.

For example, the Army uses Wi-Gig (IEEE 802.11ay), a version of Wi-Fi that operates at 60 GHz, for communication, as it creates narrow beams that allow Army command posts to evade detection. In addition, researchers at the University of California, Davis, have developed a 110-mW 39-GHz Doppler radar front end in 65-nm CMOS, which allows the radar to achieve 4-µm range accuracy.

At DARPA, participants in the Millimeter Wave Digital Arrays (MIDAS) program are pioneering digital phased arrays in the 18 to 50 GHz range to create power-efficient 4x4 dual-polarized transceiver tiles using advanced semiconductor and integration techniques. The agency intends for them to serve as foundational components for scalable arrays, supporting mobile ad-hoc networks and LEO satellite communications. The program's element-level digital beamforming aims to enable multi-beam operation, potentially reducing network discovery time and boosting throughput in high-data-rate communication systems.

The U.S. has also fielded several programs that use millimeter-wave frequencies, including the AN/APG-78 Longbow Ka-band fire-control radar for the AH-64D/E Apache attack helicopter manufactured by The Boeing Company. The radar and interferometer are housed in a dome located above the main rotor which allows target detection while the helicopter is behind obstacles. It can simultaneously track up to 128 targets and engage 16 at once with an attack initiated within 30 seconds.

The highly directional nature of millimeter-wave beams enables precise and effective jamming and counter-jamming systems. These systems can disrupt enemy communications and sensors while protecting friendly assets from similar attacks. They are designed to target specific frequencies and communication channels, effectively neutralizing enemy electronic systems. For example, millimeter-wave jammers can disrupt radar systems, rendering them less effective or inoperable. In addition to jamming, millimeter-wave technology is employed in counter-jamming systems that protect friendly communications and sensors from adversary attacks. These systems can detect and counteract jamming attempts, ensuring the integrity and reliability of military communications.

Rapid, high-capacity data transmission becomes paramount as military operations increasingly rely on real-time data for decision-making and coordination. Millimeter waves provide the bandwidth necessary to handle large volumes of data, including real-time video feeds, complex sensor information, and extensive network communications. This capability is particularly critical in network-centric warfare, where the rapid exchange of information can decisively influence battlefield outcomes.

The millimeter-wave region is immense and comparatively free of interference, unlike traditional communication frequencies, such as those in the UHF and microwave frequencies, which are congested because of their widespread use in commercial and civilian applications. The massive available bandwidth at millimeter-wave frequencies allows communications systems to achieve gigabit-per-second data rates.

Millimeter-wave frequencies benefit radars because they can obtain higher resolution of objects. This is because resolution is inversely proportional to wavelength, so shorter wavelengths allow for better resolution. For example, the angular resolution of a radar system is proportional to λ/D, where λ is the wavelength, and D is the antenna aperture size. As a result, shorter wavelengths lead to better angular resolution for a given antenna size. In addition, the range resolution is proportional to c/B, where c is the speed of light and B is the bandwidth. This means that millimeter-wave radars can employ larger bandwidths that improve range resolution. This enhanced resolution allows these radars to detect and resolve smaller objects and fine details.

Millimeter-wave radar systems can detect and track objects with remarkable accuracy. For instance, these systems can identify incoming missiles, potentially with higher precision than conventional radar. This capability is crucial for missile defense systems, which must detect, track, and intercept threats before they reach their targets. Millimeter-wave radar is also invaluable in urban warfare scenarios, where high-resolution imaging is required to distinguish between potential threats and non-threatening objects in complex environments.

Millimeter waves are also employed in **advanced imaging and sensing systems**, which benefit from their ability to penetrate certain materials while being absorbed by others. This property allows specialized sensors to detect concealed objects or substances that are otherwise difficult to identify. For example, millimeter-wave imaging systems detect concealed weapons or explosives and can differentiate between materials based on their absorption characteristics, enabling the identification of hidden threats. This capability is essential for military and homeland security operations, where detecting and neutralizing potential threats is critical.

Stealth technology aims to reduce the radar cross-section of military assets, such as aircraft, ships, and vehicles, making them less detectable by enemy radar systems. Defense researchers are creating more effective stealth coatings and shapes by understanding and manipulating the interaction of millimeter waves with various surfaces. These coatings can absorb millimeter waves, reducing the radar signature of the asset and enhancing its ability to evade detection.

30.6 MILLIMETER-WAVE FOR SOLDIER-TO-SOLDIER COMMUNICATIONS

In a work done by the U.K. Ministry of Defence assessed the feasibility of using 60 GHz millimeter-wave smart antenna technology to provide covert communications capable of meeting these stringent networking needs. The introduction of covert communications between soldiers will require the development of a bespoke directive medium access layer. A number of adjustments to the IEEE 802.11 distribution coordination function that will enable directional communications are suggested. The successful implementation of future smart antenna technologies and direction of arrival-based protocols will be highly dependent on thorough knowledge of transmission channel characteristics prior to deployment. A novel approach to simulating dynamic soldier-to-soldier signal propagation using state-of-the-art animation-based technology developed for computer game design is described using mm waves.

CHAPTER 31

UNDERSEA WIRELESS COMMUNICATION IN WARFARE

31.1 UNDERWATER WIRELESS COMMUNICATION USING EM WAVES

On June 1, 2015, China faced one of the deadliest maritime disasters in its history when a ship was traveling with 454 people on-board in the Yangtze River encountered a massive thunderstorm that capsized the ship and caused its complete destruction. About 442 people died and only 12 were rescued. After facing gusts over 118 km/h, the ship sank in approximately 15 meters of water. It took about 12 hours to provide full strength rescue support to the ship and 30 days before they could identify the correct number of people rescued. One of the major reasons for the disaster seemed to be the delay in the notification of the thunderstorm to the technical staff of the ship. These kind of incidents reminds us of the importance of underwater (seawater) research; including the design and analysis of short range, high bandwidth (MHz), higher data rates (Mbps) and high propagation velocity approximately near to free space ($\approx 3*10^7$ m/Sec with minimum latency) underwater electromagnetic (EM) communication systems.

31.1.1 Electromagnetics and the Ocean

EM propagation as it applies to the sea has many challenges. Some areas that can be examined include:

- Interaction of RF with sea surface and near surface.
- Emission of radio and microwave energy from the sea surface.
- Induced voltages caused by seawater moving in a magnetic field.

Interest in such problems stem in part from the capability of remotely measuring number of physical properties of ocean using EM radiation. Therefore, learning about certain limited but nevertheless important oceanic processes; operational needs for communication and surveillance in the marine environment; and propagation of infrared and optical wavelength energy in the sea. Physical properties or processes that can be observed via EM radiation (albeit with varying accuracies and coverage's) include the following: ocean surface wind stress (i.e., magnitude and direction), surface wave spectra, internal wave surface signatures, surface geotropic current values, sea surface topography, sea surface temperature, and sea ice cover. In addition, hydrodynamic transport rates in limited passages can be inferred from the EM potentials induced in electrodes and cables mounted on the sea floor.

31.1.2 Underwater Electromagnetic (EM) Communication

Underwater EM communication was investigated with intensity in last century up until 1970s. Underwater (seawater) conductivity σ (S/m), permittivity ϵ (F/m) along with restricted range by attenuation α (Np/m) were not helpful for significant breakthrough. Even though acoustic waves can travel long distances in seawater with little α at speed of 1500m/s; communicating through EM waves with approximately ($3*10^7$ m/sec) in seawater at higher data rate means (high bandwidth) for short range has many important applications like pinpointing submerged targets during naval operations, communication across sea/air boundary and recording of seismic waves, etc. The oldest known application of EM in a seawater environment is magnetic compass used for navigation.

31.1.3 Wireless Underwater Sensor Networks Overview

Wireless Underwater Sensor Networks (WUSN) are envisioned to operate in quite different environments, including soil medium, water medium, oil reservoirs, and underground mines and tunnels. System architecture of general WUSNs consists of large number of wireless sensor nodes buried in underwater (seawater).

Hostile underwater environments prevent direct use of most, if not all, existing wireless communication and networking solutions due to the extremely high path loss PL (dB), short communication range R (meters), and higher dynamics of EM waves while penetrating soil, sand, rock, water, and crude oil mediums. In generic WUSN architecture, there may still be some devices, such as sink nodes, deployed above water. Hence, communication between underwater sensors and above water sinks should also be considered. Most WUSNs use acoustic waves as transmission type, but the chance of getting much more out of acoustic modems is quite remote. Optical links are impractical for many underwater applications. Given modern operational requirements and digital communications technology, the time is now appropriate for re-evaluating the role of EM signals in underwater environments. Even though, in general, acoustic waves perform better and have been used for underwater communications, there are some applications like border security which display features different than those assumed for underwater communications. In cases of shallow or average depth small rivers, small lakes, water ponds and fish farms, acoustic waves do not yield the expected results in terms of price and performance.

An example is a terrestrial country border which will pass through terrains of land and rock, but also areas of where water has collected naturally (lakes, ponds, etc.) or on purpose (dams, fish farms, etc.). As EM wave propagation would be the best choice, implementing a different propagation method for limited water reservoirs will be inadequate. The optimal solution for this specific scenario will be to apply the same propagation method to the whole network. As a result the necessity arises to examine propagation characteristics of EM waves in water mediums as well and suggest a viable engineering solution that will take into consideration all the requirements. WUSN can be defined as a group of sensor nodes whose means of data transmission and reception is completely underwater. This includes situations in which a node is underwater, yet in an open space such as a cave or mine, as well as when a node is completely embedded within dense soil or rock.

WUSNs have a number of promising applications including:

(i) **Environmental monitoring,** for example WUSNs can be used to monitor and report the presence of toxic substances for soils near rivers;

(ii) **Infrastructure monitoring,** for example WUSN can be used for monitoring underwater water pipes;

(iii) **Location determination,** for example WUSNs can be used in the location of miners in underwater mines in case of an accident; border patrol and security monitoring are other areas where WUSNs can be applied.

The unique nature of channel in WUSNs possess the biggest challenge as far as communication amongst underwater wireless sensors is concerned. Wireless communication with EM waves through a dense medium such as seawater, soil or rock experiences high levels of α (dB/m) due to the absorption of the signal. Overall, underwater wireless channels for EM waves can be characterized by signal loss due to the multi-path effects caused by inhomogeneous nature of soil such as reflection, refraction, phase shifting, noise caused by electrical ground currents, and extended black-out periods after a rainfall due to wet soil. The ability to predict signal strength provided by sensor nodes in WUSNs is not only useful to researchers but also a convenient capability in practice. For instance, propagation models can be used in conception and design of communication interfaces in order to optimize their performance. They can also be used during the actual field deployment of communication systems in order to determine the coverage. Ultimate goal of propagation modeling is to determine the probability of satisfied performance of a wireless system in seawater that depends on radio wave propagation.

Using EM waves in RF, conventional radio does not work well in an underwater environment due to the conducting nature of the medium, especially in the case of seawater. However, if EM could work underwater, even in a short distance, its much faster propagation speed is definitely a great advantage for faster and efficient communication. Free-space optical (FSO) waves used as wireless communication carriers are generally limited to very short distances because the severe water absorption at the optical frequency band and strong backscatter from suspending particles. Even the clearest water has 1000 times α (dB/m) of clear air, and turbid water has more than 100 times α (dB/m) of the densest fog. Nevertheless, underwater **FSO, especially in the blue-green wavelengths, offers a practical choice for high bandwidth communication (10–150 Mbps), over moderate ranges (10–100 meters).** This communication range is much needed in harbor inspection, oil-rig maintenance, and linking submarines to land, to just name a few of the demands on this front.

Depending on the applications, we can roughly classify the targeted dense sensor **networks into two categories**.

(i) UWSNs for long-term non-time critical aquatic monitoring applications (such as oceanographic data collection, pollution monitoring/detection, and offshore oil/gas field monitoring).

(ii) UWSNs for short term time-critical aquatic exploration applications (such as submarine detection, loss treasure discovery, and hurricane disaster recovery).

The former category of UWSNs can be either mobile or static depending on the deployment of sensor nodes (buoyancy-controlled or fixed at sea floor), while the latter category of UWSNs is usually mobile since it is natural to imagine that the cost of deploying/recovering fixed sensor nodes is typically forbidden for short term time-critical applications.

31.1.4 Channel Characterization of RF Communication at MHz Frequencies

Seawater EM communication presents advantages over acoustic and optical in shallow water and deep oceans. Theoretical analysis of EM wave propagation in seawater helps us to estimate maximum distance covered in seawater at multiple depth points up to 5500 m. Mathematics of EM propagation in seawater (conducting medium) shows dependence on f (Hz), ϵ (F/m) and σ (S/m) of transmission medium.

To characterize seawater communication channels at MHz frequency, real time averaged data (averaged decades) is used for T (Co) and S (ppt) from (1955-2012) at multiple depths of oceans (Indian, Pacific, Atlantic, Southern and Arctic) up to 5500 m (https://www.nodc.noaa.gov/). The average depth of world oceans is in excess of 3500 m. Average data for these parameters taken vertically along depth of oceans at each point of latitude (angle which ranges from 0° at Equator to 90° (North or South) at the poles) and longitude (angular distance ranges from 0° to 180° in degrees, minutes and seconds; a point east or west of Prime (Greenwich) Meridian). Latitude used together with longitude to specify precise location on surface of Earth.

Challenge for industry working through international maritime organization (IMO) will be to produce unified strategy for integration and to develop systems to meet the requirements through S-mode. Concept of S-mode is default setting to bring all inputs like radar, charts, positioning for e-navigation to bring on one platform. This all needs to be achieved through acceptable cost and benefit's-navigation will be used for monitoring all kinds of activities in oceans.

31.2 US NAVY PICKS SYQWEST TO DESIGN AND BUILD UNDERSEA SONAR COMMUNICATIONS FOR SURFACE WARSHIPS AND SUBMARINES

U.S. Navy undersea warfare experts needed a new sonar communications system to replace the AN/WQC-2A underwater communications system. They found their solution from SyQwest Inc. in Cranston, R.I.

Officials of the Naval Undersea Warfare Center Division Keyport in Keyport, Wash., announced a $16.6 million contract to SyQwest earlier this month to build the first Next Generation Gertrude (NGG) underwater communication device. The NGG will replace the Navy's AN/WQC-2A underwater communications system.

The NGG will be for new-construction surface ships, as well as for back-fits to Navy Arleigh Burke-class destroyers and future Constellation-class frigates.

The AN/WQC-2A **sonar communications set** is a single sideband, general-purpose voice and continuous-wave communication set that functions as an underwater communications system to link surface ships, submarines, and shore stations.

The AN/WQC-2A sonar communications set has been in service for more than 35 years and is one of the Navy's in-service sonar underwater communication system for surface ships, submarines, and coastal based shore installations.

Navy officials are requiring its replacement with a commercially available or purpose-built product capable of providing Navy surface combatants with underwater communications between surface ships and submarines.

The NGG underwater communicator will use existing Navy transducers and established electronic interfaces necessary to transmit and receive single-sideband modulated voice, audio, and continuous-wave signals acoustically through the water. NGG will be a STANAG 1074-compliant system.

Test range sites and other coastal installations also use the AN/WQC-2A to communicate with nearby vessels. This system is installed on most U.S. Navy surface ships and submarines, and transmits and receives voice, audio, and low-speed telegraphy for short- and long-distance underwater communications.

The AN/WQC-2A also can amplify and transmit signals from external sources. Ultra Electronics Ocean Systems in Braintree, Mass., is a longtime supplier of the AN/WQC-2A, with more than 300 sonar sets produced for the U.S. Navy and foreign customers, company officials say.

The AN/WQC-2A consists of a control station, a remote-control station, a receiver-transmitter, as well as low- and high-frequency transducers.

On this NGG contract SyQwest will do the work in Cranston, R.I., and should be finished by June 2028.

31.3 INDIAN NAVY SEEKS PRIVATE SECTOR INNOVATION FOR AUTONOMOUS UNDERWATER SMART COMMUNICATION BUOY

The Indian Navy has launched the iDEX DISC 12 Challenge, inviting private sector companies to contribute to the design and development of an advanced Underwater Smart Communication Buoy. This buoy, intended for passive surveillance and data transmission in open sea environments, represents a strategic effort to bolster indigenous defense capabilities **under the Atmanirbhar Bharat initiative.** The program aims to decrease reliance on foreign technology by promoting domestic solutions for essential defense systems.

The envisioned Underwater Smart Communication Buoy will facilitate passive underwater surveillance at depths of up to 200 meters. At programmed intervals, it will surface to relay critical acoustic data to a base station via satellite communication. Designed for autonomous operation, the buoy is expected to function independently in open seas for up to 90 days, gathering and analyzing data without human intervention.

Key Features Outlined by the Indian Navy:

(i) **Smart Communication:** The buoy will surface periodically to ensure efficient, real-time transmission of data, supporting continuous situational awareness.

(ii) **Autonomous Operation:** It must operate independently, adapting to programmed depths and locations in the open sea, without the need for manual oversight.

(iii) **Data Collection and Edge Computing:** The buoy will capture acoustic data for surveillance, utilizing edge computing to process and filter critical information. Only the most relevant data will be transmitted to the base station via satellite.

(iv) **Compact and Self-Sustaining Design:** Designed for durability, the buoy must be compact, easy to deploy, and capable of sustaining itself in challenging marine environments for up to 90 days.

This initiative underscores the Indian Navy's commitment to advancing indigenous technology in defense and expanding India's self-reliance in critical maritime systems.

31.4 UNDERWATER WIRELESS COMMUNICATION MARKET POISED TO REACH VALUATION OF USD 21.06 BILLION BY 2032

The global underwater wireless communication market was valued at US $ 6.13 billion in 2023 and is anticipated to reach US $ 21.06 billion by 2032 at a CAGR of 14.7% during the forecast period 2024–2032.

The demand for underwater wireless communication is experiencing a significant surge, driven by a multitude of factors. The expansion of offshore oil and gas exploration activities has been a major contributor, with the underwater communication system market revenue in North America, largely influenced by offshore activities. Military and defense applications, such as submarine communication, surveillance, and underwater asset monitoring, have also played a crucial role in driving this demand. The U.S. and Russian navies have conducted tests of satellite-to-submarine and plane-to-submarine communication topologies, highlighting the importance of these technologies in military settings.

The Navy's focus on submarine assets and related infrastructure is closely linked to the growth of the underwater wireless communication market. As the demand for advanced underwater communication systems increases, particularly in military and defense applications, the Navy's investments in submarine capabilities are likely to drive innovation and adoption of cutting-edge technologies. The market for underwater acoustic communication, for example, is projected to reach USD 3.5 billion by 2028 and USD 9.2 billion by 2031, largely due to the expansion of defense spending and the need for reliable communication systems in naval operations.

Moreover, the rising demand for controlled underwater vehicles and the increasing deployment of unmanned devices have further fueled the growth of the underwater wireless communication market. These vehicles require high-bandwidth and high-capacity information transfer to facilitate various activities, including exploration, monitoring, and data collection. Environmental monitoring, such as pollution detection and marine life observation, has also contributed to the adoption of underwater wireless communication systems, with scientific divers using these technologies to avoid disturbing marine animals during their research.

CHAPTER 32

ELECTRO-OPTICAL SENSORS IN WARFARE

32.1 INTRODUCTION TO ELECTRO-OPTICAL SENSORS

Electro-optical sensors are electronic detectors that convert light, or a change in light, into an electronic signal. These sensors are able to detect electromagnetic radiation from the infrared down to the ultraviolet wavelengths.

They are used in many industrial and consumer applications, for example:

- Lamps that turn on automatically in response to darkness;
- Position sensors that activate when an object interrupts a light beam;
- Flash detection, to synchronize one photographic flash to another;
- Photoelectric sensors that detect the distance, absence, or presence of an object.

An optical sensor converts light rays into electronic signals. It measures the physical quantity of light and then translates it into a form readable by an instrument. An optical sensor is generally part of a larger system that integrates a source of light, a measuring device, and the optical sensor. This is often connected to an electrical trigger. The trigger reacts to a change in the signal within the light sensor. An optical sensor can measure the changes from one or several light beams. When a change occurs, the light sensor operates as a photoelectric trigger and therefore either increases or decreases the electrical output. An optical switch enables signals in optical fibers or integrated optical circuits to be switched selectively between circuits. An optical switch can operate by mechanical means or by electro-optic effects, magneto-optic effects, and other methods.

32.1.1 Types of optical sensors and switches

There are many different kinds of optical sensors, the most common types are:

(i) **Photoconductive devices** convert a change of incident light into a change of resistance.

(ii) **Photovoltaics**, commonly known as solar cells, convert an amount of incident light into an output voltage.

(iii) **Photodiodes** convert an amount of incident light into an output current.

(iv) **Phototransistors** are a type of bipolar transistor where the base-collector junction is exposed to light. This results in the same behavior of a photodiode, but with an internal gain.

(v) **Optical Switches** are usually used in optical fibers, where the electro-optic effect is used to switch one circuit to another. These switches can be implemented with, for example, microelectromechanical systems or piezoelectric systems.

32.1.2 Applications of Electro-optical Sensors

(i) Electro-optical sensors are used whenever light needs to be converted to energy. Because of this, electro-optical sensors can be seen almost anywhere. Common applications are smartphones where sensors are used to adjust screen brightness, and smartwatches in which sensors are used to measure the wearer's heartbeat.

(ii) Optical sensors can be found in the energy field to monitor structures that generate, produce, distribute, and convert electrical power. The distributed and nonconductive nature of optical fibers makes optical sensors perfect for oil and gas applications, including pipeline monitoring. They can also be found in wind turbine blade monitoring, offshore platform monitoring, power line monitoring and downhole monitoring. Other applications include the civil and transportation fields such as bridge, airport landing strip, dam, railway, airplane, wing, fuel tank and ship hull monitoring.

(iii) Among other applications, optical switches can be found in thermal methods which vary the refraction index in one leg of an interferometer in order to switch the signal, MEMS approaches involving arrays of micromirrors that can deflect an optical signal to the appropriate receiver, piezoelectric beam steering liquid crystals which rotate polarized light depending on the applied electric field and acousto-optic methods which change the refraction index as a result of strain induced by an acoustic field to deflect light.

(iv) Another important application of optical sensor is to measure the concentration of different compounds by both visible and infrared spectroscopy.

32.2 USE OF ELECTRO-OPTICAL SENSORS IN MILITARY

Electro-optical sensors are indispensable tools for modern military operations. As technology advances, these sensors will continue to evolve, providing enhanced situational awareness and lethality to military forces.

Electro-optical sensors play a crucial role in technologies beyond night vision including soil disturbance detection (indicating potential roadside bombs), missile launch detection, and locating small boats at sea. Additional ways defense is using EOs include:

(i) Surveillance and Reconnaissance

EO sensors are used extensively for gathering intelligence, monitoring enemy movements, and conducting reconnaissance missions. They provide real-time imagery of the battlefield, enabling commanders to make informed decisions.

(ii) Target Acquisition and Tracking

EO sensors help military forces locate and track targets, including vehicles, personnel, and infrastructure. They are integrated into various platforms such as unmanned aerial vehicles (UAVs), helicopters, and ground-based surveillance systems.

(iii) Precision Guided Munitions

EO sensors play a critical role in guiding precision munitions to their targets with high accuracy. Laser-guided bombs and missiles use laser designators to mark targets, while imaging sensors provide feedback for terminal guidance.

(iv) Navigation and Situational Awareness

In addition to targeting, EO sensors contribute to navigation and situational awareness for military platforms. They help pilots and vehicle operators navigate terrain, avoid obstacles, and detect threats.

32.3 FUTURE USES OF ELECTRO-OPTICAL SENSORS IN DEFENSE

While the U.S. Army has long claimed to "own the night" thanks to EO sensors – a claim that may no longer be true, according to Modern War Institute – advancements are on the horizon that may revolutionize their capabilities. Future EO sensor systems are likely to incorporate multiple sensors, including imaging, infrared, and laser sensors, to provide enhanced capabilities such as improved target detection and discrimination.

Researchers are working on extending the range of EO sensors, allowing them to detect targets from even greater distances, as well as focusing on shrinking pixel size. Smaller pixels enable higher-resolution images, providing better situational awareness and long-range performance. Infrared search-and-track systems, in particular, benefit from this advancement.

Researchers are also working to reduce the size, weight, and power consumption (SWaP) of EO sensors to allow for easier integration into various platforms including drones and soldier-worn devices. In addition, they are exploring the use of new materials that will enable sensing at higher temperatures, further expanding operational capabilities. New materials for electro-optical sensors also are coming online to enhance pixel pitch, SWaP, range and resolution. One can move from mercury-cadmium telluride and indium antimonide to a material called strained layer superlattice, also known as SLS. These are semiconductor materials that can operate at much higher temperatures. Sensors made from these materials also are called high-operating-temperature (HOT) detectors.

32.3.1 Artificial Intelligence And Machine Learning in EO Sensors

Recent advancements in EO technology have been revolutionary, especially with the inclusion of artificial intelligence (AI) and machine learning. These technologies not only enhance the precision of images, expand the range, and enable the detection of elusive objects, but they can also assist the sensors in identifying critical information.

The operator doesn't need to have constant eyes on target. AI helps to find conditions that are different than normal. It can say with 80 percent certainty that the target is a tank, a SAM site, or another kind of threat.

For the same token, AI also can help detect targets that might not be visible to the human eye, such as a small boat on a vast ocean. The ocean is enormous, and people do not get a sense for how large it is. With AI we can scan all the visible ocean surface and find objects that are not waves, whether it is a fishing trawler, commercial shipping vessel, or a navy destroyer.

With sensing in general it is about getting further and further range to see if there are threats. Third-generation sensing is about seeing further and identifying further out. This can enhance weapons sights to enable gunners to fire effectively at targets they can see.

32.3.2 Cooling and thermal management

High-performance electro-optical sensors for high resolution and long ranges today still require cooling to enhance the contrast between objects of interest and their backgrounds. Coolers tend to be large, heavy, and expensive, and can be critical single points of failure in important applications.

Innovation in cooler longevity is critical. A big part of the market is maintaining them; cooling is a big maintenance item. There is development for coolers with no moving parts, which represents a change in coolers' getting less expensive and smaller.

The need to address cooling connects directly to new generations of high-operating-temperature (HOT) detectors, which by nature require less cooling than sensors made from older-generation materials. If you operate at a higher operating temperatures, the cryogenic cooler power requirements goes down. There is an impact on cooler life expectancy.

The influence of these new HOT detectors on cooler maintenance is substantial. One can have new world-class sensors with MTBF [mean times between failures] of close to 30,000 hours. The impact of that on users in the military is enormous — whether it is applications in border patrol continuous monitoring. Now we can double the cooler life.

Not only can materials in these new HOT detectors enhance logistics and maintenance, but they also can reduce cooling requirements for sensors systems designers. A big part of this is less cooling demand, so your cooler capacity can be lower; it directly translates into power, and translates into the needs for power supplies like batteries, reducing the need for thermal heat dissipation contributes to reducing SWaP.

In HOT detectors coolers are more efficient, the heat load is less, and you have an extremely compact sensor module. Power consumption is driven down, size driven down dramatically, and the resolution is better, relative to indium antimonide. Reducing cooling demands has a big influence on military electro-optical sensors design and capability.

32.3.3 Uncooled sensors

Some electro-optical sensors applications, such as infantry rifle sights, are extremely sensitive to size, weight, power consumption, and cost (SWaP-C). These applications often must compromise on range and resolution in the interest of small size and cost. Uncooled longwave sensors represent a big market because they're cheap.

Uncooled solutions also must compensate for their relative weaknesses in range and resolution with larger lenses to enhance light sensitivity. Design tradeoffs for uncooled sensors often involve an intricate dance. Uncooled cameras are used where cost is a major consideration, as is SWaP.

You have a certain sensitivity with a lens that has an F1 aperture. In a cooled solution you could do that with an F4 lens. An uncooled camera might have a 1-Watt load, but you typically get to a limit on the focal length and how far out you can see. Uncooled sensors are for drivers aides, not for long-range systems.

Still, there are advances in uncooled detectors that are improving resolution and range. The microbolometer camera that FLIR designs for the Black Hornet 4 palm-sized unmanned helicopter, for example, has moved from a 160-by-12-pixel detector in the Black Hornet 3 to a 640-by-512-pixel camera in the Black Hornet. With those kinds of resolutions, the ability to identify objects at the same range is so much more significant.

32.3.4 Digital image processing

The embedded computing digital signal processing capability of today's electro-optical sensors is just as important — if not more so — than the sensors themselves. Advanced processing technologies such as high-performance central processing units (CPUs), field-programmable gate arrays (FPGAs), general-purpose graphics processing units (GPGPUs), and a new generation of circuit technology called 3-D Heterogeneous Integration (3DHI) are helping reduce costs, increase range, enhance resolution, and help pull out a growing amount of situational awareness information from every digital image.

Add in advanced processing techniques such as AI, machine learning, neomorphic processing, standards-based rapidly adaptable embedded computing architectures, and 3DHI, and electro-optical sensors designers can pull out more useful information from digital imagery than ever before.

Signal processing, from a modality standpoint, helps to identify and recognize the target. It can help fuse the information together, in applications such as radar and in passive sensing in SIGINT [signals intelligence] and COMINT [communications intelligence]. We want to recognize the signal not for intelligence, but for understanding.

Signal processing also helps systems designers blend information not just from electro-optical sensors, but also from RF sensors and even auditory sensors to create a rich, deep picture of the battlespace. The soldier on the battlefield has a radio, and he pushes the push-to-talk button, you can sense the RF signature. Signal processing also can help orbital electro-optical sensors such as those from Lockheed Martin make difficult predictions, such as where maneuvering hypersonic missiles will impact, and blend information from longwave infrared, mid-wave infrared, shortwave infrared, RF, radar, and other sensors.

How signal processing influences sensor data is where much technological innovation is being brought to bear today. One can see a general shift between general-purpose computers to vector processors, and now we are seeing and are designing with these SOCs [systems-on-chips] that have vector processing capability inside them. We are seeing automatic target recognition, and image-recognition software at the sensor. Some of today's SOC processors blend GPGPU, CPU, and FPGA processing all on one device. The amount of development in there is for autonomy.

Other enabling technologies in which advanced signal processing are critical are multispectral and hyperspectral sensing, in which processors blend information from different light spectra to uncover information that only one light spectrum might miss. The digital detectors that one can use in systems use 3DHI to bring multiple layers of sensors together in the same package, to build a system on a chip for our detectors.

Instead, companies like Teledyne FLIR are creating computer models and synthetic data generation for training AI systems. "We can create wire frames, and turn them into an infrared image, and use graphics engines, to look at that vehicle from every angle and every distance," Stout says. "The introduction of synthetic data to training is a significant advance."

EOs will continue to play a key role in enabling autonomy in military systems, allowing unmanned platforms to detect, track, and engage targets without human intervention. As they become more prevalent in military operations, adversaries may develop countermeasures to evade detection and targeting. Therefore, there will be a continued emphasis on developing advanced sensor technologies and tactics to maintain superiority on the battlefield.

32.4 HOW ELECTRO-OPTICAL/INFRARED PROVIDES WARFIGHTING CAPABILITIES?

As the sun sets, a speck appears on the horizon and naval watch-standers must be able to tell if it's hostile or friendly before it's too late.

New advancements in Electro-Optical and Infrared (EO/IR) capabilities help warfighters differentiate that speck, allowing them more time to react if needed.

Compared to traditional radar systems that offer a top-down view, these advancements give the surface navy a 360-degree view of their environment, increase situational awareness, and assist in detecting and identifying objects around them.

The global electro-optical/infrared systems market is projected to reach $11.68 billion by 2026. The increased development of advanced sensor technology-based systems assists in meeting end users' numerous applications and being on the cutting edge of those opportunities.

This system can be deployed on multiple Navy platforms from small unmanned surface vessels to massive aircraft carriers, using a Modular Open Systems Approach (MOSA). This approach supports the use on multiple platforms with significant reduction in engineering rework and life cycle cost by leveraging common imaging components and technologies to the largest extent possible. The goal of MOSA is to provide a flexible design strategy with a simple, more cost-effective approach to upgrade as newer technologies become available.

32.4.1 Comparing Radar to EO/IR

The U.S. military has relied on radar since the 1930s and continues to improve the system, which bounces radio waves off objects to determine their location. However, as with other active sensors, radar creates a signal that can be detected by adversaries. It is also difficult to detect small objects like Unmanned Aerial Vehicles (UAVs) and Fast Inshore Attack Craft (FIAC) with radar due to their size and surrounding clutter.

Electro-optical sensors work by converting visible light and infrared energy into electronic signals, which are then analyzed. Invisible to the naked eye, infrared radiation can be detected in total darkness. EO/IR technology combines the two capabilities into one view to create a clearer picture of threats for the warfighter. The benefit of fusing visible light with infrared signatures provides the ability to detect, recognize and identify potential threats during all conditions of visibility such as day/night and through atmospheric obscurations.

Unlike radar technology, which relies on emitting active radiation signals and waiting for a return signal, EO/IR sensors are **totally passive**. EO/IR sensors detect signals in the visible/infrared portion of the electromagnetic spectrum, naturally emitted by objects of interest. This passive operation is a key strategic advantage of EO/IR sensors by assisting Navy ships in avoiding detection and becoming targets themselves.

EO/IR technology, such as the SPEIR program, uses infrared heat signatures and algorithms that look for objects based on size, motion, characteristics and other factors. It can point out objects as small as one to two pixels on a screen – including incoming targets, such as anti-ship cruise missiles, FIAC and UAV systems.

Every object has a distinct electromagnetic fingerprint, and these EO/IR sensors provide a more complete picture of that electromagnetic spectrum. This enables better target discrimination and identification in a cluttered and congested environment.

32.5 MULTISPECTRAL ELECTRO-OPTICAL SURVEILLANCE SENSORS FOR MILITARY APPLICATIONS AT SEA

Teledyne FLIR LLC in Wilsonville, Ore., is introducing the SeaFLIR 240 and TacFLIR 240 high-definition multispectral surveillance systems for military and homeland security applications on land and at sea. These electro-optical surveillance systems each has lightweight stabilized turret, HD payload options, and inertial navigation capabilities.

These integrated sensors offer advanced image-processing and small form factor, and are for U.S. Navy, Marine Corps and Coast Guard intelligence, surveillance and reconnaissance (ISR); search and rescue; and special operations aboard combat ships, small boats, and unmanned surface vessels.

Tailored for manned and unmanned vehicle use, TacFLIR 240 can identify and track smugglers, terrorists, and similar threats during the day or at night, and over tough terrain. The system can support mid-range object and vehicle detection and assessment both for military and homeland security applications.

Key features include multi-sensor capabilities; superior image clarity; detection, recognition, and identification capability; and extensible processing control electronics unit.

These sensor systems offer high-definition medium-wave infrared sensors for thermal detection; HD daylight and HD low-light camera options; eye-safe laser range finder and laser pointer; and built-in inertial measurement unit (IMU) and GPS interface.

They offer multispectral dynamic imaging edge detail from color overlay on infrared; and advanced local area processing for detailed imagery in cold or low-contrast environments.

These surveillance sensors also provide range performance for target recognition; advanced video tracking that keeps system locked on target; built-in inertial sensors that provide stabilized imagery and target location accuracy.

The extensible processing control electronics unit provides advanced image processing capabilities, with configurations that include a high-performance central processing unit, graphics-processing unit, and digital video recorder with one terabyte solid-state hard drive supporting H.264 streaming video, plus discrete video outputs, serial interface ports, and specialty protocols like ARINC and MIL-STD-1553.

32.6 DRDO DEVELOPS ELECTRO-OPTICAL INFRARED SYSTEM FOR SURVEILLANCE

The Centre for Air Borne Systems (CABS), a laboratory of the Defence Research Development Organisation (DRDO), Govt. of India, has indigenously developed electro-optical/infrared (EO/IR) system to carry out surveillance from airborne, land and naval platforms.

Speaking on the sidelines of Aero India 2023, an official from DRDO, said, "It (EO/IR system) can be mounted on ships, aircraft and vehicles. We have evaluated it in Ladakh for the Indian Army and they are happy with it. However, they want certain modifications. It has the capability to detect objects even behind fire and smoke. The EO/IR system has a short-wave infrared (SWIR) imager, laser range finder, laser pointer and cameras. The sensors can operate simultaneously depending on daylight weather conditions and display HD images to the operator. It is also integrated with a video tracker which can track moving targets."

CHAPTER 33

FIBER-OPTIC SENSORS IN WARFARE

33.1 INTRODUCTION TO FIBER-OPTIC SENSORS

A fiber-optic sensor is a sensor that uses optical fiber either as the sensing element ("intrinsic sensors"), or as a means of relaying signals from a remote sensor to the electronics that processes the signals ("extrinsic sensors"). Fibers have many uses in remote sensing. Depending on the application, fiber may be used because of its small size, or because no electrical power is needed at the remote location, or because many sensors can be multiplexed along the length of a fiber by using light wavelength shifting for each sensor, or by sensing the time delay as light passes along the fiber through each sensor. Time delay can be determined using a device such as an optical time-domain reflectometer and wavelength shift can be calculated using an instrument implementing optical frequency domain reflectometry.

Fiber-optic sensors are also immune to electromagnetic interference, and do not conduct electricity so they can be used in places where there is high voltage electricity or flammable material such as jet fuel. Fiber-optic sensors can be designed to withstand high temperatures as well.

(i) Intrinsic fiber-optic sensors

Optical fibers can be used as sensors to measure strain, temperature, pressure and other quantities by modifying a fiber so that the quantity to be measured modulates the intensity, phase, polarization, wavelength or transit time of light in the fiber. Sensors that vary the intensity of light are the simplest, since only a simple source and detector are required. A particularly useful feature of intrinsic fiber-optic sensors is that they can, if required, provide distributed sensing over very large distances.

Temperature can be measured by using a fiber that has evanescent loss that varies with temperature, or by analyzing the Rayleigh Scattering, Raman scattering or the Brillouin scattering in the optical fiber. Electrical voltage can be sensed by nonlinear optical effects in specially-doped fiber, which alter the polarization of light as a function of voltage or electric field. Angle measurement sensors can be based on the Sagnac effect. Special fibers like long-period fiber grating (LPG) optical fibers can be used for direction recognition.

Optical fibers are used as hydrophones for seismic and sonar applications. Hydrophone systems with more than one hundred sensors per fiber cable have been developed. Hydrophone sensor systems are used by the oil industry as well as a few countries' navies. Both bottom-mounted hydrophone arrays and towed streamer systems are in use. The German company Sennheiser developed a laser microphone for use with optical fibers.

A fiber-optic microphone and fiber-optic based headphone are useful in areas with strong electrical or magnetic fields, such as communication amongst the team of people working on a patient inside a magnetic resonance imaging (MRI) machine during MRI-guided surgery.

Optical fiber sensors for temperature and pressure have been developed for downhole measurement in oil wells. The fiber-optic sensor is well suited for this environment as it functions at temperatures too high for semiconductor sensors (distributed temperature sensing).

Optical fibers can be made into interferometric sensors such as fiber-optic gyroscopes, which are used in the Boeing 767 and in some car models (for navigation purposes). They are also used to make hydrogen sensors.

Fiber-optic sensors have been developed to measure co-located temperature and strain simultaneously with very high accuracy using fiber Bragg gratings. This is particularly useful when acquiring information from small or complex structures. Fiber optic sensors are also particularly well suited for remote monitoring, and they can be interrogated 290 km away from the monitoring station using an optical fiber cable. Brillouin scattering effects can also be used to detect strain and temperature over large distances (20–120 kilometers).

Other examples of fiber-optic sensors

A fiber-optic AC/DC voltage sensor in the middle and high voltage range (100–2000 V) can be created by inducing measurable amounts of Kerr nonlinearity in single-mode optical fiber by exposing a calculated length of fiber to the external electric field. The measurement technique is based on polarimetric detection and high accuracy is achieved in a hostile industrial environment.

High frequency (5 MHz–1 GHz) electromagnetic fields can be detected by induced nonlinear effects in fiber with a suitable structure. The fiber used is designed such that the Faraday and Kerr effects cause considerable phase change in the presence of the external field. With appropriate sensor design, this type of fiber can be used to measure different electrical and magnetic quantities and different internal parameters of fiber material.

Electrical power can be measured in a fiber by using a structured bulk fiber ampere sensor coupled with proper signal processing in a polarimetric detection scheme. Experiments have been carried out in support of the technique.

Fiber-optic sensors are used in electrical switchgear to transmit light from an electrical arc flash to a digital protective relay to enable fast tripping of a breaker to reduce the energy in the arc blast.

Fiber Bragg-Grating (FBG)-based fiber-optic sensors significantly enhance performance, efficiency and safety in several industries. With FBG integrated technology, sensors can provide detailed analysis and comprehensive reports on insights with very high resolution. These type of sensors are used extensively in several industries like telecommunication, automotive, aerospace, energy, etc. Fiber Bragg gratings are sensitive to the static pressure, mechanical tension and compression and fiber temperature changes. The efficiency of fiber Bragg grating based fiber-optic sensors can be provided by means of central wavelength adjustment of light emitting source in accordance with the current Bragg gratings reflection spectra.

(ii) Extrinsic fiber-optic sensors

Extrinsic fiber-optic sensors use an optical fiber cable, normally a multimode one, to transmit modulated light from either a non-fiber optical sensor, or an electronic sensor connected to an optical transmitter. A major benefit of extrinsic sensors is their ability to reach places which are otherwise inaccessible. An example is the measurement of temperature inside aircraft jet engines by using a fiber to transmit radiation into a radiation pyrometer located outside the engine. Extrinsic sensors can also be used in the same way to measure the internal temperature of electrical transformers, where the extreme electromagnetic fields present make other measurement techniques impossible.

Extrinsic fiber-optic sensors provide excellent protection of measurement signals against noise corruption. Unfortunately, many conventional sensors produce electrical output which must be converted into an optical signal for use with fiber. For example, in the case of a platinum resistance thermometer (PRT), the temperature changes are translated into resistance changes. The PRT must therefore have an electrical power supply. The modulated voltage level at the output of the PRT can then be injected into the optical fiber via the usual type of transmitter. This complicates the measurement process and means that low-voltage power cables must be routed to the transducer.

Extrinsic sensors are used to measure vibration, rotation, displacement, velocity, acceleration, torque, and temperature.

33.1.1 Chemical sensors and biosensors

It is well-known the propagation of light in optical fiber is confined in the core of the fiber based on the total internal reflection (TIR) principle and near-zero propagation loss within the cladding, which is very important for the optical communication but limits its sensing applications due to the non-interaction of light with surroundings. Therefore, it is essential to exploit novel fiber-optic structures to disturb the light propagation, thereby enabling the interaction of the light with surroundings and constructing fiber-optic sensors. Until now, several methods, including polishing, chemical etching, tapering, bending, as well as femtosecond grating inscription, have been proposed to tailor the light propagation and prompt the interaction of light with sensing materials. In the above-mentioned fiber-optic structures, the enhanced evanescent fields can be efficiently excited to induce the light to expose to and interact with the surrounding medium. However, the fibers themselves can only sense very few kinds of analytes with low-sensitivity and zero-selectivity, which greatly limits their development and applications, especially for biosensors that require both high-sensitivity and high-selectivity. To overcome the issue, an efficient way is to resort to responsive materials, which possess the ability to change their properties, such as RI, absorption, conductivity, etc., once the surrounding environments change. Due to the rapid progress of functional materials in recent years, various sensing materials are available for fiber-optic chemical sensors and biosensors fabrication, including graphene, metals and metal oxides, carbon nanotubes, nanowires, nanoparticles, polymers, quantum dots, etc. Generally, these materials reversibly change their shape/volume upon stimulation by the surrounding environments (the target analysts), which then leads to the variation of RI or absorption of the sensing materials. Consequently, the surrounding changes will be recorded and interrogated by the optical fibers, realizing sensing functions of optical fibers. Currently, various fiber-optic chemical sensors and biosensors have been proposed and demonstrated.

33.2 SOME OPTICAL FIBER USE CASES IN DEFENCE

(i) Enhanced and Secured Communication

Fiber optics are considered a highly secure means of communication for military applications due to their unique characteristics. Unlike traditional copper wires, fiber optics use light to transmit data, which means that no electromagnetic fields are produced during the transmission. This makes it extremely difficult for an eavesdropper to intercept or tap into the communication signal.

The lack of electromagnetic emissions from fiber optics also makes them immune to interference from radio frequency (RF) devices, which are commonly used for wireless communication. This eliminates the risk of jamming or other forms of interference that can disrupt military communication systems. The use of fiber in communication systems enables longer transmission distances without the need for repeaters compared to twisted copper wires or coaxial cables. Rigorous testing conducted by all three military services has indicated that under the most battlefield conditions, optical fiber systems exhibit superiority.

(ii) Weaponry

Fiber optic technology is finding new use cases in defence weapons as the need for better communication and precision guidance increases. The FOG-M and FOG-S weapons, which use fiber optic cables are hit-to-kill weapons designed to take down helicopters, ground vehicles, and tanks. The use of fiber optic cables enables a high bandwidth data link, which is not possible with traditional copper cables. This link provides better communication between the missile and the gunner, allowing the missile to be guided more precisely to its target.

Another optical fiber use case in defence that is being considered by the Army is the PDAMS ballistic missile bus system, which carries submunitions linked to the bus by fiber optic cables. This system would be used to destroy tactical ballistic missiles within minutes of hostilities. The Indian Air Force and Navy are also considering fiber-guided weapons that can be dispensed from an aircraft. The development of these fiber optic-guided weapons represents a significant advancement in defence technology, enhancing the accuracy and efficiency of military operations.

(iii) Intelligence, Surveillance and Reconnaissance (ISR) Systems

Among the various types of sensors used in ISR, those that rely on optical technology are a major component of modern ISR systems. Specialty optics, such as night vision goggles, helmet recording camera systems, thermal imaging sights, and spotting scopes, play a crucial role in the success of soldiers on the ground. The optics in these units are designed to be durable and reliable, even in extreme conditions. They are typically made of lightweight, rugged materials that can withstand the impacts of water and dust.

Optical sensors are particularly useful in ISR operations because they can provide high-resolution imagery over large areas, even from considerable distances. This enables analysts to collect detailed information about potential targets, including their location, movement patterns, and even their composition. As technology continues to advance, the use of optical sensors in ISR is likely to become even more prevalent.

(iv) Navigation

Fiber Optic Gyroscopes (FOGs) are crucial in the military due to their exceptional performance in measuring angular rate and acceleration in addition to their superior dynamic range and reliability. They have a vast range of applications in military vehicles such as tanks, self-propelled artillery, submarines, and armored assault vehicles. In situations where electronic interference hinders satellite navigation's accuracy, FOGs can be used to enable autonomous navigation, precise guidance, and accurate target hitting for aircraft.

Moreover, FOGs are critical components in aviation fire control systems, providing stability to aiming and firing lines of weapon systems, such as armed helicopters. This capability ensures that weapons can perform search, aim, track, and shoot in motion accurately. FOGs are also the only navigation technology that works effectively underwater, making them crucial for the positioning, orientation, and navigation of submarines.

(v) Avionics

Fiber optics technology is revolutionizing avionics in defence by providing solutions to size, weight, and power (SWaP) challenges while enabling high-speed processing and data transmission. Military fighter aircraft, for instance, requires a high-speed operation to detect targets and incoming missiles. With fiber-optic-based systems, avionics can deliver real-time imaging without digital or video compression. This capability supports quick, decisive actions by pilots based on radar and camera images, which is critical in combat situations. Additionally, the next-generation military fighter platforms will require extremely high data transmission speeds with low latency, which fiber optics networks can deliver. Although the development of fully autonomous unmanned aerial vehicles (UAVs) is still in its early stages, the high-speed capability of fiber optics networks will bring this technology closer to realization. The use of fiber optics in avionics is a game-changer, enabling next-generation avionics systems to meet the demanding requirements of modern defence missions.

(vi) Others Applications of Optical fibers

Optical fibers are also used in other defence applications such as acoustic sensing, chemical sensing, and laser systems. The fibers are used for remote sensing of underwater objects and for detecting chemical agents in the air. They are also used in laser-based weapons systems for transmitting high-energy laser beams.

(vii) Fiber coupled lighting

Some submarines use fiber coupled light from centralized light sources to simulate natural circadian rhythms. They can tune the colors to imitate natural light cycles on the surface, orange in the morning to bluer, then more orange at night. It also makes changing dead bulbs easier because there's only one centralized light source, that is coupled to the rest of the submarine through the fiber.

33.3 CHALLENGES AND MYTHS OF USE OPTICAL FIBERS

While there are some issues that remain challenges in the development of fiber optics in the military context, rapid development has mitigated or even negated some of them.

(i) Initial Cost

A few years ago, the initial cost of implementing an all-fiber communications network was high. Today, with the widespread use of fiber in everything that needs to communicate, this cost is extremely low. On the other hand, the increase in the cost of copper and electronic equipment for wireless networks, makes fiber optics networks have the lowest initial implementation costs.

(ii) Specialized Maintenance

Being a passive network, the need for maintenance is very low. The robustness of the cables, especially the tactical cables (used in military environment), is such that it is very difficult to damage one of these cables. But when some kind of repair is needed, maintenance is no longer an issue. Many people are already trained, educated and comfortable with fiber optics.

(iii) Physical Vulnerability

Although fiber is resistant to electromagnetic interference, it is vulnerable to physical damage, especially in a hostile environment. Major developments have been made in this field, the use of tactical cables of high mechanical strength and the use of underground channels for cable routing and redundancy of cable paths partly solves this problem.

Fiber optic technology continues to evolve, new applications and developments will emerge, further extending its advantages in general and consequently for military applications. It is future-proof. We don't know what communication speed we will need in the near future with the development of new military solutions, but we do know that they will use fiber optics as a means to communicate.

33.4 INDIAN ARMY SELECTS HAWK'S FIBER OPTIC SENSING SYSTEM FOR PERIMETER MONITORING

HAWK's Fiber Optic Sensing System for Perimeter Security was selected by the Indian Army to detect intruders entering Indian territory.

The surrounding region consists of a thick forest with very heavy fog, making it extremely difficult to monitor and detect any activities at the borders. HAWK's Praetorian Fiber Optic Sensing System was selected and installed along the entire perimeter to immediately detect an intruder and send an alarm of their exact location. The HAWK engineering team performed testing, installation, calibration and commissioning of the entire project and it was carried out successfully. The Fiber Optic Sensing System was integrated into the Army's surveillance sensors with SCADA system and has been successfully detecting intrusions on one of world's most environmentally challenging borders for over four years.

The Praetorian Fiber Optic System is a compete perimeter security and border control monitoring system. Using a combination of Rayleigh backscatter and time of flight, Praetorian determines the presence, location, intensity and frequency of vibrations along an optical fiber in real time. HAWK's Praetorian FOS can detect amplitude, spectrum of vibrations and position of intruders walking or entering a secure zone by using buried fiber optic cable. Praetorian FOS can identify vibration due to weather effects, walking and crawling to reduce incidence of false alarms.

CHAPTER 34

MAGNETIC FIELD SENSORS IN WARFARE

34.1 WHAT IS A MAGNETIC SENSOR?

A magnetic sensor usually refers to a sensor that converts the magnitude and variations of a magnetic field into electric signals.

Magnetic fields, as exemplified by the magnetic field of the earth (earth magnetism) or magnets are familiar yet invisible phenomena. Magnetic sensors that convert invisible magnetic fields into electric signals and into visible effects have long been the subject of research.

It started decades ago with sensors using the electromagnetic induction effect and these efforts were extended to applications of the galvanomagnetic effect, magnetoresistance effect, Josephson effect and other physical phenomena.

34.1.1 Typical magnetic sensors and their applications

(i) Coils

The coil is the most classic and simple form of sensor. Although a coil cannot be used alone to directly detect a magnetic field, it can detect the variations in a magnetic field.

Bringing a magnet close to a coil will increase the magnetic flux density in the coil. The increase of magnetic flux density in the coil will also generate opposing forces in the form of **induced electromotive force and induced current.** When the coil stops moving, the magnetic flux density variations also stop and the induced electromotive force and induced current cease.

Used alone, a coil provides only limited functionality. However, when combined with other coils or magnetic materials, it can become a highly sensitive magnetic sensor.

Currently, magnetic sensors that use coils include search coils, resolvers or rotation angle sensors as well as fluxgate sensors, a type of sensor used in a broad range of applications.

(ii) Reed switches

A reed switch consists of a glass tube encapsulating two reeds, the contacts, which come from the right and left ends of the tube. The reeds are made of nickel or other magnetic material and are separated by a gap. The glass tube is filled with nitrogen or other inert gas to prevent the activation (deterioration) of the contacts.

The reed switch is normally open, but when both ends of the magnetic material are exposed to a magnetic field, the magnetic material is magnetized and the contacts are attracted to each other closing the circuit (conduction state).

Unlike semiconductor sensors such as MR sensor elements or Hall elements (see below), the reed switch operates without a power supply and is therefore often used in automobiles or other locations where power is difficult to supply.

(iii) MR sensor elements

An MR sensor element is a magnetic sensor element using the Magneto-Resistance effect (MR effect). There are a number of MR sensor types using different operating principles. The following describes the basic MR effect.

The MR effect is a phenomenon where resistance changes with changes in a magnetic field. It is an effect that occurs in magnetic materials (for example, iron, nickel or cobalt).

The MR effect requires an understanding of electron spin and how the Lorentz force operates using electron charges. When electrons move through a ferromagnetic material (a material with a certain level of magnetism) and the spinning of the electrons fluctuates, the scattering probability (of electrons) in the magnetized material rises and falls. This is what causes the MR effect.

Electrons have two important parameters: charge and spin. They have the same negative charge, but electron spin is of two kinds: up-spin and down-spin. Electron spin was verified by an experiment in 1922 and it was confirmed that electrons exhibit electronic angular momentum and magnetic moment characteristic to electrons. When electrons pass through conductive materials, they scatter (electron scattering). Electron scattering is a phenomenon caused by static electricity in the material that causes electrons to deviate from their normal trajectory. Lorentz force is a force that comes into play when mobile particles (electrons) in a conductive material are exposed to a magnetic field. It affects all charged particles and does not rely on electron spin.

In 1856, William Thomson discovered **Anisotropic Magneto-Resistance effect (AMR effect)** by observing a ferromagnetic material placed in an external magnetic field environment. When the magnetization direction in a ferromagnetic material is parallel to the current, the electron orbital becomes perpendicular to the current, which maximizes resistance. This increases the spin-dependent scattering causing electric resistance to rise. When the magnetization direction is perpendicular to the current, the electron orbital becomes horizontal to the current reducing the spin-dependent scattering, which minimizes resistance.

The rate of change in resistance caused by the state of the magnetic field is called magnetoresistive ratio (MR ratio). The MR ratio for an AMR sensor element is about 5%. The AMR sensor element is often used in magnetic switches and rotation sensors because of its simple structure.

The Giant Magneto-Resistive effect (GMR effect) was independently and simultaneously discovered by Albert Fert and Peter Grünburg in 1988 by observing a non-magnetic conductive thin film structure sandwiched between two conductive ferromagnetic material layers. The magnetization of each ferromagnetic layer is exposed to the spin-dependent scattering of electrons as they pass through the middle layer. If the spin direction of electrons passing through the ferromagnetic layer is opposite that of the magnetization of the ferromagnetic material, the interaction effect is much weaker than when the direction of spin is parallel to magnetization. As a result, when the direction of magnetization of the upper and lower ferromagnetic material is parallel, resistance to the current flowing along the boundary surface of the conductive material drops, while it increases if the direction of magnetization is anti-parallel.

The GMR sensor element is a magnetic sensor element applying the GMR effect. It has a magnetic sensitivity that is between two to five times greater than that of an AMR sensor element. This greater sensitivity allows a GMR sensor to detect minute changes in magnetic flux densities that were previously not possible. By replacing the coils in the read-write heads of a hard disk drive, the heads can be made more compact and more sensitive. This has vastly increased the storage densities of hard disks increasing their storage capacities.

The MR ratio of a GMR sensor element is about 20%. Their high sensitivity makes GMR sensor elements the device of choice for magnetic heads, rotational sensors and other devices.

The Tunnel Magneto-Resistance effect (TMR effect) at room temperature was discovered by Professor Terunobu Miyazaki at Tohoku University in 1995. A TMR sensor element is a magnetic sensor element using the TMR effect and configured from an extremely thin **nanometer** level nonmagnetic insulation layer sandwiched between two ferromagnetic layers. Electrons tunnel from one ferromagnetic layer into the other via insulation layer. This is a quantum mechanical phenomenon. Resistance decreases when the magnetization direction of the two ferromagnetic materials is parallel and increases when it is antiparallel.

The MR ratio (the rate of change in resistance depending on the state of a magnetic field) in TMR junctions can reach more than 100% in production. In lab conditions, levels upwards of 1,000% have been achieved.

The TMR sensor elements are ideal for use in hard disk magnetic heads or high sensitivity rotation angle sensors.

(iv) Hall elements

The Hall element is an application of the Hall Effect. The Hall effect discovered by Edwin H. Hall in 1879 proved that the Lorentz force generated a voltage at right angles to the direction of the current and magnetic field. This voltage is called a Hall voltage and according to Fleming's left hand rule the direction of the voltage changes with the direction of magnetic flux. The magnitude and direction (plus, minus) of the voltage make it possible to detect the magnitude and direction of the magnetic field (N-pole, S-pole).

The magnetic sensitivity of a Hall element is not as good as that of magnetic resistance sensor element. However, as a magnetic sensor that does not rely on magnetic material, it can be used in a ferromagnetic field environment or harsh environments and therefore finds application as a current sensor or as a variety of magnetic switches.

(v) SQUID

Superconducting Quantum Interference Device (SQUID) is a magnetic sensor element capable of measuring minute magnetic fields by applying the Josephson effect. SQUID, a device that combines a ring-shaped superconductor with the Josephson Junction proposed by Brian D. Josephson in 1962 is the most sensitive magnetic sensor currently available.

This sensor can detect the heart's and brain's electromagnetic fields, which are undetectable to other sensor technologies.

34.2 MEMS MAGNETIC FIELD SENSOR

A MEMS magnetic field sensor is a small-scale microelectromechanical systems (MEMS) device for detecting and measuring magnetic fields (magnetometer). Many of these operate by detecting effects of the Lorentz force: a change in voltage or resonant frequency may be measured electronically, or a mechanical displacement may be measured optically. Compensation for temperature effects is necessary. Its use as a miniaturized compass may be one such simple example application.

34.2.1 Advantages of MEMS-based sensors

A MEMS-based magnetic field sensor is small, so it can be placed close to the measurement location and thereby achieve higher spatial resolution than other magnetic field sensors. Additionally, constructing a MEMS magnetic field sensor does not require the microfabrication of magnetic material. Therefore, the cost of the sensor can be greatly reduced. Integration of MEMS sensor and microelectronics can further reduce the size of the entire magnetic field sensing system.

Lorentz-force-based MEMS sensor

This type of sensor relies on the mechanical motion of the MEMS structure due to the Lorentz force acting on the current-carrying conductor in the magnetic field. The mechanical motion of the micro-structure is sensed either electronically or optically. The mechanical structure is often driven to its resonance in order to obtain the maximum output signal. Piezoresistive and electrostatic transduction methods can be used in the electronic detection. Displacement measurement with laser source or LED source can also be used in the optical detection.

34.3 ROLE OF ACCELEROMETERS IN MILITARY/AEROSPACE APPLICATIONS

Most accelerometers are Micro-Electro-Mechanical Sensors (MEMS technology)-based. The MEMS provide low-power, ultra-low mass components which may be included into a variety of aerospace systems. Compared to traditional MEMS markets such as home computers and automobiles, the aerospace-specific MEMS are limited by the relatively small size of the aerospace vehicle market. The MEMS devices or systems have the ability to control, actuate and sense on the micro scale, and produce effects on the macro scale. These sensors could be deployed in public buildings, military camps, airports and other strategic locations. For military applications, miniaturized MEMS sensors have become popular. To provide technical solutions for the modern forces, many agencies are working on MEMS sensors for military and defence projects.

34.3.1 MEMS Technology In Military Applications

The MEMS technology in military applications is used in the following applications. These were mainly chosen as they rely heavily on cost, size and self-containment benefits of MEMS systems, and at times they appear to be viable using near-term or current developmental technologies.

(i) Chemical Attack Warning Sensor

The MEMS chemical sensors are currently being developed for the semiconductor wafer industry. It may also be possible to develop chemical sensors for the soldier, which inform him/her of the presence of noxious chemicals. Micro-chemical sensors are also being addressed by numerous other researchers. The size and cost are the major benefits that MEMS bring to this application. Currently, the chemical sensors carried by soldiers are expensive and bulky, built of distinct components. The MEMS technology would allow the systems to be micro-sized, throwaway modules, customized to the threat. The MEMS device used in military applications are able to precisely detect chemical agents below the thresholds, at which those agents become harmful to the soldiers. It also meets all the usual military requirements of long shelf life, ease of use, ruggedness, etc.

(ii) Identification of Fried or Foe (IFF)

To signal a vehicle's presence, most of discrimination aids use active beacons, reflective tapes, or transponders. Such systems, especially in fluid battlefields with intermingled forces, are extremely susceptible to interception. A concept for an IFF system was developed by a researcher based on macro-sized corner cubes mounted on the surface of a vehicle. The main benefits that MEMS provide to this application are - proliferation across the vehicle by reducing dust, mud problems and laser pointing errors. It does not require the system to be connected through armor to vehicle data/power buses. It has a low power drain because of minute actuator excursions, and has a fast response because of resulting high frequency of actuating elements and small size.

(iii) Active Surfaces

Originally developed by Defense Advanced Research Projects Agency (DARPA), a special application of active surface technology is possible with the X-wing experimental aircraft. The key aspects provided by MEMS in all active surface applications are performance, size and self-contained operations.

(iv) Distributed Battlefield Sensor Net (DBSN)

Vast sums have been invested by modern armies in fielding and developing systems to locate the enemy. There would be difficulties in detecting a number of important targets. However, MEMS could possibly help greatly in such task. A possible solution for this would be the distributed battlefield sensor net (DBSN). The basic idea behind this is to deploy a large number of disposable and cheap sensor systems over critical areas.

(v) Micro-robotic Electronic Disabling System (MEDS)

The MEDS would be present in the general target vicinity in a manner similar to DBSN. Rather than indiscriminate coverage of an area, timely and accurate placement of the MEDS devices appears to be important for payload, cost and time reasons.

34.3.2 MEMS Technology In Aerospace Applications

The use of MEMS technology in aerospace is used in a wide variety of applications. Some of them are:

(i) An active control of thin boundary layer flows with the potential to reduce drag, to eliminate conventional flight control surfaces, enhance aerodynamic performance of turbines, compressors, and low-observable intakes, provide lift-on-demand.

(ii) Compared to conventional systems, complete navigation and inertial units on a single chip, which provide main benefits in terms of weight, size and cost over conventional systems.

(iii) Arming and fusing/safety systems for torpedo applications.

(iv) Applications in harsh environments.

(v) Applications for storage environments monitoring, autonomous inventory and for service life predictions.

(vi) Extensive tests have been performed by the researchers and determined that space vacuum produces an ideal environment for some applications using MEMS devices.

34.4 NANOELECTROMECHANICAL SYSTEMS (NEMS)

Nanoelectromechanical systems (NEMS) are a class of devices integrating electrical and mechanical functionality on the nanoscale. NEMS form the next logical miniaturization step from so-called microelectromechanical systems, or MEMS devices. NEMS typically integrate transistor-like nanoelectronics with mechanical actuators, pumps, or motors, and may thereby form physical, **biological, and chemical sensors**. The name derives from typical device dimensions in the nanometer range, leading to low mass, high mechanical resonance frequencies, potentially large quantum mechanical effects such as zero-point motion, and a high surface-to-volume ratio useful for surface-based sensing mechanisms. Applications include accelerometers and sensors to detect chemical substances in the air.

In the year 2000, the first very-large-scale integration (VLSI) NEMS device was demonstrated by researchers at IBM. Its premise was an array of AFM tips which can heat/sense a deformable substrate in order to function as a memory device (Millipede memory). Further devices have been described by Stefan de Haan. In 2007, the International Technical Roadmap for Semiconductors (ITRS) contains NEMS memory as a new entry for the Emerging Research Devices section.

34.4.1 Materials for NEMS

Many of the commonly used materials for NEMS technology have been carbon based, specifically diamond, carbon nanotubes and graphene. This is mainly because of the useful properties of carbon-based materials which directly meet the needs of NEMS. The mechanical properties of carbon (such as large Young's modulus) are fundamental to the stability of NEMS while the metallic and semiconductor conductivities of carbon-based materials allow them to function as transistors.

Both graphene and diamond exhibit high Young's modulus, low density, low friction, exceedingly low mechanical dissipation, and large surface area. The low friction of CNTs, allow practically frictionless bearings and has thus been a huge motivation towards practical applications of CNTs as constitutive elements in NEMS, such as nanomotors, switches, and high-frequency oscillators. Carbon nanotubes and graphene's physical strength allows carbon-based materials to meet higher stress demands, when common materials would normally fail and thus further support their use as a major materials in NEMS technological development.

Carbon-based materials have served as prime materials for NEMS use, because of their exceptional mechanical and electrical properties.

Recently, nanowires of chalcogenide glasses have shown to be a key platform to design tunable NEMS owing to the availability of active modulation of Young's modulus.

Researchers from the California Institute of Technology developed a NEM-based cantilever with mechanical resonances up to very high frequencies (VHF). It is incorporation of electronic displacement transducers based on piezoresistive thin metal film facilitates unambiguous and efficient nanodevice readout. The functionalization of the device's surface using a thin polymer coating with high partition coefficient for the targeted species enables NEMS-based cantilevers to provide chemisorption measurements at room temperature with mass resolution at less than one attogram. Further capabilities of NEMS-based cantilevers have been exploited for the applications of sensors, scanning probes, and devices operating at very high frequency (100 MHz).

34.5 MILITARY APPLICATIONS OF NEMS

Nanoelectromechanical systems (NEMS) have many military applications, including:

(i) Sensors

NEMS-based sensors can detect chemical signals, vibrations, stresses, and forces at the atomic level. These sensors can be used in a variety of military applications, such as:

- **Weapons safing, arming, and fuzing**: MEMS can be used in these applications.
- **Chemical and biological agent detection**: NEMS-based electronic noses can detect a variety of gases, including those from chemical, biological, and explosive weapons.
- **Aircraft aerodynamic control**: MEMS can be used to control aircraft aerodynamics.
- **Engine and propulsion control**: MEMS can be used to control engines and propulsion.
- **Nanoenergetic materials**

These materials can be used in explosives, propellants, pyrotechnics, and antitamper devices. They can also be used to generate and store energy.

- **Microrobotic systems**

These systems can be deployed in swarms to increase mission flexibility and complicate countermeasures.

- **Nanosatellites**

These satellites can be powered by nanoenergetic materials to maneuver in orbit.

CHAPTER 35

MULTISPECTRAL AND HYPERSPECTRAL IMAGING IN WARFARE

35.1 MULTISPECTRAL IMAGING

Multispectral imaging captures image data within specific wavelength ranges across the electromagnetic spectrum. The wavelengths may be separated by filters or detected with the use of instruments that are sensitive to particular wavelengths, including light from frequencies beyond the visible light range (i.e. infrared and ultraviolet). It can allow extraction of additional information the human eye fails to capture with its visible receptors for red, green and blue. It was originally developed for military target identification and reconnaissance. Early space-based imaging platforms incorporated multispectral imaging technology to map details of the Earth related to coastal boundaries, vegetation, and landforms. Multispectral imaging has also found use in document and painting analysis.

Multispectral imaging measures light in a small number (typically 3 to 15) of spectral bands. **Hyperspectral imaging** (discussed below) is a special case of spectral imaging where often hundreds of contiguous spectral bands are available.

For different purposes, different combinations of spectral bands can be used. They are usually represented with red, green, and blue channels. Mapping of bands to colors depends on the purpose of the image and the personal preferences of the analysts. Thermal infrared is often omitted from consideration due to poor spatial resolution, except for special purposes.

True-color uses only red, green, and blue channels, mapped to their respective colors. As a plain color photograph, it is good for analyzing man-made objects, and is easy to understand for beginner analysts.

Green-red-infrared, where the blue channel is replaced with near infrared, is used for vegetation, which is highly reflective in near IR; it then shows as blue. This combination is often used to detect vegetation and camouflage.

Blue-NIR-MIR, where the blue channel uses visible blue, green uses NIR (so vegetation stays green), and MIR is shown as red. Such images allow the water depth, vegetation coverage, soil moisture content, and the presence of fires to be seen, all in a single image.

Many other combinations are in use. NIR is often shown as red, causing vegetation-covered areas to appear red.

35.1.1 Typical spectral bands in Multispectral imaging

The wavelengths are approximate; exact values depend on the particular instruments (e.g. characteristics of satellite's sensors for Earth observation, characteristics of illumination and sensors for document analysis):

(i) **Blue, 450–515/520 nm,** is used for atmosphere and deep-water imaging, and can reach depths up to 150 feet (50 m) in clear water.

(ii) **Green, 515/520–590/600 nm**, is used for imaging vegetation and deep-water structures, up to 90 feet (30 m) in clear water.

(iii) **Red, 600/630–680/690 nm,** is used for imaging man-made objects, in water up to 30 feet (9 m) deep, soil, and vegetation.

(iv) **Near infrared (NIR),** 750–900 nm, is used primarily for imaging vegetation.

(v) **Mid-infrared (MIR), 1550–1750 nm,** is used for imaging vegetation, soil moisture content, and some forest fires.

(vi) **Far-infrared (FIR), 2080–2350 nm,** is used for imaging soil, moisture, geological features, silicates, clays, and fires.

(vii) **Thermal infrared,** 10,400–12,500 nm, uses emitted instead of reflected radiation to image geological structures, thermal differences in water currents, fires, and for night studies.

Radar and related technologies are useful for mapping terrain and for detecting various objects.

35.1.2 Applications of Multispectral imaging

(i) Military target tracking

Multispectral imaging measures light emission and is often used in detecting or tracking military targets. In 2003, researchers at the United States Army Research Laboratory and the Federal Laboratory Collaborative Technology Alliance reported a dual band multispectral imaging focal plane array (FPA). This FPA allowed researchers to look at two infrared (IR) planes at the same time. Because mid-wave infrared (MWIR) and long wave infrared (LWIR) technologies measure radiation inherent to the object and require no external light source, they also are referred to as **thermal imaging methods.**

The brightness of the image produced by a thermal imager depends on the objects emissivity and temperature. Every material has an infrared signature that aids in the identification of the object. These signatures are less pronounced in hyperspectral systems (which image in many more bands than multispectral systems) and when exposed to wind and, more dramatically, to rain. Sometimes the surface of the target may reflect infrared energy. This reflection may misconstrue the true reading of the objects' inherent radiation. Imaging systems that use MWIR technology function better with solar reflections on the target's surface and produce more definitive images of hot objects, such as engines, compared to LWIR technology. However, LWIR operates better in hazy environments like smoke or fog because less scattering occurs in the longer wavelengths. Researchers claim that dual-band technologies combine these advantages to provide more information from an image, particularly in the realm of target tracking.

For nighttime target detection, thermal imaging outperformed single-band multispectral imaging. Dual band MWIR and LWIR technology resulted in better visualization during the nighttime than MWIR alone. The US Army reports that its dual band LWIR/MWIR FPA demonstrated better visualizing of tactical vehicles than MWIR alone after tracking them through both day and night.

(ii) Land mine detection

By analyzing the emissivity of ground surfaces, multispectral imaging can detect the presence of underground missiles. Surface and sub-surface soil possess different physical and chemical properties that appear in spectral analysis. Disturbed soil has increased emissivity in the wavelength range of 8.5 to 9.5 micrometers while demonstrating no change in wavelengths greater than 10 micrometers. The US Army Research Laboratory's dual MWIR/LWIR FPA used "red" and "blue" detectors to search for areas with enhanced emissivity. The red detector acts as a backdrop, verifying realms of undisturbed soil areas, as it is sensitive to the 10.4 micrometer wavelength. The blue detector is sensitive to wavelengths of 9.3 micrometers. If the intensity of the blue image changes when scanning, that region is likely disturbed. The scientists reported that fusing these two images increased the detection capabilities.

(iii) Ballistic missile detection

Intercepting an intercontinental ballistic missile (ICBM) in its boost phase requires imaging of the hard body as well as the rocket plumes. MWIR presents a strong signal from highly heated objects including rocket plumes, while LWIR produces emissions from the missile's body material. The US Army Research Laboratory reported that with their dual-band MWIR/LWIR technology, tracking of the Atlas 5 Evolved Expendable Launch Vehicles, similar in design to ICBMs, picked up both the missile body and plumage.

(iv) Space-based imaging

Most radiometers for remote sensing (RS) acquire multispectral images. Dividing the spectrum into many bands, multispectral is the opposite of panchromatic, which records only the total intensity of radiation falling on each pixel. Usually, Earth observation satellites have three or more radiometers. Each acquires one digital image (in remote sensing, called a 'scene') in a small spectral band. The bands are grouped into wavelength regions based on the origin of the light and the interests of the researchers.

(v) Weather forecasting

Multispectral imaging combines two to five spectral imaging bands of relatively large bandwidth into a single optical system. A multispectral system usually provides a combination of visible (0.4 to 0.7 μm), near infrared (NIR; 0.7 to 1 μm), short-wave infrared (SWIR; 1 to 1.7 μm), mid-wave infrared (MWIR; 3.5 to 5 μm) or long-wave infrared (LWIR; 8 to 12 μm) bands into a single system.

In the case of Landsat satellites, several different band designations have been used, with as many as 11 bands (Landsat 8) comprising a multispectral image. Spectral imaging with a higher radiometric resolution (involving hundreds or thousands of bands), finer spectral resolution (involving smaller bands), or wider spectral coverage **may be called hyperspectral or ultraspectral.**

(vi) Documents and artworks

Multispectral imaging can be employed for investigation of paintings and other works of art. The painting is irradiated by ultraviolet, visible and infrared rays and the reflected radiation is recorded in a camera sensitive in this region of the spectrum. The image can also be registered using the transmitted instead of reflected radiation. In special cases the painting can be irradiated by UV, VIS or IR rays and the fluorescence of pigments or varnishes can be registered.

Multispectral analysis has assisted in the interpretation of ancient papyri, such as those found at Herculaneum, by imaging the fragments in the infrared range (1000 nm). Often, the text on the documents appears to the naked eye as black ink on black paper. At 1000 nm, the difference in how paper and ink reflect infrared light makes the text clearly readable. It has also been used to image the Archimedes palimpsest by imaging the parchment leaves in bandwidths from 365–870 nm, and then using advanced digital image processing techniques to reveal the undertext with Archimedes' work. Multispectral imaging has been used in a Mellon Foundation project at Yale University to compare inks in medieval English manuscripts.

Multispectral imaging has also been used to examine discolorations and stains on old books and manuscripts. Comparing the "spectral fingerprint" of a stain to the characteristics of known chemical substances can make it possible to identify the stain. This technique has been used to examine medical and alchemical texts, seeking hints about the activities of early chemists and the possible chemical substances they may have used in their experiments. Like a cook spilling flour or vinegar on a cookbook, an early chemist might have left tangible evidence on the pages of the ingredients used to make medicines.

35.2 HYPERSPECTRAL IMAGING

Hyperspectral imaging collects and processes information from across the electromagnetic spectrum. The goal of hyperspectral imaging is to obtain the spectrum for each pixel in the image of a scene, with the purpose of finding objects, identifying materials, or detecting processes. There are three general types of spectral imagers. There are push broom scanners and the related whisk broom scanners (spatial scanning), which read images over time, band sequential scanners (spectral scanning), which acquire images of an area at different wavelengths, and snapshot hyperspectral imagers, which uses a staring array to generate an image in an instant.

Whereas the human eye sees color of visible light in mostly three bands (long wavelengths, perceived as red; medium wavelengths, perceived as green; and short wavelengths, perceived as blue), spectral imaging divides the spectrum into many more bands. This technique of dividing images into bands can be extended beyond the visible. In hyperspectral imaging, the recorded spectra have fine wavelength resolution and cover a wide range of wavelengths. Hyperspectral imaging measures continuous spectral bands, as opposed to multiband imaging which measures spaced spectral bands.

Engineers build hyperspectral sensors and processing systems for applications in astronomy, agriculture, molecular biology, biomedical imaging, geosciences, physics, and surveillance. Hyperspectral sensors look at objects using a vast portion of the electromagnetic spectrum. Certain objects leave unique "fingerprints" in the electromagnetic spectrum. Known as spectral signatures, these "fingerprints" enable identification of the materials that make up a scanned object. For example, a spectral signature for oil helps geologists find new oil fields.

Figuratively speaking, hyperspectral sensors collect information as a set of "images." Each image represents a narrow wavelength range of the electromagnetic spectrum, also known as a spectral band. These "images" are combined to form a three-dimensional (x, y, λ) hyperspectral data cube for processing and analysis, where x and y represent two spatial dimensions of the scene, and λ represents the spectral dimension (comprising a range of wavelengths).

Technically speaking, there are four ways for sensors to sample the hyperspectral cube: spatial scanning, spectral scanning, snapshot imaging, and spatio-spectral scanning.

Hyperspectral cubes are generated from airborne sensors like NASA's Airborne Visible/Infrared Imaging Spectrometer (AVIRIS), or from satellites like NASA's EO-1 with its hyperspectral instrument Hyperion. However, for many development and validation studies, handheld sensors are used.

The precision of these sensors is typically measured in spectral resolution, which is the width of each band of the spectrum that is captured. If the scanner detects a large number of fairly narrow frequency bands, it is possible to identify objects even if they are only captured in a handful of pixels. However, spatial resolution is a factor in addition to spectral resolution. If the pixels are too large, then multiple objects are captured in the same pixel and become difficult to identify. If the pixels are too small, then the intensity captured by each sensor cell is low, and the decreased signal-to-noise ratio reduces the reliability of measured features.

The acquisition and processing of hyperspectral images is also referred to as imaging spectroscopy or, with reference to the hyperspectral cube, as 3D spectroscopy.

35.2.1 Distinguishing hyperspectral from multispectral imaging

Hyperspectral imaging is part of a class of techniques commonly referred to as spectral imaging or spectral analysis. The term "hyperspectral imaging" derives from the development of NASA's Airborne Imaging Spectrometer (AIS) and AVIRIS in the mid-1980s. Although NASA prefers the earlier term "imaging spectroscopy" over "hyperspectral imaging," use of the latter term has become more prevalent in scientific and non-scientific language. Experts recommend using the terms "imaging spectroscopy" or "spectral imaging" and avoiding exaggerated prefixes such as "hyper-," "super-" and "ultra-," to prevent misnomers in discussion.

Hyperspectral imaging is related to multispectral imaging. The distinction between hyper- and multi-band is sometimes based incorrectly on an arbitrary "number of bands" or on the type of measurement. Hyperspectral imaging (HSI) uses continuous and contiguous ranges of wavelengths (e.g. 400 - 1100 nm in steps of 1 nm) whilst multiband imaging (MSI) uses a subset of targeted wavelengths at chosen locations (e.g. 400 - 1100 nm in steps of 20 nm).

Multiband imaging deals with several images at discrete and somewhat narrow bands. Being "discrete and somewhat narrow" is what distinguishes multispectral imaging in the visible wavelength from color photography. A multispectral sensor may have many bands covering the spectrum from the visible to the longwave infrared. Multispectral images do not produce the "spectrum" of an object. Landsat is a prominent practical example of multispectral imaging.

Hyperspectral imaging deals with imaging narrow spectral bands over a continuous spectral range, producing the spectra of all pixels in the scene. A sensor with only 20 bands can also be hyperspectral when it covers the range from 500 to 700 nm with 20 bands each 10 nm wide, while a sensor with 20 discrete bands covering the visible, near, short wave, medium wave and long wave infrared would be considered multispectral.

Ultraspectral imaging could be reserved for interferometer type imaging sensors with a very fine spectral resolution. These sensors often have (but not necessarily) a low spatial resolution of several pixels only, a restriction imposed by the high data rate.

35.2.2 Some Military Applications of Hyperspectral Imaging

(i) Surveillance

Hyperspectral surveillance is the implementation of hyperspectral scanning technology for surveillance purposes. Hyperspectral imaging is particularly useful in **military surveillance** because of countermeasures that military entities now take to avoid airborne surveillance. The idea that drives hyperspectral surveillance is that hyperspectral scanning draws information from such a large portion of the light spectrum that any given object should have a unique spectral signature in at least a few of the many bands that are scanned. Hyperspectral imaging has also shown potential to be used in facial recognition purposes. **Facial recognition** algorithms using hyperspectral imaging have been shown to perform better than algorithms using traditional imaging.

Traditionally, commercially available thermal infrared hyperspectral imaging systems have needed liquid nitrogen or helium cooling, which has made them impractical for most surveillance applications. In 2010, Specim introduced a thermal infrared hyperspectral camera that can be used for outdoor surveillance and UAV applications without an external light source such as the sun or the moon.

(ii) Chemical imaging

Soldiers can be exposed to a wide variety of chemical hazards. These threats are mostly invisible but detectable by hyperspectral imaging technology. The Telops Hyper-Cam, introduced in 2005, has demonstrated this at distances up to 5 km.

35.3 INDIAN DEFENCE STARTUPS BREAK INTO HYPERSPECTRAL IMAGING SMALL SATELLITES IN SPACE

A breakthrough in small and advanced multi-sensor satellites opens further ground for critical **hyperspectral imaging in Indian space**. Smallsats have the potential to radically improve and contribute to Telecommunication by providing mobile services and Internet of Things (IoT) connectivity and purpose built for specific task (ISR) in military and civilian space.

Indian defence startup breaks into coveted areas of hyperspectral imaging. SpacePixxel Technologies and GalaxEye will design and develop a 'Miniaturized satellite capable of carrying Electro-Optical, Infrared, Synthetic Aperture Radar, and Hyperspectral payloads up to 150 kg. The development is part of a contract under the aegis of Innovations for Defence Excellence (iDEX) in the area of hyperspectral imaging satellites in space.

The breakthrough is a milestone as the miniature satellites will be the world's first advanced multi-sensors (EO+SAR) satellite, designed, built, and launched from India by domestic private players. While there will be many elements of key systems and components from outside, the capability is certainly a substantial step in broadening the space clusters to domestic private players, working on existing and futuristic space tech for the military and commercial arena.

SpacePixxel plans to launch hyperspectral imaging satellites, enabling high-resolution images which is critical for advanced mapping, Intelligence, surveillance, and reconnaissance (ISR).

The breakthrough is a milestone as the miniature satellites will be the world's first advanced multi-sensors (EO+SAR) satellite, designed, built, and launched from India by domestic private players.

The small satellites will unfold such possibilities through the efficient constellations of the small satellites which can focus with much greater details and accuracy to provide detailed earth observation data. Moreover, the smallsat has the potential to radically improve and contribute to Telecommunication by providing mobile services and Internet of Things (IoT) connectivity in military and civilian space.

In communication, small satellites facilitate global internet coverage, especially in remote and underserved regions. These satellites are instrumental in establishing IoT networks for real-time data collection from sensors on the ground, which is crucial for industries like logistics, utilities, and smart grids. In space technologies, the miniaturization of the satellites— small satellites — is no small breakthrough as it further opens great new possibilities in space-based tech capabilities.

Defence applications of multi-sensor small satellites are equally diverse and critical. They enhance situational awareness on the battlefield through constant surveillance and reconnaissance. These satellites can provide critical threat assessment data to the armed forces in wartime and conflict scenarios. The versatility of multi-sensor small satellites allows for rapid reconfiguration and adaptation to new missions, making them a strategic asset in modern warfare.

GalaxEye has formally signed an agreement with iDEX to build systems in a Defspace Challenge of designing & develop a miniaturised satellite capable of carrying multiple payloads up to 150 kg for the Indian Air Force.

35.4 BENGALURU-BASED PIXXEL EYES INDIA'S DEFENCE FOR APPLICATION OF ITS HYPERSPECTRAL IMAGERY SATELLITES

Last week, United States' National Reconnaissance Office (NRO) awarded Pixxel a 5-year contract for providing them with hyperspectral imagery (HSI)- enabled remote sensing capabilities via modeling and simulation and data evaluation.

Bengaluru-based space tech start-up Pixxel, which marked its first foray into the surveillance and reconnaissance sector after it recently won a United States defence contract, is keen to use its hyperspectral imagery satellites for the Indian defence sector.

"We are testing internally around strategic and intelligence initiatives. We are surrounded by two countries that have been quite hostile to us for quite a few decades. So, knowing exactly what is happening there (is important); and hyperspectral (imagery) can help unlock new things that one won't be able to see otherwise (sic)," Awais Ahmed, CEO and co-founder of Pixxel told.

Hyperspectral imaging takes a spectrum of light and divides it into hundreds of narrow spectral bands so one can better understand and discern anything on the planet. Pixxel is aiming to build a constellation of such hyperspectral earth imaging satellites and analytical tools to mine insights from space data.

The start-up has already deployed three demo satellites, with the latest, called Anand, being launched from an Indian Space Research Organisation rocket last year. In their vision for the company, the start-up had said that their satellites will be used to build a 'health monitor' for Earth, by focusing on climate, environment, forestry and agriculture use cases.

CHAPTER 36

SONAR SYSTEMS IN WARFARE

36.1 INTRODUCTION TO SONAR

Sonar (sound navigation and ranging or sonic navigation and ranging) is a technique that uses sound propagation (usually underwater, as in submarine navigation) to navigate, measure distances (ranging), communicate with or detect objects on or under the surface of the water, such as other vessels.

"Sonar" can refer to one of two types of technology: passive sonar means listening for the sound made by vessels; active sonar means emitting pulses of sounds and listening for echoes. Sonar may be used as a means of acoustic location and of measurement of the echo characteristics of "targets" in the water. Acoustic location in air was used before the introduction of radar. Sonar may also be used for robot navigation, and sodar (an upward-looking in-air sonar) is used for atmospheric investigations. The term sonar is also used for the equipment used to generate and receive the sound. The acoustic frequencies used in sonar systems vary from very low (infrasonic) to extremely high (ultrasonic). The study of underwater sound is known as underwater acoustics or hydroacoustics.

The first recorded use of the technique was in 1490 by Leonardo da Vinci, who used a tube inserted into the water to detect vessels by ear. It was developed during World War I to counter the growing threat of submarine warfare, with an operational passive sonar system in use by 1918. Modern active sonar systems use an acoustic transducer to generate a sound wave which is reflected from target objects.

36.2 ACTIVE SONAR

Active sonar uses a sound transmitter (or projector) and a receiver. When the two are in the same place it is monostatic operation. When the transmitter and receiver are separated it is bistatic operation. When more transmitters (or more receivers) are used, again spatially separated, it is multistatic operation. Most sonars are used monostatically with the same array often being used for transmission and reception. Active sonobuoy fields may be operated multistatically.

Active sonar creates a pulse of sound, often called a "ping", and then listens for reflections (echo) of the pulse. This pulse of sound is generally created electronically using a sonar projector consisting of a signal generator, power amplifier and electro-acoustic transducer/array. A transducer is a device that can transmit and receive acoustic signals ("pings"). A beamformer is usually employed to concentrate the acoustic power into a beam, which may be swept to cover the required search angles. Generally, the electro-acoustic transducers are of the Tonpilz type and their design may be optimized to achieve maximum efficiency over the widest bandwidth, in order to optimize performance of the overall system. Occasionally, the acoustic pulse may be created by other means, e.g. chemically using explosives, air-guns or plasma sound sources.

To measure the distance to an object, the time from transmission of a pulse to reception is measured and converted into a range using the known speed of sound. To measure the bearing, several hydrophones are used, and the set measures the relative arrival time to each, or with an array of hydrophones, by measuring the relative amplitude in beams formed through a process called beamforming. Use of an array reduces the spatial response so that to provide wide cover multibeam systems are used. The target signal (if present) together with noise is then passed through various forms of signal processing, which for simple sonars may be just energy measurement. It is then presented to some form of decision device that calls the output either the required signal or noise. This decision device may be an operator with headphones or a display, or in more sophisticated sonars this function may be carried out by software. Further processes may be carried out to classify the target and localize it, as well as measuring its velocity.

The pulse may be at constant frequency or a chirp of changing frequency (to allow pulse compression on reception). Simple sonars generally use the former with a filter wide enough to cover possible Doppler changes due to target movement, while more complex ones generally include the latter technique. Since digital processing became available pulse compression has usually been implemented using digital correlation techniques. Military sonars often have multiple beams to provide all-round cover while simple ones only cover a narrow arc, although the beam may be rotated, relatively slowly, by mechanical scanning.

Particularly when single frequency transmissions are used, the Doppler effect can be used to measure the radial speed of a target. The difference in frequency between the transmitted and received signal is measured and converted into a velocity. Since Doppler shifts can be introduced by either receiver or target motion, allowance has to be made for the radial speed of the searching platform.

One useful small sonar is similar in appearance to a waterproof flashlight. The head is pointed into the water, a button is pressed, and the device displays the distance to the target. Another variant is a "fishfinder" that shows a small display with shoals of fish.

When active sonar is used to measure the distance from the transducer to the bottom, it is known as echo sounding. Similar methods may be used looking upward for wave measurement.

Active sonar is also used to measure distance through water between two sonar transducers or a combination of a hydrophone (underwater acoustic microphone) and projector (underwater acoustic speaker). When a hydrophone/transducer receives a specific interrogation signal it responds by transmitting a specific reply signal. To measure distance, one transducer/projector transmits an interrogation signal and measures the time between this transmission and the receipt of the other transducer/hydrophone reply. The time difference, scaled by the speed of sound through water and divided by two, is the distance between the two platforms. This technique, when used with multiple transducers/hydrophones/projectors, can calculate the relative positions of static and moving objects in water.

In combat situations, an active pulse can be detected by an enemy and will reveal a submarine's position at twice the maximum distance that the submarine can itself detect a contact and give clues as to the submarine's identity based on the characteristics of the outgoing ping. For these reasons, active sonar is not frequently used by military submarines.

A very directional, but low-efficiency, type of sonar (used by fisheries, military, and for port security) makes use of a complex nonlinear feature of water known as non-linear sonar, the virtual transducer being known as a parametric array.

36.3 MILITARY APPLICATIONS OF SONAR

Modern naval warfare makes extensive use of both passive and active sonar from water-borne vessels, aircraft and fixed installations. Although active sonar was used by surface craft in World War II, submarines avoided the use of active sonar due to the potential for revealing their presence and position to enemy forces. However, the advent of modern signal-processing enabled the use of passive sonar as a primary means for search and detection operations. In 1987 a division of Japanese company Toshiba reportedly sold machinery to the Soviet Union that allowed their submarine propeller blades to be milled so that they became radically quieter, making the newer generation of submarines more difficult to detect.

The use of active sonar by a submarine to determine bearing is extremely rare and will not necessarily give high quality bearing or range information to the submarines fire control team. However, use of active sonar on surface ships is very common and is used by submarines when the tactical situation dictates that it is more important to determine the position of a hostile submarine than conceal their own position. With surface ships, it might be assumed that the threat is already tracking the ship with satellite data as any vessel around the emitting sonar will detect the emission. Having heard the signal, it is easy to identify the sonar equipment used (usually with its frequency) and its position (with the sound wave's energy). Active sonar is similar to radar in that, while it allows detection of targets at a certain range, it also enables the emitter to be detected at a far greater range, which is undesirable.

Since active sonar reveals the presence and position of the operator, and does not allow exact classification of targets, it is used by fast (planes, helicopters) and by noisy platforms (most surface ships) but rarely by submarines. When active sonar is used by surface ships or submarines, it is typically activated very briefly at intermittent periods to minimize the risk of detection. Consequently, active sonar is normally considered as a backup to passive sonar. In aircraft, active sonar is used in the form of disposable sonobuoys that are dropped in the aircraft's patrol area or in the vicinity of possible enemy sonar contacts.

Passive sonar has several advantages, most importantly that it is silent. If the target radiated noise level is high enough, it can have a greater range than active sonar, and allows the target to be identified. Since any motorized object makes some noise, it may in principle be detected, depending on the level of noise emitted and the ambient noise level in the area, as well as the technology used. To simplify, passive sonar "sees" around the ship using it. On a submarine, nose-mounted passive sonar detects in directions of about 270°, centered on the ship's alignment, the hull-mounted array of about 160° on each side, and the towed array of a full 360°. The invisible areas are due to the ship's own interference. Once a signal is detected in a certain direction (which means that something makes sound in that direction, this is called broadband detection) it is possible to zoom in and analyze the signal received (narrowband analysis). This is generally done using a Fourier transform to show the different frequencies making up the sound. Since every engine makes a specific sound, it is straightforward to identify the object. Databases of unique engine sounds are part of what is known as acoustic intelligence or ACINT.

Another use of passive sonar is to determine the target's trajectory. This process is called target motion analysis (TMA), and the resultant "solution" is the target's range, course, and speed. TMA is done by marking from which direction the sound comes at different times, and comparing the motion with that of the operator's own ship. Changes in relative motion are analyzed using standard geometrical techniques along with some assumptions about limiting cases.

Passive sonar is stealthy and very useful. However, it requires high-tech electronic components and is costly. It is generally deployed on expensive ships in the form of arrays to enhance detection. Surface ships use it to good effect; it is even better used by submarines, and it is also used by airplanes and helicopters, mostly to a "surprise effect", since submarines can hide under thermal layers. If a submarine's commander believes he is alone, he may bring his boat closer to the surface and be easier to detect, or go deeper and faster, and thus make more sound.

Examples of sonar applications in military use are given below.

(i) Variable depth sonar and its winch

Until recently, ship sonars were usually made with hull mounted arrays, either amidships or at the bow. It was soon found after their initial use that a means of reducing flow noise was required. The first were made of canvas on a framework, then steel ones were used. Now domes are usually made of reinforced plastic or pressurized rubber. Such sonars are primarily active in operation. An example of a conventional hull mounted sonar is the SQS-56.

Because of the problems of ship noise, towed sonars are also used. These have the advantage of being able to be placed deeper in the water, but have limitations on their use in shallow water. These are called towed arrays (linear) or variable depth sonars (VDS) with 2/3D arrays. A problem is that the winches required to deploy/recover them are large and expensive. VDS sets are primarily active in operation, while towed arrays are passive.

An example of a modern active-passive ship towed sonar is Sonar 2087 made by Thales Underwater Systems.

(ii) Torpedoes

Modern torpedoes are generally fitted with an active/passive sonar. This may be used to home directly on the target, but wake homing torpedoes are also used. An early example of an acoustic homer was the Mark 37 torpedo.

Torpedo countermeasures can be towed or free. An early example was the German Sieglinde device while the Bold was a chemical device. A widely used US device was the towed AN/SLQ-25 Nixie while the mobile submarine simulator (MOSS) was a free device. A modern alternative to the Nixie system is the UK Royal Navy S2170 Surface Ship Torpedo Defence system.

(iii) Mines

Mines may be fitted with a sonar to detect, localize and recognize the required target. An example is the CAPTOR mine.

(iv) Mine countermeasures

Mine countermeasure (MCM) sonar, sometimes called "mine and obstacle avoidance sonar (MOAS)", is a specialized type of sonar used for detecting small objects. Most MCM sonars are hull mounted but a few types are VDS design. An example of a hull mounted MCM sonar is the Type 2193 while the SQQ-32 mine-hunting sonar and Type 2093 systems are VDS designs.

(v) Submarine navigation

Submarines rely on sonar to a greater extent than surface ships as they cannot use radar in water. The sonar arrays may be hull mounted or towed. Information fitted on typical fits is given in Oyashio-class submarine and Swiftsure-class submarine.

(vi) Aircraft

Helicopters can be used for antisubmarine warfare by deploying fields of active-passive sonobuoys or can operate dipping sonar, such as the AQS-13. Fixed wing aircraft can also deploy sonobuoys and have greater endurance and capacity to deploy them. Processing from the sonobuoys or dipping sonar can be on the aircraft or on ship. Dipping sonar has the advantage of being deployable to depths appropriate to daily conditions. Helicopters have also been used for mine countermeasure missions using towed sonars such as the AQS-20A.

(vii) Underwater communications

Dedicated sonars can be fitted to ships and submarines for underwater communication.

(viii) Ocean surveillance

The United States began a system of passive, fixed ocean surveillance systems in 1950 with the classified name Sound Surveillance System (SOSUS) with American Telephone and Telegraph Company (AT&T), with its Bell Laboratories research and Western Electric manufacturing entities being contracted for development and installation. The systems exploited the SOFAR (Sound Fixing and Ranging) channel, also known as the deep sound channel, where a sound speed minimum creates a waveguide in which low frequency sound travels thousands of miles. Analysis was based on an AT&T sound spectrograph, which converted sound into a visual spectrogram representing a time–frequency analysis of sound that was developed for speech analysis and modified to analyze low-frequency underwater sounds. That process was Low Frequency Analysis and Recording and the equipment was termed the Low

Frequency Analyzer and Recorder, both with the acronym LOFAR. LOFAR research was termed Jezebel and led to usage in air and surface systems, particularly sonobuoys using the process and sometimes using "Jezebel" in their name. The proposed system offered such promise of long-range submarine detection that the Navy ordered immediate moves for implementation.

Between installation of a test array followed by a full scale, forty element, prototype operational array in 1951 and 1958 systems were installed in the Atlantic and then the Pacific under the unclassified name Project Caesar. The original systems were terminated at classified shore stations designated Naval Facility (NAVFAC) explained as engaging in "ocean research" to cover their classified mission. The system was upgraded multiple times with more advanced cable allowing the arrays to be installed in ocean basins and upgraded processing. The shore stations were eliminated in a process of consolidation and rerouting the arrays to central processing centers into the 1990s. In 1985, with new mobile arrays and other systems becoming operational the collective system name was changed to Integrated Undersea Surveillance System (IUSS). In 1991 the mission of the system was declassified. The year before IUSS insignia were authorized for wear. Access was granted to some systems for scientific research.

A similar system is believed to have been operated by the Soviet Union.

(ix) Anti-frogman techniques § Detection

Sonar can be used to detect frogmen and other scuba divers. This can be applicable around ships or at entrances to ports. Active sonar can also be used as a deterrent and/or disablement mechanism. One such device is the Cerberus system.

(x) Intercept sonar

This is a sonar designed to detect and locate the transmissions from hostile active sonars. An example of this is the Type 2082 fitted on the British Vanguard-class submarines.

36.4 USHUS (SONAR), NPOL, DRDO, INDIA

USHUS is an Integrated Submarine Sonar System developed by the Naval Physical and Oceanographic Laboratory (NPOL) of the Defence Research and Development Organisation (DRDO) of India and manufactured by Bharat Electronics Limited (BEL). It has been developed for use in submarines of the Indian Navy, especially for Sindhughosh-class submarines. Some reports also suggest that Arihant-class nuclear ballistic missile submarines are also equipped with USHUS system. USHUS replaces Russian systems like MGK-400 and MGK-519 sonars on Indian submarines.

36.4.1 Design and description of USHUS

USHUS is a multipurpose integrated sonar suite used for detecting and tracking enemy submarines, surface vessels, and torpedoes as well as for underwater communication and obstacles avoidance operations. The sonar capabilities of includes active and passive sonar surveillance, underwater communication and is capable of interception and. The sub-systems of the sonar includes Active Sonar

Array, Passive Sonar Array, Intercept Sonar Array (low to medium frequency), Obstacle Avoidance System and Underwater Telephony. The USHUS-2 variant has a detection range of 30 km.

36.4.2 Production of USHUS

The production of the sonar is done by Bharat Electronics (BEL) at its Bengaluru unit, after transfer of technology from NPOL. NPOL continues to provide technical consultancy and support. The Indian Ministry of Defence signed a contract worth ₹167 crore with BEL for the delivery of the sonars for four Kilo-class submarines between 2003 and 2007. Initially one sonar system was installed and integrated in Russia and other system was installed on board an Indian submarine.

36.4.3 Current status of USHUS

By April 2013, five Sindhughosh class submarines of the navy were upgraded to include the USHUS system. They are, in order of their upgrade: INS Sindhuvir (S58), INS Sindhudhvaj (S56), INS Sindhuvijay (S62) and INS Sindhurakshak (S63). INS Sindhukirti (S61) was upgraded in India at Hindustan Shipyard. The remaining submarines of the class are expect to follow.

As of 2024, USHUS-1 has been installed on 5 of the Sindhughosh-class submarines whereas USHUS-2 variant of the sonar system is expected to be installed on INS Sindhughosh (S55), INS Sindhuraj (S57) and INS Sindhuratna (S59).

USHUS-2 is a state of art upgrade of NPOL designed sonar USHUS in terms of the technology and sonar capabilities. USHUS is operational onboard five of the nine frontline EKM submarines of Indian Navy. USHUS-2 is a world class sonar suite, tailored for the remaining four EKM class of submarines.

36.5 NEW SONAR COULD TURN ANY SHIP INTO SUBMARINE HUNTER

A sonar system being installed on new U.S. Navy Constellation-class frigates could also protect merchant ships during a conflict and give them the ability to search for submarines.

The towed-sensor system is already being used by U.S. allies and could be quickly installed on non-military ships. It's a modular system that can be placed on vessels of opportunity.

The company believes the sonar system could be loaded on a military cargo plane and quickly flown to a ship that needs it.

This is not a concept that is new or developmental. It's a concept of how to rapidly address [anti-submarine warfare] capacity from a DOD or navy perspective.

The US Navy said it would install Advanced Acoustic Concepts' CAPTAS anti-submarine sonar on its Constellation-class frigates over similar Raytheon-made technology.

The sonar system is made up of a long cable that tows a sensor in the water that pings for submarines. Unlike the Navy's current bow-mounted sonars, the so-called variable-depth sonar can be "placed at a depth that has the highest probability of getting a target acquisition," Bock said.

Thales acquired Advanced Acoustic Concepts, a undersea technology joint venture it previously had with Leonardo DRS, in July. The acquisition, Thales said at the time, was to "increase its engineering and industrial footprint in the U.S. defense market, with reinforced U.S.-based teams and capabilities."

The anti-submarine technology is already used on British, French, Spanish, and Chilean ships.

The system is mature, has a track record, past performance record against targets.

CHAPTER 37

BIOTECHNOLOGY IN WARFARE

37.1 BIOTECHNOLOGY: THE DEADLY WEAPONIZATION IN WARFARE

Using biological agents as weapons is illegal, but what is banned depends on what an actor plans to do rather than the actual pathogens. While it's hard to assess the intentions of a malicious actor, international consensus leans toward not using biological agents as weapons.

Biotechnology is a technique employed to develop a weapon that a malign actor could use to obtain a hazardous agent.

With the emergence of biotechnology, these considerations become murkier. Efficient manipulation of an organism's DNA can produce more lethal agents leveraging recent scientific advances. In lieu of this, the International Committee of the Red Cross (ICRC), an international organization, urged all countries to renounce this potential path to novel weapons of mass destruction.

Leveraging biotechnology could be a new, more terrorizing form of adopting technologies for greater destruction. Nevertheless, for one specialist, this is only another iteration of the age-old weaponization of life.

Malign actors sought to weaponize organisms, including anthrax, in the past, but intentions lie far from achievement, according to one researcher, as scaling production poses different challenges from developing a pathogen on paper or in digital form or as a small culture in a lab.

Developing an aggressive organism or toxin is only one part of the equation. Production, storage, transport and deployment require knowledge specific to one agent and cannot be transferred to others. To produce a biological weapon, you don't need just one person; you need a team who has skills in different stages of processing the agent. Therefore, if a terrorist group manipulates a specific virus, it needs specialists that will most likely only know how to work with that specific pathogen. In case this malicious actor chooses another microbe, they should find different willing specialists.

37.2 BIOLOGICAL WARFARE

Biological warfare, also known as germ warfare, is the use of biological toxins or infectious agents such as bacteria, viruses, insects, and fungi with the intent to kill, harm or incapacitate humans, animals or plants as an act of war. Biological weapons (often termed "bio-weapons", "biological threat agents", or "bio-agents") are living organisms or replicating entities (i.e. viruses, which are not universally considered "alive"). Entomological (insect) warfare is a subtype of biological warfare.

Biological warfare is subject to a forceful normative prohibition. Offensive biological warfare in international armed conflicts is a war crime under the 1925 Geneva Protocol and several international humanitarian law treaties. In particular, the 1972 Biological Weapons Convention (BWC) bans the development, production, acquisition, transfer, stockpiling and use of biological weapons. In contrast, **defensive biological research for prophylactic, protective or other peaceful purposes is not prohibited by the BWC.**

Biological warfare is distinct from warfare involving other types of weapons of mass destruction (WMD), including nuclear warfare, chemical warfare, and radiological warfare. None of these are considered conventional weapons, which are deployed primarily for their explosive, kinetic, or incendiary potential.

Biological weapons may be employed in various ways to gain a strategic or tactical advantage over the enemy, either by threats or by actual deployments. The biological weapons may also be useful as area denial weapons. These agents may be lethal or non-lethal, and may be targeted against a single individual, a group of people, or even an entire population. They may be developed, acquired, stockpiled or deployed by nation states or by non-national groups. In the latter case, or if a nation-state uses it clandestinely, it may also be considered **bioterrorism**.

A biological attack could conceivably result in large numbers of civilian casualties and cause severe disruption to economic and societal infrastructure. A nation or group that can pose a credible threat of mass casualty has the ability to alter the terms under which other nations or groups interact with it. The biological weapons possess destructive potential and loss of life far in excess of nuclear, chemical or conventional weapons. Accordingly, biological agents are potentially useful as strategic deterrents, in addition to their utility as offensive weapons on the battlefield.

As a tactical weapon for military use, a significant problem with biological warfare is that it would take days to be effective, and therefore might not immediately stop an opposing force. Some biological agents (smallpox, pneumonic plague) have the capability of person-to-person transmission via aerosolized respiratory droplets. This feature can be undesirable, as the agent(s) may be transmitted by this mechanism to unintended populations, including neutral or even friendly forces. Worse still, such a weapon could escape the laboratory where it was developed, even if there was no intent to use it – for example, by infecting a researcher who then transmits it to the outside world before realizing that they were infected. Several cases are known of researchers becoming infected and dying of Ebola, which they had been working with in the lab (though nobody else was infected in those cases) – while there is no evidence that their work was directed towards biological warfare, it demonstrates the potential for accidental infection even of careful researchers fully aware of the dangers. While containment of biological warfare is less of a concern for certain criminal or terrorist organizations, it remains a significant concern for the military and civilian populations of virtually all nations.

37.3 BIOLOGICAL WARFARE AND BIOTERRORISM: A HISTORICAL REVIEW

37.3.1 Early Use of Biological Warfare

Infectious diseases were recognized for their potential impact on people and armies **as early as 600 BC.** The crude use of filth and cadavers (corpse), animal carcasses, and contagion had devastating effects and weakened the enemy. Polluting wells and other sources of water of the opposing army was a common strategy that continued to be used through the many European wars, during the American Civil War, and even into the 20th century.

Military leaders in the Middle Ages recognized that victims of infectious diseases could become weapons themselves. During the siege of **Caffa**, a well-fortified Genoese-controlled seaport (now Feodosia, Ukraine), in 1346, the attacking Tartar force experienced an epidemic of plague. The Tartars, however, converted their misfortune into an opportunity by hurling the cadavers of their deceased into the city, thus initiating a plague epidemic in the city. The outbreak of plague followed, forcing a retreat of the Genoese forces. The plague pandemic, also known as the Black Death, swept through Europe, the Near East, and North Africa in the 14th century and was probably the most devastating public health disaster in recorded history. The ultimate origin of the plague remains uncertain: several countries in the Far East, China, Mongolia, India, and central Asia have been suggested.

The Caffa incident was described in 1348 or 1349 by Gabriel de Mussis, a notary born in Piacenza north of Genoa. De Mussis made two important claims: plague was transmitted to the citizens of Caffa by the hurling of diseased cadavers into the besieged city, and Italians fleeing from Caffa brought the plague into the Mediterranean seaports. In fact, ships carrying plague-infected refugees (and possibly rats) sailed to Constantinople, Genoa, Venice, and other Mediterranean seaports and are thought to have contributed to the second plague pandemic. However, given the complex ecology and epidemiology of plague, it may be an oversimplification to assume that a single biological attack was the sole cause of the plague epidemic in Caffa and even the 14th-century plague pandemic in Europe. Nonetheless, the account of a biological warfare attack in Caffa is plausible and consistent with the technology of that time, and despite its historical unimportance, the siege of Caffa is a powerful reminder of the terrible consequences when diseases are used as weapons.

During the same 14th-century plague pandemic, **which killed more than 25 million Europeans** in the 14th and 15th centuries, many other incidents indicate the various uses of disease and poisons during war. For example, bodies of dead soldiers were catapulted into the ranks of the enemy in Karolstein in 1422. A similar strategy using cadavers of plague victims was utilized in 1710 during the battle between Russian troops and Swedish forces in Reval. On numerous occasions during the past 2000 years, the use of biological agents in the form of disease, filth, and animal and human cadavers has been mentioned in historical recordings.

Another disease has been used as an effective biological weapon in the New World: **smallpox**. Pizarro is said to have presented South American natives with variola-contaminated clothing in the 15th century. In addition, during the French-Indian War (1754–1767), Sir Jeffrey Amherst, the commander of the British forces in North America, suggested the deliberate use of smallpox to diminish the native Indian population hostile to the British. An outbreak of smallpox in Fort Pitt led to a significant generation of fomites and provided Amherst with the means to execute his plan. On June 24, 1763, Captain Ecuyer, one of Amherst's subordinate officers, provided the Native Americans with **smallpox-laden blankets** from the smallpox hospital. He recorded in his journal: "I hope it will have the desired effect". As a result, a large outbreak of smallpox occurred among the Indian tribes in the Ohio River Valley. Again, it has to be recognized that several other contacts between European colonists and Native Americans contributed to such epidemics, which had been occurring for over 200 years. In addition, the transmission of smallpox by fomites was inefficient compared with respiratory droplet transmission.

The description of these historical attempts of using diseases in biological warfare illustrates the difficulty of differentiating between a naturally occurring epidemic and an alleged or attempted biological warfare attack—a problem that has continued into present times.

37.3.2 Biological Warfare in the 19th and 20th Centuries

The use of biological warfare became more sophisticated during the 19th century. The conception of Koch's postulates and the development of modern microbiology during the 19th century made possible the isolation and production of stocks of specific pathogens.

(i) World War I

Substantial evidence suggests the existence of an ambitious biological warfare program in Germany during World War I. This program allegedly featured covert operations. During World War I, reports circulated of attempts by Germans to ship horses and cattle inoculated with disease-producing bacteria, such as Bacillus anthracis (anthrax) and Pseudomonas pseudomallei (glanders), to the USA and other countries. The same agents were used to infect Romanian sheep that were designated for export to Russia. Other allegations of attempts by Germany to spread cholera in Italy and plague in St. Petersburg in Russia followed. Germany denied all these allegations, including the accusation that biological bombs were dropped over British positions.

In 1924, a subcommittee of the Temporary Mixed Commission of the League of Nations, in support of Germany, found no hard evidence that the bacteriological arm of warfare had been employed in war. However, the document indicated evidence of use of the chemical arm of warfare. In response to the horror of chemical warfare during World War I, international diplomatic efforts were directed toward limiting the proliferation and use of weapons of mass destruction, i.e., biological and chemical weapons. On June 17, 1925, the "Protocol for the Prohibition of the Use in War of Asphyxiating, Poisonous or Other Gases and of Bacteriological Methods of Warfare," commonly called the Geneva Protocol of 1925, was signed. Because viruses were not differentiated from bacteria at that time, they were not specifically mentioned in the protocol. A total of 108 nations, including eventually the 5 permanent members of the United Nations (UN) Security Council, signed the agreement. However, the Geneva Protocol did not address verification or compliance, making it a "toothless" and less meaningful document. Several countries that were parties to the Geneva Protocol of 1925 began to develop biological weapons soon after its ratification. These countries included Belgium, Canada, France, Great Britain, Italy, the Netherlands, Poland, Japan, and the Soviet Union. The USA did not ratify the Geneva Protocol until 1975.

(ii) World War II

During World War II, some countries began a rather ambitious biological warfare research program. Various allegations and countercharges clouded the events during and after World War II. Japan conducted biological weapons research from approximately 1932 until the end of World War II. The program was under the direction of Shiro Ishii (1932–1942) and Kitano Misaji (1942–1945). Several military units existed for research and development of biological warfare. The center of the Japanese biowarfare program was known as "Unit 731" and was located in Manchuria near the town of Pingfan. The Japanese program consisted of more than 150 buildings in Pingfan, 5 satellite camps, and a staff of more than 3000 scientists. Organisms and diseases of interest to the Japanese program were B. anthracis, Neisseria meningitidis, Vibrio cholerae, Shigella spp, and Yersiniapestis. More than 10,000 prisoners are believed to have died as a result of experimental infection during the Japanese program between 1932 and 1945. At least 3000 of these victims were prisoners of war, including Korean, Chinese, Mongolian, Soviet, American, British, and Australian soldiers. Many of these prisoners died as a direct effect of experimental inoculation of agents causing gas gangrene, anthrax, meningococcal infection, cholera, dysentery, or plague. In addition, experiments with terodotoxin (an extremely poisonous fungal toxin) were conducted. In later years, Japanese officials considered these experiments as "most regrettable from the view point of humanity".

In addition to the experiments conducted on prisoners in the camps of Unit 731, the Japanese military developed plague as a biological weapon by allowing laboratory fleas to feed on plague infected rats. On several occasions, the fleas were released from aircraft over Chinese cities to initiate plague epidemics. However, the Japanese had not adequately prepared, trained, or equipped their own military personnel for the hazards of biological weapons. An attack on the city of Changteh in 1941 reportedly led to approximately 10,000 casualties due to biological weapons. During this incident, 1700 deaths were reported among Japanese troops. Thus, "field trials" were terminated in 1942.

In December 1949, a Soviet military tribunal in Khabarovsk tried 12 Japanese prisoners of war for preparing and using biological weapons. Major General Kawashima, former head of Unit 731's First, Third, and Fourth Sections, testified in this trial that no fewer than 600 prisoners were killed yearly at Unit 731. The Japanese government, in turn, accused the Soviet Union of experimentation with biological weapons, referring to examples of B. anthracis, Shigella, and V. cholerae organisms recovered from Russian spies.

Although German medical researchers infected prisoners with disease-producing organisms such as Rickettsia prowazekii, hepatitis A virus, and malaria, no charges were pressed against Germany regarding experimentation with agents of biological warfare. Allegedly Hitler issued orders prohibiting the development of biological weapons, referring to his own devastating experience with the effects of chemical agents used during World War I. However, with the support of other high-ranking Nazi officials, German scientists began biological weapons research. Despite these efforts, which clearly lagged behind those of other countries, a German offensive biological weapons program never materialized.

On the other hand, German officials accused the Allies of using biological weapons: Joseph Goebbels accused the British of attempting to introduce **yellow fever into India by importing infected mosquitoes from West Africa**. This was in fact believable by many, because the British were actually experimenting with at least one organism of biological warfare: B. anthracis. Bomb experiments of weaponized spores of B. anthracis were conducted on Gruinard Island near the coast of Scotland. These experiments lead to heavy contamination of the island with persistence of viable spores. In 1986, the island was finally decontaminated by using formaldehyde and seawater.

In the USA, an offensive biological warfare program was begun in 1942 under the direction of a civilian agency, the War Reserve Service. The program included a research and development facility at Camp Detrick, Maryland (renamed Fort Detrick in 1956 and known today as the US Army Medical Research Institute of Infectious Diseases [USAMRIID]), testing sites in Mississippi and Utah, and a production facility in Terra Haute, Indiana.

Initially, organisms of interest were B. anthracis and Brucella suis. Although about 5000 bombs filled with B. anthracis spores were produced at Camp Detrick, the production facility lacked adequate engineering safety measures, precluding a large-scale production of biological weapons during World War II.

(iii) Bio-Warfare after World War II

During the years immediately after World War II, newspapers were filled with articles about disease outbreaks caused by foreign agents armed with biological weapons. During the Korean War, the Soviet Union, China, and North Korea accused the USA of using agents of biological warfare against North Korea. In later years the USA admitted that it had the capability of producing such weapons, although it denied having used them. However, the credibility of the USA was undermined by its failure to ratify the Geneva Protocol of 1925, by public acknowledgment of its own offensive biological warfare program, and by suspicions of collaboration with former Unit 731 scientists.

In fact, the US program expanded during the Korean War (1950–1953) with the establishment of a new production facility in Pine Bluff, Arkansas. In addition, a defensive program was launched in 1953 with the objective of developing countermeasures, including vaccines, antisera, and therapeutic agents, to protect troops from possible biological attacks. By the late 1960s, the US military had developed a biological arsenal that included numerous biological pathogens, toxins, and fungal plant pathogens that could be directed against crops to induce crop failure and famine.

At Fort Detrick, biological munitions were detonated inside a hollow 1-million-liter, metallic, spherical aerosolization chamber known as the "eight ball". Volunteers inside this chamber were exposed to Francisella tularensis and Coxiella burnetii. The studies were conducted to determine the vulnerability of humans to certain aerosolized pathogens. Further testing was done to evaluate the efficacy of vaccines, prophylaxis, and therapy. During the offensive biological program (1942–1969), 456 cases of occupational infections acquired at Fort Detrick were reported at a rate of < 10 infections per 1 million hours worked. This rate of infection was well within the contemporary standards of the National Safety Council and below the rate reported from other laboratories. Three fatalities due to acquired infections were reported from Fort Detrick during this period: 2 cases of anthrax occurred in 1951 and 1958, and one case of viral encephalitis was reported in 1964. In addition, 48 occupational infections were reported from the other testing and production sites, but no other fatalities occurred.

Between 1951 and 1954, several studies were conducted to demonstrate the vulnerability of US cities. Cities on both coasts were surreptitiously used as laboratories to test aerosolization and dispersal methods when simulants were released during covert experiments in New York City, San Francisco, and other sites. Aspergillus fumigatus, Bacillus subtilis var globigii, and Serratia marcescens were selected for these experiments. Organisms were released over large geographic areas to study the effects of solar irradiation and climatic conditions on the viability of organisms. Concerns regarding potential public health hazards were raised after outbreaks of urinary tract infections caused by nosocomial S. marcescens at Stanford University Hospital between September 1950 and February 1951. The outbreak followed covert experiments using S. marcescens as a simulant in San Francisco.

In addition to these efforts in the USA, many other countries continued their biological weapons research, including Canada, Britain, France, and the Soviet Union. In the United Kingdom, the Microbiological Research Department was established in 1947 and expanded in 1951. Plans for pilot biological warfare were made, and research continued on the development of new biological agents and weapons design. Britain conducted several trials with biological warfare agents in the Bahamas, in the Isles of Lewis, and in Scottish waters to refine these weapons. However, in 1957, the British government decided to abandon the offensive biological warfare research and to destroy stockpiles. At that time, a new emphasis was put on further development of biological defensive research. At the same time, the Soviet Union increased its efforts in both offensive and defensive biological warfare research and development. Reports regarding offensive research repeatedly occurred in the 1960s and 1970s, although officially the Soviet Union claimed not to possess any biological or chemical weapons.

(iv) Other allegations occurred during the post—World War II period(F

The Eastern European press stated that Great Britain had used biological weapons in Oman in 1957.

The Chinese alleged that the USA caused a cholera epidemic in Hong Kong in 1961.

In July 1964, the Soviet newspaper Pravda asserted that the US Military Commission in Columbia and Colombian troops had used biological agents against peasants in Colombia and Bolivia.

In 1969, Egypt accused the "imperialistic aggressors" of using biological weapons in the Middle East, specifically causing an epidemic of cholera in Iraq in 1966.

37.3.3 The 1972-Biological Weapons Convention

During the late 1960s, public and expert concerns were raised internationally regarding the indiscriminate nature of, unpredictability of, epidemiologic risks of, and lack of epidemiologic control measures for biological weapons. In addition, more information on various nations-biological weapons programs became evident, and it was obvious that the 1925-Geneva Protocol was ineffective in controlling the proliferation of biological weapons. In July 1969, Great Britain submitted a proposal to the UN Committee on Disarmament outlining the need to prohibit the development, production, and stockpiling of biological weapons. Furthermore, the proposal provided for measures for control and inspections, as well as procedures to be followed in case of violation. Shortly after submission of the British proposal, in September 1969, the Warsaw Pact nations under the lead of the Soviet Union submitted a similar proposal to the UN. However, this proposal lacked provisions for inspections. Two months later, in November 1969, the World Health Organization issued a report regarding the possible consequences of the use of biological warfare agents.

Subsequently, the 1972 "Convention on the Prohibition of the Development, Production, and Stockpiling of Bacteriological (Biological) and Toxin Weapons and on Their Destruction," known as the BWC, was developed. This treaty prohibits the development, production, and stockpiling of pathogens or toxins in "quantities that have no justification for prophylactic, protective or other peaceful purposes". Under the BWC, the development of delivery systems and the transfer of biological warfare technology or expertise to other countries are also prohibited. It further required the parties to the BWC to destroy stockpiles, delivery systems, and production equipment within 9 months of ratifying the treaty. This agreement was reached among 103 cosigning nations, and the treaty was ratified in April 1972. The BWC went into effect in March 1975. Signatories that have not yet ratified the BWC are obliged to refrain from activities that would defeat the purpose of the treaty until they specifically communicate to the UN their intention not to ratify the treaty. Review conferences to the BWC were held in 1981, 1986, 1991, and 1996. Signatories to the BWC are required to submit the following information to the UN on an annual basis: facilities where biological defense research is being conducted, scientific conferences that are held at specified facilities, exchange of scientists or information, and disease outbreaks.

However, like the 1925 Geneva Protocol, the BWC does not provide firm guidelines for inspections and control of disarmament and adherence to the protocol. In addition, there are no guidelines on enforcement and how to deal with violations. Furthermore, there are unresolved controversies about the definition of "defensive research" and the quantities of pathogens necessary for benevolent research. Alleged violations of the BWC were to be reported to the UN Security Council, which may in turn initiate inspections of accused parties, as well as modalities of correction. The right of permanent members of the Security Council to veto proposed inspections, however, undermines this provision. More recent events in 2003 and 2004 again illustrated the complexity and the enormous difficulties the UN faces in enforcing the statutes of the BWC.

In the USA, the offensive biological weapons program was terminated by President Nixon by executive orders in 1969 and 1970. The USA adopted a policy to never use biological weapons, including toxins, under any circumstances. National Security Decisions 35 and 44, issued in November 1969 (microorganisms) and February 1970 (toxins), mandated the cessation of offensive biological weapons research and production and the destruction of the biological weapons arsenal. However, research efforts continued to be allowed for the purpose of developing countermeasures, including vaccines and antisera. The entire arsenal of biological weapons was destroyed between May 1971 and February 1973 under the auspices of the US Department of Agriculture, the US Department of Health, Education, and Welfare, and the Departments of Nature Resources of Arkansas, Colorado, and Maryland. After the termination of the offensive program, USAMRIID was established to continue research for development of medical defense for the US military against a potential attack with biological weapons. The USAMRIID is an open research institution, and none of the research is classified.

37.3.4 After BWC

Despite the agreement reached in 1972, several of the signatory nations of the BWC participated in activities outlawed by the convention. These events clearly demonstrate the ineffectiveness of the convention as the exclusive approach for eradicating biological weapons and preventing further proliferation. The number and identity of countries that have engaged in offensive biological weapons research is largely still classified information. However, it can be accurately stated that the number of state sponsored programs of this type has increased significantly during the past 30 years. In addition, several assassination attempts and attacks, as well as non—state-sponsored terrorist attacks, have been documented.

During the 1970s, biological weapons were used for covert assassinations. In 1978 a Bulgarian exile named Georgi Markov was attacked and killed in London, England. This assassination later became known as the "umbrella killing," because the weapon used was a device disguised as an umbrella. This weapon discharged a tiny pellet into the subcutaneous tissue of Markov's leg while he was waiting at a bus stop in London. The following day, he became severely ill, and he died only 3 days after the attack. On autopsy, the pellet, cross-drilled as if it was designed to contain another material, was retrieved. As it was revealed in later years, this assassination was carried out by the communist Bulgarian secret service, and the technology to commit the crime was supplied to the Bulgarians by the Soviet Union. Only 10 days before the assassination of Markov, an attempt to kill another Bulgarian exile, Vladimir Kostov, had occurred in Paris, France. Kostov said that one day when he was leaving a metro stop in Paris, he had felt a sharp pain in his back. When he turned around, he saw a man with an umbrella running away.

Two weeks later, after he had learned of Markov's death, Kostov was examined by French doctors. They removed a similar pellet, which was made from an exotic alloy of iridium and platinum and contained the toxin ricin.

In the late 1970s, allegations were made that planes and helicopters delivering aerosols of different colors may have attacked the inhabitants of Laos and Kampuchea. People who were exposed became disoriented and ill. These attacks were commonly described as "yellow rain." In fact it was highly controversial whether these clouds truly represented biological warfare agents. Some of these clouds were believed to comprise trichothecene toxins (e.g., T-2 mycotoxin). Some scientists believed that the yellow rains were most likely the fecal matter of wild honeybees dropped during their "cleansing flights." The controversy over the yellow rain incidents remains unresolved.

During April 1979, an epidemic of anthrax occurred among the citizens of Sverdlovsk (now Ekaterinburg), Russia. The epidemic occurred among people who lived and worked near a Soviet military microbiology facility (Compound 19) in Sverdlovsk. In addition, many livestock died of anthrax in the same area, out to a distance of 50 km. European and US intelligence suspected that this facility conducted biological warfare research and attributed the epidemic to an accidental release of anthrax spores. Early in February 1980, the widely distributed German newspaper Bild Zeitung carried a story about an accident in a Soviet military settlement in Sverdlovsk in which an anthrax cloud had resulted. When this story was published, other major Western newspapers and magazines began to take an interest in the anthrax outbreak in Sverdlovsk, a city of 1.2 million people, 1400 km east of Moscow. Later that year several articles occurred in Soviet medical, veterinary, and legal journals reporting an anthrax outbreak among livestock. Human cases of anthrax were attributed to the ingestion of contaminated meat.

In 1986, Matthew Meselson (Department of Molecular and Cellular Biology, Harvard University, Cambridge, Massachusetts) renewed previously unsuccessful requests to Soviet officials to bring independent scientists to Sverdlovsk to investigate the incident. This request finally resulted in the invitation to come to Moscow to discuss the incident with 4 Soviet physicians who had gone to Sverdlovsk to deal with the outbreak. The impression after these meetings was that a plausible case had been made, and further investigation of the epidemiologic and pathoanatomical data was needed. The Soviet Union maintained that the anthrax outbreak was caused by consumption of contaminated meat that was purchased on the black market. However, after the collapse of the Soviet Union, Boris Yeltsin, then the president of Russia, directed his counselor for ecology and health to determine the origin of the epidemic in Sverdlovsk. In May 1992, Yeltsin admitted that the facility had been part of an offensive biological weapons program and that the epidemic was caused by an accidental release of anthrax spores. He was quoted as saying, "The KGB admitted that our military developments were the cause." Meselson and his team returned to Russia to aid in these further investigations. Among the evidence reviewed were a private pathologist's notes from 42 autopsies that resulted in the diagnosis of anthrax. Demographic, ecologic, and atmospheric data were also reviewed. The conclusion was that the pattern of these 42 cases of fatal anthrax bacteremia and toxemia were typical of inhalational anthrax as seen in experimentally infected nonhuman primates. In summary, the narrow zone of human and animal anthrax cases extending downwind from Compound 19 indicated that the outbreak resulted from an aerosol that originated there.

A 1995-report stated that the Russian program continued to exist after the 1979 incident and had temporarily increased during the 1980s. In 1995, the program was still in existence and employed 25,000 to 30,000 people. At the same time, several high-ranking officials in the former Soviet military and Biopreparat had defected to Western countries. The information provided by these former employees gave further insight into the biological weapons program of the former Soviet Union. After the anthrax incident in Sverdlovsk, the research was continued at a remote military facility in the isolated city of Stepnogorsk in Kazakhstan, producing an even more virulent strain of anthrax. In 1980, the former Soviet Union expanded its bioweapons research program and was eventually able to weaponize smallpox. This research was conducted at remote facilities in Siberia, and very little information is available about the extent and outcome of this research and where it was conducted.

During Operation Desert Shield, the build-up phase of the Persian Gulf War (Operation Desert Storm) after Iraq had invaded and occupied Kuwait in the fall and winter of 1990, the USA and the coalition of allied countries faced the threat of biological and chemical warfare. The experience gained from observations during the first Persian Gulf War in the late 1980s supported the information on biological and chemical weapons available to the Western intelligence community. In fact, Iraq had used chemical warfare against its own people on many occasions in the 1980s. Intelligence reports from that time suggested that the Iraqi regime had sponsored a very ambitious biological and chemical warfare program.

Coalition forces prepared in 1990–1991 for potential biological and chemical warfare by training in protective masks and equipment, exercising decontamination procedures, receiving extensive education on possible detection procedures, and immunizing troops against potential biological warfare threats. Approximately 150,000 US troops received a Food and Drug Administration—licensed toxoid vaccine against anthrax, and 8000 received a new botulinum toxoid vaccine. For further protection against anthrax spores, 30 million 500-mg oral doses of ciprofloxacin were stockpiled to provide a 1-month course of chemoprophylaxis for the 500,000 US troops that were involved in the operation.

At the end of the Persian Gulf War in August 1991, the first UN inspection of Iraq's biological warfare capabilities was carried out. Representatives of the Iraqi government announced to representatives from the UN Special Commissions Team 7 that Iraq had conducted research into the offensive use of B. anthracis, botulinum toxins, and Clostridium perfringens. Iraq had extensive and redundant research facilities at Salman Pak, Al Hakam, and other sites, only some of which were destroyed during the war. Despite these elaborate efforts by the UN, the struggle with enforcement of the BWC continued throughout the late 1990s and into the 21st century. As the recent developments in Iraq have shown, development of biological and chemical weapons is a real threat, and efforts to control its proliferation are limited by logistical and political problems. As long as there are no concrete provisions for enforcement, the BWC will remain a toothless instrument in the hands of the UN Security Council.

In addition to these state-sponsored and military-related biowarfare programs, private and civilian groups have attempted to develop, distribute, and use biological and chemical weapons. One incident was the intentional contamination of salad bars **in restaurants in Oregon by the Rajneeshee cult** during late September 1984. A total of 751 cases of severe enteritis were reported, and Salmonella typhimurium was identified as the causative organism. Forty-five victims were hospitalized during this outbreak. Although the Rajneeshees were suspected, the extensive research and investigation conducted by the Oregon Health Department and the Centers for Disease Control could not conclusively identify the origin of the epidemic. However, in 1985, a member of the cult confirmed the attack and identified the epidemic as a deliberate biological attack.

Unfortunately, recent examples of the intentional use of biological weapons are not difficult to find. In the mid-1990s, large amounts of botulinum toxin were found in a laboratory in a safe house of the Red Army Faction in Paris, France. Apparently, the toxin was never used. The bioterrorism threat resurfaced then on March 18, 1995, after the Aum Shinrikyo attacked the Tokyo subway system with sarin gas. The investigations after this incident disclosed evidence of a rudimentary biological weapons program. Allegedly before March 1995, the cult had attempted 3 unsuccessful biological attacks in Japan using anthrax and botulinum toxin. In addition, cult members had attempted to acquire Ebola virus in Zaire during 1992. However, only a small portion of the entire program was discovered by Japanese police and intelligence, and only fragments of evidence have been made available to the public. Until the present time, the full extent of the biological weapons program by the Aum Shinrikyo, as well as its present condition, remains unknown.

37.4 POSSIBLE BIOLOGICAL THREATS BY 2030

The biological threats of the future are increasingly severe. The individual threats are often incomprehensible for nonexperts, as biological warfare can be carried out by using viruses, bacteria, fungi, insects, or plants. Almost all animals are possible vectors, and in the far future, even mechanical products or highly manipulated organisms could also be possible vectors. In addition, synthetic biology, nanotechnology, and DNA manipulation open up a whole new range of possibilities for modifying or even completely rebuilding or recreating viruses and bacteria. The latter are called designer pathogens. These technological advances were foreseeable for some time, and yet they only came to public attention because of the global pandemic. But as complex and diverse as the possible types of biological weapons are, so are the techniques to enhance the efficacy of biological weapons through biological engineering. A 2013-report in the Dartmouth Undergraduate Journal of Science lists the possible techniques for weaponizing biological materials. These include the manipulation of bacteria; the aforementioned designer pathogens; the destruction or replacement of individual genes in the context of misused gene therapy; stealth viruses that only unfold their effect in the body after external or internal activation; host swapping diseases that, for example, specifically jump from domestic cats to humans; designer diseases that, for example, cause artificial cancer; and personalized biological weapons. The latter spread approximately asymptomatically in the population and only have an effect on certain genetic characteristics of a person or group of people.

In his 2002-contribution to The Counterproliferation Papers of the U.S. Air Force Counterproliferation Center at Air University, Michael J. Ainscough describes the threats that could become reality by 2030. Based on findings of the JASON Defense Advisory Panel in 1997, Ainscough describes six future threats. First, he talks about binary biological weapons that can be used for extortion or safe handling. For this, a harmless host bacterium and a virulent plasmid would be isolated separately and threatened with the release of the associated second component, which would then interact to produce its effect. As far as designer genes are concerned, the researcher concludes that these have long been state of the art with simple modifications at the time of the study. Future designer pathogens will have far more complex capabilities and will be able to exhibit a whole range of modified characteristics. Regarding gene therapy, he writes:

There are two general classes of gene therapy: germ-cell line (reproductive) and somatic cell line (therapeutic). Changes in DNA in germ cells would be inherited by future generations. Changes in DNA of somatic cells would affect only the individual and could not be passed on to descendants. Manipulation of somatic cells is subject to less ethical scrutiny than manipulation of germ cells.

Already 25 years ago, viruses were used as vectors to insert genes into mammalian cells. This genetically engineered virus was successfully used to prevent rabies in wildlife. Likewise, viruses were successfully used as vectors for mousepox viruses 25 years ago. This allowed vaccination of mice to be circumvented, which died shortly afterwards. The concept of stealth viruses is not new in nature. In this case, an initially unnoticed virus could enter human cells and wait for an external or internal signal. One related example are oncogenes, which are mutated genes that cause cancer as soon as they are activated. Some viruses have segments of DNA that mimic oncogenes. Other substances, bioregulators, physical processes, or external influences such as ultraviolet light could thus activate the virus. Ainscough also writes about host-swapping diseases and designer diseases. In the future of 2030, it could be possible to create the suitable pathogens for a certain disease pattern. This would make it possible, for example, to temporarily shut down the immune system or induce cell death in certain cells. Twenty years later, Ainscough's prognoses are all proving to be increasingly technically feasible.

Although some of the possible applications mentioned have not yet been achieved in practice, thanks to the aforementioned CRISPR (Clustered regularly interspaced short palindromic repeats - Cas9 gene-editing technology and the general progress in the field, it is only a matter of time before the biological weapons mentioned are successfully tested within military or civil dual-use research. Another extremely problematic aspect is that CRISPR is not a high-tech technology that is only available in secure laboratories. At the current rate, it is foreseeable that in the world of 2030, manipulated and synthetic biological substances could take on an almost everyday character. But how difficult is it really for future actors to actually develop and deploy one of these methods themselves?

It is possible that a biohacker could lose control of a potentially dangerous agent as a result of an accident, since generally weaker standards of safety are observed in amateur labs. In future, it will probably be even easier to circumvent the barriers as, for example, the aforementioned small flow reactors and CRISPR-Cas9 applications become widely marketable.

All in all, synthetic or DNA-engineered biological weapons can potentially cause enormous damage, but a closer look reveals that, at least for nonstate actors, production is currently not as easy as it might seem. By 2030, however, some of the current barriers are expected to be significantly lower. Although it is possible to learn the fundamentals via internet courses, in most cases a solid academic education is needed to gain practical experience with the laboratory equipment. Compared to genetically modified agents, existing natural pathogens may pose an even greater danger, as slightly less experience is required to weaponize them. There is also more publicly available research and potential natural source sites for such pathogens. In 2014, for example, a Tunisian jihadist did not even attempt to produce complicated pathogens, but instead records were found on their laptop of how the causative pathogen of plague (Yersinia pestis) can be isolated from infected animals and subsequently weaponized. The chemist and physicist would presumably have had the theoretical prerequisites for creating his own strain, but it seems the costs were too high compared to the benefits. He was caught without carrying out an attack.

It would also be relatively easy for nonstate actors to take advantage of a natural outbreak to infect themselves and then infect as many other people as possible. Breaking an imposed quarantine during a disease outbreak for political reasons could also be classified as terrorism, as people could be killed indirectly. Such intentions, as well as acting as a so-called superspreader, are entirely possible, as already described in the section on SARS-CoV-2. However, it is relatively difficult to deliberately infect oneself with a naturally occurring virus as the first carrier. Another comparatively simple biological weapon that could be used for attacks in the future is the mass breeding of insects. This can lead to effective attacks on crops, but as soon as the insects are to be used as vectors for diseases against humans, a greater effort might be required, although it might still be much less than that of producing a synthetic pathogen. The use of insectoid vectors proved to be very effective in the operations carried out by the Japanese during the Second World War. Other biological agents already used in the past, such as anthrax and ricin toxin, might also potentially be used in the future again. Currently, the Centers for Disease Control and Prevention lists more than 20 dangerous bioterrorism agents, which they subdivided into three categories.63

37.5 LIST OF SOME BIOLOGICAL WARFARE AGENTS

Some biological warfare agents include:

(i) **Anthrax:** A disease caused by the bacterium Bacillus anthracis that can be transmitted through contact with infected animals or contaminated animal products. Anthrax spores can survive in water or soil for up to 40 years.

(ii) **Botulism:** A disease caused by the toxin Clostridium botulinum that is often associated with consuming preserved foods. Botulinum toxins are a common biological weapon used by terrorists.

(iii) **Plague:** A disease characterized by high fevers, painful lymph nodes, and bacteremia.

(iv) **Smallpox:** A biological warfare agent.

(v) **Tularemia:** Also known as "rabbit fever", this disease is caused by the bacterium Francisella tularensis. Humans can become infected through tick bites, handling infected animals, or inhaling the bacteria.

(vi) **Viral hemorrhagic fevers:** A group of illnesses caused by several distinct families of viruses, including Ebola, Marburg, Lassa, Machupo, Bolivian hemorrhagic fever, Korean hemorrhagic fever, and Dengue hemorrhagic fever.

Biological agents that are considered Tier 1 present the greatest risk of deliberate misuse and have the potential for mass casualties. Risk Group 4 (RG4) agents are likely to cause serious or lethal human disease, and there are usually no preventive or therapeutic interventions available.

37.6 DEFENDING AGAINST BIOLOGICAL WEAPONS

To prevent international terrorists and rogue states from easily acquiring the agents for biological weapons from the United States (for example), the Commerce Department employs export controls. The amended Export Administration Act of 1979 prohibits the export, to certain proscribed countries, of any substance or technology that facilitates development or delivery of a biological weapon. Any company or individual that knowingly exports materials used for biological weapons to one of these countries faces severe civil and criminal penalties. The Chemical and Biological Weapons Control Act of 1991 also establishes an elaborate system of economic sanctions and export controls to curb the proliferation of biological weapons. This act was enacted in an attempt to curb the transfer of biological weapons to rogue nations.

Export control through export licensing is an essential element in the overall strategy to limit the spread of biological weapons. Yet export controls are difficult to administer, in part because of the dual nature of the agents and equipment that might be used for producing biological weapons. Before the Gulf War, the Department of Commerce had approved the shipment of cultures of B. anthracis to Iraq for peaceful biomedical and diagnostic purposes, but these cultures may subsequently have been misused by Iraq and incorporated into its biological weapons program. Modifications of existing export control systems will be needed to maximize their usefulness for limiting the threat of biological warfare and bioterrorism. Given that there are hundreds of culture collections around the world—many in nations with no export controls—the utility of export controls will be limited. Even if international accords establish uniform regulations for the transfer of dangerous pathogens, rogue arms dealers supplying these agents could remain a threat.

To curb the domestic acquisition of biological weapons, Congress passed the Biological Weapons Act of 1989 and the Anti-Terrorism Act of 1996. These statutes impose severe criminal penalties for the possession, manufacture, and use of biological weapons. They authorize the federal government to seize any pathogens or other materials used to develop a biological weapon or its delivery system. The federal government need only establish that someone is attempting to acquire a biological agent that can be used as a biological weapon with no apparent legitimate purpose. The 1996 Anti-Terrorism Act permits the federal government to act quickly to prevent the potential use of a biological weapon.

The Anti-Terrorism Act of 1996 also established a regulatory framework for controlling the distribution of agents that could be used as biological weapons. Congress directed the Centers for Disease Control and Prevention (CDC) to develop a regime to identify those biological agents that could pose a high threat to human health as bioweapons and to institute a procedure that controls the transfer and use of those agents. CDC's regulations governing hazardous biological agents went into effect 15 April 1997. CDC identified 30 infectious agents and 12 toxins (Table 6) and established procedures that must be followed for tracking the transfer of these agents. Some transfers, namely those involving some vaccine strains and toxins for medical use and those to diagnostic laboratories, were exempted from regulation. Some law enforcement officials have proposed limiting possession of dangerous pathogens to a few researchers and laboratories licensed to have them. Congress and CDC are still assessing whether additional oversight for possession of certain agents is necessary to enhance the protective network against domestic bioterrorism.

37.7 ABOUT BIOLOGICAL WARFARE IN INDIA

India is a signatory to the BWC of 1972. India has a well-developed biotechnology infrastructure that includes numerous pharmaceutical production facilities bio-containment laboratories (including BL-3) for working with lethal pathogens. It also has qualified scientists with expertise in infectious diseases. Some of India's facilities are being used to support research and development for BW defense purposes. These facilities constitute a substantial potential capability for offensive purposes as well.

The Defence Research and Development Establishment (DRDE) at Gwalior is the primary establishment for studies in toxicology and biochemical pharmacology and development of antibodies against several bacterial and viral agents. Work is in progress to prepare responses to threats like Anthrax, Brucellosis, cholera and plague, viral threats like smallpox and viral hemorrhage fever and bio-toxic threats like botulism. Researchers have developed chemical/biological protective gear, including masks, suits, detectors and suitable drugs.

CHAPTER 38

CHEMICAL WARFARE

38.1 INTRODUCTION TO CHEMICAL WARFARE

Chemical warfare (CW) involves using the toxic properties of chemical substances as weapons. This type of warfare is distinct from nuclear warfare, biological warfare and radiological warfare, which together make up CBRN, the military acronym for chemical, biological, radiological, and nuclear (warfare or weapons), all of which are considered "weapons of mass destruction" (WMDs), a term that contrasts with conventional weapons.

Chemical warfare is different from the use of conventional weapons or nuclear weapons because the destructive effects of chemical weapons are not primarily due to any explosive force. The offensive use of living organisms (such as anthrax) is considered biological warfare rather than chemical warfare; however, the use of nonliving toxic products produced by living organisms (e.g. toxins such as botulinum toxin, ricin, and saxitoxin) is considered chemical warfare under the provisions of the **Chemical Weapons Convention (CWC)**. Under this convention, any toxic chemical, regardless of its origin, is considered a chemical weapon unless it is used for purposes that are not prohibited (an important legal definition known as the General-Purpose Criterion).

The use of chemical weapons in international armed conflicts is prohibited under international humanitarian law by the 1925 Geneva Protocol and the Hague Conventions of 1899 and 1907. The 1993 Chemical Weapons Convention prohibits signatories from acquiring, stockpiling, developing, and using chemical weapons in all circumstances except for very limited purposes (research, medical, pharmaceutical or protective).

About 70 different chemicals have been used or were stockpiled as chemical warfare agents during the 20th century. The entire class, known as Lethal Unitary Chemical Agents and Munitions, has been scheduled for elimination by the CWC.

Under the convention, chemicals that are toxic enough to be used as chemical weapons, or that may be used to manufacture such chemicals, are divided into three groups according to their purpose and treatment:

Schedule 1 – Have few, if any, legitimate uses. These may only be produced or used for research, medical, pharmaceutical or protective purposes (i.e. testing of chemical weapons sensors and protective clothing). Examples include nerve agents, ricin, lewisite and mustard gas. Any production over 100 grams (3.5 oz) must be reported to the Organisation for the Prohibition of Chemical Weapons (OPCW) and a country can have a stockpile of no more than one tonne of these chemicals.

Schedule 2 – Have no large-scale industrial uses, but may have legitimate small-scale uses. Examples include dimethyl methyl-phosphonate, a precursor to sarin also used as a flame retardant, and thiodiglycol, a precursor chemical used in the manufacture of mustard gas but also widely used as a solvent in inks.

Schedule 3 – Have legitimate large-scale industrial uses. Examples include phosgene and chloropicrin. Both have been used as chemical weapons but phosgene is an important precursor in the manufacture of plastics, and chloropicrin is used as a fumigant. The OPCW must be notified of, and may inspect, any plant producing more than 30 tons per year.

Chemical weapons are divided into three categories:

Category 1 – based on Schedule 1 substances,

Category 2 – based on non-Schedule 1 substances,

Category 3 – devices and equipment designed to use chemical weapons, without the substances themselves.

38.2 CHEMICAL WEAPONS CONVENTION (CWC)

The Convention on the Prohibition of the Development, Production, Stockpiling and Use of Chemical Weapons and on their Destruction (the Chemical Weapons Convention or CWC), is comprised of a Preamble, 24 Articles, and 3 Annexes — the Annex on Chemicals, the Verification Annex, and the Confidentiality Annex.

The Convention aims to eliminate an entire category of weapons of mass destruction by prohibiting the development, production, acquisition, stockpiling, retention, transfer or use of chemical weapons by States Parties. States Parties, in turn, must take the steps necessary to enforce that prohibition in respect of persons (natural or legal) within their jurisdiction.

All States Parties have agreed to chemically disarm by destroying any stockpiles of chemical weapons they may hold and any facilities which produced them, as well as any chemical weapons they abandoned on the territory of other States Parties in the past. States Parties have also agreed to create a verification regime for certain toxic chemicals and their precursors (listed in Schedules 1, 2 and 3 in the Annex on Chemicals) in order to ensure that such chemicals are only used for purposes not prohibited under the Convention.

A unique feature of the Convention is its incorporation of the 'challenge inspection', whereby any State Party in doubt about another State Party's compliance can request a surprise inspection. Under the Convention's 'challenge inspection' procedure, States Parties have committed themselves to the principle of 'anytime, anywhere' inspections with no right of refusal.

Article I – General Obligations

Article I sets out the general obligations of each State Party under the Convention

Article II – Definitions and Criteria

Article II sets out the definitions and criteria to be used in implementing the Convention

Article III – Declarations

Article III requires each State Party to submit declarations to the OPCW within 30 days after the Convention enters into force for that particular State Party

Article IV – Chemical Weapons

Article IV sets out the requirements for States Parties to destroy their chemical weapons

Article V – Chemical Weapons Production Facilities

Article V relates to the requirement for States Parties to destroy and/or convert their Chemical Weapons Production Facilities (CWPFs)

Article VI – Activities Not Prohibited under this Convention

Article VI covers "activities not prohibited under this Convention", otherwise known as the non-proliferation or industry verification regime

Article VII – National Implementation Measures

Article VII covers national implementation of the Convention and requires each State Party to enact implementing legislation at the national level

Article VIII – The Organization

Article VIII establishes the OPCW as the implementing body of the Convention

Article IX – Consultations, Cooperation and Fact-Finding

Article IX provides for the consultation and clarification if concerns about possible non-compliance arise

Article X – Assistance and Protection against Chemical Weapons

Article X provides for assistance and protection to a State Party if it is attacked or threatened with attack by chemical weapons

Article XI – Economic and Technological Development

Article XI provides international cooperation for the economic and technological development of States Parties

Article XII – Measures to Redress a Situation and to Ensure Compliance, Including Sanctions

Article XII deals with measures to ensure compliance, including sanctions against a State Party that fails to uphold its treaty obligations

Article XIII – Relation to Other International Agreements

Article XIII deals with the relations with other international treaties

Article XIV – Settlement of Disputes

Article XIV deals with the settlement of disputes that may arise concerning the application or the interpretation of the Convention

Article XV – Amendments

Article XV deals with the amendments to the Convention

Article XVI – Duration and Withdrawal

Article XVI deals with the duration of the Convention and the States Parties' the right to withdraw from the Convention

Article XVII – Status of the Annexes

Article XVII relates to the status of the Convention's annexes

Article XVIII – Signature

Article XVIII deals with the signature of the Convention

Article XIX – Ratification

Article XIX related to the ratification of the Convention

Article XX – Accession

Article XX relates to the accession to the Convention

Article XXI – Entry into Force

Article XXI deals with the entry into force of the Convention

Article XXII – Reservations

Article XXII deals with reservations

Article XXIII – Depositary

Article XXIII relates to the Depositary of the Convention

Article XXIV – Authentic Texts

Article XXIV covers the authenticity of texts

38.3 WHICH COUNTRIES OUTLAW CHEMICAL WEAPONS?

(i) Syria

In 2015, the Conference of the States Parties to the Chemical Weapons Convention declared November 30 the Day of Remembrance for all Victims of Chemical Warfare. Although the underlying convention entered into force as early as April 29, 1997, attacks with chemical weapons have verifiably been carried out well into the 21st century.

One well-known example is the Syrian Civil War, where thousands of people were killed and injured due to the use of chemical weapons. The country acceded to the convention in September 2013, only a couple of weeks after a sarin gas attack in Damascus' Ghouta district. Reuters reports that a coalition of various rights groups and legal experts has now unveiled its plan to create a tribunal at The Hague able to persecute perpetrators of such attacks worldwide. Our map shows that as of 2018, almost every country in the world either signed and ratified or acceded to the convention.

Only four countries haven't ratified or signed the treaty yet. While Israel signed on to the convention in January 1993, it hasn't enshrined any of the measures in its national legislation at the time of writing. **Egypt, South Sudan and North Korea** aren't classified as states parties at all, which leads some spectators to believe these countries still possess illegal chemical weapons.

As for the 193 states and regions partially or fully passing legislation to adhere to the Chemical Weapons Convention, according to verified data by the Organisation for the Prohibition of Chemical Weapons, 100 percent of all declared stockpiles have been destroyed and all of the production plants for chemical weapons have either been dismantled or converted to a peaceful purpose.

The Convention on the Prohibition of the Development, Production, Stockpiling and Use of Chemical Weapons and on their Destruction or Chemical Weapons Convention is meant to be an addendum to the Geneva Convention, which already prohibits the use of biological and chemical weapons in war but not their production, possession or trade.

The convention text was drafted in September 1992. The treaty was open for signatures from January 13, 1993 to April 29, 1997. Key obligations for states parties of the conventions are not to "develop, produce, otherwise acquire, stockpile or retain chemical weapons, or transfer, directly or indirectly, chemical weapons to anyone", "to use chemical weapons" and "to assist, encourage or induce, in any way, anyone to engage in any activity prohibited to a State Party under this Convention."

38.4 LIST OF CHEMICAL WARFARE AGENTS

A chemical weapon agent (CWA), or chemical warfare agent, is a chemical substance whose toxic properties are meant to kill, injure or incapacitate human beings. **About 70 different chemicals** have been used or stockpiled as chemical weapon agents during the 20th century. These agents may be in liquid, gas or solid form.

In general, chemical weapon agents are organized into **several categories** (according to the physiological manner in which they affect the human body). They may also be divided by tactical purpose or chemical structure. The names and number of categories may vary slightly from source to source, but, in general, the different types of chemical warfare agents are listed below.

(i) Harassing agents

These are substances that are not intended to kill or injure. They are often referred to as Riot Control Agents (RCAs) and may be used by civilian police forces against criminals and rioters, or in the military for training purposes. These agents also have tactical utility to force combatants out of concealed or covered positions for conventional engagement, and preventing combatants from occupying contaminated terrain or operating weapons. In general, harassing agents are sensory irritants that have fleeting concentration dependent effects that resolve within minutes after removal. Casualty effects are not anticipated to exceed 24-hours nor require medical attention.

(ii) Tear agents

These sensory irritants produce immediate pain to the eyes and irritate mucous membranes (aka lachrymatory agent or lachrymator).

Benzyl chloride

Benzyl bromide

Bromoacetone (BA)

Bromobenzyl cyanide (CA)

Bromomethyl ethyl ketone

Capsaicin (OC)

Chloracetophenone (MACE; CN)

Chloromethyl chloroformate

Dibenzoxazepine (CR)

Ethyl iodoacetate

Ortho-chlorobenzylidene malononitrile (Super tear gas; CS)

Trichloromethyl chloroformate

Xylyl bromide

(iii) Vomiting agents

These sensory irritants are also termed sternators or nose irritants. They irritate the mucous membranes to produce congestion, coughing, sneezing, and eventually nausea.

Adamsite (DM)

Diphenylchloroarsine (DA)

Diphenylcyanoarsine (DC)

Cyanide.

(iv) Malodorants

These are compounds with a very strong and unpleasant smell, which produce powerfully aversive effects without the toxic effects of tear agents or vomiting agents.

(v) Incapacitating agents

These are substances that produce debilitating effects with limited probability of permanent injury or loss of life. The casualty effects typically last over 24 hours, and though medical evacuation and isolation is recommended, it is not required for complete recovery. These, together with harassing agents, are sometimes called nonlethal agents. There may be as high as 5% fatalities with the use of these agents.

(vi) Psychological agents

These are substances that produce casualty effects through mental disturbances such as delirium or hallucination.

3-Quinuclidinyl benzilate (BZ)

Phencyclidine (SN)

Lysergic acid diethylamide (LSD)

(vii) Other incapacitating agents

These substances have also been investigated as incapacitants, though they operate more through interactions outside the central nervous system.

KOLOKOL-1 (tranquilizer)

(viii) Lethal agents

These substances are for producing chemical casualties without regard to long-term consequences or loss of life. They cause injuries that require medical treatment.

Blister agents: A blister agent is a chemical compound that irritates and causes injury to the skin. These substances also attack the eyes, or any other tissue they contact.

Vesicants: The vesicants are substances that produce large fluid-filled blisters on the skin.

(ix) Nitrogen mustards

Bis(2-chloroethyl)ethylamine (HN1)

Bis(2-chloroethyl)methylamine (HN2)

Tris(2-chloroethyl)amine (HN3)

(x) Sulfur mustards

1,2-Bis(2-chloroethylthio) ethane (Sesquimustard; Q)

1,3-Bis(2-chloroethylthio)-n-propane

1,4-Bis(2-chloroethylthio)-n-butane

1,5-Bis(2-chloroethylthio)-n-pentane

2-Chloroethylchloromethylsulfide

Bis(2-chloroethyl) sulfide (Mustard gas; HD)

Bis(2-chloroethylthio) methane

Bis(2-chloroethylthiomethyl) ether

Bis(2-chloroethylthioethyl) ether (O Mustard; T)

(xi) Arsenicals

Ethyldichloroarsine (ED)

Methyldichloroarsine (MD)

Phenyldichloroarsine (PD)

2-Chlorovinyldichloroarsine (Lewisite; L)

(xii) Urticants

The urticants are substances that produce a painful wheal on the skin. These are sometimes termed skin necrotizers and are known as the most painful substances produced.

Phosgene oxime (CX)

(xiii) Blood agents

These substances are metabolic poisons that interfere with the life-sustaining processes of the blood.

Cyanogen chloride (CK)

Hydrogen cyanide (AC)

Arsine (SA)

(xiv) Choking agents

These substances are sometime referred to as pulmonary agent or lung irritants and cause injury to the lung-blood barrier resulting in Asphyxia.

Chlorine (CL)

Chloropicrin (PS)

Diphosgene (DP)

Phosgene (CG)

carbon monoxide

(xv) Nerve agents

Nerve agents are substances that disrupt the chemical communications through the nervous system. One mechanism of disruption, utilized by the G, GV, and V series of chemicals is caused by blocking the acetylcholinesterase, an enzyme that normally destroys and stops the activity of acetylcholine, a neurotransmitter. Poisoning by these nerve agents leads to an accumulation of acetylcholine at the nerve axon, producing a perpetual excited state in the nerve (e.g. constant muscle contraction). The eventual exhaustion of muscles leads to respiratory failure and death. A separate class of nerve agents are related to Tetrodotoxin, frequently abbreviated as TTX, is a potent neurotoxin with no known antidote. Tetrodotoxin blocks action potentials in nerves by binding to the voltage-gated, fast sodium channels in nerve cell membranes, essentially preventing any affected nerve cells from firing by blocking the channels used in the process.

G series

These are high volatility nerve agents that are typically used for a nonpersistent to semipersistent effect.

Tabun (GA)

Sarin (GB)

Soman (GD)

Cyclosarin (GF)

GV series

These agents have a volatility between the V and G agents and are typically used for a semi-persistent to persistent effect.

Novichok agents

GV (nerve agent)

V series

These agents have low volatility and are typically used for a persistent effect or liquid contact hazard.

VE

VG

VM

VX

T series

These agents are related to the puffer fish

Tetrodotoxin

Saxitoxin (TZ)

Other agents

Botulinum toxin

38.5 DEFENSE AGAINST CHEMICAL WEAPONS OR AGENTS

Tokyo subway attack of 1995

Workers cleaning a train car after members of AUM Shinrikyo released sarin in the Tokyo subway system, March 1995.

Until the 1990s, terrorists had rarely possessed or employed chemical weapons. However, several states that have sponsored terrorism have also possessed chemical weapons—**Libya, Iran, and Iraq**—and there is a concern that they and groups they sponsor might use chemical weapons in the future.

An example of an organization that learned to produce and use chemical weapons is the AUM Shinrikyo sect in Japan, members of which used sarin nerve agent to kill 12 people and injure more than 1,000 in a March 1995 chemical weapons attack inside the Tokyo subway system. Members of this same group had killed seven and injured more than 300 in a June 1994 attack in Matsumoto, Japan. They also assassinated one opponent using VX nerve agent in Osaka and injured another by the same means in Tokyo in early 1995. Finally, in May and July 1995, members of the AUM Shinrikyo used hydrogen cyanide in two follow-up strikes in the Tokyo subway that injured four persons. Altogether, the several attacks with three different types of chemical weapons killed 20 people, injured some 1,300, and sent more than 5,600 to the hospital for examination. Casualties would likely have been much higher had the Japanese police not intervened when they did.

Al-Qaeda leaders have shown an interest in acquiring and employing chemical weapons, as indicated by experiments testing the use of hydrogen cyanide on animals in al-Qaeda camps in Afghanistan prior to the September 11 attacks on the United States in 2001. In addition to other documents showing ongoing research on chemical weapons, al-Qaeda planned and then aborted a chemical attack on the New York City subway system in 2005. Furthermore, al-Qaeda of Mesopotamia (also known as al-Qaeda in Iraq) initiated chlorine attacks in Iraq in 2007. It is believed by some Western analysts that al-Qaeda leaders would not hesitate to use any chemical, biological, radiological, or nuclear weapons that they might acquire. For example, al-Qaeda of Mesopotamia openly issued a public invitation for Muslim chemists, biologists, and physicists to join their cause.

Unfortunately, a substantial amount of information on how to manufacture chemical weapons already exists in the public domain, particularly on the Internet, which is within reach of individuals and groups worldwide.

Since World War I the military organizations of all the great powers have acquired defensive equipment to cope with emerging offensive chemical weapons.

The first and most important line of defense against chemical agents is the individual protection provided by gas masks and protective clothing and the collective protection of combat vehicles and mobile or fixed shelters. Filters for masks and shelters contain specially treated activated charcoal, to remove vapors, and paper membranes or other materials, to remove particles. Such filters typically can reduce the concentration of chemical agents by a factor of at least 100,000. Masks can be donned in less than 10 seconds and can be worn for long periods, even in sleep.

Modern protective overgarments are made of fabric containing activated charcoal or other adsorptive forms of carbon. A complete suit typically weighs about 2 kg (4.4 pounds). The fabric can breathe and pass water-vapour perspiration. In warm weather, periods of heavy exertion in full protective gear would have to be limited in order to avoid heat stress. Also, removing such gear in a contaminated environment would raise the risk of becoming a casualty or fatality, and so gear must be removed within toxic-free shelters after following decontamination procedures at the shelter entrance.

Chemical detectors have been developed to help identify levels and places of contamination. These include chemically treated litmus paper used to determine the presence of chemical agents. Other sensors may include handheld assays, vehicles equipped with scoops and laboratory analysis tools, and both point and standoff sensors. Automatic field alarm systems are employed by some military forces to alert personnel to the presence of chemical agents.

Well-equipped troops are supplied with hypodermic needles filled with antidotes to be administered in the event of toxic poisoning from nerve agents. For example, atropine shots can be injected to fight the effects of nerve gas exposures, and different medicines are available to treat casualties.

A number of methods have been found useful in decontaminating areas and people covered with chemical agents, including spraying with super tropical bleach (chlorinated lime) or washing contaminated surfaces or garments with warm soapy water. The challenge is finding and using a decontamination solution that is strong enough to neutralize the chemical agent without damaging the equipment or harming the personnel.

In some military forces, modular field hospitals have been developed that are stocked with resuscitation devices for respiratory support and other necessary equipment, decontamination solutions, and staff trained to decontaminate chemical warfare casualties. Collective protective shelters, complete with filters for airflow systems, have been provided to shield personnel in an otherwise contaminated area. Such shelters can provide a toxic-free area for personnel to change clothes, get medical attention, sleep, and take care of bodily functions with less danger of exposure to lethal chemicals.

Chemical agents used against unprotected forces can cause high casualties, fear, and confusion. Thus, personnel facing adversaries equipped with chemical weapons must be trained to don individual protective equipment, seek cover in collective protection shelters, avoid contaminated areas, and rapidly decontaminate personnel and equipment that have been exposed. However, such measures, while necessary to protect against chemical attacks, may expose protected forces to greater casualties from conventional weapons fire and lead to a loss of conventional combat effectiveness. Indeed, exercises have shown that conventional combat effectiveness can be decreased by 25 percent or more for military forces compelled to operate in masks, protective overgarments, special gloves, and boots. This is especially true if temperatures are high and forces are required to stay sealed in their gear for many hours or days without relief. Prolonged wearing of individual protective equipment can lead to stress, fatigue, disorientation, confusion, frustration, and irritability. Also, heat can build up and lead to dehydration. Thus, there is generally a trade-off between protecting one's force through chemical-protection gear and maintaining conventional fighting effectiveness.

While most military forces have at least some defense against chemical attack, this is not the case for most civilian populations, which typically have no individual protective equipment (masks, overgarments, boots, or gloves) or collective protection shelters. One notable exception is Israel, which has been at war numerous times since its independence in 1948. Israeli citizens are assigned their own gas masks, and new buildings in Israel must contain a reinforced shelter. Israel also conducts civil defense exercises on a regular basis in order to prepare its citizens for attack.

A further problem for almost every country is the presence in most urban centres of storage or manufacturing facilities that contain toxic industrial chemicals and other toxic materials. A conventional attack on such a site would be the functional equivalent of a chemical weapons attack. Most countries do not have adequate security around such areas.

One response to the threat of a chemical weapons attack on civilian society has been the creation of active, well-trained emergency response teams that know how to identify chemical agents, decontaminate areas and people exposed to chemical weapons, and coordinate rescue operations. Cognizant of the growing risk posed by weapons of mass destruction (WMD), the United States in 1998 authorized the creation of 10 National GuardWMD Civil Support Teams (WMD-CST) within its territory; each team was organized, trained, and equipped to handle chemical emergencies in support of local police, firefighters, medical personnel, and other first responders. In subsequent years, dozens of new WMD-CST were authorized, with plans for eventually certifying units for every state and some U.S. protectorates. In addition, the U.S. Centers for Disease Control and Prevention maintains the Strategic National Stockpile, which contains medical supplies and equipment positioned around the country to provide medical help in emergencies, including a chemical weapons attack.

CHAPTER 39

NUCLEAR WAR, ITS LIKELIHOOD, AND DETERRENCE

"We are now in a world where we're facing multiple nuclear competitors, multiple states that are growing, diversifying and modernizing their nuclear arsenals and also, unfortunately, prioritizing the role that nuclear weapons play in their national security strategies," said Richard C. Johnson, the Deputy Assistant Secretary of Defense for Nuclear and Countering Weapons of Mass Destruction Policy (N-CWMD).

As the security environment evolves, adjustments to the 2022 Nuclear Posture Review may be required to sustain the ability to achieve **nuclear deterrence**, in the light of enhanced nuclear capabilities of China and Russia and possible lack of nuclear arms control agreements after February.

39.1 NUCLEAR WEAPONS IN THE 21ST CENTURY, LIKELIHOOD AND DETERRENCE

The aggressive war that the Russian Federation is waging in Ukraine has rehashed fears about nuclear weapons 75 years since the first atomic bomb was used.

The Ukraine war is the first conventional war fought since WWII In Europe under the shadow of nuclear threats. The Russian Federation has the largest nuclear arsenal globally, and its nuclear doctrine envisages the possibility of using nuclear weapons first against enemies preparing to invade the country with conventional capabilities.

The global nuclear order had been in trouble for quite some time before the Ukraine war began. Since 2006, the North Korean regime has steadily advanced its nuclear capabilities to strike within the United States territory. China has been working on expanding its arsenal and dominating in critical technology fields, including space and artificial intelligence. And doubts linger on the military dimension of the Iran nuclear program. But the Russian threat over nuclear weapons in Ukraine underscores a much more severe and entrenched problem: how fragile the global nuclear balance is.

For decades, the best way to manage potential nuclear escalations has been by seeking to design **nuclear deterrence postures** that were credible to discourage the adversary from undertaking a surprise attack but equally intentional and cautious so as not to trigger inadvertent escalations.

The urgency of nuclear threats today raise an inevitable question: Is nuclear deterrence still appropriate in the 21st century?

On May 10, 2024, the Belfer Center Project on Managing the Atom launched the Research Network on Rethinking Nuclear Deterrence (RNDn) to address two fundamental and interrelated questions:

- Given the deteriorating geopolitical and security landscape, how do we make sure nuclear deterrence does not fail? And,
- What alternatives can replace nuclear deterrence, and what conditions should exist to materialize them?

There is much at stake if nuclear deterrence fails. We are determined not to let it happen.

The nuclear deterrence theories that continue to dominate and inform strategic thinking today were wholly designed to manage the risks of a bipolar nuclear order consisting of two blocs of internally homogeneous and ideologically distinct states. These theories were also conceived in isolation from considerations of a broader nature, including the quest for human security, social justice, inclusion, and equality.

Three-quarters of a century after the first nuclear detonations, nuclear weapons and the threat of nuclear conflict continue to cast a long shadow on global affairs. **But the current nuclear age is radically different than the era in which nuclear deterrence theories were originally developed.** It consists of many more players bound together by complex economic and technological interdependencies. And it interplays with other global challenges—climate change, social justice issues, violent nationalism, global pandemics in particular—of unprecedented scale.

Since the end of the Cold War, legitimate questions have been raised about how universally appropriate deterrence is as the answer to existential security risks in a world of multiple state and potentially non-state nuclear actors, extensive nuclear entanglement with non-nuclear technologies and military arrangements, terrorist groups with nuclear ambitions, and hybrid/gray zone attack that seeks to defeat or disable the adversary without necessarily going to war.

Since 2011, a coalition of non-nuclear weapons states launched a powerful initiative to explore the humanitarian consequences of nuclear weapons, claiming that any use of nuclear weapons anywhere in the globe will have catastrophic consequences on the food chain and water supply, agricultural and health infrastructures of all. The initiative later culminated in the adoption of the Treaty on the Prohibition of nuclear weapons that entered into force last February 2021. It is essential to point out that none of the nuclear weapons states or their allies participated in any negotiating sessions.

The divide has been growing between countries relying for their security on nuclear weapons and those who believe that security can only be guaranteed through a world free of nuclear weapons.

Nuclear deterrence is not a condition that you meet or a status you reach. Instead, it is a relationship between two or more nuclear states and a tool employed by states to achieve specific goals. Accordingly, it is subject to changing strategic considerations and global circumstances. It is dynamic and ever-evolving.

The concept of deterrence has deep historical roots and, thanks to the nuclear age, has been deeply developed and entrenched, especially in the existing nuclear-weapon states. **Deterrence, in broad terms, is about influencing decision-making through the threat of force to discourage an opponent from taking an undesired action. This can be achieved through the threat of retaliation (deterrence by punishment) or by denying the opponent's aims (deterrence by denial), or both.**

The concept of nuclear deterrence—equally theoretically parsimonious and morally perplexing—has been historically contested and highly disputed. States rely on nuclear deterrence when, to intimidate an enemy, they threaten the use of weapons of mass and indiscriminate destructive power—nuclear weapons—and potentially on a scale that could produce catastrophic global carnage.

During the period of intense Soviet-U.S. rivalry, some questioned whether the benefits of nuclear deterrence were worth the terrifying risk that nuclear weapons could be launched, either intentionally or by accident. On balance, however, the recent memory of a catastrophic great power war, within an international system marked by a bitter ideological clash, deep mistrust, and intense security competition made the possibility that nuclear deterrence could provide stability and decrease if not eliminate the prospect of total war appealing.

Contestation to the concept of nuclear deterrence also led to the creation of powerful schools of thought regarding nuclear weapons. The first urged the United States to seek to maintain a qualitative technological edge and to prepare for nuclear warfighting. The second advocated for a more restrained nuclear posture and bilateral and multilateral agreements to constrain dangerous forms of nuclear competition. The third argued for complete nuclear disarmament given human fallibility, the catastrophic impacts of nuclear weapons, and the precarity of nuclear deterrence.

Even though unsatisfactory and at least debatable, nuclear deterrence offered a way to navigate through dangerous nuclear war inclinations and uncompromising disarmament aspirations. And however frightening and risky, in fact, nuclear deterrence had, according to many policymakers and scholars, some positive effects in reducing the risk of war during the Cold War. Relying on nuclear deterrence to maintain peace is like skating on thin ice. The fact that you have done it before does not mean you should expect to be able to do it safely forever.

39.2 NUCLEAR DETERRENCE

Nuclear deterrence refers to a principle in international relations where the retaliatory potential and destructive force of nuclear weapons prevents nations from launching a nuclear attack. However, there are questions as to whether nuclear deterrence is sufficient, effective, just, or ethical.

China is dramatically expanding its nuclear arsenal, and North Korea now boasts of the "accelerating development" of land-based tactical nuclear weapons. The rising risk of limited nuclear attacks makes up an important element of the future threat from China and North Korea. Meanwhile, Russia, with its stockpile of more than 4,300 nuclear warheads, continues its nuclear saber-rattling over Ukraine, while Iran stands on the cusp of building its own nuclear weapons.

The difference is stark. While the United States has sought to reduce its nuclear weapons stockpile, China, North Korea, and Russia are accelerating strategic weapons development and doctrine for employment at the operational level of war.

This shortfall will not be resolved by the modernization of US Air Force and Navy nuclear forces alone. The US Army, in particular, requires much greater focus in this area, though it has made some progress. As part of its broader pivot to large-scale combat operations, the Army has taken recent steps to increase conventional-nuclear integration, for example. But overall, the US Army is not adapting quickly enough to the real and growing threat of nuclear weapons use on a battlefield. The US Army's ability to fight and win on a battlefield threatened by an adversary's use of strategic and nuclear weapons must catch up to current and near-future threats.

It won't be easy. The Army must overcome a generation of neglect toward its readiness to counter such weapons of mass destruction. This neglect began after the Presidential Nuclear Initiatives terminated ground-launched nuclear artillery in the 1990s, and it accelerated after the 9/11 attacks and the Army's subsequent divorce from large-scale combat operations.

Most important, US military and defense thinkers must overcome two major misconceptions that for too long have shaped US actions, delaying both support for integrating conventional maneuver and US readiness to counter weapons of mass destruction. The first misconception is that the Army developing—and training based on—nuclear doctrine is itself provocative to the extent that it should be curtailed. The second misconception, related to the first, is that the likelihood of nuclear use on the battlefield by either the United States or its adversaries is so remote as to be unthinkable. In reality, both of these misconceptions imperil the Army's readiness to wage all-domain, full-spectrum large-scale combat operations in the current environment.

(i) Misconception one: Countering weapons of mass destruction readiness is provocative

Proponents of this misconception argue that because the consequences of nuclear warfare are potentially catastrophic, the prudent course is to avoid actions that provoke nuclear-armed adversaries or encourage employment strategies sooner in warfare. This view reveals that US military thinkers tend to assess escalation dangers more pointedly than force benefits. In a 2020 evaluation, Center for Strategic and International Studies researchers noted, "Policymakers were highly attuned to the escalatory risk. . . often weighing their concerns about the potential provocational risks to be more important than the. . . benefit that capabilities may provide." Some think tank recommendations for military planning argue for less provocative means of establishing joint readiness. In some instances, the provocation argument further leads to pacifist calls for unilateral disarmament.

This action-reaction dynamic—that if we only don't provoke, we can reduce or contain risk—has not held true for adversary nuclear expansion and alliance building. The adversary does what it wants when it wants to provide advantage as it views strategic stability. The United States must be able to do the same. One's provocation is another's defensive readiness. For example, the Russian and Chinese nuclear bomber flight in July near Alaska was viewed as provocative by US officials. Russian experts claimed that they viewed their actions merely "as part of. . . military cooperation plan for 2024" and a response to a US B-52 flight in Finland.

If the provocation logic prevails, the ground force cedes maneuver advantage because it does not develop a counter to a weapon the adversary possesses, has employment doctrine for, and trains to have survivability and resiliency from. A ground force that cannot maneuver is one that can neither seize terrain nor force objectives. It is, in other words, one that cannot win battles. Ceding maneuver is a sure path to battlefield defeat.

(ii) Misconception two: Nuclear use is unthinkable

Another perspective, and a plank of US policy, states that nuclear war cannot be won and must never be fought. As a broad statement of hope and as a diplomatic position, this is noble. Others posit that preparing for doomsday strategies is wasteful or that nuclear weapons are unlikely to be used in any near-future conflict. With this logic, the highest priority for the land component of the US armed forces is to assure conventional superiority.

Is this realistic? North Korea's declared strategy and priority military investment is in nuclear weapons. Is there a credible scenario in which the regime doesn't employ nuclear (and chemical) weapons if it feels existentially threatened? Or take China, which isn't revealing its nuclear end game. Is it parity or something more? In a possible major war with China, how confident can the United States be, as the blood of many thousands is being spilled and national treasuries drained, that nuclear weapons won't come to the forefront? Russia, too, threatens nuclear weapons use if adversaries cross red lines that impact its ability to achieve its objectives.

It is hard to believe large-scale combat operations against a nuclear-armed adversary do not go nuclear if victory is being sought. The calculus that the United States employed in August 1945 to win the war with Imperial Japan will be the same that combatants use today. Admittedly, one other course is possible, and it is the outcome that the United States accepted in the Korean War. The decision then was not to employ nuclear weapons, and thereby not to win the war. Not winning may not be a viable option in the next major state-versus-state war.

39.2.1 The time for action is now

If these misconceptions persist, the United States is heading toward unilateral vulnerability. This outcome is the least viable, most dangerous way to mitigate the use of nuclear weapons at the operational level of war. If provocation is the United States' concern, it is on the wrong side of the escalation curve. If the United States believes that nuclear weapons are tools too terrible to employ, then it must think more like its adversaries, who express that they don't fear the use of nuclear weapons—and build arsenals to back up their claims.

Now is time for the US Army to shatter these misconceptions and develop forward-looking institutional strategies and operational plans. Army preparation to survive nuclear attack and remain on the offensive constitutes a necessary defensive reaction to adversary capabilities. It is certainly possible, and almost likely, that a land force will be required to operate in a nuclear environment against adversaries who possess nuclear weapons. Against the most formidable state adversaries outlined by US leaders, Russia and China each possess large and extensive landpower capabilities, including flexible and intermingled nuclear capabilities.

It will take dedicated commitment over time for the US Army to address this issue in full, but three steps should be taken as soon as possible.

39.3 NUCLEAR WARS CANNOT BE WON: AN ARGUMENT FOR STRATEGIC DETERRENCE

The increasing polarization of international politics indicates that the future will be characterized by an intensification of conflicts. And, as the tensions intensify between the world's two largest nuclear weapons states (NWSs), Russia and the US, and potential nuclear near peer, China, the underpinning principle of the nuclear revolution theory will only grow in salience: nuclear wars cannot be won, and therefore should not be fought. However, the return of great power competition has also galvanized sceptics of the nuclear revolution theory, who reject the logic and tenability of nuclear deterrence and advocate instead for nuclear superiority and a shift toward war-winning nuclear postures.

An understanding of nuclear deterrence, as engendered by the nuclear revolution theory, is critical in preserving peace and strategic stability. States may choose to pursue superiority, but the nuclear revolution theory can more adeptly provide the requisite insights for policymakers and scholars alike to better navigate the challenges inherent in the return of great power competition between China, Russia and the US.

Nuclear deterrence can be defined as the threat of nuclear retaliation against an adversary for an attack on a state's vital interests, thereby imposing costs that would significantly outweigh any potential gains. The material reality of nuclear weapons, that is their immense destructive potential, informs the principle that nuclear wars cannot be won. Bernard Brodie stated in 1946 that 'the factor of increase of destructive efficiency is so great that there arises at once the strong presumption that the experience of the past concerning eventual adjustment might just as well be thrown out the window.'

The significance and veracity of mutual assured destruction (MAD) increases when two or more nuclear adversaries possess mutual second-strike capabilities. Additionally, when nuclear adversaries possess larger, qualitatively robust nuclear forces, the reality and risk inherent in MAD intensify. In other words, pre-nuclear-age major power war strategies became obsolete with the advent of nuclear weapons. Whereas once, superiority in the size and breadth of forces had a decisive effect on results of warfighting, in the nuclear age, the size of an NWS's forces is of less significance. Instead, the efficacy of deterrence rests in the credibility of the threat of retaliation, as opposed to the size and breadth of a deterring actor's forces.

Hence, nuclear superiority is of secondary importance with regards to nuclear deterrence, as even states with small nuclear forces can inspire restraint in larger NWSs. Because all it takes to cause unacceptable levels of damage is for one nuclear weapon to penetrate a state's defenses, the utility of nuclear weapons is limited. As Susan Martin puts it: 'nuclear warfighting is not a strategy for survival'. This irrefutable reality will remain important in understanding the geopolitical dynamics of the future. Indeed, this outlook is collectively echoed by the five possessor states.

A key concept in this logic is nuclear danger, which argues that the risk of nuclear war, rather than the nuclear balance of forces, reinforces the value of deterrence: since having more nuclear weapons or more nuclear options than the adversary cannot provide much assistance in terminating war, this posture should not provide a great peacetime advantage. This dynamic was observable in South Asia, when Pakistani forces crossed the Line of Control into India-controlled parts of contested Kashmir in 1999 (**commonly known as the Kargil War**). Yet, instead of waging all-out war, **India restrained from retaliating with nuclear force despite its advantage over Pakistan.** Concomitantly, while Pakistan also possessed the capability to credibly retaliate with nuclear force, the Kargil War eventually de-escalated, and a reversion to crisis stability was achieved.

Because nuclear war would yield no victors, nuclear powers historically loathe to wage direct war against each other. The Cold War provides compelling evidence for this assertion. The Cuban Missile Crisis, arguably the closest the US and USSR came to fighting a nuclear war, eventually ended in compromise between the superpowers. The resulting decision to undertake confidence-building measures, such as establishing a direct hotline between the states' leaders, is emblematic of the aversion both superpowers had toward risking nuclear war. The rising tensions in East Asia bring the concept of nuclear danger and the irrelevance of nuclear superiority into sharp focus. **China's nuclear capability is roughly 10 times smaller than the US's, yet the nuclear deterrence relationship between them remains stable.**

The Cuban Missile Crisis, the Kargil War, and an understanding of nuclear revolution theorists' arguments regarding the material reality of nuclear weapons reinforce the idea that superiority is not required for nuclear weapons to have a deterrent effect, and is also not effective at compelling the opponent to yield.

Indeed, superiority is not the decisive factor in crisis de-escalation; rather, it is the possibility of nuclear retaliation even by the weaker opponent that inspires restraint. That over 50 nuclear threats were made in the 20 years following the end of the Cold War, with none of them resulting in all-out war, is testament to the deterrent value of nuclear weapons and to the restraint the nuclear revolution inspires. This pattern suggests that the next 20 years could be no different, if superiority is not pursued. However, if superiority is asymmetrically pursued by an NWS, this would not give them an advantage, instead precipitating an inconclusive arms race, intensifying tensions and plunging the global security landscape into an indefinite state of danger and instability.

Nuclear weapons have served most effectively as a deterrent, preventing nuclear powers from engaging one another in direct conflict. Nuclear weapons dissuade states from going to war. The historical evidence demonstrates that nuclear deterrence has facilitated peace and stability. As China and Russia strive to challenge the US for global predominance, the underpinning principle of the nuclear revolution theory will become ever more important: nuclear wars cannot be won, and therefore should not be fought.

CHAPTER 40

ENVIRONMENTAL WARFARE OR ECO-TERRORISM

40.1 DEFINITION OF ENVIRONMENTAL WARFARE AND ECO-TERRORISM

Environmental Warfare and Eco-Terrorism is defined as the direct manipulation or destruction of ecological resources as either a political threat or for actual military advantage.

Environmental warfare means waging warfare by means of deliberate environmental destruction or alteration, in order to repel enemy assault, as well as to hinder, hamper or injure the opponent. Environmental welfare operations combine hydrogeological, physical, and chemical processes. The design of the environmental warfare operation process starts with selecting the damage effects required. then selecting suitable chemical agents or microbiological agents for these applications. then selecting a method for introduction through water, soil, food, or air.

Environmental warfare is a war whose goal is for generating partial health damage for the enemy country/group. Environmental warfare was known a long time ago.

Some of the steps in the Environmental warfare are:

- Estimating the damage effects required from this operation of environmental warfare.
- Choosing chemical or biological compounds and environmental carriers for these compounds.
- Estimating the time required for this process and the time required for damage effects.

Environmental warfare is not visible to many of us. but it is classified as cooled chemical and microbiological warfare. The main target of any war is to defeat the enemy. Environmental warfare is a serious threat in this century.

Ecoterrorism, destruction, or the threat of destruction, of the environment is by states, groups, or individuals in order to intimidate or to coerce governments or civilians. The term also has been applied to a variety of crimes committed against companies or government agencies and intended to prevent or to interfere with activities allegedly harmful to the environment.

Ecoterrorism has been practiced by groups engaged in "anti-system" violence (i.e., violence against existing political structures). This kind of terrorism, also known as bioterrorism, includes, for example, threats to contaminate water supplies or to destroy or disable energy utilities, as well as practices such as the deployment of anthrax or other biological agents.

Nevertheless, such destruction has occurred with some regularity. In the 1960s and '70s the U.S. military used the **defoliant Agent Orange** to destroy forest cover in Vietnam, and in 1991 Iraqi military forces retreating during the Persian Gulf War set fire to **Kuwaiti oil wells**, causing significant **environmental damage**. The Rome Statute of the International Criminal Court, adopted in 1998, defines such modification or destruction as a **war crime**.

40.2 USE OF ENVIRONMENTAL WARFARE: THE KUWAITI OIL FIRES

The use of environmental warfare is not unheard of in history. Neither is the destruction of oil facilities. During World War II, an intensive American airstrike focused on the Ploiești petrol perimeter in Romania for months in 1944, in order to deprive Germany of its ally´s highly needed oil resources. Approximately 25% of all ships that the US lost during World War II were oil tankers, releasing about 5.5 million cubic meters of oil into the ocean.

During the Iran-Iraq War, both parties reciprocally destroyed oil facilities. Nevertheless, due to its military lack of necessity and thus tactical irrelevance and unprecedented objective and impact, the Gulf War remains unique in comparison to previous cases of using environmental warfare and especially attacking oil facilities. After invading Kuwait on 2 August 1990, Iraq threatened to destroy the Kuwaiti oil wells if the US and allied forces would force Iraq to retreat from Kuwait. The United Nations reacted immediately, the Security Council adopting multiple resolutions in order to condemn the invasion, sanction Iraq for its actions and enable military action against Iraq. Since Kuwait has a high number of oil wells (ca 1270), apart from oil storage depots, oil refineries and petrochemical factories, the possibility of a global environmental crisis was predicted already in November 1991 during the Second World Climatic Conference in Geneva by King Hussein of Jordan, who, primarily used the report made by his science advisor, Dr. Abdullah Toukan. Dr. Toukan estimated that the fires´ emissions of soot (term used by Dr. Toukan for the particles of unburned organic matter given off in the fire) and hazardous gases would have devastating and maybe far-reaching effects, disrupting the normal weather patterns in Asia, especially the monsoon rains, on which hundreds of millions of people are dependent. A "petroleum winter effect" was feared as well by some ecologists, but with the help of mathematical models, this theory was dismissed.

An oil well is made of a pipeline that connects the surface with the oil producing stratum underneath, extending for several thousand meters. A wellhead is usually made of steel and built to fit and form a seal at the end of this pipeline in order to prevent oil or gas from blowing or leaking at the surface. The nature of the damage made to the wellheads, pressure inside the well, exit velocity of the jet and heat radiation determine the extent of smoke plume height and its lateral spread. Emissions from the oil fires took two major forms: the liquid one due to the damaged oil wells, pipelines, storage tanks and oil tankers (disrupting the marine and terrestrial environment) and as smoke from the oil fires as soot and other gases (affecting the atmosphere and later the terrestrial and marine environment as fallout). Iraq already started using explosives on wellheads and tried out their effectiveness on 6 wells in December 1990.

It is estimated that during the allied forces´ airstrike in January 1991, 60 wells were blown up by Iraqis and 34 wells (loaded with explosives) by the allied troops. The use of environmental warfare by the allied troops is often neglected and a complete depiction of the facts would only be possible by revealing the allies´ implication in the environmental damage done during the Gulf War, even if their acts fade before the Iraqi troops´ actions. Iraq packed wellheads with plastic explosives and linked them with electrical and mechanical detonators after the tryouts in December 1990. Finally, before the conclusion of the Gulf War (in February 1991), as a result of Saddam Hussein´s scorched earth policy, 730 wells exploded (from the 800 wells detonated with explosives that were ignited by Iraq), out of which 656 were in flames for several months and 74 gushed oil, forming lakes. It is estimated that because the explosive charges on 76 wellheads were unable to set them ablaze, these gushed crude oil without fire, forming more than 100 oil lakes covering an area of 16 km^2 Additionally, the accumulation of unburned oil in low-lying areas led to the formation of 39 oil lakes covering an area of 3,08 km^2. All these pools and lakes contained 24-40 million barrels of oil. A total amount of 1-1,5 billion barrels of crude oil was lost during the oil fires. However, all these estimates vary depending on the consulted sources.

Apart from the oil leaks, the above-mentioned smoke plumes played a significant role in heavily polluting the atmosphere and consequently the marine and the terrestrial environment, as proven by empirical data. Characteristics of the smoke plume have a direct impact on the severity of the environmental damage. The location of the rupture and the extent of the damage to the wellhead determined the extent of the smoke cloud and coverage in the Kuwaiti oil fields. Oil emerged as a high-pressure jet in the form of smoke from wells under high pressure, resulting in smoke emitting vertically upward in 40% of the cases. In the cases where the wellheads were completely blown out, the smoke cloud formation was much closer to the surface, with a few cases of smoke jet escaping horizontally. Wells with wellheads covered with coke emitted black smoke due to the incomplete combustion process. On the other hand, pool fires were a source of black smoke as well. This plume was composed of smoke levels of ca. 500 $\mu g/m^3$, according to the analysis of a UK meteorology flight in 1992, being characterized by a large number of carbon chain agglomerates, commonly known as soot. Apart from soot there were various combustion gases included.

The concentration of the smoke particles decreased when further from the fire source, with the proportion of aggregates increasing with the age of the plume. Additionally, the soot/smoke emission was dependent on the type of fire, black smoke plumes containing higher amounts of soot/smoke particles, white smoke plumes being richer in salts and lake fires producing more sooty masses than oil well fires. Meteorological conditions in the region are highly relevant, because the rise of these plumes was dependent on the wind speed and atmospheric stability conditions.

The most difficult task was to extinguish all these fires, to cap the oil wells and to restore the damaged infrastructure. Extinguishing oil well fires has always been extremely problematic, as previous examples in history show (for example during the Iran-Iraq conflict, when two oil wells and one gas well were capped after 3 years of continuous fire). Because extinguishing and capping the sabotaged wells proportionally decreases the emissions, the duration of this process is in a strict correlation with the amount of lost oil and gas. In the case of the Kuwaiti oil wells, this process was even more challenging (at least in the beginning) due to the lack of water to fight fires, near destruction of the Kuwaiti infrastructure, delay in transporting necessary equipment from North America, presence of numerous unexploded ordnance in the fields, restricted access due to oil pool formation and delays in receiving necessary expenditure authorizations from the Kuwaiti government. However, by October 1991, 730 wells were extinguished. The last fire was extinguished in November 1991, leaving ca. 300 oil lakes behind and a layer of soot and oil that after falling from the sky and mixed with sand and gravel formed the so-called "tarcrete" on about 5% of Kuwait´s territory. The capping of the oil wells was a very slow and difficult procedure due to the above-mentioned challenges, so that at the end of July 1991, only 15% of the burning wells were capped, especially because in the beginning, the focus was on low pressured wells and gushing wells without fire. These results were achieved by an international team of workers, who extinguished the fires, capped the wells and reconstructed oil fields.97 Even if these workers´ speed and efficiency surprised everyone,98 the emissions of oil and smoke still massively disrupted the environmental balance in the region.

40.3 ENVIRONMENTAL CONSEQUENCES OF OIL FIRES

As previously explained, the Kuwaiti oil well fires emitted immense quantities of oil as well as smoke. These fires mainly produced solid particles and gases. The massive amounts of inhalable particles and noxious gases that appeared from the exposure of oil to the natural environment are all potentially harmful for human health and vegetation growth: sulphur dioxide, carbon monoxide, carbon dioxide, hydrogen sulphide, oxides of nitrogen and particulate matter containing partially burned hydrocarbons and metals. The combined oil and gas emissions equaled the consumption of ca. 4.6 million barrels of oil per day. Initial predictions about an intensified global warming due to the high emissions of CO_2 turned out to be false since a total amount of 133 million tons of CO_2 resulted from the oil well fires, which is the equivalent of 1.5% of the global annual emissions from fossil fuels and biomass burning and could therefore only slightly influence global warming on a short-term basis. The burning oil wells emitted daily about 20 Gg of sulphur dioxide, 1.5 Gg of particulate matter, 250 Mg of carbon monoxide and 500 Mg of oxides of nitrogen to the atmosphere. High amounts of sulphur (which, absorbed through breathing, produces increased air resistance in the lungs and their severe impairment when exposed to high doses) were expected to be released from the fires, especially because Kuwaiti crude oil has a relatively high sulphur concentration. In the end, it is estimated that 2.3 million tonnes of SO_2 were emitted. However, concentrations of SO_2 remained in Kuwait, Saudi Arabia, Bahrain and other nearby locations below the air quality guidelines.110 There was a relatively rapid removal of SO_2 in the smoke from the fires (within hours), oxidation to SO_4 being seen as a plausible cause, so that SO_2 levels in the upper atmosphere had no global impact, but may have increased acidification on a local scale. On the ground level, however, the SO_2 concentration was high near the fires, exceeding the air quality guidelines (and therefore affecting plant growth). About 80,000 metric tonnes of oxides of nitrogen (NOx) were emitted (NOx can produce inactive forms of haemoglobin, ultimately reducing the capability to carry oxygen to body tissues), but were present consistently below the proposed air quality limit and rapidly removed, hence having insignificant effects on the environmental conditions in Kuwait and other Gulf countries. However, NOx was involved in depleting the ozone in the vicinity of the fire sources. Hydrocarbons represent a major proportion of crude oil and their emission rate is dependent on the combustion efficiency because in the case of complete combustion hydrocarbons are turned to water and CO_2 and during incomplete combustion they are partially released in their initial form or as an intermediary combustion product. The combustion efficiency in Kuwait was generally over 83%, which would however not exclude the emission of volatile and non-volatile organic hydrocarbons. Nonetheless, the levels of nonmethane hydrocarbons (NMHC) were comparable with clean to moderately polluted atmosphere and the emission of methane is thought to be insignificant for the local, regional or global environment.

Conclusively, although the concentration of gaseous pollutants was usually below the permitted values, it was significantly increased during the oil fires compared to previous levels: there was an increase in SO_2 of 71%, in NO of 98%, in NO_2 of 25% and in non-methane hydrocarbon of 200% during the oil fires.

The smoke´s solid particulates included inorganic salt crystals, organic chemicals and soot particles. Although it was initially prognosticated that due to the large emissions of soot (initial estimates of 50,000 tons per day) a high amount of sun radiation would be absorbed and thus the regional and even global climate would change, ulterior detailed airborne measurements showed that only 0.45% (soot emission factor) of the mass of fuel burned turned into fine particle soot, emitting 3,400 tons of soot per day. The cause was the efficient combustion, with 96% of the carbon burned emitted as CO_2 and with very low emissions of CO and soot. The estimates concerning the soot emission factor significantly vary, but all reported values that are based on actual and/or airborne measurements report soot emission factors of less than 3%. Based on a factor of 2%, it was estimated that emissions of about 2.5 million tons of soot/smoke, assuming that 1.12 billion barrels burned with a combustion rate of 90% in the Kuwaiti oil fires. This amount was enough to lead to an absorption of over 90% of solar radiation in the thickest part of the smoke plume and about 75-80% in the super-composite part. The decrease in solar radiation from March to May 1991 was caused by the smoke cover from the Kuwaiti fires, in June to August 1991 by the combined effects of smoke and dust and in September the difference was caused mainly by dust. Solar radiation reduction was registered in Kuwait, the northeastern part of Saudi Arabia and Bahrain from March to August 1991. Increases in particulate and smoke density led to a significant decrease in temperature: the temperature dropped by 3-4°C at midday during smoky days in Kuwait, on average by 10°C between February and October 1991 in Kuwait and northern Saudi-Arabia, by 5-8°C about 250 km south and southwest of the fires and by 1-2°C 750 km away from the fires. The solid particulates found in the smoke include heavy metals like nickel, small amounts of vanadium, iron and trace quantities of aluminium, beryllium, cadmium, calcium, chromium, arsenic, silicon, zinc and lead, all of which are highly hazardous for the human health, because their inhalation can induce chronic health effects. Toxic metals were found in precipitation samples in Dhahran, Saudi Arabia in March 1991 (about 400 km south from Kuwait). Metal concentrations in air particulates increased rapidly from April to July 1991 and relatively higher concentrations of nickel, vanadium and chromium were found in the regional inhalable particulates in Dhahran in 1991 and 1992. Similarly, the levels of arsenic, cadmium, cobalt, copper, chromium and molybdenum increased as well. The total of 7224 tons of vanadium, 2084 tons of nickel, 917 tons of aluminum, 747 tons of iron, 432 tons of zinc, 77 tons of chromium and 28 tons of cadmium released from the Kuwaiti oil fires is especially relevant for the assessment of the environmental effects. The levels of cadmium and vanadium exceeded the permissible limits in Kuwait, Saudi Arabia and Bahrain.

The prevailing north-westerly wind (shamal) near the fire sources led the plume towards the eastern part of the Kingdom of Saudi Arabia over the Persian Gulf coast, covering the sky and turning day into night. Therefore, the smoke was reported to have affected especially neighboring countries. The dispersion of the smoke led to black rainfalls in Qatar, Turkey, Iran, the Himalayas and black snow in the Indian state of **Kashmir** (2600 km to the east). An unusually high proportion of soot containing particles (53%) was found in April 1991 at 7.5 km altitude above Japan, which could easily be explained by the Kuwaiti oil fires (due to the presence of vanadium and a likely upper troposphere origin from oil combustion, enabled by cloudy cyclones). It showed a strong connection between the Kuwaiti oil fires and the sharp increase (five times higher than the normal value) of black carbon concentration, respectively deposition of black carbon on ice cores during 1991 at the Muztagh Ata Mountain, in northern Tibet, leading to increased snow melting on the northern Tibetan Plateau. In some cases, where the evidence establishing the extent and nature of pollution due to the Kuwaiti oil fires was insufficient, although the pollution was confirmed, there was no compensation granted by the United Nations Compensation Commission (UNCC), such as in the case of Iran, who claimed compensation for the damage inflicted by the smoke from the oil fires on its national cultural heritage. Inorganic and organic fractions as well as microbiotic crusts based on samples from Iranian archeological sites in Khuzestan and Fars provinces in 2003, where monument surfaces and bas-reliefs were blackened and soiled. The inorganic fraction analyses concluded that the effect of an episodic event could not be seen in the samples, the organic fraction analyses identified PAHs and petroleum molecular markers only in negligible amounts and it was concluded that the cause of the blackening and soiling of Iranian monuments was colonization of Nostocales (a species of cyanobacteria). Another study shows that the Kuwaiti smoke plumes influenced the formation and properties: clouds formed at the top of the smoke plumes, containing giant cloud condensation nuclei formed by the salt particles from the plume, which caused coalescence in the highly polluted clouds close to the source, reduced growth with altitude and diminished the water and ice precipitation forming process farther away. The plume remained between ca. 500 and 4,000 m and was never detected above 5,500 m, leading to a minimized global spread in the upper atmosphere. During the British Meteorological Office´s Meteorological Research Flight´s (MRF) missions in March 1991 high amounts of smoke were observed within 5 km altitude. Nonetheless, this did have a significant local climatic effect by reducing the area surface temperature by 12° F in 1991, even if increased global warming or effects on the global climate have to be excluded. Although the atmospheric pollution was heavy and the pollution levels at regional population centres were higher than in normal cases, they were much lower than initially predicted because, according to the World Meteorological Organisation (WMO) and United Nations Environmental Programme (UNEP), the local meteorological and geographic conditions limited the materials in the plumes to a horizontal cloud and prevented them from mixing with higher or lower layers of air. According to MRF, in March 1991 the plume formed two layers: an upper layer between 3,600 m and 4,600 m and a lower layer between 1,000 m and 3,200 m, separated by a highly moist layer at 3000-m altitude. Depletion of ozone was observed at the levels where the highest concentration of gaseous pollutants was found, between 1,000 m and 1,500 m altitude in the bottom layer and at 4,500 m altitude in the top layer. Global ozone depletion is one of the biggest environmental concerns today and was already feared at the beginning of the Gulf crisis. During later measurements, it was discovered that ozone was depleted near the fire sources probably because of the reaction with NOx within a few kilometres from the source. The ozone was also depleted farther downwind due to its reaction with alkanes and possible adsorption on the soot particles. It was also observed that in the diffused region of the smoke, where ultraviolet radiation was more intense, the ozone was already regenerated. This phenomenon is explained by the fact that the conversion of O_3 to NOx is a photochemical reaction so that the smoke cloud limiting the radiation levels was a critical factor in the depletion of O_3 by NOx. Nonetheless, the plume didn´t rise to the stratosphere (it remained in the troposphere, not exceeding 5,500 m altitude) and is therefore believed by some researchers not to pose any global environmental problems. However, it is well known that disturbances in the ozone layer lead to radiation imbalances which impact the ecosystem by increasing temperatures, changing the monsoon winds and causing inequal distribution of rainfall pattern, affecting living organisms, intensifying the greenhouse effect etc. The fact that between 1992 and 1998 higher than average rainfalls were recorded in Kuwait confirms the idea that the ozone depletion did indeed have at least local environmental effects, even if this can be seen as a positive effect, due to the remarkable vegetation growth afterwards. The unburned amounts of oil that managed to pass through the flame caused oil rains which in return formed new oil lakes, which were a source of emissions to the air, particularly volatile organic compounds. The particulate matter and the gushed oil (from the oil wells that didn´t explode) contributed to the oil lake formation. Additionally, this oil and soot fallout covered large areas in the desert with a black, tar-like coating and the surface of the Persian Gulf water, killing or damaging some vegetation.

40.4 ENVIRONMENTAL WAR BY THE ISLAMIC STATE

Of the tactics employed by the Islamic State to resist counterinsurgent forces and subjugate civilians, few produced such far-reaching damage as its deliberate annihilation of the natural environment in Syria and Iraq. In 2016, Islamic State fighters set fire to sulfur plants and oil fields in Qayyarah, near Mosul, releasing dense clouds of pollutant smoke to obscure their positions from American drones and warplanes. The fires resulted in humanitarian, environmental, and health crises: toxins seeped into the surrounding soil and water sources, contaminating croplands and causing serious respiratory and other medical conditions among the civilian population.

In wielding environmental degradation as a weapon, the Islamic State drew on a wartime tactic with a long and devastating history—one that stretches from Cyrus' diversion of the Euphrates to overtake Babylon, to Sherman's scorched-earth campaign across Georgia and Virginia, to Saddam Hussein's destruction of Kuwaiti oil wells during the Gulf War. All wars entail environmental damage, often significant and sometimes irreversible. Yet there has been little effort by social scientists to identify the conditions under which conflict actors make use of environmental warfare tactics in either conventional or irregular warfare; the limited social science scholarship on this topic typically makes use of single case studies rather than cross-case analysis.

In some cases, there is a clear strategic or tactical logic that governs the treatment of nature in war. While counterinsurgents in Vietnam, Indonesia, Colombia, and Turkey carried out extensive defoliation in pursuit of guerrilla targets, their Marxist opponents adopted conservationist practices to conceal strategic bases in densely forested terrain. Often, however, the intentional destruction of the natural landscape is difficult to explain by reference to operational strategy alone.

Consider the following examples:

- Conflict actors sometimes engage in widespread environmental destruction far beyond the threshold of military necessity. After the 1991 Gulf War, Saddam Hussein's government retaliated against Shi'a rebels by draining the marshes of southern Iraq. Cynically described as an agricultural development project, Saddam's actions obliterated the largest wetland ecosystem in the region and devastated the Ma'dan population that had inhabited the marshlands for over five thousand years.

- In other cases, conflict actors show restraint in their treatment of the natural environment even when military strategy might dictate otherwise. Despite extensive research into the development of cloud seeding during the Vietnam War, the United States agreed to prohibit the use of weather modification techniques in war as part of the 1977 Convention on the Prohibition of Military or Any Other Hostile Use of Environmental Modification Techniques (ENMOD) treaty. More recently, the US military declined to bomb Islamic State-controlled oil wells, citing concerns about local environmental damage.
- The intentional and widespread destruction of the natural landscape often conflicts with the efforts of insurgents and counterinsurgents alike to secure the support of civilian populations. During the Vietnam War, the US military persisted in its use of Agent Orange even as the strategy alienated South Vietnamese farmers, whose hearts and minds it hoped to win, by inadvertently spraying their crops. The toxic chemicals proved a valuable propaganda tool for North Vietnamese forces competing with the United States for civilian loyalties.

Deliberate attacks on the natural landscape, the frequency with which these tactics are deployed, and the diverse forms they take are conditioned by context-specific incentives, constraints, and intervening variables.

Demonstrating intentionality of action (such that we may distinguish between direct targeting and collateral damage) is intrinsically challenging, especially as any action in wartime may have more than one intent. That said, the empirical efforts to determine intent are essential to the long-term protection of the environment from wartime damage: even though it is unlikely that warfare can ever be cleansed of its passive effects on the environment, we must work to eliminate the intentional use of the environment as a weapon in order to safeguard against some forms of severe degradation.

Given that all wars produce environmental damage, pinning down a clear definition of environmental warfare is a particularly tricky task. Some weapons and tactics are difficult to classify. For instance, landmines, underwater mines, cluster bombs, and other area denial devices target enemy combatants, rather than the terrain itself; but we might think of landmines as an example of a weapon that uses mined terrain as a medium through which to inflict harm.

40.5 INTERNATIONAL LAW ON ENVIRONMENTAL WARFARE

Articles 35(3) and 55(1) of Additional Protocol I: Purpose and Applicability

On 8 June 1977 the Diplomatic Conference on the Reaffirmation and Development of International Humanitarian Law Applicable in Armed Conflicts adopted two protocols additional to the Geneva Conventions of 12 August 1949. The Additional Protocol I included two provisions specifically dealing with the question of environmental damage during international armed conflicts: Art 35 and Art 55. Additional Protocol I and Environmental Modification (ENMOD)-Convention were designed to achieve different purposes and therefore do not substantially overlap. The scope of Additional Protocol I is more limited than that of the ENMOD-Convention, since it only refers to international armed conflicts. Because of the overflowing and transnational character of environmental damage, Additional Protocol I doesn´t differentiate between the territory of the belligerent state causing the environmental damage and the territory of the state suffering the environmental damage, but even protects non-belligerent states as well, according to the ICRC Commentary. Additional Protocol I protects the environment as a victim against both intentional infliction of damage during war as well as incidental or unintentional damage, if it may be expected to occur. Consequently, belligerent states must assess the possibility of environmental damage that may be inflicted by their actions. The threshold imposed by the terms "widespread, long-lasting and severe" will be discussed in the following sections. This thesis focuses on Art 35(3) and 55(1), which together contain multiple prohibitions and one positive obligation of environmental protection, establishing different standards of harm, as the following shall demonstrate. Art 35(3) Additional Protocol I proclaims:

"It is prohibited to employ methods or means of warfare which are intended, or may be expected, to cause widespread, long-term and severe damage to the natural environment." Article 55(1) Additional Protocol I stipulates: "Care shall be taken in warfare to protect the natural environment against widespread, long-term and severe damage. This protection includes a prohibition of the use of methods or means of warfare, which are intended or may be expected to cause such damage to the natural environment and thereby to prejudice the health or survival of the population." Art 35 (3) Additional Protocol I prohibits the employment of methods or means of war that are intended or expected to cause widespread, long-term and severe damage to the natural environment. Art 55 (1) additionally mentions harm to the health or survival of the population as a consequence of damage to the natural environment, therefore emphasizing the anthropogenic nature and relevance of the natural environment but making the term even more elaborate on the other hand. These provisions offer no definition whatsoever for the concept of environmental damage and a systematic or teleological interpretation of the treaty regarding such a complex term would lead to ambiguity. Some consider this ambiguity in the interest of environmental protection and preferable to use the general understanding of this concept. Because the subsequent practice is very limited and was practically inexistent until the Gulf War and the activity of the UNCC, the travaux préparatoires to Art 35 (3) and 55(1) Additional Protocol I have to be consulted according to Art 32 VCLT. The official discussion sessions for Additional Protocol I lasted from 1974 until 1977 and were marked by difficulties concerning the use and definition of the terms "environment" and "damage". Although there is no definition for the term "natural environment" in Additional Protocol I, the ICRC understands it "in the widest sense to cover the biological environment in which a population is living", consisting of objects indispensable to survival, forests, vegetation, fauna, flora and other biological or climatic elements, when commenting Art 55(1). An earlier draft from 1975 of Art 33 explains damage to the environment "in such a way that the stability of the ecosystem is disturbed" and a draft of Art 55 from the same year understands harm to the natural environment as a disturbance of the stability of the ecosystem. "Alteration" or "disturbance" were some of the alternative terms debated to refer to "damage" and although they do not have an intrinsic negative connotation, the discussions suggest only a negative impact was taken into consideration. A disturbance of the ecosystem is therefore a threshold beyond which damage is prohibited. A better definition of environmental damage is however found by using certain criteria: negative consequences, human causation of harm, the level of impact required for the qualification as damage and the wider effects of environmental damage. These criteria are flexible and depend on the objective of the treaty/instrument used. The term "damage" implies a negative effect, but this can be problematic especially in complex cases where human behavior causes a change in an environmental component, which consequently impacts other environmental components in both positive and negative ways, as it is the case with the burning oil wells in Kuwait. However, one must consider that positive changes themselves may negatively affect certain environmental elements, although increased rainfalls caused by the meteorological changes inflicted by the oil well fires benefitted the local vegetation, they intensified the seepage of oil into the ground, affecting the groundwater) or even lead to the evolution of certain species, which would imply the extinction of other species. These challenging situations generate many questions: When is a change to the environment positive and when is it negative? When is a mixed result of both positive and negative effects considered environmental damage? Is the question of reversibility of the ecosystem relevant for the evaluation of damage? Can even the reparation of environmental damage be classified as damage, if it inevitably involves certain negative effects, the saltwater used during the firefighting actions contaminated the groundwater)? These questions cannot be easily answered but will be further analyzed in correlation to the ENMOD Convention, respectively Art 35 (3) and Art 55 (1) Additional Protocol I in the following sections. Art 55(1) has the highest threshold of environmental damage within Additional Protocol I because in the second sentence it requires environmental damage capable of inflicting human injury. The first sentence specifies the obligation of environmental protection, accepting the peacetime norm of environmental protection into the law of armed conflict and essentially implying taking reasonable steps to protect the natural environment against widespread, long-term and severe damage. The second sentence of Art 55(1) essentially replicates Art 35(3), although it adds the term "includes" and human-centric elements. Therefore, the verb "includes" suggests that this prohibition is just an example for the environmental protection guaranteed in the first sentence and not a definition of it. This provision generally refers to the "health or survival of the population", without using the adjective "civilian" and hence including the entire population, without regard to its combatant status, which is reasonable since "long-lasting" environmental effects are relevant for any category of population. Furthermore, the word "health" expands the range of applicability for Art 55(1) by not being reduced to mere survival of the population, but covering acts that seriously prejudice health on a long-term basis (e.g. congenital defects, degenerations, deformities).

CHAPTER 41

CYBER WARFARE

41.1 INTRODUCTION TO CYBER WARFARE

Cyberwarfare is the use of cyber-attacks against an enemy state, causing comparable harm to actual warfare and/or disrupting vital computer systems. Some intended outcomes could be espionage, sabotage, propaganda, manipulation or economic warfare.

There is significant debate among experts regarding the definition of cyberwarfare, and even if such a thing exists. One view is that the term is a misnomer since no cyber-attacks to date could be described as a war. An alternative view is that it is a suitable label for cyber-attacks which cause physical damage to people and objects in the real world.

While the majority of scholars, militaries, and governments use definitions that refer to state and state-sponsored actors, other definitions may include non-state actors, such as terrorist groups, companies, political or ideological extremist groups, hacktivists, and transnational criminal organizations depending on the context of the work.

Many countries, including the United States, United Kingdom, Russia, China, Israel, Iran, and North Korea, have active cyber capabilities for offensive and defensive operations. As states explore the use of cyber operations and combine capabilities, the likelihood of physical confrontation and violence playing out as a result of, or part of, a cyber operation is increased. However, meeting the scale and protracted nature of war is unlikely, thus ambiguity remains.

The first instance of kinetic military action used in response to a cyber-attack resulting in the loss of human life was observed on 5 May 2019, when the Israel Defense Forces targeted and destroyed a building associated with an ongoing cyber-attack.

41.1.1 Cyberwarfare vs. cyber war

The term "cyberwarfare" is distinct from the term "cyber war". Cyberwarfare includes techniques, tactics and procedures that may be involved in a cyber war, but the term does not imply scale, protraction or violence, which are typically associated with the term "war", which inherently refers to a large-scale action, typically over a protracted period of time, and may include objectives seeking to utilize violence or the aim to kill. A cyber war could accurately describe a protracted period of back-and-forth cyber-attacks (including in combination with traditional military action) between warring states. To date, no such action is known to have occurred. Instead, armed forces have responded with tit-for-tat military cyber actions. For example, in June 2019, the United States launched a cyber-attack against Iranian weapons systems in retaliation to the shooting down of a US drone in the Strait of Hormuz.

41.1.2 Cyberwarfare and cyber sanctions

In addition to retaliatory digital attacks, countries can respond to cyber-attacks with cyber sanctions. Sometimes, it is not easy to detect the attacker, but suspicions may focus on a particular country or group of countries. In these cases, unilateral and multilateral economic sanctions can be used instead of cyberwarfare. For example, the United States has frequently imposed economic sanctions related to cyber-attacks. Two Executive Orders issued during the Obama administration, specifically focused on the implementation of the cyber sanctions. Subsequent US presidents have issued similar Executive Orders. The US Congress has also imposed cyber sanctions in response to cyberwarfare. For example, the Iran Cyber Sanctions Act of 2016 imposes sanctions on specific individuals responsible for cyber-attacks.

41.2 TYPES OF CYBER WARFARE

Cyber warfare can present a multitude of threats towards a nation. At the most basic level, cyber-attacks can be used to support traditional warfare. For example, tampering with the operation of air defenses via cyber means in order to facilitate an air attack. Aside from these "hard" threats, cyber warfare can also contribute towards "soft" threats such as espionage and propaganda.

(i) Espionage

Traditional espionage is not an act of war, nor is cyber-espionage, and both are generally assumed to be ongoing between major powers. Despite this assumption, some incidents can cause serious tensions between nations, and are often described as "attacks". For example:

- **Massive spying by the US on many countries**

After the NSA's spying on Germany's Chancellor Angela Merkel was revealed, the Chancellor compared the NSA with the Stasi.

- The NSA recording nearly every cell phone conversation in the Bahamas, without the Bahamian government's permission, and similar programs in Kenya, the Philippines, Mexico and Afghanistan.
- The "Titan Rain" probes of American defense contractors computer systems since 2003.
- The Office of Personnel Management data breach, in the US, widely attributed to China.
- The security firm Area 1 published details of a breach that compromised one of the European Union's diplomatic communication channels for three years.

Out of all cyber-attacks, 25% of them are espionage based.

(ii) Sabotage

Computers and satellites that coordinate other activities are vulnerable components of a system and could lead to the disruption of equipment. Compromise of military systems, such as C4ISTAR components that are responsible for orders and communications could lead to their interception or malicious replacement. Power, water, fuel, communications, and transportation infrastructure, all may be vulnerable to disruption.

In mid-July 2010, security experts discovered a malicious software program called *Stuxnet* that had infiltrated factory computers and had spread to plants around the world. It is considered the first attack on critical industrial infrastructure that sits at the foundation of modern economies,

Stuxnet, while extremely effective in delaying Iran's nuclear program for the development of nuclear weaponry, came at a high cost. For the first time, it became clear that not only could cyber weapons be defensive but they could be offensive. The large decentralization and scale of cyberspace makes it extremely difficult to direct from a policy perspective. Non-state actors can play a part in the cyberwar space as state actors, which leads to dangerous, sometimes disastrous, consequences. Small groups of highly skilled malware developers are able to as effectively impact global politics and cyber warfare as large governmental agencies. A major aspect of this ability lies in the willingness of these groups to share their exploits and developments on the web as a form of arms proliferation. This allows lesser hackers to become more proficient in creating the large-scale attacks that once only a small handful were skillful enough to manage. In addition, thriving black markets for these kinds of cyber weapons are buying and selling these cyber capabilities to the highest bidder without regard for consequences.

(iii) Denial-of-service attack

In computing, a denial-of-service attack (DoS attack) or distributed denial-of-service attack (DDoS attack) is an attempt to make a machine or network resource unavailable to its intended users. Perpetrators of DoS attacks typically target sites or services hosted on high-profile web servers such as banks, credit card payment gateways, and even root nameservers. DoS attacks often leverage internet-connected devices with vulnerable security measures to carry out these large-scale attacks. DoS attacks may not be limited to computer-based methods, as strategic physical attacks against infrastructure can be just as devastating. For example, cutting undersea communication cables may severely cripple some regions and countries with regards to their information warfare ability.

(iv) Electrical power grid

The federal government of the United States admits that the electric power grid is susceptible to cyberwarfare. The United States Department of Homeland Security works with industries to identify vulnerabilities and to help industries enhance the security of control system networks. The federal government is also working to ensure that security is built in as the next generation of "smart grid" networks are developed. In April 2009, reports surfaced that China and Russia had infiltrated the U.S. electrical grid and left behind software programs that could be used to disrupt the system, according to current and former national security officials. The North American Electric Reliability Corporation (NERC) has issued a public notice that warns that the electrical grid is not adequately protected from cyber-attack. China denies intruding into the U.S. electrical grid. One countermeasure would be to disconnect the power grid from the Internet and run the net with droop speed control only. Massive power outages caused by a cyber-attack could disrupt the economy, distract from a simultaneous military attack, or create a national trauma.

Iranian hackers, possibly Iranian Cyber Army pushed a massive power outage for 12 hours in 44 of 81 provinces of Turkey, impacting 40 million people. Istanbul and Ankara were among the places suffering blackout.

It's possible that hackers have gotten into administrative computer systems of utility companies, but says those aren't linked to the equipment controlling the grid, at least not in developed countries.

In June 2019, Russia said that its electrical grid has been under cyber-attack by the United States. The New York Times reported that American hackers from the United States Cyber Command planted malware potentially capable of disrupting the Russian electrical grid.

(v) Propaganda

Cyber propaganda is an effort to control information in whatever form it takes, and influence public opinion. It is a form of psychological warfare, except it uses social media, fake news websites and other digital means. In 2018, this kind of attack took place from actors such as Russia. It is a form of system warfare that seeks to de-legitimize the political and social system on which our military strength is based".

The propaganda is the deliberate, systematic attempt to shape perceptions, manipulate cognitions, and direct behavior to achieve a response that furthers the desired intent of the propagandist. The internet is the most important means of communication today. People can convey their messages quickly across to a huge audience, and this can open a window for evil. Terrorist organizations can exploit this and may use this medium to brainwash people. It has been suggested that restricted media coverage of terrorist attacks would in turn decrease the number of terrorist attacks that occur afterwards.

(vi) Economic disruption

In 2017, the WannaCry and Petya cyber-attacks, masquerading as ransomware, caused large-scale disruptions in Ukraine as well as to the U.K.'s National Health Service, These attacks are also categorized as cybercrimes, specifically financial crime because they negatively affect a company or group.

(vii) Surprise cyber attack

The idea of a "cyber-Pearl Harbor" has been debated by scholars, drawing an analogy to the historical act of war. Others have used "cyber 9/11" to draw attention to the nontraditional, asymmetric, or irregular aspect of cyber action against a state.

41.3 EXAMPLES OF CYBER WARFARE OPERATIONS

Here are some examples of cyber warfare in recent times:

(i) Sony Pictures Hack

An attack on Sony Pictures followed the release of the film "The Interview", which presented a negative portrayal of Kim Jong Un. The attack is attributed to North Korean government hackers. The FBI found similarities to previous malware attacks by North Koreans, including code, encryption algorithms, and data deletion mechanisms.

(ii) Bronze Soldier

In 2007, Estonia relocated a statue associated with the Soviet Union, the Bronze Soldier, from the center of its capital Tallinn to a military cemetery near the city. Estonia suffered a number of significant cyber-attacks in the following months. Estonian government websites, media outlets, and banks were overloaded with traffic in massive denial of service (DoS) attacks and consequently were taken offline.

(iii) Fancy Bear

CrowdStrike claims that the Russian organized cybercrime group Fancy Bear targeted Ukrainian rocket forces and artillery between 2014 and 2016. The malware was spread via an infected Android application used by the D-30 Howitzer artillery unit to manage targeting data.

Ukrainian officers made wide use of the app, which contained the X-Agent spyware. This is considered to be a highly successful attack, resulting in the destruction of over 80% of Ukraine's D-30 Howitzers.

(iv) Enemies of Qatar

Elliott Broidy, an American Republican fundraiser, sued the government of Qatar in 2018, accusing it of stealing and leaking his emails in an attempt to discredit him. The Qataris allegedly saw him as an obstacle to improving their standing in Washington.

According to the lawsuit, the brother of the Qatari Emir was alleged to have orchestrated a cyber warfare campaign, along with others in Qatari leadership. 1,200 people were targeted by the same attackers, with many of these being known "enemies of Qatar", including senior officials from Egypt, Saudi Arabia, the United Arab Emirates, and Bahrain.

(v) North Korea's Lazarus Group

Another cyber warfare example comes from North Korea's Lazarus Group, a sophisticated hacker organization known for using its technical prowess to steal millions of dollars' worth of cryptocurrency. Crypto platform cyber-attacks make it possible to steal money, disable critical systems, and even disrupt economies. The cyber-attacks executed by Lazarus Group over the past 10+ years have caused substantial financial losses and exposed just how vulnerable new financial technologies are to cyber war.

(vi) Operation Orchard

Still, attacks are not limited to the digital realm, as evidenced by Israel's *Operation Orchard.* Operation Orchard, which targeted a suspected Syrian nuclear complex, demonstrated how cyber-attacks can be integrated with conventional action. Israel disabled Syrian air defenses while showing the world how types of cyber warfare can align with military action.

41.4 PREVENTING AND COUNTERING CYBER WARFARE

From hacking to cyber espionage, cyber warfare tactics target the military, utility companies, businesses, financial institutions, and government agencies. Governments and multinational corporations must prioritize cybersecurity to prevent cyber attackers from accessing private information, damaging computer systems, and disrupting operations. Similarly, public agencies should coordinate with domestic security firms to safeguard industrial infrastructure against cyber threats.

NATO-style international partnerships combine cyber-attack prevention and mitigation strategies. The Department of Defense runs cyber centers to coordinate security and counter cyber threats.

Government Strategies

Cyber warfare attacks represent a growing problem. Governments worldwide have enacted laws and taken other preventative measures to defend against a myriad of cyber attackers. Their critical industrial infrastructure and national security depend on robust cybersecurity. Financial systems also generally incorporate robust security to protect against cyber espionage and other potential threats.

Governments set the precedent when it comes to defending digital networks against cyber warfare attacks. Compliance and cybersecurity laws continue to evolve, often accounting for the unique needs of the financial (payment processing systems), utilities (power grid), and telecommunications industries. Laws and other regulations are key for protecting infrastructure, national security, and modern economies.

Governments are also taking steps to educate their citizens about cyber hazards. With phishing, ransomware, and other external attacks on the rise, corporations and the general public must stay informed to protect themselves. Governments may launch cyber warfare attacks and run dedicated command centers to combat targeted hacking and giant ransomware attacks.

In addition, governments take offensive and defensive cyber actions, most notably, to defend their countries against cyber warfare acts executed by North Korea. National and international organizations, like NATO, routinely exchange sensitive information electronically. When conflicts arise, nations must fight a hybrid war, which combines cyber and conventional warfare efforts.

Sharing knowledge and best practices helps nations anticipate and respond to cyberattacks, while simultaneously boosting global cybersecurity.

41.5 CYBER WARFARE IN RUSSIA-UKRAINE WAR

Russia's invasion of Ukraine is significant for cyber warfare because it shows how cyber warfare can be used in conjunction with conventional military assets. While cyber warfare was largely overshadowed by other aspects of Russia's invasion like the movements of armor units and use of artillery, the Russians used it throughout as part of their overall war plan. This includes some notable operations that had effects beyond Ukraine.

For example:

- The Russians targeted *Viasat*, an American satellite communications company that provided support to the Ukrainian military, with malware designed to erase its data before disabling it. Because the Russians did not limit the malware's scope, it ended up affecting other ground satellite components, causing hundreds of thousands of people outside of Ukraine to lose electrical power and their connection to the Internet.
- A cyberattack against the City Council of Odessa, a major Ukrainian port city situated on the Black Sea, was timed to coincide with a cruise missile attack that was meant to disrupt Ukraine's response to Russian forces attacking in the South.
- Cyberattacks have also been launched against many parts of Ukraine's infrastructure and government and civilian networks, including hospitals.

These actions show that cyber operations are not limited to the military forces of combatants and, like World War II strategic bombing efforts, often extend to strike at infrastructure and areas of economic significance. The Russians continued to use cyber in Ukraine in 2023, reusing a malware program called *Cadet Blizzard* in February that was used originally in cyber-attacks in 2020.

41.6 DEFENCE CYBER AGENCY OF INDIA

The Defence Cyber Agency (DCyA) is an integrated tri-services agency of the Indian Armed Forces. Headquartered in New Delhi. The agency is tasked with handling cyber security threats. The DCyA draws personnel from all three branches of the Armed Forces. The head of the DCyA is an officer of two-star rank, and reports to the Chief of Defence Staff (CDS) through the Integrated Defence Staff (IDS).

Indian Navy's Rear Admiral Mohit Gupta was appointed in May 2019 as the first head of the DCyA. The DCyA was expected to be operational by November 2019. As of 2021, DCyA was fully operational with Army, Air Force, and Navy establishing their respective Cyber Emergency Response Teams (CERT).

The creation of the Defence Cyber Agency (DCyA), the Defence Space Agency (DSA), and the Armed Forces Special Operations Division (AFSOD) was approved by Prime Minister Narendra Modi during the Combined Commanders' Conference at Jodhpur Air Force Station on 28 September 2018. The existing Defence Information Assurance and Research Agency was upgraded to form the new Defence Cyber Agency.

The Week reported that the DCyA would have the capability to hack into networks, mount surveillance operations, lay honeypots, recover deleted data from hard drives and cellphones, break into encrypted communication channels, and perform other complex objectives. According to Lieutenant General Deependra Singh Hooda, the DCyA would have the responsibility of framing a long-term policy for the security of military networks, including eliminating the use of foreign hardware and software in the Indian Armed Forces, and preparing a cyberwarfare doctrine.

41.7 ADAPTATION OF CYBER WARFARE ACTIVITY BY THE INDIAN MILITARY

Within the Indian Armed Forces, the Army remains at the forefront of adaptation to cyber warfare. Recently, the Indian army raised new cyber units called **'Command Cyber Operations and Support Wings (CCOSW)'**. Moreover, to hire skilled human resources, the Indian army conducted a maiden **hackathon** in 2021, which lasted three consecutive months. The hackathon included participants from across India, whose numbers exceeded 15,000, demonstrating various skills, such as offensive cyber operations, coding, and intercepting encrypted communications. Similarly, the Indian Air Force (IAF) and the Indian Navy are also working to strengthen their cyber capabilities.

While most of these efforts remain shrouded in secrecy, statements from respective service chiefs indicate that significant progress has been made in the cyber domain. For instance, the IAF chief publicly announced the importance of cyber capability by stating that the IAF considers cyber operations an integral part of modern military operations and works constantly to enhance its cyber capabilities.

More recently, the military unveiled its maiden Joint Doctrine for Cyberspace Operations. The new framework is a key document because it aims to amalgamate each service's cyber warfighting capabilities in the future. The release of the doctrine is a continuation of India's attempts to consolidate its cyber prowess under one roof to enhance operational effectiveness.

India is also actively engaged in boosting its indigenous cyber capabilities by signing multiple cyber partnerships with various countries, including the United States, Japan, United Kingdom, Israel, Russia, Japan, Australia, and France. The focus of these partnerships is providing cyber training, transferring knowledge, and collaborations with Indian law enforcement and military services.

Publicly available information also indicates that India has built sophisticated cyber warfare skills with a specific focus on targeting **Pakistan**. Its 'Advanced Persistent Threats' (APT) groups, including the SideWinder, Bahamut, and PatchWork, have targeted key domestic government organizations in recent years. The Indian cyber teams are targeting **web traffic in Pakistan**. In 2020, a cyber team tried to hack Pakistan's military and civilian government official's cellular and computer devices. Actions to disrupt and hack Pakistan's state machinery are likely to expand in the future, necessitating appropriate protective measures.

Pakistan has undertaken concrete measures to address such cyber threats. Notably, the establishment of a Computer Emergency Response Team (CERT) marked a significant step in countering cyber-attacks. The CERT initiative is designed to protect the country's critical infrastructure from offensive cyber actions and to identify and mitigate existing vulnerabilities, thereby enhancing the nation's cybersecurity posture.

The Indian military is expected to continue enhancing its warfighting capabilities and spearheading cyber-attacks **against Pakistan**. This necessitates that Pakistan not only improves its cyber capabilities but also develops and maintains robust 'cyber deterrence' to prevent India from initiating cyber-attacks. This can be accomplished by increasing the capacity of the national CERT team to counter threats and strengthen cyber security, as well as creating cyber awareness among personnel working in government institutions. Importantly, Pakistan also needs to set up a dedicated national cyber command to improve and synergize its response, making it more effective and efficient.

India is facing growing threat to its national security from adversaries resorting to a new kind of cyber-warfare and, therefore, an offensive approach involving "super cyber force" and "surgical strikes" is required, according to a report by non-government organization Prahar.

The report, titled 'The Invisible Hand', projected that if unchecked, cyberattacks on India was likely to rise to one trillion per annum by 2033, reaching 17 trillion by 2047.

Stating that cyberspace is the new battlefield, the report suggested that India must go on the offensive. "Other interventions include sophisticated tech infrastructure, skill improvement, whitelisting digital apps and platforms, and educating citizens. Until India creates a holistic cyber policy and execution strategy, restricting identifiable legitimate platforms could push citizens into the hands of dark web operators," it said.

CHAPTER 42

ELECTRONIC WARFARE

42.1 INTRODUCTION TO ELECTRONIC WARFARE (EW)

Electromagnetic warfare or electronic warfare (EW) is warfare involving the use of the electromagnetic spectrum (EM spectrum) or directed energy to control the spectrum, attack an enemy, or impede enemy operations. The purpose of electronic warfare is to deny the opponent the advantage of—and ensure friendly unimpeded access to—the EM spectrum. Electronic warfare can be applied from air, sea, land, or space by crewed and uncrewed systems, and can target communication, radar, or other military and civilian assets.

Military operations are executed in an information environment increasingly complicated by the electromagnetic spectrum. The electromagnetic spectrum portion of the information environment is referred to as the electromagnetic environment (EME). The recognized need for military forces to have unimpeded access to and use of the electromagnetic environment creates vulnerabilities and opportunities for electronic warfare in support of military operations.

Within the information operations construct, EW is an element of information warfare; more specifically, it is an element of offensive and defensive counterinformation.

NATO has a different and arguably more encompassing and comprehensive approach to EW. A Military Committee Transformation Concept for Future NATO Electronic Warfare, recognized the EME as an operational maneuver space and warfighting environment/domain. In NATO, EW is considered to be warfare in the EME. NATO has adopted simplified language which parallels those used in other warfighting environments like maritime, land, and air/space. For example, an electronic attack (EA) is offensive use of EM energy, electronic defense (ED), and electronic surveillance (ES). The use of the traditional NATO EW terms, electronic countermeasures (ECM), electronic protective measures (EPM), and electronic support measures (ESM) has been retained as they contribute to and support electronic attack (EA), electronic defense (ED) and electronic surveillance (ES). Besides EW, other EM operations include intelligence, surveillance, target acquisition and reconnaissance (ISTAR), and signals intelligence (SIGINT). Subsequently, NATO has issued EW policy and doctrine and is addressing the other NATO defense lines of development.

Primary EW activities have been developed over time to exploit the opportunities and vulnerabilities that are inherent in the physics of EM energy. Activities used in EW include electro-optical, infrared and radio frequency countermeasures; EM compatibility and deception; radio jamming, radar jamming and deception and electronic counter-countermeasures (or anti-jamming); electronic masking, probing, reconnaissance, and intelligence; electronic security; EW reprogramming; emission control; spectrum management; and wartime reserve modes.

42.1.1 Subdivisions of EW

Electronic warfare consists of three major subdivisions: electronic attack (EA), electronic protection (EP), and electronic warfare support (ES).

(i) Electronic Attack (EA)

Electronic attack (EA), also known as electronic countermeasures (**ECM**), involves the offensive use of electromagnetic energy weapons, directed energy weapons, or anti-radiation weapons to attack personnel, facilities, or equipment with the intent of degrading, neutralizing, or destroying enemy combat capability including human life. In the case of electromagnetic energy, this action is most commonly referred to as "**jamming**" and can be performed on communications systems or radar systems. In the case of anti-radiation weapons, this often includes missiles or bombs that can home in on a specific signal (radio or radar) and follow that path directly to impact, thus destroying the system broadcasting.

In November 2021, Israel Aerospace Industries announced a new electronic warfare system named **Scorpius** that can disrupt radar and communications from ships, UAVs, and missiles simultaneously and at varying distances.

On 8 September 2024, Russian drones entered both Romanian and Latvian airspace. Romania scrambled two F-16s to monitor the drone's progress, it landed "in an uninhabited area" near Periprava, according to the Romanian Ministry of Defence. This comes as the ISW noted increased success in Ukrainian Electronic Warfare against Russian drones that resulted in "several Russian Shahed drones (that) recently failed to reach their intended targets for unknown reasons." Two Kh-58s also reportedly failed to reach their targets.

(ii) Electronic protection

Electronic protection (EP), also known as an electronic protective measure (**EPM**) or electronic counter-countermeasure (**ECCM**) are a measure used to protect against an electronic enemy attack (EA) or to protect against friendly forces who unintentionally deploy the equivalent of an electronic attack on friendly forces. (sometimes called EW fratricide). The effectiveness of electronic protection (EP) level is the ability to counter an electronic attack (EA).

Flares are often used to distract infrared homing missiles into missing their target. The use of flare rejection logic in the guidance (seeker head) of an infrared homing missile to counter an adversary's use of flares is an example of EP. While defensive EA actions (jamming) and EP (defeating jamming) both protect personnel, facilities, capabilities, and equipment, EP protects from the effects of EA (friendly and/or adversary). Other examples of EP include spread spectrum technologies, the use of restricted frequency lists, emissions control (EMCON), and low observability (stealth) technology.

Electronic warfare self-protection (EWSP) is a suite of countermeasure systems fitted primarily in the aircraft for the purpose of protecting the host from weapons fire and can include, among others: directional infrared countermeasures (DIRCM, flare systems and other forms of infrared countermeasures for protection against infrared missiles; chaff (protection against radar-guided missiles); and DRFM decoy systems (protection against radar-targeted anti-aircraft weapons).

An electronic warfare tactics range (EWTR) is a practice range that provides training for personnel operating in electronic warfare. There are two examples of such ranges in Europe: one at RAF Spadeadam in the northwest county of Cumbria, England, and the Multinational Aircrew Electronic Warfare Tactics Facility Polygone range on the border between Germany and France. EWTRs. are equipped with ground-based equipment to simulate electronic warfare threats that aircrew might encounter on missions. Other EW training and tactics ranges are available for ground and naval forces as well.

Antifragile EW is a step beyond standard EP, occurring when a communications link being jammed actually increases in capability as a result of a jamming attack, although this is only possible under certain circumstances such as reactive forms of jamming.

(iii) Electronic warfare support

Electronic warfare support (ES) is a subdivision of EW involving actions taken by an operational commander or operator to detect, intercept, identify, locate, and/or localize sources of intended and unintended radiated electromagnetic (EM) energy. These Electronic Support Measures (ESM) aim to enable immediate threat recognition focuses on serving military service needs even in the most tactical, rugged, and extreme environments. This is often referred to as simply reconnaissance, although today, more common terms are intelligence, surveillance and reconnaissance (ISR) or intelligence, surveillance, target acquisition, and reconnaissance (ISTAR). The purpose is to provide immediate recognition, prioritization, and targeting of threats to battlefield commanders.

Signals intelligence (**SIGINT**), a discipline overlapping with ES, is the related process of analyzing and identifying intercepted transmissions from sources such as radio communication, mobile phones, radar, or microwave communication. SIGINT is broken into two categories: electronic intelligence (**ELINT**) and communications intelligence (**COMINT**). Analysis parameters measured in signals of these categories can include frequency, bandwidth, modulation, and polarization.

The distinction between SIGINT and ES is determined by the controller of the collection assets, the information provided, and the intended purpose of the information. Electronic warfare support is conducted by assets under the operational control of a commander to provide tactical information, specifically threat prioritization, recognition, location, targeting, and avoidance. However, the same assets and resources that are tasked with ES can simultaneously collect information that meets the collection requirements for more strategic intelligence.

42.2 ELECTRONIC WARFARE IS A GAME OF CAT AND MOUSE

The story of electronic warfare is one of intrigue, secrecy and technological innovation at the cutting edge. Throughout its history, EW has played a significant role in helping military leaders maintain a strategic edge in a battlespace experiencing rapid technological advances.

As nations learned to exploit the electromagnetic spectrum for military advantage – in areas like communications, navigation and radar – military strategists and scientists simultaneously engineered ways to deny their adversaries those similar advantages. A cat-and-mouse dynamic emerged in the competition for spectrum superiority – one that continues to define the advancement of the field today.

Global technologies and developments in EW are leveling the playing field. The proliferation and affordability of commercial electronics and computing power means that EW is no longer the exclusive province of wealthy nations; it is now a battlefield for smaller states and even non-state actors. EW helps sort through this complexity, making sure our systems are able to communicate, identify and combat enemy radar.

Electronic warfare (EW) systems can be configured for a variety of different missions and use a host of different subsystems. But despite this incredible sophistication and diversity, there are three main capabilities common to most electronic warfare systems – sensing the environment (receiver sensor), analyzing the environment (signal analysis), and responding to the environment (technique generation and high-power transmission).

42.3 BASIC CONSTRUCTION OF EW SYSTEMS

An Electronic Warfare (EW) system comprises of several interconnected components, each serving a vital function. Antennas collect and transmit electromagnetic signals, and the RF front-end filters and amplifies these signals for processing. The signal processing unit analyzes and interprets the data, identifying enemy emissions and threats. Electronic attack components, such as jammers, disrupt enemy radar and communication systems, while deception systems create false targets to confuse adversaries.

Defensive elements include sensors that detect enemy signals, with electronic protection systems designed to prevent jamming and spoofing. Power supplies and cooling systems ensure the system remains operational under stressful conditions, and operators manage the system through user interfaces that display real-time data and control EW operations. The entire setup is orchestrated by a command and control (C2) system that ensures effective coordination between components.

Together, these components enable EW systems to perform critical functions like signal interception, jamming, intelligence gathering, and protection of military assets, ensuring operational effectiveness across diverse environments.

42.4 APPLICATIONS OF ELECTRONIC WARFARE (EW)

(i) **Suppression of Enemy Air Defenses (SEAD)**: EW is crucial in SEAD missions, which aim to disable or disrupt enemy air defense systems like radar-guided surface-to-air missiles (SAMs) and anti-aircraft artillery. By jamming radar and communication systems, EW protects aircraft from detection, ensuring safer air operations for strikes and reconnaissance.

A notable real-world example is Israel's 2007 Operation **Orchard**. During this covert airstrike on a Syrian nuclear facility, Israeli jets employed advanced EW tactics to neutralize Syrian air defenses. By jamming and disrupting radar and missile systems, Israeli forces effectively rendered Syria's air defenses useless, allowing the mission to proceed undetected. This operation highlighted the critical role of EW in modern warfare, showcasing how it can disable enemy defenses and ensure mission success.

(ii) **Command and Control Warfare (C2W):** EW plays a significant role in disrupting an enemy's command and control systems. By jamming or spoofing communications, it impairs the enemy's ability to coordinate and control forces, leading to disorganized and ineffective operations.

(iii) **Electronic Intelligence (ELINT) and Signals Intelligence (SIGINT):** EW is a key tool for gathering intelligence. ELINT and SIGINT systems intercept enemy radar signals and communications, providing vital information on enemy locations, capabilities, and plans. This intelligence supports tactical and strategic decision-making.

(iv) **Protection of Friendly Assets (Electronic Defense):** EW provides defense for critical military systems by countering enemy electronic attacks. Techniques like frequency hopping and stealth technology protect aircraft, ships, and ground units from detection and jamming, ensuring operational effectiveness in hostile environments.

(v) **Cyber-Electronic Warfare Integration:** Modern EW integrates with cyber capabilities to disrupt both electromagnetic and digital domains. This coordination allows forces to simultaneously attack enemy networks, communication infrastructure, and radar systems, maximizing the impact of EW and cyber operations.

42.5 FUTURE OF ELECTRONIC WARFARE

The future is about ensuring spectrum-wide superiority at all times, in all domains. Like today, the capabilities of the future will involve using the spectrum to attack, defend and understand battlefield dynamics. But all this will be done faster, more covertly and more autonomously than today.

Electromagnetic spectrum maneuvers will be more agile, and forces better equipped to pivot in real time to changes on the battlefield, even employing the spectrum to mount new kinds of attacks on hostile infrastructure and networks.

Electronic warfare systems must be capable of deploying overwhelming power with minute precision, of deploying countermeasures to counter never-before-encountered threats, of intelligent spectrum management and sharing with joint and allied forces.

Achieving this will require incremental improvements and ingenious innovations, new ways of addressing old problems and solutions to problems that don't yet exist. This is where we're headed:

42.5.1 Key Concepts and Capabilities of Electronic Warfare

(i) **Smaller and lighter, more power with less energy:** It's clear that small-size, weight and power (SWAP) systems will drive the near-term evolution of EW. Powerful digital capabilities in smaller packages will be critical to extending spectrum superiority to small, unmanned systems and rotary platforms where every ounce counts, as well as to dismounted forces constantly on the move. Small, light and affordable systems will enable a whole range of new solutions that bring once-costly EW capabilities to single-use or expendable platforms like missiles and decoys.

(ii) **Multifunctional and reprogrammable:** Giving systems the ability to pivot in real time from one function to another is the cornerstone of agile and adaptable EW. Whether through manual on-the-fly reprogramming or intelligent, autonomous reconfiguration, a system's ability to perform multiple functions like EA, EP and ESM, as needs evolve, is all but expected. Innovations will allow systems to transmit and receive simultaneously without unwanted interference or signal fratricide. Machine learning will usher in a new generation of cognitive EW systems that learn how to categorize and respond to new threats and reduce the burden on the warfighter.

(iii) **Modular, open and scalable:** Systems that function independently, or that can be connected together to handle a greater diversity of threats simultaneously, will help enable future forces to control the spectrum. The ability to add new functions to a system or to create a networked "system of systems" will make EW operations more flexible, resilient and easier to maintain. Modular, software-defined and scalable systems will enable powerful capabilities like distributed jamming against multiple threats and facilitate cooperation between manned and unmanned systems in the battlespace.

(iv) Future systems will not only be able to use RF transmitters to jam enemy radar and communications, but also to insert viruses and other destructive computer code into enemy systems to spoof or disable them.

This extension of EW definition is a real Spectrum Warfare that starts from traditional EW, but adds cyber warfare optical warfare, navigation warfare and includes also the tactical use of signal intelligence.

This new scenario creates a new discipline, CEMA (Cyber Electro-Magnetic Activities) that can be defined as "The synchronization and coordination of cyber and electromagnetic activities, delivering operational advantage thereby enabling freedom of movement, and effects, while simultaneously, denying and degrading adversaries' use of the electromagnetic environment and cyberspace". This new concept introduces a new order of battle oriented at disrupting or degrading enemy functionalities in military operations which are more and more net-centric and computer-centric.

42.6 INDIA PROCURES INDIGENOUS ELECTRONIC WARFARE SYSTEMS FOR $364M

The Indian Ministry of Defence has awarded Bharat Electronics Limited (BEL) a 30-billion-rupees ($364 million) contract to deliver two integrated Electronic Warfare Systems (EWS) to the Indian Army. Based on a Defence Electronics Research Laboratory (DLRL) design, BEL developed the Himshakti EWS system for mountainous regions.

According to Defence Update, the system can jam electromagnetic frequencies emitted by mobile phones, wireless phones, satellite phones, and radio receivers over 10,000 square kilometers (3,861 square miles). The wheeled system is ideal for mechanized forces, protecting them from electronic attack, the outlet added.

42.6.1 Features of Electronic Warfare System

The system can be used for signal intelligence, surveillance, analysis, interception, direction finding, and position fixing, listing, prioritizing and jamming of all communication and radar signals from HF (high frequency) to MMW (millimeter wave).

It will also protect our own electronic assets on the battlefield. Addition of such a system will paralyze the enemy and degrade its will to fight the war.

The contract has been awarded under the Indigenously Designed Developed and Manufactured category, expecting to generate approximately three lakh man-days over a period of two years.

42.7 INDIAN ARMY PURSUING NEW ELECTRONIC WARFARE ARCHITECTURE

The Indian Army is looking to devise an EW architecture separate from the highly classified processes associated with signals intelligence. After deciding to split up its integrated signals intelligence and electronic warfare platform, the Army is pursuing a new architecture for its EW suite.

Following operational demonstrations, the Army determined that the concept for the Terrestrial Layer System-Brigade Combat Team (TLS_BCT) was not going to work the way it was intended or gain the efficiencies desired.

TLS-BCT was designed as the first integrated signals intelligence, cyber and electronic warfare platform, devised roughly six years ago. It has been described as a key enabler of Army priorities — considering the service has been without a program-of-record jammer for decades — that will support multi-domain operations. As initially conceived, it was to be mounted on Strykers and then Army Multi-Purpose Vehicle variant prototypes.

Outside experts had always voiced concern with such a setup given the highly classified nature and authorities that come with signals intelligence and the issues associated with putting that on the same platform as electronic warfare tools.

Now, the Army has decided to split the system up — along with the TLS-Echelons Above Brigade, designed primarily for divisions, corps and Multi-Domain Task Forces to sense across greater ranges than its brigade counterpart — into two separate systems.

TLS-BCT, specifically, this program was birthed with the concept that you could have one system where you had EW and SIGINT soldiers on board the same platform, operating at the same time, supporting different battlefield operating systems.

Although SIGINT and EW are very similar — they use very similar hardware, software — they have different mission threads and they often need to be at different places on the battlefield. They have different requirements in terms of timeliness and data sharing. The operational demonstration wasn't going to work the way that it was envisioned. This has now forced the Army to relook and reset its electronic warfare architecture as to not be tied to the signals intelligence production chain that requires a different hardware and software approach, Strayer said.

One of the key lessons being learned in conflicts like Ukraine is the need for speed. Classification is often a barrier to moving fast, forcing the Army in other portfolios to loosen the reins with concepts such as secure but unclassified-encrypted communications, which reduces overall network complexity and has had huge benefits in terms of interfacing with partner nations and eschewing the need for liaisons. That increases the pace of operations.

Moreover, as the Army shifts to the division as the unit of action instead of the brigade of the last 20 years during the global war on terror, higher classified networks such as signals intelligence will be pushed to higher echelons while smaller units such as brigade and below will need to be unburdened and empowered to share with coalition forces on faster timelines, such as near real-time.

This really needs to be pushed down to the unclassified level, so it can share with coalition partners and rapidly feed the fires targeting cycle without having to go up through national SIGINT chains.

As a result, the program office is gearing up to release a request for information to industry to inquire on the availability of commercially or government-owned, preferably off-the-shelf electronic warfare hardware and software ecosystems or architectures that the Army could leverage to tailor for the unique requirements of each echelon to converge on a common hardware/software architecture that's scalable across echelons.

We don't want to go trying to reinvent something. We think the commercial marketplace is really caught up based on the work that's going on with other services and throughout industry. We're excited to hear what industry has to offer in the coming months. That's going to drive our acquisition strategy for EW moving forward.

The new architecture it's pursuing is supposed to allow for the rapid collection, dissemination and reprogramming of signals in the field at the speed of war. The service wants the ability to have a computed architecture of standard CPUs and GPUs that can be purchased to facilitate the ability to pull out classification and identification of signals in the environment, and plug in third-party capabilities such as artificial intelligence and machine learning to keep pace with threats by being able to identify and classify signals.

All the while, the Army wants to be able to purchase commercial sensors given the rich marketplace that exists in the private sector now, rather than spending money to develop its own, unique, fit-for-purpose sensors. As the service is conducting market research on this evolving electronic warfare architecture, it's still pursing the signals intelligence system development, which is more mature than the electronic warfare portions.

The Army was "pretty close" on the Stryker-based signals intelligence system configuration, formerly TLS-BCT, at the operational demonstration last year. Officials will look to continue refining that effort with a customer test this year and a follow-on demonstration next year before getting it into the field.

On TLS-EAB, the Army started bifurcating the electronic warfare and signals intelligence capabilities based on lessons learned from Europe, despite initially envisioning it as one system that does it all. That program initiated after the brigade version. As a result, that effort is also going down two pathways: signals intelligence and electronic warfare.

Given the new approach with the electronic warfare architecture, the Army has asked the vendor to prioritize the signals intelligence portion of the system. The vendor is working on a hands-on physical integration of the desired signals intelligence architecture that could be a 20-foot container for the first prototype, allowing it to be mounted in different ways.

That prototype is scheduled to be delivered to the Army in early calendar 2025 but could slip to second quarter.

On the electronic warfare aspect, officials are using this system as the main component for defining and demonstrating the initial EW architecture, given the BCT portion of the program went so far with a demonstration and design on an integrated platform.

Once developed, the architecture will have the ability to tailor for the need at each echelon.

Next year will be a focal point for the effort, with demonstrations planned and the goal of selecting a common architecture by the end of 2025. It will then be instantiated in some physical prototypes in 2026 for both the EAB and hopefully BCT as well, according to Army plans.

42.8 SAMYUKTA ELECTRONIC WARFARE SYSTEM OF INDIA

Samyukta (lit. 'United') is a mobile integrated electronic warfare system. Touted to be the largest electronic warfare system in India, it was developed jointly by DRDO, Bharat Electronics Limited, Electronics Corporation of India Limited, and Corps of Signals of Indian Army. The System is fully mobile and is meant for tactical battlefield use. It covers wide range of frequencies and coverage of electromagnetic spectrum is handled by the communication segment and the non-communication segment. Its functions include various ELINT, COMINT and electronic attack (ECM) activities.

Each system operates on 145 ground mobile vehicles which has three communication and two non-communication segments and can cover an area of 150 x 70 km^2. The system has the capability for surveillance, analysis, interception, direction finding, and position fixing, listing, prioritizing and jamming of all communication and radar signals from HF to MMW.

The development of the system was led by Defence Electronics Research Laboratory, Bharat Electronics Limited, Electronics Corporation of India Limited, Corps of Signals of Indian Army and private companies like Data Patterns India Ltd (Chennai), CMC and Tata Power Company Ltd. Strategic Electronics Division (Tata Power SED). Around 40 companies also contributed by producing various components indigenously. The challenge was to tackle the sanctions imposed by the United States after 1998 nuclear tests conducted by India which banned the import of advanced electronic components. CMC and Tata Power SED jointly developed Command and Control Software having 10 million lines of code even though project was not attractive commercially.

42.9 CHINA'S ELECTRONIC WARFARE SURGE SHOCKS US IN SOUTH CHINA SEA

China's cutting-edge electronic warfare (EW) capabilities are transforming the balance of power in the South China Sea, as shown by a recent encounter between US and Chinese forces.

Recently, the **South China Morning Post (SCMP)** reported on China's enhanced EW capabilities by shedding light on a December 2023 incident between a US EA-18 Growler carrier-based EW aircraft and China's Type 055 cruiser Nanchang in the contested South China Sea.

In December 2023, the US Navy dismissed William Coulter, commander of US Electronic Attack Squadron 136 (VAQ-136), stationed on the USS Carl Vinson, citing a loss of confidence in his ability to command. The report says that a month later, the People's Liberation Army (PLA) recognized the Nanchang's crew for their actions against a US carrier fleet. It also notes that Chinese media highlighted an encounter involving an EA-18G, believed to be from Coulter's squadron, and the Nanchang cruiser. The report mentions that PLA scientists recently disclosed in a Radar & ECM journal article that AI-enhanced radar gave the Nanchang an advantage over the EA-18G's jamming capabilities. It claims that the EA-18G, manufactured by Boeing, has been upgraded since 2021 for future warfare but faces new challenges from the PLA–Navy's (PLA-N) integrated radar systems and communication strategies.

SCMP notes that these advancements allow PLA-N warships to form a "kill web" to counter the EA-18G's attacks.

It also says that the Nanchang's reported proactive tactics and successful engagement with US forces illustrate a shift in the PLA-N's EW approach. Much-improved Chinese EW capabilities developed after then-US Speaker of the House of Representatives Nancy Pelosi's controversial August 2022 visit to Taiwan may have enabled a feat. SCMP noted that the PLA failed to track and surveil the US Air Force transport plane carrying Pelosi during her visit despite deploying Type 055 cruisers and J-16D EW aircraft. The source says that almost all of the PLA's EW equipment failed to function because of electronic interference from Pelosi's escorting aircraft force. On Pelosi's aircraft escort, John Tkacik says in an August 2022 article for Taipei Times that it could have been a massive force of US F-15s that flew out of Kadena Air Force Base in Japan supported by the USS Ronald Reagan carrier strike group and the USS Tripoli with embarked F-35s stationed in the Philippine Sea.

From that experience, China may have improved its EW capabilities quickly by investing in new technologies and placing them in a more extensive kill web consisting of kinetic and non-kinetic elements. SCMP reported in February 2024 that Chinese scientists have invented a new class of EW equipment that can reportedly rapidly detect, decode and suppress enemy signals. The new system, SCMP says, allows the PLA to seamlessly monitor signals into the gigahertz zone, encompassing frequencies used by amateur radio and even Elon Musk's Starlink satellites. It notes that the equipment includes innovative signal processing chips and AI integration, enhancing China's ability to counter enemy jamming and maintain communication flow.

Furthermore, SCMP claims that in encounters with US Navy ships with EW activity, China has used electromagnetic-emitting equipment, including high-power phased array radars, to lock on to multiple targets including US carrier-based aircraft.

Aside from developing new tech, China may already have elevated EW into a strategic capability, integrating it into its multi-domain operations alongside other kinetic and non-kinetic capabilities in a complex kill web.

Asia Times noted in April 2024 that the rebranding of China's PLA Strategic Support Force (PLA-SSF) into the PLA-Information Support Force (PLA-ISF) highlights China's strategic shift towards technology-driven "intelligentized warfare."

The PLA-ISF is designed to integrate emerging AI, quantum and other technologies into China's multi-domain operational strategy against potential adversaries like the US and its allies. The rebranding reflects an evolution in Chinese military thought, transitioning from "informationized wars" to "intelligentized warfare" that includes EW, cyber operations and signals intelligence (SIGINT).

EW is also a key component of China's Multi-Domain Precision Warfare (MDPW) concept, which leverages AI and big data to identify and exploit weaknesses in US operational systems.

China's MDPW seeks to dismantle and destroy US kill chains by targeting critical information nodes such as aircraft and satellites through physical attacks and targeting information networks by using EW and cyberattacks.

While the US arguably still has the edge in EW, near-peer adversaries like China and Russia may be closing the gap.

Army Technology reported in May 2024 that the US spent US $5 billion on EW capabilities in 2024, accounting for 45% of global EW spending from 2021-2023, compared to just 14% by Russia and 13% by China. However, Army Technology says that the US' dominant position in the EW market is being challenged, as Russia, China and India's share is projected to increase by the next decade. The report notes that over the past two decades, Russia has exploited alleged US complacency in EW strategies, which have focused on counterinsurgency versus non-state actors.

It notes that in Ukraine, Russia has used EW to disrupt adversary battlefield networks, support conventional assault forces through SIGINT and jamming attacks, and secure captured territory against counterattacks. The source also mentions that Russia has used EW to disrupt regional civilian services such as GPS and telecoms.

Likewise, Army Technology mentions that China has mirrored Russia's use of EW and has equated information dominance with electromagnetic dominance. It says that in addition to shipborne EW equipment, China has installed such equipment and more in its occupied features in the South China Sea. In line with that, Matthew Funaiole and other writers highlight in a December 2021 CSIS article the expansion of China's facilities on Hainan Island, Subi Reef, and Fiery Cross Reef, which now includes satellite tracking, communication platforms, and systems potentially used in EW and SIGINT.

CHAPTER 43

SPACE WARFARE

43.1 INTRODUCTION TO SPACE WEAPONS

Space weapons are weapons used in space warfare. They include weapons that can attack space systems in orbit (for example, anti-satellite weapons), attack targets on the earth from space or disable missiles travelling through space. In the course of militarization of space, such weapons were developed mainly by the contesting superpowers during the Cold War, and some remain under development today.

(i) Space-to-space weapons

The Soviet **Almaz** secret military space station program was equipped with a fixed 23mm autocannon to prevent hostile interception or boarding by hostile forces. This was the first and so far the only space-to-space weapon to be fired in orbit.

The Soviet uncrewed **Polyus** weapons platform was designed to be equipped with a megawatt **carbon-dioxide laser** and a self-defense cannon.

(ii) Earth-to-space weapons

Anti-satellite weapons, which are primarily surface-to-space and air-to-space missiles, have been developed by the United States, the USSR/Russia, **India** and the People's Republic of China. Multiple test firings have been done as part of recent Chinese and U.S test programs that involved destroying an orbiting satellite. In general, the use of explosive and kinetic kill systems is limited to relatively low altitudes due to space debris issues and so as to avoid leaving debris from launch in orbit.

Strategic Defense Initiative (SDI)

On March 23, 1983, the then US President Ronald Reagan proposed the Strategic Defense Initiative, a research program with a goal of developing a defensive system which would destroy enemy ICBMs. The defensive system was nicknamed ***Star Wars***, after the movie, by its detractors. Some concepts of the system included Brilliant Pebbles, which were Kinetic Kill Vehicles, essentially small rockets launched from satellites toward their targets (a warhead, warhead bus, or even an upper stage of an ICBM). Other aspects included satellites in orbit carrying **powerful laser weapons**, plasma weapons, or particle beams. When a missile launch was detected, the satellite would fire at the missile (or warheads) and destroy it. Although no real hardware was ever manufactured for deployment, the military did test the use of **lasers mounted on Boeing 747s to destroy missiles in the 2000s**, however these were discontinued due to practical limitations of keeping a constant fleet airborne near potential launch sites due to the **lasers range limitations** keeping a small number from being sufficient. The tests took place at Edwards Air Force Base.

(iii) Space-to-Earth weapons

Orbital weaponry is any weapon that is in orbit around a large body such as a planet or moon. As of December 2022, there are no known operative orbital weapons systems, but several nations have deployed orbital surveillance networks to observe other nations or armed forces. Several orbital weaponry systems were designed by the United States and the Soviet Union during the Cold War. During World War II, Nazi Germany was also developing plans for an orbital weapon called the Sun gun, an orbital mirror that would have been used to focus and weaponize beams of sunlight.

Development of orbital weaponry was largely halted after the entry into force of the Outer Space Treaty and the SALT II treaty. These agreements prohibit weapons of mass destruction from being placed in space. As other weapons exist, notably those using kinetic bombardment, that would not violate these treaties, some private groups and government officials have proposed a Space Preservation Treaty which would ban the placement of any weaponry in outer space.

Orbital (Kinetic) bombardment

Orbital bombardment is the act of attacking targets on a planet, moon or other astronomical object from orbit around the object, rather than from an aircraft, or a platform beyond orbit. It has been proposed as a means of attack for several weapons systems concepts, including kinetic bombardment and as a nuclear delivery system.

During the Cold War, the Soviet Union deployed a Fractional Orbital Bombardment System from 1968 to 1983. Using this system, a nuclear warhead could be placed in low Earth orbit, and later de-orbited to hit any location on the Earth's surface. While the Soviets deployed a working version of the system, they were forbidden by the Outer Space Treaty to place live warheads in space. The fractional orbital bombardment system was phased out in January 1983 in compliance with the SALT II treaty of 1979, which, among other things, prohibited the deployment of systems capable of placing weapons of mass destruction in such a partial orbit.

Orbital bombardment systems with conventional warheads are permitted under the terms of SALT II. Some of the proposed systems rely on large tungsten carbide/uranium cermet rods dropped from orbit and depend on kinetic energy, rather than explosives, but their mass makes them prohibitively difficult to transport to orbit.

As of 2020 the only true orbital bombardment in history has been executed for scientific purposes. On 5 April 2019 the Japanese Hayabusa2 robotic space probe released an explosive device called an "impactor" from space onto the surface of asteroid 162173 Ryugu, in order to collect debris released by the explosion. The mission was successful and Hayabusa2 retrieved valuable samples of the celestial body which it brought back to Earth.

43.2 SIGNIFICANCE OF SPACE FOR MODERN WARFARE

The general population, including the armed forces, is becoming phenomenally more dependent on space-based technologies. Therefore, The primary objective has been to enhance their technological capabilities in space, minimize their susceptibilities, and elevate their significance as a strategic sphere. Space systems are crucial in various areas, such as navigation, communication, and remote piloting. Additionally, they are indispensable for operating GPS-guided weapons, surveillance, and information warfare. GPS signals have become essential to our day-to-day activities for various purposes, such as banking, travel, navigation, agriculture, and communication facilities, using the internet access. Reliance on satellite navigation accounts for approximately 6-7% of Western countries' Gross Domestic Product (GDP). Communications satellites are utilized not only for direct broadcast television but also to facilitate numerous terrestrial networks. In isolated regions of the globe, they could serve as the sole method of communication. In the foreseeable future, global broadband internet access could be facilitated by communications satellites. Satellites facilitate the acquisition of weather forecasts and enhance agricultural productivity. In addition, they assist us in strategizing for disaster relief efforts, locating and extracting natural resources, monitoring environmental well-being, and various other applications. The first Gulf War, which took place in 1991, is often denoted as the first space war, although it was not fought in space. Instead, the United States and coalition forces heavily depended on GPS and other satellite technology to further their interests in that conflict.

(i) ASATs

The proliferation of anti-satellite weapons (ASATs) is gradually becoming challenging. In 2007, China utilized a ground-based missile to obliterate an inactive communication satellite, resulting in the creation of approximately 3,000 fragments of debris that can be monitored. Similarly, the United States launched a missile to target a reconnaissance satellite during its descent from orbit in 2008. It is worth observing that Russia has also conducted various tests with ground-based missiles. **India conducted a successful ASAT test**, codenamed 'Mission Shakti', on March 27, 2019. The test successfully targeted and destroyed a live satellite in the low earth orbit using indigenous technology. India's space programmes have achieved substantial growth. like other nations,

India possesses space assets that must be protected and preserved. This achievement brought India to join the select group of nations, the US, Russia, and China, that possessed this capability. The use of Kinetic ASAT weapons has caused significant international concern due to the resulting creation of debris, which poses a potential threat to all space systems. Debris resulting from ASAT tests has penetrated orbits at higher altitudes than initially predicted, with a portion persisting in space. Currently, the potential for global disapproval has prevented kinetic ASATs from posing a more significant risk to security. Subsequently, space-based resources have facilitated enhanced capacity for land, naval and air forces. Given the dual use of numerous satellites, an armed conflict in space could have catastrophic effects on modern life. These satellites are a crucial element of ballistic missile defence, capable of detecting missiles immediately after launch and tracking their paths. The establishment of the United States Space Force has evoked a plethora of imaginative conjectures about the prospect of warfare within the Space realm. This created suspicion about the conduct of operations in the observers' minds.

(ii) Space Warfare vs Conventional Warfare

In contrast to terrestrial warfare, when opposing troops strive to control a specific physical area, satellites in orbit do not occupy invariably a singular position. Satellites are commonly employed in circular orbits exhibit velocities ranging from 3km/s to 8km/s, dependent upon their respective altitudes. In contrast, an average bullet travels approximately 0.75 kilometres per second. They appear quickly and then disappear. The volume of the section between the low earth orbit as well as the geostationary orbit is almost 200 trillion cubic kilometres. The volume is 190 times greater than that of earth. Placing a satellite in orbit requires significant time and delta-V (change in velocity) to execute phasing maneuvers. Space operations, therefore, including moves and actions, necessitate meticulous pre-planning. any conflict in space will exhibit a significantly reduced pace and heightened intentionality. Throughout history, warfare has invariably ventured into unexplored regions to achieve the element of surprise. It might not be possible to precisely quantify the prospective role that space could play in future conflicts.

Future space conflicts will primarily revolve around satellites. Several nations presently possess military satellites deployed within the space domain. The United States, Russia, and China are widely recognized as the prominent triumvirate in global power dynamics. However, it is noteworthy that nations such as France, India, Israel, and the United Kingdom have also made significant explorations to enter the competitive domain of military aerospace development.

Could we anticipate the emergence of conflict solely within the dimensions of outer space? It is highly improbable that we shall witness any conflict exclusively transpiring within the celestial world in the foreseeable future. Space warfare remains predominantly theoretical today, as no tangible instance of armed conflict in outer space has transpired. nevertheless, numerous concepts and technologies have been explored for implementation in space combat. as the existence of space technology continues to proliferate among countries, its implementation in various world conflicts is set to increase.

There are two main strands through which satellites can be drawn into a war in space: "cyber-attacks" and "anti-satellite missiles."

The fundamental nature of space warfare is expected to differ significantly from the typical forms of terrestrial warfare. However, the physical limitations of space may render its use impossible. In the huge expanse that is outer-space, there is no air resistance or gravity making it an environment where weapons can be used as much as one wishes.

Besides some specific weaponry tactics, what else would distinguish space combat, are:

- Space warfare would have an extraordinary swiftness and span our long distances, taking place within seconds. Space amplifies weapon platforms by increasing velocities for missile trajectories, thus, enhancing speed and accuracy.
- Getting weapons into space would create tremendous precision, making it difficult for enemy spacecraft and satellites to try to hide. It is also important to note that aerial equipment cannot be shielded in the empty void of space, making aerial assets in the air highly vulnerable to potential acts of aggression.

- Ships in the sea and satellites in space are highly vulnerable to attacks due to a lack of relative shielding. Moreover, wars in space will be strongly interconnected with ongoing events on earth.

43.3 TYPES OF WEAPONS FOR SPACE WARFARE

The weapons used in space warfare may differ greatly from those on earth. In the great emptiness of space, there is no atmospheric pressure to resist the motion of solid particles flying out of a gun and no medium to slow down these projectiles.

(i) Kinetic Weapons

Conventional kinetic armaments such as projectiles and missiles would multiply their destructive potential, greatly increasing their potential. Moreover, with no atmosphere to resist their movement through space, we could **use high-powered laser and particle beam technologies to neutralize or utterly destroy enemy ships.**

(ii) Non-Kinetic Weapons

Non-kinetic weapons like cyber warfare and electromagnetic pulses (EMPs) are also being used in space warfare. EMPs have the capability to render electronic systems inoperable, leading to the possibility of substantial disorder and disruption. Cyberwarfare has the potential to intrude and interrupt adversary's communication networks, command and control systems, satellite navigation and other systems.

(iii) Strategic Considerations

Space warfare would necessitate a completely different set of strategic considerations than conventional land warfare. rapid response to attacks in space would be challenging due to the immensity and severity of the environment and the speed of movements. It would be difficult to conceal spacecraft from detection because there would be no air resistance to contend with. consequently, any space war in the future would probably be waged using long-range weaponry and emphasis on hitting enemy infrastructure instead of directly confronting enemy forces.

43.4 CHINA'S SPACE CAPABILITY: A CONCERN FOR US

The People's Republic of China (PRC) has rapidly advanced in the space domain on several fronts simultaneously. Gen. Stephen Whiting, commander of U.S. Space Command testified in early 2024 that, "The PRC is moving breathtakingly fast in space. America must rapidly increase the timeliness, quality and quantity of our critical national space and missile defense systems to match China's speed and maintain our advantage." The revelation in February 2024 that Russia was developing a nuclear-enabled anti-satellite weapon led to speculation about whether it was a fission reactor-powered electronic warfare satellite or a nuclear detonation device, with the latter seeming to be the more likely, despite Russia's treaty commitments. Such developments have also stoked anxieties about China's intentions in space, including after its demonstrations of a so-called "fractional orbital bombardment system" and the deployment of a reusable spaceplane. PRC researchers have themselves studied the effects of upper atmospheric nuclear detonation on LEO, modeling radiological effects based on altitude for optimal strategic utility. While Russia has garnered much of the spotlight with its employment or threatened employment of novel weapons systems, China's development of similar emerging technologies deserves as much or more scrutiny. On the other hand, there have been signs of internal frictions and underlying problems within the PRC defense and aerospace establishment.

The PRC remains a hard analytic target requiring scrutiny of both its triumphs and its points of weakness. Understanding China Aerospace Studies Institute June 2024, this mix of factors is key for assessing how to best cooperate and compete with this rising space power, and the various implications for space domain awareness. A Challenging Adversary Foremost in the minds of many in the community are the achievements and capabilities that PRC military space programs have developed in recent years. These include development and maturation of reusable space plane operations, the possibility of an alarming return to the Cold War with a potential Fractional Orbital Bombardment hypersonic vehicle, ongoing rendezvous and proximity operations particularly targeting the GEO belt, the prospect of using nuclear weapons in an anti-satellite capacity, and cyber-electronic warfare (EW) threats.

U.S. defense leadership has also sounded the alarm at the steady rise in the number of intelligence satellites China has placed in orbit: "As of January 2024, the PRC has deployed a fleet of 359 intelligence satellites, [Gen. Stephen Whiting] said, 'more than tripling its on-orbit collection presence since 2018." After the establishment of the People's Liberation Army (PLA) Strategic Support Force in 2015, it was clear that space as a domain would be an important strategic focus for future developments, and as the PRC builds out its satellite operational support networks along with the projects below, the military continues to play a key role in driving space development. Researchers at institutions affiliated with the Aerospace Force and former Strategic Support Force have called for using a "combination of soft and hard kill methods" to hold at risk orbital assets, including commercial satellites like the Starlink constellation.

Reusable Experimental Spacecraft PRC scientists, defense industry engineers, and military analysts have long been fascinated by the concept of a reusable spaceplane, particularly after NASA retired the space shuttle and the U.S. Air Force operationalized the long dormant concept of an uncrewed reusable spaceplane in the X-37B project. When first launched, PRC analyst speculated about whether it could serve as a platform for grappler arms, high-power microwave weapons, or **laser weapons**. China Aerospace Science and Technology Corporation China Aerospace Studies (CASC) first launched the "Reusable Experimental Spacecraft in 2020 and has launched subsequent missions in 2022 and 2023, with the turnaround between recovery in May 2023 and launch in October demonstrating rapid redeployment capabilities. Little has been revealed in public about the payloads of the missions, other than their use as an experimental platform to test reusable aerospace technologies and to promote "the peaceful uses of outer space." Each of the three missions to date, however, have released payloads in orbit, assessed by Western observes to be RPO (rendezvous and proximity operations)-capable microsatellites that were able to maintain station with their mothership. The latest spaceplane launch resulted in at least six objects in orbit, with two that "gave off radio signals." While the applications of this or any other PRC spaceplane platform have been kept confidential, there are implications for development of hypersonic flight technology, RPO, directed energy, satellite-to-satellite ISR (intelligence, surveillance, and reconnaissance), signals intelligence, and more. The tacit message from the PRC seems to be, whatever the X-37B vehicle can do, we can do, as well, including keeping mission details private, making independent tracking and observation of the vehicle a vital mission. For space situational awareness purposes, maintaining observations of a maneuverable vehicle with long endurance will necessitate greater investment in both terrestrial and on-orbit tracking and characterization assets across the geographic and spatial domain.

Fractional Orbital Bombardment System

PRC efforts to revive space technology concepts like a reusable spaceplane may be harkening back to the Space Shuttle or Buran, but China has seemingly reached even further back to revive the Soviet-era concept of a "fractional orbital bombardment system" (FOBS), in the form of a hypersonic boost-glide system. The original rationale for launching a nuclear weapon on a fractional orbital trajectory was to evade early warning radars by taking a different approach vector than traditional ICBM launchers. Public details on the capabilities of the PRC's system are still closely guarded secrets and the Foreign Ministry simply indicated the test was of a hypersonic spaceplane, but may have been referring to a different test altogether that occurred around the same time as the July 2021 launch. PRC military doctrinal writings indicate that developing an orbital bombardment capability could be part of a deterrence and compellence strategy aimed at holding assets at risk in expanded regions: The use of an orbital bombardment system could increase PLA power projection capabilities against bases and territories globally, including targets in the 50 states. The use of an orbital bombardment system can complicate U.S. missile defenses by forcing the U.S. to defend against joint and combined arms attacks from multiple directions.

If the PRC hypersonic vehicle is intended to carry a nuclear payload and conducts a full orbit, such a system would be in violation of the 1967 Outer Space Treaty, to which the PRC is a signatory. Demonstrating such a system would be a sharp break from both PRC diplomatic commitments as well as longstanding positions it has taken in the Conference on Disarmament and the Committee on the Peaceful Uses of Outer Space, where both China and Russia have consistently held that weaponizing the space domain was undesirable. The likelihood that the PRC would invest prestige and sustained funding into a new weapons system that undid those previous international commitments seems incongruous and uncharacteristically risky.

An alternative explanation could be found in the strange operational behavior of the reentry vehicle observed by U.S intelligence and as reported in the media. U.S. analysts were apparently surprised enough to characterize the test as having "defied the laws of physics" when the primary vehicle appeared to have fired a secondary munition—possibly a countermeasure, air-to-air missile, or something else—while traveling at hypersonic speeds. This would be a capability no other nation has demonstrated or has claimed to be pursuing:

Experts at DARPA, the Pentagon's advanced research agency, remain unsure how China managed to fire countermeasures from a vehicle travelling at hypersonic speeds. Military experts have been poring over data related to the test to understand how China mastered the technology. They are also debating the purpose of the projectile, which was fired by the hypersonic vehicle with no obvious target of its own, before plunging into the water. Through the nuclear lens, such a secondary payload could represent a MIRV/MaRV (multiple independently targetable reentry vehicle/maneuverable reentry vehicle) warhead, an atmospheric countermeasure, or a self-defense capability. Yet, Secretary Kendall admitted in his initial disclosure that this interpretation was colored by Cold War history. The original Soviet FOBS did not include a maneuvering hypersonic glide vehicle, much less a secondary vehicle.

43.5 'ANTARIKSHA ABHYAS – 2024' BY INDIA FOR READINESS IN SPACE WARFARE

Defence Space Agency of Headquarters Integrated Defence Staff successfully conducted the Space Table Top Exercise Antariksha Abhyas-2024 from 11 - 13 Nov 2024; a significant milestone aimed at bolstering the strategic readiness of the Indian Armed Forces in the domain of **space warfare**. This pioneering event marked a crucial step in strengthening India's space-based operational capabilities, enhancing tri-services integration for space security.

Key components of the exercise included focused discussions on emerging space technologies, space situational awareness and India's space programmes. The discussions highlighted the importance of monitoring and protecting critical assets and maintaining situational awareness in the increasingly contested space environment.

Throughout the three-day event, participants engaged in scenario-based exercises, facilitated by subject experts from various Ministries and Departments of the Government of India, besides military, scientific and academia. The experts provided valuable insights into the present and future landscape of military space capabilities and technologies, elucidating specific challenges faced in defence space operations and also the evolving nature of space safety, security, and international space laws.

Antariksha Abhyas 2024 successfully met its objectives of improving interoperability, fostering mutual understanding and enhancing cohesion between the tri-services and Defence Space Agency. Key outcomes included refined strategies for operational preparedness, a robust framework for future collaboration and a clear road map for advancing India's Space doctrine and capabilities in the line with National Security objectives. This exercise marks a pivotal point in India's journey to secure its interests in space, reaffirming the nation's advancement and strategic focus on this critical domain.

43.6 IS INDIA PREPARED FOR EMERGING SPACE WARFARE?

The development of non-kinetic weapons and a diversified space defence strategy is critical for safeguarding our assets, says Lt Gen AK Bhatt (Retd), Director General of Indian Space Association (ISpA).

In an episode of the State of Economy Podcast, Business-line's Dalip Singh spoke with Lt Gen AK Bhatt (Retd), Director General of the Indian Space Association (ISpA), who shared valuable insights into India's evolving space defence strategies. With the militarization of space accelerating globally, General Bhatt discussed how space is no longer just a domain for peaceful exploration but a vital strategic asset for national security.

The episode started with how nations are increasingly using space for military purposes, such as satellite communication, intelligence gathering, surveillance, and navigation. General Bhatt also addressed India's space defence roadmap, emphasizing the government's strategic investments, including satellite-based surveillance programs and growing collaboration with the private sector.

The discussion touched on the competition with global powers, particularly China. "China is aspiring not to compete in space with India, but to compete with USA and their concerns for ASR, for communication, for navigation are global. They want to protect and have visibility in the first chain of islands and second chain of islands."

General Bhatt emphasized the importance of space situational awareness (SSA), a system that tracks satellites and debris to prevent accidents and potential threats from adversarial actions. This growing need for global cooperation in space defence, as highlighted by Bhatt, is crucial for maintaining space security.

General Bhatt went on to explain how the new form of space warfare can disrupt communication systems and affect the overall operational integrity of a nation's space infrastructure. General Bhatt pointed out a critical shift: Space warfare is no longer confined to government-controlled entities, and private players have become essential in providing support during conflicts.

CHAPTER 44

HYBRID WARFARE

44.1 WHAT IS HYBRID WARFARE?

Hybrid warfare is a theory of military strategy, first proposed by Frank Hoffman, which employs political warfare and blends conventional warfare, irregular warfare, and cyberwarfare with other influencing methods, such as fake news, diplomacy, lawfare, regime change, and foreign electoral intervention. By combining kinetic operations with subversive efforts, the aggressor intends to avoid attribution or retribution.

Hybrid warfare can be used to describe the flexible and complex dynamics of the battlespace requiring a highly adaptable and resilient response. There are a variety of terms used to refer to the hybrid war concept: hybrid war, hybrid warfare, hybrid threat, or hybrid adversary (as well as non-linear war, non-traditional war or special war). US military bodies tend to speak in terms of a hybrid threat, while academic literature speaks of a hybrid warfare.

44.1.1 Relation of Hybrid Warfare to the Grey-Zone

The concept of grey-zone conflicts or warfare is distinct from the concept of hybrid warfare, although the two are intimately linked as in the modern era states most often apply unconventional tools and hybrid techniques in the grey-zone. However many of the unconventional tools used by states in the grey-zone such as propaganda campaigns, economic pressure, and the use of non-state entities do not cross over the threshold into formalized state-level aggression.

44.1.2 Effectiveness of Hybrid Warfare

Traditional militaries find it hard to respond to hybrid warfare since it is hard to agree on the source of the conflict. An article published in Global Security Review, "What is Hybrid Warfare?" compares the notion of hybrid warfare to the Russian concept of "non-linear" warfare, which it defines as the deployment of "conventional and irregular military forces in conjunction with psychological, economic, political, and cyber assaults." The article partially attributes the difficulty to the "rigid" or static military taxonomy used by NATO to define the very concept of warfare.

To counter a hybrid threat, hard power is often insufficient. Often, the conflict evolves under the radar, and even a "rapid" response turns out to be too late. Overwhelming force is an insufficient deterrent. Many traditional militaries lack the flexibility to shift tactics, priorities, and objectives constantly.

44.2 HISTORY OF HYBRID WARFARE

The combination of conventional and irregular methods is not new and has been used throughout history. A few examples of that type of combat are found in the American Revolutionary War (a combination of George Washington's Continental Army with militia forces) and the Napoleonic Wars (British regulars co-operated with Spanish guerrillas).

One can find examples of hybrid warfare in smaller conflicts during the 19th century. For instance, between 1837 and 1840, Rafael Carrera, a Conservative peasant rebel leader in Guatemala, waged a successful military campaign against the Liberals and the federal government of Central America by using a strategy that combined classical guerrilla tactics with conventional operations. Carrera's hybrid approach to warfare gave him the edge over his numerically-superior and better-armed enemies.

The Soviet Union engaged in an early case of hybrid warfare in 1944. When the Tuvan Army was away in Europe, fighting along the Red Army against the Third Reich, Moscow annexed the Tuvan People's Republic by pressing the Tuvan government to ask for joining the Soviet Union.

(i) After 1945

The Vietnam War saw hybrid warfare tactics employed by both sides, with the US using the CIA to support civil war parties in Laos and the Cambodian Civil War as well as ethnic groups inside Vietnam for its cause and the Soviet Union supported the Viet Cong militia.

(ii) After 1989

The end of the Cold War created a unipolar system with a preponderant American military power. Though that has tempered traditional conflicts, regional conflicts and threats that leverage the weaknesses of conventional military structures are becoming more frequent.

At the same time, the sophistication and the lethality of non-state actors has increased. They are well-armed with technologically advanced weapons, which are now available at low prices. Similarly, commercial technologies such as cellphones and digital networks are being adapted to the battlefield. Another new element is the ability of non-state actors to persist within the modern system.

44.3 EXAMPLES OF HYBRID WARFARE

(i) 2006-Israel-Hezbollah Conflict

One of the most often quoted examples of a hybrid war is the 2006-conflict between Israel and the Hezbollah. The Hezbollah is a sophisticated non-state actor sponsored by Iran. While the group often acts as a proxy for Iran, it has its own agenda. It was Hezbollah policy, rather than Iran's, that led to the kidnapping of Israeli troops that was the impetus for the war. The war featured about 3,000 Hezbollah fighters embedded in the local population attacked by about 30,000 Israeli regular troops.

The group used decentralized cells composed of guerrillas and regular troops armed with weaponry that nation states use such as precision missiles, rockets, armed unmanned aerial vehicles, and advanced improvised explosive devices. Hezbollah cells downed Israeli helicopters, damaged Merkava-IV tanks, communicated with encrypted cell phones, and monitored Israeli troops movements with night vision and thermal imaging devices. Iranian Quds Force operatives acted as mentors and suppliers of advanced systems.

Hezbollah leveraged mass communication immediately distributing battlefield photos and videos dominating the perception battle throughout the conflict. Israel did not lose the war on the battlefield but lost the information battle as the overwhelming perception at the time was of Israeli defeat.

However, despite the perceived successes on the tactical level and propaganda, Hezbollah casualties were heavy (at least 600 killed and an unknown number wounded) and in retrospect, Israel achieved its political goal of deterring Hezbollah attacks - from summer 2000 to summer 2006 Hezbollah conducted approximately 200 attacks on Israel, but during the following six years Hezbollah refrained completely from attacking Israel and distanced itself vociferously from attacks on Israel by other parties based in Lebanon.

(ii) 2014-Islamic State advance into Iraq

Islamic State is a non-state actor utilizing hybrid tactics against the conventional Iraqi military. Islamic State has transitional aspirations, it uses irregular and regular tactics, and it uses terror as a part of its arsenal. In response, the state of Iraq itself turned to hybrid tactics utilizing non-state and international actors to counter the IS advance. United States likewise is a hybrid participant through a combination of traditional air power, advisers to Iraqi government troops, Kurdish peshmerga, and sectarian militias, and training opposition forces within Syria. The Iraq–Syria hybrid war is a conflict with an interconnected group of state and non-state actors pursuing overlapping goals and a weak local state.

(iii) Russian involvement with Middle-Eastern refugees

General Philip Breedlove, in a US Senate hearing February 2016, claimed that Russia is using refugees to weaken Europe, directing the influx of refugees in the continent to destabilize areas and regions in terms of economy and to create social unrest. On 10 February 2016, Finnish Defence Minister Jussi Niinistö told a meeting of NATO Defence Ministers that Finland expects Russia to open a second front, where as many as 1 million migrants may arrive over the Finnish/Russian border. A similar statement was made by Ilkka Kanerva, Finland's former foreign minister and now chairman of the country's parliamentary Defense Committee.

(iv) United States on Russian activities

Moscow has accused Washington of conducting hybrid warfare against Russia during the colour revolutions. Its perception of being at war or in a permanent state of conflict with the US and its allies was furthered by the 2014 Maidan uprising in Ukraine.

Speaking at the Valdai Discussion Club in November 2014, Russian Foreign Minister Sergey Lavrov said:

"It is an interesting term, but I would apply it above all to the United States and its war strategy – it is truly a hybrid war aimed not so much at defeating the enemy militarily as at changing the regimes in the states that pursue a policy Washington does not like. It is using financial and economic pressure, information attacks, using others on the perimeter of a corresponding state as proxies and of course information and ideological pressure through externally financed non-governmental organizations. Is it not a hybrid process and not what we call war?"

44.4 FUTURE OF HYBRID WARFARE FOR NATO

In the future, NATO should embrace the concept of hybrid warfare and incorporate it into how it deters and defends. Starting with resiliency as the first step, NATO can use its inherent strengths to put adversaries at a disadvantage and strengthen its strategic approach to hybrid warfare.

Warfare evolves and belies the Western tendency to make black and white distinctions, such as that between conventional and irregular warfare. This reflects a rigid categorization rather than a continuum of conflict and a variety of methods and modes of warfare. However, in spite of this tendency, NATO should more fully integrate hybrid warfare in its defense strategies. To do this, it must overcome natural disadvantages while using its own strengths, interests, and values to establish resilience against adversaries' hybrid activities.

The original perspective on hybrid threats reflected a violent blend of regular capabilities and irregular tactics. This mode of conflict was defined as an adversary that "simultaneously and adaptively employs a fused mix of conventional weapons, irregular tactics, catastrophic terrorism, and criminal behavior in the battlespace to obtain desired political objectives." Violent conflict is increasingly likely in the future, and it is very likely to be hybrid in its character due to the desire of great powers to avoid direct confrontation.

Perhaps because of its intensive but stalled battle with Kyiv, Moscow is starting to expand its active measures and hybrid conflict tactics throughout Europe. This includes sabotage, arson, and cyberattacks. NATO's concern about the dark side of Russian hybrid conflict is warranted. NATO publicly warned Russia about its hybrid activities identified in Czechia, Estonia, Germany, Latvia, Lithuania, Poland, and the United Kingdom. Regardless of how it fares in Ukraine, it is clear that Russia will continue to try to destabilize the West, interfering through funding friendly media outlets and subverting political parties. As noted by Mark Galeotti, there is nothing novel in Russia's use of these techniques, most of which are as old as time. NATO countries will have to be prepared to deflect this "war in the shadows" while also preparing its collective defense capabilities.

The fusion of advanced capabilities with irregular forces and tactics is also key, as borne out repeatedly over the last decade in conflicts involving Hezbollah, the conflict in Syria, and Russian campaigns in Georgia and Ukraine. The Syrian case study is concerning, as the civil war in Syria has evolved into a hybrid conflict where Russian military support, Syrian irregulars, and mercenaries are fighting against an equally blended mix of local irregulars supported by U.S. air and artillery support.

Hybrid conflict takes an indirect approach via proxy forces to increase deniability. Autocratic powers would prefer to advance their interests as sponsors of proxy forces. Proxy wars, appealing to some powers as "warfare on the cheap," are historically ubiquitous. Sponsorship from a major power can generate hybrid threats more readily by the provision of advanced military capabilities to the proxy. Russian private military companies like the Wagner Group operate as subsidiaries of the Kremlin and have operated in Syria, Ukraine, and Libya. Putin has provided advanced weapons, intelligence, and various "volunteers" to these conflicts. Although the private military companies have a rather mixed track record, Russia's use of the companies will likely continue. More important to major states, the lack of direct contact helps defuse the potential for attribution and escalation. Proxy wars have taken many forms in the past, and the employment of cyberattacks and drones may continue to expand the use of proxy actors.

Chinese thinking about more unconventional approaches to conflict includes the "three types of warfare" (psychological, political, and lawfare) to achieve its goals. For example, its investment in the United Front department serves to control the behavior and narrative of Chinese diasporas and foreign communities, while using front organizations to maximize Chinese influence overseas. China is also increasing its use of commercial security operations to protect its economic and political interests abroad. The probability of indirect and proxy Chinese intervention should be anticipated as Chinese interests expand in depth and reach. Lastly, China's continued use of blurred tactics and quasi-military forces in the maritime domain manifestly demonstrates Chinese use of coercive force in innovative ways. China has been carefully adapting its maritime assets and extending its influence, conducting hybrid or grey zone actions with "Chinese characteristics." As NATO countries deploy forces to the Pacific region to defend international norms and order, they will face Chinese hybrid warfare activities there as well as in other regions, as China's reach expands to the Middle East and possibly the Arctic.

In response, NATO has adopted an expansive definition of hybrid conflict that includes political interference, sabotage, subversion, and malign influence activities. That perspective captures the alliance's persistent challenge from Moscow, which has long favored indirect methods or "Active Measures," and now includes mass media disinformation. In response, NATO places resiliency as the "first line" of deterrence and defense. This emphasis is a key shift but is also difficult to do, or do quickly. The very things that make NATO and its democracies strong in the face of autocracies—laws, principles, and values—also make collective action and decisions difficult in the grey zone. It is relatively easier to focus on nuclear and conventional military forces than to discuss societal resiliency in the context of an alliance.

The future of hybrid warfare for NATO is likely to see the most consequential form of armed conflict in the lower-to-middle portion of the conflict spectrum, likely with surrogates and proxies. Fortunately, NATO has invested time and thought into this problem and has slowly integrated countering surrogates and proxies into its strategies for collective defense and integrated deterrence. As Russia continues to antagonize the West and China looks to gain advantages across the globe, NATO will have to reinforce its ability to detect, deter, and respond to this type of malign assault on its values and freedoms.

44.5 HYBRID WARFARE: CONCEPT & IMPLICATIONS FOR INDIA

Discussions on hybrid conflicts have existed in India since at least the era of Kautilya (the ancient Indian polymath also known as Chanakya). However, India's current military system is a colonial legacy and has developed from Europe's set-piece battlefields. A major challenge for the Indian armed forces is to adapt to the desired threshold of the complex and hybrid battlespace.

Given the threats posed by its adversaries (chiefly, **Pakistan and China**), India has prioritized sub-conventional and conventional warfare, albeit under the shadow of nuclear war. Still, future threats will not be simplistic and will give rise to complex and hybrid structures, necessitating a complete overhaul of force application and warfighting doctrines and strategies. To coherently define a specific objective and the desired end status of any war or conflict in compliance with the national and military aim—which will remain paramount and non-negotiable—it is necessary to establish an overall campaign 'grand strategy' with multiple sub-strategies. Countries require a strong political will, reinforced by a firm and decisive joint services capability (that goes beyond the land, air, and sea domains) to protect their core national interest. This mandates a strategic transition from the present focus on territorial integrity towards a more interest-based capability. As such, India must reorient existing political thought towards the development of a cohesive national mindset that does not rely solely on soft power and status quo options.

Kautilya's *Arthashastra*, a third-century treatise, and *Nitisara* by Kamandaki (a lesser-known military strategist) are essential readings on the Indian history of warfare. The key lesson in both is that a complex **and hybrid mix of elements of national power** is necessary to influence the battlefield before any actual force is used during conflict or thereafter. While Kautilya expounds on diplomacy, dissent, economic coercion, and the continuum of force before, during, and after conflict, Kamandaki adds deception, psychological warfare, and benign neglect to these tools of intimidation. When applied to the present context of the information age, these concepts will mean weaponizing every modern, niche, and disruptive tool, technology, and ecosystem.

To successfully strategize its joint warfare doctrines and commensurately address the 'threat and capability' paradigm of war, a country and its military will need to adopt an integrated multipronged approach to achieve 'maximum gain', which can only emerge from professional military education and the civil-military fusion. As the adversaries' game plan could comprise a diverse range of threats—from low-cost and low-tech to high cost and niche technologies—it is incumbent that all ingredients of national power are coalesced into the nation's strategized joint operational plans. The political leadership must encourage politico-military diplomacy and astute statecraft policies that push the optimization of soft power strategic communications to deter conflicts, while progressively developing and preserving a robust and credible military capability to safeguard the country's vital interests.

The Government of India must utilize inter-ministerial and inter-agency planning and incorporation processes by coordinating civil-military fusion with the appropriate civilian agencies, and engage the expertise of the private sector, including non-governmental organizations and academia. Such an augmentation will enhance the ability to address the non-military aspects of conflict. At the operational level, however, India must review and further refine its integrated joint warfighting principles to engage with the complex and hybridized character of warfare. The military must reassess its ideologies as they pertain to operational art and attempt to impeccably integrate new pillars of national power into the canvas of its strategic plans. Crucial to achieving this critical target is preparing the military's highest leaders with a thorough understanding of how strategizing in the military profession is organized and conducted, and the ways in which it is related to strategy and policy. Under the agile leadership of the Chief of Defence Staff (CDS), India's joint and integrated military planning can become doctrinally more perspicuous, adaptive, synergized, and combat-responsive in the strategic context.

India's ability to grasp the new signposts of evolving non-contact, non-kinetic, and hybrid warfare precepts, in addition to recognizing that warfare has now expanded beyond the conventional dimensions of land, sea, and air to that of space and information, is vital to achieve success in the emergent environment of **hybrid conflicts**. As India engages, either directly or indirectly, in such multispectral warfare, it will need professionals who can outperform the enemy even in conditions of volatility, uncertainty, complexity, and ambiguity (popularly referred to as a VUCA environment). The complexity of hybrid threats makes it imperative for future leaders to understand the implications, spin effects, and trade-offs between preparing for and carrying out counterinsurgency, partner-building, and stability operations, and enhancing cyber and information capabilities, while maintaining a conventional ascendency in combat. As a result, capability development is essential for India's armed forces, who can then decide what levels of knowledge and capability fusion are required to achieve effective operational readiness for a multifront scenario in a hybrid combat environment.

44.5.1 On the Western Front: Pakistan-Centric Threats

Pakistan continues to engage in antagonistic rhetoric against India even amid unprecedented domestic turmoil. The Pakistan military is currently focused on maintaining a hybrid strategy against the Taliban on the country's Western front (Afghanistan) and India on its Eastern front. It is important to understand Pakistan's use of **hybrid warfare against India** in the context of its origins as an independent country; in the aftermath of Partition, the Pakistani Army, unlike the Indian Army, became politicized and emerged as a predominant power in the nation's ruling structure (which now includes the government, judiciary, and military).

To understand the threats posed by Pakistan, it is crucial to examine some of the geopolitical and strategic imperatives that define the India-Pakistan relationship. Jammu and Kashmir (J&K) is important to Pakistan for geostrategic, socioeconomic, cultural, and demographic reasons. Pakistan has long suffered from the fear of a lack of 'strategic depth'. At the time of Partition, Pakistan felt that this non-adequacy of depth would force it to surrender in the event of a bold military offensive by India. An additional concern was that the sizeable Indian strike formations could threaten the epicentre of Pakistan's Punjab, a prosperous state. The proximity of J&K to Pakistan's Khyber Pakhtunkhwa province and its large Pashtun population also posed a demographic and geopolitical challenge for the Pakistani authorities. Hydro-politics is another issue for Pakistan, which is the most canal-irrigated country globally and has the largest network of inundation systems. Waters of the Indus River and its tributaries, which flow from J&K to Pakistan, are crucial to the country's economy. As a riparian state, India has the option to openly control the waterways and use 'water as a weapon'.

In the aftermath of the 1971 Bangladesh Liberation War, Pakistan unified based on religious ideology and fundamentalism to prevent any potential further division within the country. The philosophy of Islamization, initiated by former President Muhammad Zia-ul-Haq, went beyond the imposition of strict rules on adhering to Islamic practices to include transforming and building an Islamic army inspired by Quranic teachings. This has been a significant aspect of Pakistan's doctrinal strategy of covert warfare.

Pakistan began pursuing a strategy of asymmetric/proxy war against India as it entered an alliance with the US during the Cold War, resulting in a massive sale of arms and transfer of aid to Islamabad. Additionally, China has long provided Pakistan with major military, technical, and economic support, as well as nuclear material, including sensitive nuclear technology and equipment.

Pakistan has also exploited Sikh separatism in India's Punjab state to promote the idea of Khalistan, a sovereign Sikh state, as a means to annex J&K. In the late 1980s, Pakistan's covert war in J&K began to concretize in the form of an insurgency in the Kashmir Valley. Pakistan's Inter-Services Intelligence (ISI) agency and the army have calibrated the insurgency in J&K to keep it below India's war threshold, thereby preventing covert operations from escalating into a full-fledged conflict. The aim of Pakistan's infiltration into India through terrorists with large amounts of counterfeit money and narcoterrorism is to create an imbalance in J&K and the country. **India views Pakistan's hybrid warfare as a well-planned and organized strategy, focused on support from local (Pakistani) population and infrastructure, and waged covertly via proxies.**

Notably, Pakistan has conducted anti-terrorist campaigns, ostensibly to placate the US and continue receiving military aid and replacements for its diminishing hardware. At the same time, it has also continued using jihadi outfits, such as the Taliban, Lashkar-e-Taiba (LeT), Jaish-e-Mohammed (JeM), and Hizbul Mujahideen (HM), as strategic assets to advance its interests in India and Afghanistan. Anticipating the forms and scope of Pakistan's potential **plans for hybrid warfare** to draw India into conflict will help New Delhi articulate more efficient responses.

Islamabad's illegitimate control over Pakistan-occupied Kashmir, and the ongoing political unrest and unstable security environment in the country have fomented conditions for terrorists to obtain access to nuclear equipment—admittedly one of Pakistan's inherent apprehensions—and launch a chemical, biological, radiological, and nuclear attack in India, by either triggering an industrial accident or using chemical and biological agents (such as mustard gas, VX nerve gas, or anthrax).

In another scenario, Pakistan—through proxies such as the LeT, HM, and the crime syndicates based in some Arab countries—may attempt to destabilize the Indian economy by hijacking a large ship (possibly carrying liquefied natural gas) and setting off a massive explosion with collateral damage in the constricted areas of one of India's chief commercial ports.

Since its earlier attempts to alter the status of J&K failed, Pakistan shifted to proxy war and other forms of unequal warfare, but these efforts were thwarted by India's resolve to stay the course in the region. The emphasis on violence has subsequently changed to include intifada-style mass protests, the use of media, and infiltration into legal and judicial systems to put the Indian security forces on the defensive. The form of terrorist attacks in India over the last few years indicates a likelihood that various jihadi groups have shifted their attention from J&K-centric rural insurgence to a pan-Indian urban terrorism. It is also likely that Pakistan will look to use India's homegrown groups (such as the Indian Mujahideen and Students' Islamic Movement of India) through subversion strategies to carry out terrorist strikes in India. Additionally, given the geographic spread of the Naxal threat across India—and its potential to destabilize the country—Pakistan could look to extend support to such outfits.

Further, given that India has now fenced the entire stretch of the border with Pakistan and the Line of Control (LoC) and conducts constant surveillance of the surrounding waters, it is likely that the porous India-Nepal border will become the route of access for any future terrorist attack in India. There is also the potential for cyberattacks on India's financial, military, and other critical networks, as well as a greater influx of fake Indian currency notes and narcotics to generate greater turmoil.

44.5.2 India's Countermeasures against Pakistan

Pakistan is unlikely to change its confrontational stance towards India despite global opprobrium and the Afghanistan-Pakistan region now often being referred to as the 'epicenter of terrorism'. As such, India must be prepared to safeguard its security interests by adopting a proactive hardline approach using smart power, and establishing a set of response options using all available tools (such as diplomatic, economic, informational, infrastructural, cultural, social, and politico-military). This will need to be supplemented by hard-hitting covert operations with designated special forces at appropriate pre-identified targets. The larger aim should be to prevent India from becoming a passive state that absorbs terror attacks without retaliation.

There are two possible approaches to achieve this goal. First, through effective measures aimed at generating a paradigm shift in India's strategy towards Pakistan and its asymmetric warfare manifestations, from reactive to proactive. This would mean foreseeing crises scenarios and developing multiple strategies to overcome them, including incorporating high-end autonomous combat technologies and creating an immediate preemptive 'response options matrix'. Second, in a potential situation where Pakistan is found to be responsible for a major terrorist strike in India, the best possible course of action will be to carry out military posturing of pre-planned strategic components and undertake swift retaliatory actions, including accurate multidimensional strikes, confirmed through post-damage assessment.

Pakistan's methods of hybrid warfare incorporate psychological and social combat. As such, it is imperative for India to practice a calibrated offensive-defence strategy using unconventional warfare methods, such as coercive tactics and guerrilla warfare through a combination of exhaustion, denial, and subversion, alongside conventional means of military actions. Furthermore, India's efforts are likely to endure by remaining just below the threshold of conventional war. Although a hybrid war can break out unexpectedly in peaceful circumstances, India has several alternatives to such warfare, which will require some degree of politico-military coordination. Another ongoing threat is that some (irregular) elements (such as non-state actors) could suddenly shift alliances to obtain their respective goals, such as destabilizing the nation or harming its peace and security. Therefore, alternative wars and non-linear conflicts require a very robust command-and-control structure and well-defined limits of operation to avoid an overexposure and any compromise of national interests.

For the most part, India's response to Pakistan's asymmetrical and hybrid war against it has been conventional. The deployment of conventional forces in hybrid conflicts demonstrates a lack of strategy, the inefficient use of resources, and the uninspired use of force. **India must create a division dedicated to hybrid alternative war, with both offensive and defensive abilities.** The various elements of the division should be able to operate covertly in areas where vital national interests are consistently threatened. The infantry division should have three brigades, two of which should be for offensive and defensive operations consisting of elements of cyber, electronic, information, space, communication, special forces, intelligence, surveillance, and psychological warfare, alongside geographically-specific air, naval, and logistical components. Further, the Ministry of External Affairs and Ministry of Home Affairs should realign the military structures to enable a special force-style structure to combat any hybrid threat. Operational research and support should fall under the purview of the third brigade. This brigade should develop effective psychological war themes, gather intelligence on culture, identify fault-lines in target areas, exploit cyber loopholes to launch attacks against adversaries, and serve as a hub for artificial intelligence (AI) and as the custodian of operational data. It should also oversee the impact assessment of operations, operational research, and the net assessment of future areas of operations. The entire structure should be confidential in nature and should be run under strict politico-military control. The structure should be agile, adaptive, and futuristic to conduct offensive and defensive hybrid war operations during both peacetime and conflict. The organization should be adaptable enough to combine the civilian and military components for 'plug-and-fight' functionality.

While a politico-military oversight must be exercised throughout an operation, a clear distinction between the strategic and tactical levels of control is necessary. At the strategic level, the actions will mostly be to formulate plans. The space, cyber, special forces, and alternative war divisions at the CDS office should be responsible for the formulation of policies and strategies. Further, the Department of Military Affairs under the defence ministry can embed a special forces headquarters, with the liberty of choosing the required tool for attaining the desired result. Notably, **when strategizing for hybrid attacks, a pre-emptive scheme is key due to the criticality of time in such situations.**

The divisional commander should be given tactical responsibility for the day-to-day coordination of hybrid and counter-hybrid war campaigns and operations. A three-star military commander can be designated as a commander-in-chief of the new force (that could be termed the 'Information and Dynamic Support Force'). The commander-in-chief should report to the CDS. India currently has a skeletal structure in the form of the Armed Forces Special Forces Division, which needs to be reorganized given its limited capability and negligible mandate. There are several policy options for India to consider when developing its 'Information and Dynamic Support Force' to counter the hybrid threats emanating from its neighbourhood.

44.5.3 Joint Military Planning and Countermeasures

Indian defence forces must analyze and co-opt each hybrid threat emerging from Pakistan into their planning parameters if they are to effectively respond with tenacity, lethality, and intensity to eradicate the risk. These conceptual foundations have the potential to influence or affect the counterstrategies, which can be further improved for the purposes of defence policy, strategy, and capability development.

Possible responses by the Indian military can range from swift covert operations by special forces, abrogation of the current ceasefire along the LoC, and multidimensional surgical/precision strikes with drones on identified terrorist camps, to a limited/all-out war in a nuclear backdrop. Given the threat of escalation, it is imperative to be operationally ready. **Modern means of mass communications and drone technologies have significantly increased the success of hybrid warfare.** As such, drone technologies have the potential to become a specialized force rather than simply a force multiplier in the present environment. The complex and flexible nature of the warfare perpetrated by Pakistan calls for an adaptive, carefully considered professional response from the Indian defence forces. Given that India is a frequent casualty of Pakistan's hybrid machinations, it needs to take proactive offensive measures as part of integrated joint strategic initiatives by the various agencies. This should be undertaken in accordance with a well-formulated national hybrid warfare strategy, which should form an integral part of the national security strategy.

44.5.4 Politico-Military Diplomacy by India

Politico-military diplomatic responses can include the suspension of confidence-building measures, cross-border rail and bus services, sporting events, cultural exchanges, Track-II channels, and one-on-one interactions between the two countries. Additionally, given the persistent risk of terrorist organizations obtaining nuclear weapons—a situation that will have implications for regional and global security—the global community should only provide Pakistan with military aid and economic assistance in exchange for the country cracking down on terrorist groups. Pakistan should also be prevented from using military aid and funds to augment its conventional military capability against India, and there should be an insistence on the need for appropriate benchmarks for accountability. India must intensify the campaign that states Pakistan will join the ranks of failed states unless stern measures are adopted by the international community.

As the largest troop contributor to UN peacekeeping operations, India must reiterate that 'terror havens' like Pakistan should not contribute troops to UN operations. Any such cutback will hurt the Pakistan economy, as there will be a shortfall in remuneration to Pakistani banks from savings made by the military peacekeeping forces. India can consider a strategic engagement through Track-II military diplomacy. Such a dialogue will present ample room for discussion, given that the military will be broadly represented on both sides. Additionally, India needs the space and time to focus its resources and consolidate its military modernisation to counter the mounting pressure from China, but it cannot accomplish this goal while preparing for a potential crisis along the western borders. This is where a Track-II dialogue can be beneficial.

44.5.5 On the Northern and Northeastern Front: The China Factor

China is attempting to achieve unilateral superpower status by dislodging the US in the military, economic, digital, and infrastructure spheres through its expansionist, augmentation, and transformation policies. These are based on a strategized civil-military fusion ideology, focus on socioeconomic growth, reforms to increase agricultural productivity, transcendence of high-end scientific technology, and infusion of renewable energy, all to eventually enhance its comprehensive national power. Despite recent economic slowdown, setbacks to the banking sector and infrastructure and housing projects, and severe droughts, China's single-party regime has enabled the State to continue with its expansionist goals.

China's approach to hybrid conflict includes standoffs with India's Intelligence Bureau; transgressions, intrusions, rising assertiveness on the political and military fronts, forays into the Indian Ocean region and threatening island territories; economic colonization through debt traps; colluding with like-minded countries; and a general attitude of 'coercive gradualism'. China furthers its national objectives through the use of non-contact warfare, including cyber warfare, electronic warfare, integrated network electronic warfare, information operations, the 'three warfare strategy' (psychological, media, and legal warfare), political and diplomatic negotiations, economic warfare, and demographic warfare.

There are several geopolitical and geostrategic factors **driving China's hybrid warfare activities against India.**

First, Beijing's expansionist efforts into Indian territory and actions in the Xinjiang Autonomous Region and in Taiwan have fostered disputes with New Delhi and with countries and non-state actors seeking to protect the Uyghurs.

Second, China's Southeast coastal territories—which have over half of the country's economic, commercial, and industrial activities—were made particularly vulnerable by the historic maritime inertia induced by the defensive policies of the erstwhile Chinese emperors (due to which the West was able to gain a technological advantage in marine engineering). Additionally, China considers Taiwan as strategically important because of its location. Controlling Taiwan will allow China to dominate the major sea lines of communication (SLOCs) in the South China Sea and increase its naval presence in those waters to compete with the US. India has been enhancing its engagement with Taiwan in political and economic terms, and, as such, Taipei has become a key partner for New Delhi.

Third, China is seeking alternatives to the Malacca Strait, via which 80 percent of its trade passes, through Belt and Road Initiative (BRI) projects to avoid an overdependence on this vital SLOC. At the same time, India has the option of using the Sunda Strait—situated between the Indonesian islands of Sumatra and Java, and which connects the Java Sea and the Indian Ocean to the waters of East Asia—as an alternative route for its maritime SLOCs in the event of a blockade of the Malacca Strait. Although the BRI was envisioned to offset the vulnerability of Chinese trade through the Malacca Strait, it has so far only incurred investments with little return. Studies on the economic viability of goods transportation have shown that even if all three alternatives—land, sea, and air—are available, water transport is still the most cost-effective and commercially viable. As such, it is unlikely that SLOCs will ever be supplanted.

Fourth, China has 17 percent of the world's population but only 7 percent of arable land (due to significant industrialization and urbanization), leading to food shortages and increased dependence on Africa and Europe for its requirements of food and energy security. Additionally, the Chinese economy is slowing down, with projections of it remaining below 7 percent over the medium term. Stagnation due to the initial advantages of low input costs (especially labor) that are no longer fruitful has forced China to look for ways and means to solve its mounting economic problems.

44.5.6 India's Options against China

China sees India as an adversary to its ambition for regional hegemony and supremacy. India's geostrategic location and credibility as a strong military force also works against Chinese desires. China's belligerence has become more pronounced in light of its imperialist aspirations, and it has started modernizing the PLA. Although India is cognizant of China's military buildup, its response is not cogent and commensurate enough to balance the power dynamics in the region. Until it can adequately leverage its military and economic capacities to match China, India must prepare interim focused strategies to counter the Chinese threat.

India can consider the following options to create hostile conditions in China and prevent it from implementing its expansionist plans.

(i) Supporting Minorities

China faces internal security problems due to its various agitated minority groups, including the Han, Zhuang, Hui, and Manchu. Minorities frequently experience discrimination and exclusion, and struggle to access their basic rights, even under conditions of full and unquestioned citizenship. Denying them citizenship can increase their vulnerability and potentially result in mass expulsion. Any demand for self-determination by ethnic minorities is often perceived as an attempt to secede from the mainland and typically receives a negative response from the State. Supporting the rights of ethnic minorities in Xinjiang and the Inner Mongolia Autonomous Region has the potential to create conditions of domestic instability, complicate Beijing's relations with its neighbors, and threaten its territorial integrity.

(ii) Supporting Greater Autonomy for Tibet

The campaign for greater autonomy for Tibet and its 5.2 million inhabitants continues amid China's subjugation of the region. India's advocacy for the Dalai Lama and outreach to Tibet's restive, discontented, and oppressed youth will certainly harm China. Its status as a permanent member of the UN Security Council has allowed China to persist with its claim of Tibet being a domestic issue that does not need outside resolution. Still, using social media, India can galvanize a massive global movement under the banner of 'revolution for Tibetan autonomy' with the help of the US and other friendly foreign nations to demonstrate solidarity for the oppressed Tibetan people.

(iii) Aligning with ASEAN Countries

The Association of Southeast Asian Nations (ASEAN) region is of key economic and strategic importance to China. If India were to enter into a strategic partnership and free trade agreement with the US, it will counter China's attempts to establish regional hegemony through various BRI plans and mechanisms, including the China-Myanmar Pipeline Corridor. The ASEAN countries are wary of the Indo-Pacific alignment in the US's geopolitical outlook. While they remain suspicious of China's intentions and strategy in the South China Sea and its overall commitment to freedom of navigation and a rules-based order, they do not want to displease Beijing because they are unsure of how far the US will go to advance their shared interests in the region. The Quad (a strategic security dialogue between India, Australia, Japan, and the US) lacks confidence because of its long history of alterations and the sluggish growth of its security route. Perhaps this was one of the driving forces behind the creation of the AUKUS (a security pact between Australia, the UK, and the US), a sort of add-on arrangement meant to inspire greater confidence and convey the American message of relying on its many partnerships. In any case, it is apparent that ASEAN does not want the conflict between superpowers to bleed into its territory. While ASEAN retains its own interests, many of its member countries are hesitant in communicating their will. This power play prevents the ASEAN countries from developing relations with India since they may be seen as going against Chinese interests. Still, India must focus on growing its ties with the ASEAN region to capitalize on their joint economic opportunities and as a counter to China.[19]

(iv) Establishing Integrated Joint Cyber Warfare Capabilities

The PLA and the Indian Army recognize the significance of network-centric combat. The PLA has undergone significant reforms in organization and commands and control, but the Indian Army has not. The Indian Army is yet to fully recognize the extent to which electronic warfare complements cyber warfare. The Indian Army has significant organic cyber and electronic warfare capabilities at all levels, and so more personnel should be trained in these areas. The territorial army can be a good incubator and manning agency for this purpose. India lacks a dedicated information warfare service that can be used in service-specific missions and military goals, while China has the Strategic Support Force. Indeed, India's information warfare capabilities are fragmented and lack a defined command structure, and its electronic warfare competencies have not acquired the same degree of miniaturization as China. This necessitates the integration of cyber and electronic warfare assets under the control of a single operational commander. Additionally, the Indian armed forces have yet to develop anything remotely resembling the Chinese integrated network electronic warfare approach. At the same time, India has established the defence cyber and space agencies and the Armed Forces Special Operations Division. A probable reason for this is that there is insufficient communication between the Armed Forces and the National Technical Reconnaissance Organisation. If India were to establish integrated theatre commands, the inter-services theatre commander will need to be given autonomy to coordinate cyber and electronic warfare for greater network centricity of the tri-services down to the tactical level.

(v) Converting Infantry Units to Special Forces

A greater willingness and mandate to embrace risk, friction, and uncertainty is required to position Indian forces for longer missions, deeper into disputed territory, and against a significantly more powerful foe. On a wider level, effective covert action centres on a strong understanding of the strategic importance of special operations, rather than a focus on short-term tactical gains. The Indian Army's idea of how to operate in case of an occurrence along the Line of Actual Control (LAC) is also not conventional or flexible. On the contrary, it places emphasis on quickly reclaiming the initiative of launching surgical strikes deep inside the adversary's territory and horizontal escalation across multiple sections of the border. It remains to be seen whether India's political establishment is willing to sign off on such plans. Even if India's current administration is determined to send a message that risk appreciation and use of force are indiscrete, much will depend on the specifics of the conflict and the exact nature of Chinese aggression.

(vi) Steady Engagement with the US

Ensuring a reliable US presence and long-term strategy for the region is another way to neutralize the China challenge. There are some expectations of the US's Biden administration engaging more closely with the South Asia countries on issues such as vaccines, emerging technologies, and climate—three areas where China has supported South Asia until now. At the same time, connectivity and infrastructure projects are now a key area of cooperation between China and several South Asian countries because these are essential to the region's development. The G7's 'Build Back Better World' initiative and the Blue Dot Network have the potential to offer South Asia the same level of engagement as China.

Additionally, the Biden administration has released an Indo-Pacific strategy document that focuses on countering China's growing economic and military strength, and empowering India to do so. The document discusses the potential for a major defence partnership with India and expresses support for its role as a net security provider in the region. It also states that the US will continue to work with India bilaterally and through groups to support New Delhi's rise and regional leadership, and India must commit to doing so as well.

The US can also provide South Asian countries with technical advice on debt management. This does not have to mean formal debt relief as envisioned in programmes like the Multilateral Debt Relief Initiative. Instead, the US can offer to support countries before and after they accept Chinese loans. It can assist countries like Sri Lanka and the Maldives in setting terms and conditions that will strengthen their negotiating position with Chinese actors, as well as by helping them manage and restructure their overall debt. The US Department of Treasury's Office of Technical Assistance helps the finance ministries and central banks of developing and transitioning countries to reinforce their ability to manage public finances effectively and safeguard their financial sectors, and it can extend such an engagement to countries in South Asia as well. Additionally, the US could use its influence on Western multilateral organizations to ease the transition and help these states in creating alternative channels for fiscal restraint and constancy.

(vii) Build Modern Infrastructure and Sustainable Capacities

Over time, building the capacities of India's neighboring countries could enable more stakeholders in those nations to recognize the risks associated with China's opaque deals. These stakeholders will likely want their political administrations to only work with entities that share their commitment to international standards of sustainability and accountability. As a relatively new source of development aid, China depends heavily on project finance and direct investment. The China International Development Cooperation Agency or its proxies do not accept most of the Chinese development in Sri Lanka, Nepal, Maldives, and Bangladesh, which are instead accepted by the arms of various Chinese state-owned enterprises that have been established to only focus on a few specific regions.

(viii) Boosting Military's Air and Naval Capabilities

China's ongoing military build-up aims to overcome the PLA's own shortfalls, increase the PLA's combat capability for conflict along the LAC, and eliminate India's airpower advantage over Tibet. Building multidimensional deterrence against China with a very short-term perspective will be a serious strategic flaw. India must develop a plan that reinforces its position as a credible deterrent. This should be supported by a strong military capability designed to carry out retaliatory strikes along the LAC. India must focus on its capabilities that seek to impose painful consequences in the event of a conflict. Since China does not have many forward bases in the region, and since the harsh climate of the Tibetan plateau makes soldier deployment and transportation extremely challenging, air superiority is required to ensure operational dominance in a high altitude. Moreover, India needs to develop its naval might in the Indian Ocean region. India must go beyond traditional means and develop the capability to execute costs beyond the immediate area of conflict through long-range missiles, cyber warfare, and space weapons.

CHAPTER 45

ASYMMETRIC WARFARE

45.1 WHAT IS ASYMMETRIC WARFARE?

Asymmetric warfare is war between belligerents whose relative military power differs significantly, or whose strategy or tactics differ significantly. This is typically a war between a standing, professional army and an insurgency or resistance movement.

Asymmetric warfare can describe a conflict in which the resources of two belligerents differ in essence and in the struggle, interact and attempt to exploit each other's characteristic weaknesses. Such struggles often involve strategies and tactics of unconventional warfare, the weaker combatants attempting to use strategy to offset deficiencies in quantity or quality. Such strategies may not necessarily be militarized. This is in contrast to symmetric warfare, where two powers have similar military power and resources and rely on tactics that are similar overall, differing only in details and execution.

The term 'Asymmetric warfare' is also frequently used to describe what is also called "guerrilla warfare", "insurgency", "terrorism", "counterinsurgency", and "counterterrorism", essentially violent conflict between a formal military and an informal, less equipped and supported, undermanned but resilient opponent. Asymmetric warfare is a form of irregular warfare.

Victory in war does not always go to the militarily superior force. Indeed, colonial powers have contended with asymmetrical threats since the rise of empires. In the 6th century BCE, Darius I of Persia, at the head of the largest and most powerful army in existence at the time, was checked by the Scythians, who possessed a smaller but far more mobile force. As recounted by Herodotus in Book IV of his History, the Scythians retreated before the main body of the Persian army, drawing it deeper into Scythian territory, only to launch lethal mounted strikes on Persian encampments. Darius was forced to retire, leaving the Scythians in command of the lands beyond the Danube River.

In the modern era, Western powers fighting in developing countries have sometimes been defeated by local forces despite massive asymmetries in terms of conventional military strength. Colonial powers were forced to withdraw from Algeria, Indochina, and other areas not necessarily as a result of defeat in battle but because of their lack of will to sustain the war. In Vietnam, a crushing defeat at the Battle of Dien Bien Phu in 1954 sapped the will of the French military, and, after some two decades of U.S. involvement in the Vietnam War, the social and political environments at home forced the United States to concede defeat and withdraw its forces. The insurgents in colonized countries often did not need to defeat the sometimes-long-established colonizer but merely persuaded it to withdraw from the region. Asymmetries of both power and will were operating: the colonial powers possessed superior military resources but were sometimes reluctant or unable to bring them to bear.

The value of asymmetrical tactics can be seen most clearly in guerrilla warfare—indeed, guerrilla means "little war" in Spanish. Guerrilla fighters are generally fewer in number and possess fewer and less-powerful weapons than the opposing force. Guerrilla tactics include ambush, avoiding open battle, cutting communication lines, and generally harassing the enemy. Guerrilla warfare has been practiced throughout history, and it includes both military operations carried out against the rear of an enemy's army and operations carried out by a local population against an occupying force. The aim of the guerrilla fighter is erosion of the enemy's will to sustain the costs of continuing the war. Henry Kissinger observed that "the guerilla wins if he does not lose. The conventional army loses if it does not win."

Although usually exercising a smaller force, guerrilla fighters, especially in urban areas, can be formidable adversaries to a conventional military. Guerrilla fighters typically do not inhabit large, well-established bases, making it impossible for their enemy to exploit technological advantages such as aerial bombardment to destroy personnel and infrastructure. If the guerrillas are in an urban area, their opponents cannot use powerful conventional weapons unless they are willing to inflict large numbers of civilian casualties and risk increasing popular support for the guerrillas. Small guerrilla or insurgent groups also tend to be less hierarchical, meaning that a force cannot be neutralized by the capture or death of a handful of leaders.

Groups lacking the ability to take power either militarily or politically may resort to terrorist attacks within the heart of a state. Terrorist attacks in cities attract more media coverage than those in rural areas; car bombs, assassinations, and bombs left in crowded public places are common tactics of urban terrorism. As long as the survival of its state is not at risk, the nation under attack may be politically unable to use its full military power and thus may have to fight a limited war while terrorists commit themselves and their resources to total war. Terrorist groups are willing to rely on tactics that the states they attack are unlikely or unwilling to use, such as suicide bombings or targeting civilians.

45.2 EXAMPLES OF ASYMMETRIC WARFARE IN 21ST CENTURY

(i) Israel/Palestine

The ongoing battle between the Israelis and some Palestinian organizations (such as Hamas and Islamic Jihad) is a classic case of asymmetric warfare. Israel has a powerful army, air force and navy, while the Palestinian organizations have no access to large-scale military equipment with which to conduct operations; instead, they utilize asymmetric tactics, such as: small gunfights, cross-border sniping, rocket attacks, and suicide bombing.

(ii) Sri Lanka

The Sri Lankan Civil War, which raged on and off from 1983 to 2009, between the Sri Lankan government and the Liberation Tigers of Tamil Eelam (LTTE) saw large-scale asymmetric warfare. The war started as an insurgency and progressed to a large-scale conflict with the mixture of guerrilla and conventional warfare. The LTTE used suicide bombing and with the use of male/female & child suicide bombers both on and off battlefield, mainly targeting innocent civilians; use of explosive-filled boats for suicide attacks on military shipping; use of light aircraft targeting military installations.

(iii) Iraq

The victory by the US-led coalition forces in the 1991 Persian Gulf War and the 2003 invasion of Iraq, demonstrated that training, tactics and technology can provide overwhelming victories in the field of battle during modern conventional warfare. After Saddam Hussein's regime was removed from power, the Iraq campaign moved into a different type of asymmetric warfare where the coalition's use of superior conventional warfare training, tactics and technology was of much less use against continued opposition from the various partisan groups operating inside Iraq.

(iv) Syria

Much of the 2012–2017 Syrian civil war has been fought asymmetrically. The Syrian National Coalition along with the Mujahideen and Kurdish Democratic Union Party, have been engaging with the forces of the Syrian government through asymmetric means. The conflict has seen large-scale asymmetric warfare across the country, with the forces opposed to the government unable to engage symmetrically with the Syrian government and have to resort to other asymmetric tactics such as suicide bombings and targeted assassinations.

45.3 WHAT IS THE DIFFERENCE BETWEEN HYBRID WARFARE AND ASYMMETRIC WARFARE?

Hybrid warfare refers to the combination of military (violent force) and non-military means (disinformation campaigns, economic sanctions); the integration of conventional (regular uniformed troops, armored battalions etc.) and unconventional forces (irregular guerrillas, proxies), usually to exploit strategic ambiguity below the threshold of all-out armed conflict.

Asymmetric warfare refers to a weaker party refusing to engage symmetrically/head-on on the enemy's terms, instead exploiting the weaknesses (e.g. whittling away its political-psychological resolve) of its superior as well as changing the design of system. Here, it consists of:

(i) **System** = Masses (Huge number of people),

(ii) **Input** = data which we are consuming from internet, and

(iii) **Output** = Reactions such as anger, fear and anxiety etc. revealing on the internet.

How does it work?

Asymmetric warfare has **3 pillars** i.e.

- Information warfare;
- Public Opinion; and
- Legal warfare.

A boy with a laptop is more dangerous than a division of Soldiers.

We will understand all the three pillars one by one and how they're coordinated to yield desired result.

45.3.1 Information warfare

Information warfare is the domain which determines the input of the system. How?

In this domain, different inputs are thrown to the masses and their reactions are recorded. Now, data scientists have what they needed to make deductions. With the help of these massive data, which we have been leaking since a decade goes into deep analysis and then they come out with a mathematical model/tool which determines "Which set of data is required to drive certain behavior on a large scale."

Now, they have formed a weapon which can use us against any country

45.3.2 Public Opinion

In a democratic country like India, Public opinion/sentiment matters a lot. This is the strength as well as the weakness of the India. With the help of above tool, It is possible to mold public opinion and pressurize the democratic government to overturn or stall key decisions.

How public opinion is formed?

It is formed by bombarding certain set of information to the particular group of people many times to make them behave in a certain way.

45.3.3 Legal warfare

This domain of Asymmetric warfare uses loopholes in the agreement to strangle you with your own systems. They can stall your key defence, strategic projects by using your very own law.

When all these 3 pillars are coordinated well, they impact directly on the Economic growth, social fabric and Military Capabilities of the Nation which results into a Civil war or Economic breakdown.

45.4 ASYMMETRIC WARS IN THE INDIAN CONTEXT

45.4.1 India-Pakistan

India is the world's largest democracy with vibrant democratic institutions, voluntary armed forces, an active media and above all a historical culture that has strong religious and moral beliefs. Given its strong democratic ideals, India's political vulnerability is high.

Pakistan, on the other hand, is a weak democracy propped up by a strong and powerful military and a fanatic religious and radical lobby. With weak democratic institutions, a stifled judiciary, a feudal system where people's rights are subdued and a media that is controlled, the state borders on the authoritarian model. Indeed, Pakistan has been under military rule for much of its history. Pakistan's political vulnerability is therefore low. Time and delays are irrelevant for such a regime as long as the elite remain in power.

In the 1965 and 1971 wars Pakistan applied the direct-direct approach strategy and consequently failed to achieve its strategic aim. Subsequently, it reverted to the indirect approach, unleashing terrorism/GWS (Guerrilla warfare strategy), first in the state of Punjab and then in Jammu and

Kashmir. Initially, this strategy achieved results by tying down the Indian Army and increasing the cost to India. It required the Indian military to change its strategy to one of combating GWS before bringing the situation well under control. The transformation strategy included the raising of the *Rashtriya Rifles* (a force detached from the conventional army), changes in the methodology of training, tactics, equipment and force structure. In addition, strategies such as WHAM (Winning Hearts and Minds) and population control measures (denying support and sanctuary to terrorists) brought about a significant change in negating Pakistan`s strategy. In short, once the Indian Army adopted the indirect approach to combat Pakistan`s indirect approach, the stronger (India) won. Resultantly, there is a relative peace in the state of Jammu and Kashmir.

According to Arreguin-Toft, direct-direct strategies also result in victory for the strong actor. Therefore, the obvious approach for India is to transfer the gains of the indirect approach strategy (GWS) to the Para Military Forces (PMFs) till they attain a certain degree of expertise in combating GWS. At the same time, in order to decisively defeat Pakistan militarily, India should prepare for a direct-direct approach (conventional war).

45.4.2 India-China

China is a major power in the making. With a GDP four times that of India's and an official military budget three times that of India's, China is way ahead in terms of power. However, it is a single party-based regime. Authority is vested in the Party, society is somewhat closed, the media is controlled, religious beliefs are shunned, although there are a few democratic institutions. China typifies an authoritarian strong actor.

While applying the `Strategic Interaction theory` to the India-China scenario, it would be correct to state that China's interest is low but being a power in the making it cannot be seen to `lose face` in the eyes of its public to a weak adversary. Thus China's vulnerability is high.

China views India as an adversary in its quest for regional hegemony and dominance in Asia. Naturally, with economic power grow military aspirations. In recent years, Chinese belligerence has become more pronounced. China has begun modernizing the PLA on a massive scale, expanding its reach and aspirations beyond its borders. India has taken cognizance of China`s military expansion, but its response has been both slow and tardy. In the interim, till India builds adequate capacities to match up to China, there is a need to prepare for strategies to `deter` China.

Applying the Arreguin-Toft theory, India as the weak actor needs to adopt indirect strategies vis-a vis China. It must be remembered that such strategies are a function of time. An authoritarian regime may not tolerate indirect acts for long and, when the elite or public get restive, the response may be that of barbarism or direct confrontation, both of which do not suit the Indian state in its present state of preparedness.

Some of the interim indirect strategies that have been suggested can best be explained as vigorous applications of India`s soft power. These are:-

- Support the Dalai Lama`s cause and reach out to the restive, discontented and oppressed Tibetan population, particularly the youth, in Tibet.

- Support the cause of ethnic minorities in Xinjiang Autonomous Region (XUAR) and Inner Mongolia Autonomous Region (IMAR).
- Develop superior Cyber Warfare capabilities to defend against cyber-attacks, as also support the cause of free expression and use of internet in Tibet.
- Develop superior Electronic Warfare capabilities to thwart propaganda and assist own communications in border areas.
- Develop Special Forces (SF) in a big way to consolidate gains.

The urgency in time and necessity of developing capabilities and capacities vis-a-vis China cannot be overemphasized, particularly if India is forced into a direct confrontation with China in the near future.

CHAPTER 46

IRREGULAR WARFARE

46.1 DIFFERENCE BETWEEN IRREGULAR, ASYMMETRIC, AND HYBRID WARFARE

Irregular warfare, asymmetric warfare, and hybrid warfare are all types of warfare that differ in their tactics and goals:

(i) Irregular warfare

Irregular warfare is a type of warfare that involves population-centered attacks. Irregular forces can be a challenge for conventional militaries.

(ii) Asymmetric warfare

Asymmetric warfare is a broad term that describes conflicts where the enemy combatants are not regular military forces. Asymmetric warfare is a type of irregular warfare.

(iii) Hybrid warfare

Hybrid warfare is a type of warfare that combines conventional and irregular tactics. Hybrid warfare is characterized by the use of blended instruments, ambiguity, and attribution.

Here are some more details about these types of warfare:

Irregular warfare involves military, paramilitary, political, economic, psychological, and civic actions.

Hybrid warfare is often used to degrade or disrupt democratic societies. Hybrid warfare can involve the use of mass communication for propaganda, including fake-news websites.

Asymmetric warfare is a euphemism for avoiding acknowledging that the opponent isn't playing to your strengths.

46.2 INTRODUCTION TO IRREGULAR WARFARE

Irregular warfare (IW) is defined as a violent struggle among state and non-state actors for legitimacy and influence over the relevant populations. Concepts associated with irregular warfare are older than the term itself.

Irregular warfare favors indirect warfare and asymmetric warfare approaches, though it may employ the full range of military and other capabilities in order to erode the adversary's power, influence, and will. It is inherently a protracted struggle that will test the resolve of a state and its strategic partners.

Irregular warfare is a fundamentally European concept that arose in the 18th century to refer to the reverse of a set of practices and representations established as norms for armed conflict. Whether consciously or not, irregular combatants defy modern military categories, in tactical and strategic terms, as well as in political and legal ones. In its various forms, irregular warfare has resulted not only in guerrilla warfare campaigns against occupying armies but also in clandestine or subversive actions aimed at surprising or circumventing a more powerful foe by attacking it logistically or politically behind the lines. With the reduction in the classical battle model immediately after World War II, irregular warfare has gradually imposed itself as the dominant mode of conflictuality, through both guerrilla warfare and terrorism. This prevalence is the result both of indirect confrontations between major powers and of asymmetrical conflicts following decolonization.

46.2.1 Irregular Warfare through the Ages

It has been amply demonstrated that, far from being a form of absolute violence, war is a socially constructed activity reflecting both the values and the political and economic organization of the societies involved. The European concept of conflictuality is no exception to that rule: it meets the standards corresponding to representations that have been forged over the course of its history. It juxtaposes different heritages: Greco-Roman and Judeo-Christian founding myths, Medieval traditions for regulating violence (e.g. the "Peace of God"), evolutions tied to the emergence of political modernity from the 15th to the 18th centuries (state apparatuses, standing armies), social, legal and economic evolutions from the 19th and 20th centuries (nationalism, industrialization, rise of international humanitarian law). These have led to a kind of "ideal type" of armed violence that is sometimes known as "regular" or "conventional" warfare, and which dominates our representation of war. In opposition to that model, a form of conflictuality that is described – in a normative and ethnocentric manner – as "irregular" has emerged. The term refers to military practices that don't comply with the representation of warfare that dominates in the western world.

In political and legal terms, irregular combatants do not recognize the principle of a state defined exclusively by its monopoly on the legitimate use of force. Whether they are revolutionary or reactionary, independentist or separatist, irregular combatants' actions, in and of themselves, contest the legitimacy of the established regime. So they situate themselves outside of positive law, which often leads to them being described as bandits or criminals. If they refer to transnational political or religious solidarity, as they sometimes do, they expose themselves to the accusation of being foreign agents. Lastly, irregulars tend to reject the distinction between combatants and non-combatants. That fundamental norm in human rights – present as far back as the Middle Ages in Thomas Aquinas's writings, then more centrally in the philosophy of Hugo Grotius, the father of modern international law – took on a particular materiality during the Renaissance, with military uniforms, for instance. So the absence of uniforms is one of irregular combatants' defining characteristics: by mingling with the civilian population, they expose that population to their adversary's indiscriminate repression.

Norms apply on a tactical level as well. While conventional armies have focused their efforts on a continuous increase in fire power, irregular fighters focus on mobility, compensatory tactics: ambushes, surprise attacks, concealment and evanescence.

In contrast to the classical tradition, based on the destruction of the enemy army, irregular strategy is more indirect, endeavoring to erode the other side's determination. "The guerrilla wins if he does not lose," Henry Kissinger wrote in 1969, referring to America's experience in Vietnam. In order to survive while wearing their opponent down, irregulars aim for small victories through harassment, sabotage and sapping power relays in the territories where they become established. In so doing, they resort to kinds of economic or social mobilization to build — by whatever means necessary — popular support, which grants them political legitimacy, as well as human and material resources.

Irregular warfare has undoubtedly always existed, motivated by the difference in weapons and manpower between two sides: Julius Caesar's Commentaries on the Gallic War is bursting with accounts of asymmetrical confrontations with Roman legions. While the methods continued into the modern era, the establishment of standing armies in the 17th century and the growth in their size and logistical needs explains the development of harassment tactics aimed at supply lines. The 18th century saw a professionalization of these "skirmishes" or "small wars." They are also known in French as "party warfare," because small detachments, or parties, lead attacks against the backlines. That is where the term "partisan" comes from.

With the emergence of national sentiments in the 19th century, skirmishes became more generalized and more and more often took the shape of popular resistance to a foreign enemy. That was the case during the Napoleonic Wars with the Spanish guerrilla, the 1812 Russian partisans, and even the 1813 Prussian Landwehr.

The advent of total war at the turn of the 20th century launched a new transformation of irregular warfare. During World War I, the strategic importance of the "home front" (mobilization of the civilian population for the war effort) increased the fear of subversion from the inside. Dropping defeatist leaflets and spreading "fake news" introduced the idea of propaganda as a weapon of war. Guerrilla warfare was part of that: here and there, it was a way of getting around a battlefield paralyzed by siege warfare. German support for the 1916 Irish insurrection and British support for the 1917 Great Arab Revolt are two examples of that. World War II accelerated that logic, notably due to the influence of Great Britain, which established an organization for "political warfare" and "special operations" aimed at bolstering resistance in territories occupied by the Axis powers.

After 1945, Cold War logic reinforced the indirect strategy, and, therefore, irregular warfare. Propaganda, subversion, clandestine actions in support of armed groups became necessary tools. In Europe, that could be seen in both the Greek Civil War and the Anglo-American secret services' attempts to destabilize Albania.

In many nations in Africa and Asia, the colonial era led into civil wars, separatist rebellions and other forms of armed struggle, which has extended irregular conflictuality into the present. Finally, terrorism emerged in the early 1970s as a new weapon in irregular warfare. First employed by revolutionary movements (the Red Brigades in Italy, and the Red Army Faction in Germany) and nationalist ones (Palestinian, Irish, Basque), it was adopted in the early 1980s by a new cause: Islamic jihadism. Since the September 11, 2001 attacks, that movement has gradually taken a central position in irregular conflictuality.

46.3 BLEED INDIA WITH A THOUSAND CUTS

'Bleed India with a Thousand Cuts' is a military doctrine **followed by the Pakistani military** against India. It consists of waging covert war against India using insurgents at multiple locations. This view was put forward in various studies by the Pakistani military, particularly in its Staff College, Quetta.

In a 1965 speech to the UN Security Council, Zulfikar Ali Bhutto, former Prime Minister and former President of Pakistan, declared a thousand-year war against India. Former Pakistani Army Chief General Zia-ul-Haq gave form to Bhutto's "thousand years war" with the 'bleeding India through a thousand cuts' doctrine using covert and low-intensity warfare with militancy and infiltration. This doctrine was first attempted during the Punjab insurgency and then in Kashmir insurgency using India's western border with Pakistan. Pakistan has devised a malevolent strategy aimed at instigating religio-political turmoil in India's border states of Jammu and Kashmir and Punjab. This strategy involves supporting and fueling acts of terror-induced violence, with the explicit intention of causing significant problems for India.

The origins of the strategic doctrine against India are attributed to Zulfikar Ali Bhutto, then a member of the military regime of the General Ayub Khan, who declared a thousand-year war against India during his speech to the United Nations Security Council in 1965. His plans for the 1971 war included severing the entire eastern India and making it a "permanent part" of East Pakistan, occupying Kashmir, and turning East Punjab into a separate 'Khalistan'. After the war ended with Pakistan's own dismemberment, he laid down the doctrine of continuing the conflict by "inflicting a thousand cuts" on India.

On 5 July 1977, Bhutto was deposed by his army chief General Zia-ul-Haq in a military coup before being controversially tried and executed.

Zia then assumed the office of President of Pakistan in 1978 and the thousand cut policy began taking shape. After the defeat of Pakistan in the 1971 war, Pakistan was divided and Bangladesh was created. The war clarified that Kashmir could no longer be taken from India by a conventional war. Zia implemented Bhutto's "thousand years war" with 'Bleed India Through A Thousand Cuts' doctrine using covert and low intensity warfare with militancy and infiltration.

46.3.1 Insurgency in Punjab, India by Pakistan

Pakistan had been helping the Sikh secessionist movement in the Indian Punjab since the 1970s. Since the early 1980s Pakistan's intelligence agency ISI created a special Punjab cell in its headquarters to support the militant Sikh followers of Bhindranwale and supply them with arms and ammunitions. Terrorist training camps were set up in Pakistan at Lahore and Karachi to train the young Sikhs. Hamid Gul (who had led ISI) had stated about Punjab insurgency that "Keeping Punjab destabilized is equivalent to the Pakistan Army having an extra division at no cost to the taxpayers." On February 26, 2023, Punjab Chief Minister Bhagwant Mann asserted that Khalistan supporters were receiving financial support from Pakistan and various other nations. This statement came against the backdrop of the prevailing unrest in the state, fueled by the recent actions of Khalistani sympathizer Amritpal Singh and his followers.

46.3.2 Insurgency in Kashmir, India by Pakistan

After the conclusion of the Soviet–Afghan War, the fighters of the Sunni Mujahideen and other Islamic militants had successfully removed the Soviet forces from Afghanistan. The military and civil government of Pakistan sought to utilize these militants in the Kashmir conflict against the Indian Armed Forces in accordance with the "thousand cuts" doctrine so as to "bleed India", using Pakistan's nuclear arsenal as a shield. In the 1980s cross-border terrorism started in the Kashmir region as armed and well-trained groups of terrorists were infiltrated into India through the border. Pakistan officially maintained that the terrorism in Kashmir was "freedom struggle" of Kashmiris and Pakistan only provided moral support to them. But this turned out to be inaccurate as Director-General of Inter-Services Intelligence (ISI) stated in the National Assembly of Pakistan that the ISI was sponsoring this support in Kashmir. Pakistan used the jihadist militias to conduct an asymmetric and irregular warfare with India.

According to a general involved with the "bleed India" strategy of infiltrating jihadists into Kashmir:

It kept 700,000 Indian troops and paramilitary forces in Kashmir at very low cost to Pakistan; at the same time, it ensured that the Indian Army could not threaten Pakistan, created enormous expenditures for India, and kept it bogged down in military and political terms.

In May 1998, India tested its nuclear weapons at Pokhran-II followed by Pakistani nuclear tests. The Infiltration of Pakistani soldiers disguised as Kashmiri militants into positions on the Indian side of the LOC, resulted in a geographically limited Kargil War, during which the Pakistani Foreign Secretary Shamshad Ahmed issued a veiled nuclear threat that, 'We will not hesitate to use any weapon in our arsenal to defend our territorial integrity.'

After the Kargil War in 1999, the Kargil Review Committee came out with a report which took reference to the concept of Pakistan bleeding India. The report said that if the "Siachenisation" of Kargil had happened prior to the war, that if troops had been stationed there all year round along a wider area, it would have resulted in huge costs "and enabled Pakistan to bleed India".

On 13 December 2001 a terrorist attack occurred on the Indian Parliament (during which twelve people, including the five terrorists who attacked the building, were killed) and the Jammu and Kashmir Legislative Assembly on 1 October 2001. India claimed that the attacks were carried out by two Pakistan-based terror groups fighting Indian administered Kashmir, the Lashkar-e-Taiba and Jaish-e-Mohammad, both of whom India has said are backed by Pakistan's ISI, a charge that Pakistan denied. The military buildup was initiated by India in response to the twin attacks leading to the 2001–02 India–Pakistan standoff between India and Pakistan. Troops were amassed on either side of the border and along the Line of Control (LoC) in the region of Kashmir. International media reported the possibility of a nuclear war between the two countries and the implications of the potential conflict on the American-led "Global War on Terrorism" in nearby Afghanistan. Tensions de-escalated following international diplomatic mediation which resulted in the October 2002 withdrawal of Indian and Pakistani troops from the international border.

In spite of grave provocations, the lack of military retaliation by India was seen as evidence of successful deterrence of India by Pakistan's nuclear capability. The nuclear deterrence has encouraged certain Pakistani elements to further provoke India. An asymmetric nuclear escalation posture of Pakistan has deterred conventional military power of India and in turn has enabled Pakistan's "aggressive strategy of bleeding India by a thousand cuts with little fear of significant retaliation.

Presently, the militant Islamists extremists in Bangladesh and Pakistan, through designated terrorist groups such as Harkat-ul-Jihad al-Islami (HuJI), have joined forces to carry out terrorist attacks on India. In 2015, a Pakistani High Commission staff member who was an ISI agent in Bangladesh had to be withdrawn by Pakistan after his involvement in financing of terrorism activities and fake Indian currency note racket was reported by intelligence sources. He was involved in financing of terrorist organizations, Hizb ut-Tahir, Ansarullah Bangla Team and Jamaat-e-Islami. In December 2015, another Pakistani diplomat, a second secretary in the high commission, was withdrawn for links with Jama'atul Mujahideen Bangladesh. The arrest of Pakistani citizens in Bangladesh with fake Indian currency was a "common phenomenon". In Bangladesh, HuJI-B provides a safe zone for training as well as help in crossing the border into India.

"Pakistan has decided to bleed India with thousand cuts. It's the policy of Pakistan. The creation of Bangladesh, which happened with the help of India, was a very humiliating defeat for them, and they feel that this is one way of avenging that defeat. They are avenging this defeat by causing casualties to our security forces and creating mayhem amongst the people."

— Chief of the Indian Army General Bipin Rawat in September 2018

In an interview in May 2016, Husain Haqqani, Pakistan's former ambassador to America, said:

"Pakistan sees jihad as a low-cost option to bleed India. The security apparatus views terrorism as **irregular warfare**. Islamabad feels this is the only way to ensure some form of military parity.

CHAPTER 47

INFORMATION WARFARE

47.1 INTRODUCTION TO INFORMATION WARFARE

Information warfare, or information operations, refers to the concepts of 21st-century warfare that are driven by electronic and information systems. It encompasses actions taken to achieve information superiority in support of national military strategy by affecting adversary information and information systems, while also leveraging and protecting our own information and information systems. It includes offensive actions to deny, corrupt, destroy, or exploit an adversary's information, defensive actions to safeguard ourselves and allies, and exploitative actions to enhance our decision/action cycle and disrupt the adversary's cycle. This form of warfare extends beyond strictly military targets and can involve civilian communities and industries.

The federal government's definition of information warfare can be divided into three general categories: offensive, defensive, and exploitative.

1. **Offensive**—Deny, corrupt, destroy, or exploit an adversary's information, or influence the adversary's perception.
2. **Defensive**—Safeguard ourselves and allies from similar actions (also known as information warfare hardening).
3. **Exploitative**—Exploit available information in a timely fashion to enhance our decision/action cycle and disrupt the adversary's cycle.

In addition, the military views information warfare to include electronic warfare (e.g., jamming communications links), surveillance systems, precision strike (e.g., if you bomb a telecommunications switching systems, it is considered information warfare), and advanced battlefield management (e.g., using information and information systems to provide information on which to base military decisions when prosecuting a war).

The rapid development of networks has turned each automated system into a potential target of invasion. The fact that information technology is increasingly relevant to people's lives determines that those who take part in information war are not all soldiers and that anybody who understands computers may become a "fighter" on the network. Think tanks composed of nongovernmental experts may take part in decision making; rapid mobilization will not just be directed to young people; information-related industries and domains will be the first to be mobilized and enter the war.

One can look at information warfare as being a factor in information, systems, and telecommunications protection. The information warriors, cyber warriors, techno-spies, Internet terrorists, or whatever one wants to call them, present additional increased challenges to all those on, attached to, or concerned with making the Internet a safe place to travel, visit, and maintain as a free space for sharing information and learning.

Remember that these armies of information warriors are looking at the targets presented along the Internet, and the first of these are commercial, or nonmilitary. The weapons that they will use can be categorized as attack, protect, exploit, and support weapons systems. They have the funding and identification of targets, and they are developing plans and sophisticated application programs to attack a nation's information infrastructure, which includes those on the Internet.

- Information Warfare (IW) is war normally conducted via computers. This perspective treats IW as the expansion of military action into the virtual battlefield of cyberspace. The emergent issues concern cyberspace as a new venue for offensive and defensive operations. Terms associated with this perspective include computer warfare, cyberwar, datawar, hackerwar, information age warfare, information systems warfare, information war, infowar, netwar, network warfare, and silicon warfare.

- IW is military operations empowered by information and knowledge. This perspective addresses the manner in which information technologies (IT) can empower and revolutionize military operations as we now know them. The emergent issues concern the means by which IT can make existing military operations more efficient or effective. Terms associated with this perspective include command and control warfare, information-based warfare, infowar, intelligence-based warfare, knowledge war, knowledge-based warfare, and third-wave warfare.

- IW is a new evolutionary stage of military evolution. This perspective addresses the notion that the previous themes indicate a new epoch in military history. The emergent issues concern how military theory and practice will be revolutionized as a result of the IT revolution. Terms associated with this perspective include cyberwar, hyperwar, post-modern warfare, post-industrial warfare, and third-wave warfare.

- IW is an emerging mode of low-intensity societal conflict. This theme treats IW as a progressively pervasive phenomenon not limited to the international politico-military actions traditionally considered wars. The emergent issues concern sociological and cultural impacts of widespread IT exploitation for the sake of causes and crimes. Terms associated with this perspective include cybercrime, cyberterrorism, information war, infowar, netwar, and post-modern warfare.

During the mid-1990s, information warfare was commonly discussed with specific regard to one or a few of these perspectives. By the beginning of the 21st century, the label had progressively grown to subsume most if not all of these themes and labels.

47.1.1 Relationship between Information Warfare and War

The predominantly military connotations attributed to information warfare suggest the subject is a topic of purely military significance, but this is not the case. Modern life can be threatened by types of warfare other than full-blown wars involving national military forces. The scale of the violence and casualties in the September 11, 2001 attacks on New York and Washington correspond to historical acts of conventional war, even though the attack employed unconventional means and the attackers (the al-Qaida terrorist network) were not a national entity. Even among national entities, conflicts with significance approximating that of historical wars may be prosecuted without military involvement. Considerable damage can be wrought on adversary nations through means other than military force (e.g., economic manipulations and trade sanctions). As such, the historical correlation of threat magnitude, national scale, and military force is a deteriorating criterion for identifying situations in which IW may be conducted. Distinguishing IW's general consequences from its specifically military ones will require terminology more discriminating than is evident in the literature to date.

One approach, more common in sociological than in military IW analyses, is to treat warfare as something of more than purely military scope. In this approach, warfare is treated as a course of deliberate action through which an actor pursues advantage over another (e.g., a competitor or adversary). The means for such an action are taken to be extraordinary (i.e., outside the scope of ordinary daily interactions), and the notion of advantage includes adverse effects on the other regardless of benefits accruing to the actor. This broad definition of warfare subsumes nonpolitical and nonmilitary actions that are unilateral in execution, constrained in scope, anonymous, surreptitious, and prosecuted without lethal means. This approach typically reserves the term war for bilateral, unlimited, explicit warfare through any and all means (including lethal). To speak of IW in this sense allows analysts to address the phenomenon as a concern for organizations other than governments and the military. IW can be attributed to activities conducted outside the scope of those overt international or civil political conflicts we associate with military operations—for example, activities conducted by nongovernmental or commercial organizations and even individuals.

The aforementioned approach is not widely acknowledged in the military community, for whom war and warfare are canonical concepts not subject to redefinition. However, even the military community recognizes the need to distinguish between information activities of general and military significance. For example, during the 1990s the U.S. military had come to focus its attention on the general concept of information operations (IO), defined in Joint Publication 3-13 (Joint Doctrine for Information Operations, 1998) as "actions undertaken to influence an adversary's information and information systems while maintaining the integrity of one's own information and information systems." By virtue of not being contingent on an overt war, this construct has certain resemblances to the liberal interpretation of information warfare noted above. In contrast, this doctrinal document then defines information warfare as "information operations (IO) undertaken in the context of crisis or conflict to pursue and accomplish particular goals and objectives relative to a specific adversary or set of adversaries." As a more specific category linked to overt conflict, this IW definition resembles what the other approach relegates to activities in the context of war.

47.2 INFORMATION WARFARE (A US-PERSPECTIVE)

The information age offers new challenges and opportunities. Cyberspace, advanced computing, mobile networks, unmanned systems, and social media present a military revolution in information warfare. To leverage its full potential, the U.S. military requires a cultural change to reconcile institutional aversion toward non-lethal information warfare. To aggressively shape, influence, control, and manipulate information, change is needed in U.S. military attitudes toward information warfare. This can be realized through better training and education, and deliberate integration of information operations across the military services during planning and operations.

Information is now the seventh joint function. Its efficacy depends upon the ability to transition practical concepts into fundamental underlying logic. The information operations is defined as the "integrated employment, during military operations, of information-related capabilities in concert with other lines of operation to influence, disrupt, corrupt, or usurp the decision making of adversaries and potential adversaries while protecting our own." As U.S. Secretary of Defense James Mattis has pointed out, the addition of information as a joint function recognizes its power to support military operations, particularly in the wake of modern technology and social media. As a joint function, information will garner additional advocacy among the joint force, and greater incorporation into plans and operations.

Information is both a resource and a weapon. The battlefield of tomorrow is compressed in time and space. When enemy actors are simultaneously operating in space, cyberspace, and across the electromagnetic spectrum, reacting in real time will become a challenge on an interconnected battlefield. The addition of space and cyberspace as warfighting domains has expanded exponentially and greatly complicated information operations.

The application of information as an instrument of war is not new, but the American military struggles with information operations. Information is ubiquitous, yet frustrates American military planners and practitioners. Information operations range from deception and cyber activities to public affairs, but traditional American military strength often relies upon overwhelming air power, fire, and maneuver. The so-called American way of warfare is characterized by the application of joint fires within a highly integrated command and control network to rapidly destroy enemy formations and command and control nodes. The American way of warfare has a successful track record in battle, but also belies a persistent inability to ruthlessly exploit the information environment at the strategic, operational, and tactical level.

U.S. military campaign strategy is not comfortable with maneuver in the information space. The U.S. military has a penchant to find it easier—from a capabilities, policies, and authorities perspective—to drop ordnance on a target rather than generating other less-lethal, less-irreversible, and less-transparent effects. Yet, capturing the perceptions of foreign audiences will replace seizing terrain as the new high ground for the future joint force. Discomfort in the realm of information operations is a challenge that must be overcome.

U.S. military discomfort with information operations extends beyond tactics and operations This heightens the perception of a persistent bifurcation in American strategic thinking, in which military professionals concentrate on winning battles and campaigns, while policymakers focus on the diplomatic struggles that precede and influence, or are influenced by, the actual fighting. The limited reference to the diplomatic instrument of national power belies the need for close cooperation. Given the power of modern communications to shape the information environment, even tactical information operations can create strategic effects.

Thus, as the U.S. military begins to consider information as a distinct warfighting function, it cannot neglect the critical linkage to grand strategy and the other instruments of national power. If war is driven by policy, which is in turn influenced by information, the military should anticipate and address underlying assumptions regarding U.S. civil-military relations. For instance, the video and imagery showing the devastation on the Iraqi highway of death by the media during the first Gulf War is considered a significant factor in the cessation of coalition attacks. In other words, while the military develops its capacity for information operations, senior leaders must contend with the need to ensure plans and operations support and complement national objectives.

Exploitation of the information environment has occurred throughout the history of warfare. The First World War featured the extensive use of information operations, including the first use of electronic warfare through the interception of wireless communications. When the war began, Great Britain, then the hub of global communications, cut telegraph cables to Germany. This led to a communications blackout for the Germans. German telegraphs had to be routed through British-controlled lines, leading to the discovery of the Zimmerman telegram.

During the First World War, the Germans also leveraged information in novel ways. In April of 1917, Germany accelerated the fall of the Russian monarchy by loading a train bound for Petrograd with human malware; a virus of sorts in the form of Vladimir Lenin. As the German army fought the Russians along the eastern front, Lenin hastened the collapse of a Russian monarchy already reeling from inept leadership and repeated military defeat. The German use of Lenin as a weapon to influence Russian behavior appears almost quaint by the scale of later efforts.

During the Second World War, Operation Bodyguard used extensive physical and electronic deception to conceal the location of the Allied landings at Normandy. The allied deception operations during the Second World War were significantly more complicated, requiring extensive coordination between military and political instruments of national power in multiple domains simultaneously.

In the first Gulf War, the U.S.-led coalition made a concerted effort to control and influence Iraqi decision making through information operations. U.S. Marine Corps forces afloat created a strategic distraction for the Iraqis, while the U.S. ensured France's Satellite Pour l Observation de la Terre (SPOT) system was unable to provide satellite imagery to Saddam Hussein, blinding his ability to observe coalition activities. More recently, the Russians have employed so-called hybrid warfare in their conflicts in Ukraine. Russia seized Crimea without firing a shot due in part to a successful information campaign. Since it annexed Crimea, Russia has continued its information operations, evidenced by the establishment of Russian-filtered internet service via underwater cable. Indeed, the so-called Gerasimov doctrine claims the "role of nonmilitary means of achieving political and strategic goals has grown, and, in many cases, they have exceeded the power of force or weapons in their effectiveness."

Generally, the U.S. military gravitates towards the use of information to support kinetic solutions. The emergence of the so-called information revolution in military affairs in the 1990s led the services to use information to strike targets faster and more accurately through concepts such as net-centric warfare and effects-based operations through networking, sensors, computers, and satellites. The American military is less inclined to leverage information with cunning and guile before or during operations. Many have characterized the use of deception, for example, as ad hoc at best.

In the information realm, the U.S. often operates at a disadvantage. American combat arms traditionally dominate warfare and war planning, while non-lethal and diplomatic integration is challenged. American military culture is also averse to deceit and falsehoods, favoring standard operating procedure over imaginative psychological means. While adversaries comfortably blend political and military operations, the American military tradition studiously avoids political considerations in campaigns. Military success is presumed to lead inexorably to a favorable political outcome overseas. Hence some have concluded "power and diplomacy to occupy separate spheres."

Used to its full potential, information operations can prepare the battlefield and set the conditions for victory. Information can soften the enemy's will to fight, deceive, and pollute his or her decision-making cycle. Given the variety of media capable of using information to influence, coerce, or deceive, it is conceivable that information operations may surpass fire and maneuver in importance at times. Yet influencing a foreign audience cannot be accomplished ad hoc without study and preparation of culture, history, and language. New institutional arrangements may be in order to fully resource requirements for successful information operations.

The explosion of the information landscape and technological information-related capabilities presents myriad challenges. The violence, like Twitter, is a means of communication. Whole populations can be reached through social media and influenced in real-time. However, the American style of warfare failed to internalize the contention that war was the continuation of politics by other means. This artificial separation between war and politics presents myriad obstacles in the execution of information operations.

In the early phases of combat operations, web-connected devices can be scraped for metadata and surveilled. Military and interagency collectors can map the information space, to include information systems, public networks, and other sensors and emitters. Commanders will require a visualization of the information space through a common informational picture. Commanders and planners will need real-time access to this information. Ethical dilemmas will arise given the need to target enemy information systems embedded within civilian infrastructure. The speed of a non-lethal operation to influence, degrade, suppress, or deny the enemy access to information will be measured in minutes or hours, not days or weeks.

The tools and authorities needed for these actions at lower tactical levels are a largely uncharted territory. The growth of cyberspace has increased the surface area of the battlespace. Potential attack vectors have multiplied and grown in complexity. There are five categories of cyber actions: stealing, spying, disrupting, destroying, and deceiving. These cyber threats can be brought to bear on U.S. forces in large scale combat operations in varying degrees. The sheer speed of a cyber action reduces an effect to mere seconds. Given the scale of networked devices on a battlefield and on the home front, the scope of attacks to defend increase exponentially.

The long-term method to close the gap in the political-military divide is professional military education. Many historians argue the quality-of-service members, and their ability to think and adapt, are leading indicators of success in war. The acid test of military effectiveness was whether one could handle not the expected but the unexpected elements thrown up in war. Notably, the National Defense Strategy states U.S. military professional military education has stagnated. Hence, professional military education for both enlisted personnel and officers is critical.

Similarly, educational opportunities should be made available to civilian policy makers and diplomats. An even better solution would provide an opportunity for mid-grade and senior officers and civilians to learn from one another. More than one thousand students are educated annually at the U.S. Army Command and General Staff Officer Course at Fort Leavenworth. Only a small fraction consist of civilian and interagency students.

The benefits to training against realistic threats in live field exercises are self-evident. The theory facilitates practice. Thus, the practice must apply the theory in a realistic manner. Results must then be recorded, analyzed, and improved upon. The consequences for inattentiveness are dire. During the First World War, the British Expeditionary Force did not possess a system for frequent re-examination of tactical methods and operational doctrine. As a result, improvement was haphazard and incoherent.

Ubiquitous media is a significant blind-spot for an open society. The largest study of fake news found that falsehoods penetrated further, faster, and deeper into the social network than accurate information on Twitter. An erroneous tweet by a state government employee set off panic and alarm across the Hawaiian Islands. Evocative information travels fast, anything that arouses emotions will rapidly pick up speed. What if a media-owned or third-party drone or satellite broadcasts dead and dying Americans to domestic audiences in high definition? This presents significant informational questions military strategists and operational planners need to consider.

While professional military education is one way to address present challenges, the services and combatant commands must also better organize themselves to act in the information domain. The U.S. Marine Corps has made a concerted effort of late, establishing an Information Group with each Marine Expeditionary Force (MEF). However, organization is only the first step. Each Marine Expeditionary Force Information Group (MIG) must develop new tactics, techniques, and procedures to conduct information operations to achieve objectives beyond simply closing with and destroying an enemy.

General Robert Neller, the Commandant of the Marine Corps, has said future conflict is "...going to be a land, air, sea operation, but it's going to involve space, it's going to involve information, it's going to involve the electromagnetic spectrum; all the things that we haven't had to think about in the past 15 to 20 years." The West in general and the U.S. military in particular are nowhere near as agile as their adversaries in the information sphere. Tactical, operational, and strategic success requires a cultural change to reconcile institutional aversion and reluctance toward non-lethal information warfare. To dominate the information domain before, during, and after the next conflict, significant change is required in the U.S. military's approach toward training and education of information as a warfighting function, and information operations as a discipline.

47.3 INFORMATION WARFARE AND INDIA'S LEVEL OF PREPAREDNESS

"Repeat a lie often enough, and it becomes the truth", said Nazi Joseph Goebbels. This maxim is quite relevant for countries that are actively involved in information and psychological warfare.

India's strategic thinkers realise well that the intensity of the information war mounted against India is increasing. Stoke White, a UK-based firm, released a forty-one pages report, titled "India's War Crimes in Kashmir: Violence, Dissent and the War on Terror", on 14 January 2022. The report claims to have gathered testimonies of two thousand residents of Jammu and Kashmir, primarily Kashmiri Muslims and alleges that the Indian state puts them under coercion time and again. The report has termed the United Nations listed terror groups like Jaish-E-Mohammad (JeM) and others like the United States and the European Union sanctioned Hizb-ul-Mujahideen as "non-state armed groups" and calls for the UK's universal jurisdiction on "war crimes under the Geneva convention" to be invoked against Indian officials and Indian government functionaries. Uncannily enough, the theme of the report aligns well with Pakistan's long-standing narrative. Even the terms in use are akin to a 2021 dossier, released by Pakistan, titled "Indian Human Rights Violations in Indian Illegally Occupied Jammu and Kashmir (IIOJK)". Interestingly, Stoke White has a link with Turkey, which, as one may be aware, has colluded with Pakistan in recent years and has been aggressively running a disinformation campaign against India. The two countries, with the assistance of their world-over network, roll out reports and documentaries containing fake information on a sustained regular basis. That they rigorously cite each other's works, with the origin of information often being flaky, is a typical feature of information warfare. The objective is not hard to decipher here. Reports like these wish to authenticate the disinformation being spread, aimed at setting global perception against India. In 2021, an international relations researcher, analyzed a Disinformation Lab Report titled "The Unending War: From Proxy War to Info War", revealing that Pakistan and Turkey are aligned in their bid to create an anti-India image using the constructivist concept of 'perception management'. Perception is the key and its management a continuous work-in-progress, as propagators of Information Warfare understand only too well.

India is aware of disinformation campaigns against it and has recently begun confronting these at international diplomatic forums by presenting timely rebuttals and with the counter use of information technology. Also, the Ministry of Information and Broadcasting, in coordination with intelligence agencies, has chosen to block channels on YouTube and other websites that are found to be spreading anti-India propaganda and false news. In December 2021, twenty YouTube.com channels and two websites belonging to a Pakistan-based disinformation network were blocked for posting divisive content on Kashmir, the Indian Army, minority communities in India, Ram Mandir, General Bipin Rawat and more. So far, India has responded to these smearing campaigns, primarily to defend itself. However, there is a dire need for a holistic approach to curb the adverse effects of these malign campaigns on India's global image and its internal social and cultural fabric.

47.3.1 How India can fight back Information Warfare?

Disinformation narratives should be watched out for, rather expected, in current times of robust information technology. Think of narratives as a stream of running water. The current never stops. Similarly, countering such narratives too should be a sustained effort, preferred over ad hoc and episodic forms of confrontation. The flow of information should be tracked, and possibly capped while in its initial stages. That way the disinformation narratives would not fructify for debates on national and international forums. Other times, a 'wait and watch' approach is advisable before the narrative takes its full form. This would help formulate the counter tactic to be employed. Tracing the origin of information would help, specifically for sensitive topics pertaining to India – Kashmir, the North-East, Human Rights and religious minorities to name a few. Additionally, field studies in sensitive areas should be undertaken by Indian think tanks, and the reports generated be widely publicized. Furthermore, research centres should be made operational in the said areas, these being of keen interest for the anti-India dissemination networks, thereby whetting the information being generated becomes of paramount importance.

Many security advisers or academicians make the mistake of assuming that information-led techniques to counter disinformation networks (that float rumors and false news to instigate the civilians) can replace temporary internet shutdowns in affected areas. The two are mutually independent of each other. Shutting down internet services is a handy and short-term tactic to quell frayed rumors arising from an unexpected event. Seemingly thwarting democratic rights, its advantages far outweigh the cons, one being its swiftness. On the other hand, countering rumors through the release of rebuttal information can be a time-consuming process requiring days and weeks before compete dissemination can take place. However, the Indian authority must think beyond internet shutdowns, especially when its use is prolonged, as happened after the abrogation of the provisions of articles 370 and 35A when the citizens in Jammu and Kashmir were debarred from using the internet for more than 145 days. [8] Such prolonged restrictions on the internet impact the economy. Also, it creates a persistent sense of alienation amongst the civilians. It is a known fact that internet penetration is higher in Jammu and Kashmir. Therefore, the government should utilize the internet, specifically the semantic part of it, i.e., Social Media. Instead of internet shutdowns every time, the government must spread awareness.

For spreading awareness (including on Social Media platforms), a network of good Samaritans must be stimulated. The perception of Kashmiris, who encourages development and integration, must be well-reflected on social media platforms. Likewise, India must create awareness and have a good slogan such as "India Against Disinformation" or "Stop the Fake News" – to facilitate our citizens to post correct information online.

There is no way India could stop disinformation flow, but through improving communication strategy in the international system, India can reduce the adverse outcomes of the disinformation. To be credible, India also needs to formulate narratives supported by videos, images, write-ups, etc., concise and detailed forms by having an independent team with specialists and professionals who will proactively ensure its projections. This could be enhanced by creating fact-checking news portals of the international level along with this Indian statesmen should regularly contribute op-eds for widely read international newspapers. As well, in search engines such as Google and Yahoo, India must ensure the presence of counter-narratives using keywords that are used by adversaries in their offensive disinformation campaigns.

CHAPTER 48

STEALTH TECHNOLOGY IN WARFARE

48.1 INTRODUCTION TO STEALTH TECHNOLOGY

Stealth technology, also termed low observable technology (LO technology), is a sub-discipline of military tactics and passive and active electronic countermeasures. The term covers a range of methods used to make personnel, aircraft, ships, submarines, missiles, satellites, and ground vehicles less visible (ideally invisible) to radar, infrared, sonar and other detection methods. It corresponds to military camouflage for these parts of the electromagnetic spectrum (i.e., multi-spectral camouflage).

Development of modern stealth technologies in the United States began in 1958, where earlier attempts to prevent radar tracking of its U-2 spy planes during the Cold War by the Soviet Union had been unsuccessful. Designers turned to developing a specific shape for planes that tended to reduce detection by redirecting electromagnetic radiation waves from radars. Radiation-absorbent material was also tested and made to reduce or block radar signals that reflect off the surfaces of aircraft. Such changes to shape and surface composition comprise stealth technology as currently used on the Northrop Grumman B-2 Spirit "Stealth Bomber".

The concept of stealth is to operate or hide while giving enemy forces no indication as to the presence of friendly forces. This concept was first explored through camouflage to make an object's appearance blend into the visual background. As the potency of detection and interception technologies (radar, Infrared search and tracking, surface-to-air missiles, etc.) have increased, so too has the extent to which the design and operation of military personnel and vehicles have been affected in response. Some military uniforms are treated with chemicals to reduce their infrared signature. A modern stealth vehicle is designed from the outset to have a chosen spectral signature. The degree of stealth embodied in a given design is chosen according to the projected threats of detection.

48.2 RADAR CROSS-SECTION (RCS) REDUCTIONS

Almost since the invention of radar, various methods have been tried to minimize detection. Rapid development of radar during World War II led to equally rapid development of numerous counter radar measures during the period; a notable example of this was the use of chaff. Modern methods include radar jamming and deception.

The term stealth in reference to reduced radar signature aircraft became popular during the late eighties when the Lockheed Martin F-117 stealth fighter became widely known. The first large scale (and public) use of the F-117 was during the Gulf War in 1991. However, F-117A stealth fighters were used for the first time in combat during Operation *Just Cause*, the United States invasion of Panama in 1989. Stealth aircraft are often designed to have radar cross sections that are orders of magnitude smaller than conventional aircraft. The radar range equation meant that all else being equal, detection range is proportional to the fourth root of RCS; thus, reducing detection range by a factor of 10 requires a reduction of RCS by a factor of 10,000.

48.3 VEHICLE SHAPE FOR STEALTH

(i) Aircraft

The F-35 Lightning II offers better stealthy features (such as this landing gear door) than prior American multi-role fighters, such as the F-16 Fighting Falcon.

The possibility of designing aircraft in such a manner as to reduce their radar cross-section was recognized in the late 1930s, when the first radar tracking systems were employed, and it has been known since at least the 1960s that aircraft shape makes a significant difference in detectability. The Avro Vulcan, a British bomber of the 1960s, had a remarkably small appearance on radar despite its large size, and occasionally disappeared from radar screens entirely. It is now known that it had a fortuitously stealthy shape apart from the vertical element of the tail. Despite being designed before a low radar cross-section (RCS) and other stealth factors were ever a consideration, a Royal Aircraft Establishment technical note of 1957 stated that of all the aircraft so far studied, the Vulcan appeared by far the simplest radar echoing object, due to its shape: only one or two components contributing significantly to the echo at any aspect (one of them being the vertical stabilizer, which is especially relevant for side aspect RCS), compared with three or more on most other types. While writing about radar systems, authors Simon Kingsley and Shaun Quegan singled out the Vulcan's shape as acting to reduce the RCS. In contrast, the Tupolev 95 Russian long-range bomber (NATO reporting name 'Bear') was conspicuous on radar. It is now known that propellers and jet turbine blades produce a bright radar image; the Bear has four pairs of large 18-foot (5.6 m) diameter contra-rotating propellers.

Another important factor is internal construction. Some stealth aircraft have skin that is radar transparent or absorbing, behind which are structures termed re-entrant triangles. Radar waves penetrating the skin get trapped in these structures, reflecting off the internal faces and losing energy. This method was first used on the Blackbird series: A-12, YF-12A, Lockheed SR-71 Blackbird.

The most efficient way to reflect radar waves back to the emitting radar is with orthogonal metal plates, forming a corner reflector consisting of either a dihedral (two plates) or a trihedral (three orthogonal plates). This configuration occurs in the tail of a conventional aircraft, where the vertical and horizontal components of the tail are set at right angles. Stealth aircraft such as the F-117 use a different arrangement, tilting the tail surfaces to reduce corner reflections formed between them. A more radical method is to omit the tail, as in the B-2 Spirit. The B-2's clean, low-drag flying wing configuration gives it exceptional range and reduces its radar profile. The flying wing design most closely resembles a so-called infinite flat plate (as vertical control surfaces dramatically increase RCS), the perfect stealth shape, as it would have no angles to reflect back radar waves.

In addition to altering the tail, stealth design must bury the engines within the wing or fuselage, or in some cases where stealth is applied to an extant aircraft, install baffles in the air intakes, so that the compressor blades are not visible to radar. A stealthy shape must be devoid of complex bumps or protrusions of any kind, meaning that weapons, fuel tanks, and other stores must not be carried externally. Any stealthy vehicle becomes un-stealthy when a door or hatch opens.

Parallel alignment of edges or even surfaces is also often used in stealth designs. The technique involves using a small number of edge orientations in the shape of the structure. For example, on the F-22A Raptor, the leading edges of the wing and the tail planes are set at the same angle. Other smaller structures, such as the air intake bypass doors and the air refueling aperture, also use the same angles. The effect of this is to return a narrow radar signal in a very specific direction away from the radar emitter rather than returning a diffuse signal detectable at many angles. The effect is sometimes called "glitter" after the very brief signal seen when the reflected beam passes across a detector. It can be difficult for the radar operator to distinguish between a glitter event and a digital glitch in the processing system.

Stealth airframes sometimes display distinctive serrations on some exposed edges, such as the engine ports. The YF-23 has such serrations on the exhaust ports. This is another example in the parallel alignment of features, this time on the external airframe.

The shaping requirements detracted greatly from the F-117's aerodynamic properties. It is inherently unstable, and cannot be flown without a fly-by-wire control system.

Similarly, coating the cockpit canopy with a thin film transparent conductor (vapor-deposited gold or indium tin oxide) helps to reduce the aircraft's radar profile, because radar waves would normally enter the cockpit, reflect off objects (the inside of a cockpit has a complex shape, with a pilot helmet alone forming a sizeable return), and possibly return to the radar, but the conductive coating creates a controlled shape that deflects the incoming radar waves away from the radar. The coating is thin enough that it has no adverse effect on pilot vision.

(ii) Ships

Ships have also adopted similar methods. Though the earlier Arleigh Burke-class destroyer incorporated some signature-reduction features. the Norwegian Skjold-class corvette was the first coastal defence and the French La Fayette-class frigate the first ocean-going stealth ship to enter service. Other examples are the Dutch De Zeven Provinciën class frigates, the Taiwanese Tuo Chiang stealth corvette, German Sachsen-class frigates, the Swedish Visby-class corvette, the USS San Antonio amphibious transport dock, and most modern warship designs.

(iii) Materials

Dielectric composite materials are more transparent to radar, whereas electrically conductive materials such as metals and carbon fibers reflect electromagnetic energy incident on the material's surface. Composites may also contain ferrites to optimize the dielectric and magnetic properties of a material for its application.

(a) Radar-absorbent material

(b) Radiation-absorbent material (RAM)

Often as paints, are used especially on the edges of metal surfaces. While the material and thickness of RAM coatings can vary, the way they work is the same: absorb radiated energy from a ground- or air-based radar station into the coating and convert it to heat rather than reflect it back. Current technologies include dielectric composites and metal fibers containing ferrite isotopes. Ceramic composite coating is a new type of material systems which can sustain at higher temperatures with better sand erosion resistance and thermal resistance. Paint comprises depositing pyramid-like colonies on the reflecting superficies with the gaps filled with ferrite-based RAM. The pyramidal structure deflects the incident radar energy in the maze of RAM. One commonly used material is called iron ball paint. It contains microscopic iron spheres that resonate in tune with incoming radio waves and dissipate most of their energy as heat, leaving little to reflect back to detectors. FSS are planar periodic structures that behave like filters to electromagnetic energy. The considered frequency-selective surfaces are composed of conducting patch elements pasted on the ferrite layer. FSS are used for filtration and microwave absorption.

48.3.1 Radar stealth countermeasures and limits

(i) Low-frequency radar

Shaping offers far fewer stealth advantages against low-frequency radar. If the radar wavelength is roughly twice the size of the target, a half-wave resonance effect can still generate a significant return. However, low-frequency radar is limited by lack of available frequencies (many are heavily used by other systems), by lack of accuracy of the diffraction-limited systems given their long wavelengths, and by the radar's size, making it difficult to transport. A long-wave radar may detect a target and roughly locate it, but not provide enough information to identify it, target it with weapons, or even to guide a fighter to it.

(ii) Multiple emitters

Stealth aircraft attempt to minimize all radar reflections, but are specifically designed to avoid reflecting radar waves back in the direction they came from (since in most cases a radar emitter and receiver are in the same location). They are less able to minimize radar reflections in other directions. Thus, detection can be better achieved if emitters are in different locations from receivers. One emitter separate from one receiver is termed **bistatic radar**; one or more emitters separate from more than one receiver is termed **multistatic radar**. Proposals exist to use reflections from emitters such as civilian radio transmitters, including cellular telephone radio towers.

48.3.2 Moore's law

By Moore's law the processing power behind radar systems is rising over time. This will eventually erode the ability of physical stealth to hide vehicles.

(i) Ship wakes and spray

Synthetic aperture side-scan radars can be used to detect the location and heading of ships from their wake patterns. These are detectable from orbit. When a ship moves through a seaway it throws up a cloud of spray which can be detected by radar.

(ii) Acoustics

Acoustic stealth plays a primary role for submarines and ground vehicles. Submarines use extensive rubber mountings to isolate, damp, and avoid mechanical noises that can reveal locations to underwater passive sonar arrays.

Early stealth observation aircraft used slow-turning propellers to avoid being heard by enemy troops below. Stealth aircraft that stay subsonic can avoid being tracked by sonic boom. The presence of supersonic and jet-powered stealth aircraft such as the SR-71 Blackbird indicates that acoustic signature is not always a major driver in aircraft design, as the Blackbird relied more on its very high speed and altitude.

One method to reduce helicopter rotor noise is modulated blade spacing. Standard rotor blades are evenly spaced, and produce greater noise at a given frequency and its harmonics. Using varied spacing between the blades spreads the noise or acoustic signature of the rotor over a greater range of frequencies.

(iii) Visibility

The simplest technology is visual camouflage; the use of paint or other materials to color and break up the lines of a vehicle or person.

Most stealth aircraft use matte paint and dark colors, and operate only at night. Lately, interest in daylight Stealth (especially by the USAF) has emphasized the use of gray paint in disruptive schemes, and it is assumed that Yehudi lights could be used in the future to hide the airframe (against the background of the sky, including at night, aircraft of any colour appear dark) or as a sort of active camouflage. The original B-2 design had wing tanks for a contrail-inhibiting chemical, alleged by some to be chlorofluorosulfonic acid, but this was replaced in the final design with a contrail sensor that alerts the pilot when he should change altitude and mission planning also considers altitudes where the probability of their formation is minimized.

In space, mirrored surfaces can be employed to reflect views of empty space toward known or suspected observers; this approach is compatible with several radar stealth schemes. Careful control of the orientation of the satellite relative to the observers is essential, and mistakes can lead to detectability enhancement rather than the desired reduction.

(iv) Infrared

An exhaust plume contributes a significant infrared signature. One means to reduce IR signature is to have a non-circular tail pipe (a slit shape) to minimize the exhaust cross sectional area and maximize the mixing of hot exhaust with cool ambient air Often, cool air is deliberately injected into the exhaust flow to boost this process. The Stefan–Boltzmann law shows how this results in less energy (Thermal radiation in infrared spectrum) being released and thus reduces the heat signature. In some aircraft, the jet exhaust is vented above the wing surface to shield it from observers below, as in the Lockheed F-117 Nighthawk, and the unstealthy Fairchild Republic A-10 Thunderbolt II. To achieve infrared stealth, the exhaust gas is cooled to the temperatures where the brightest wavelengths it radiates are absorbed by atmospheric carbon dioxide and water vapor, greatly reducing the infrared visibility of the exhaust plume. Another way to reduce the exhaust temperature is to circulate coolant fluids such as fuel inside the exhaust pipe, where the fuel tanks serve as heat sinks cooled by the flow of air along the wings.

Ground combat includes the use of both active and passive infrared sensors. Thus, the United States Marine Corps (USMC) ground combat uniform requirements document specifies infrared reflective quality standards.

(v) Reducing radio frequency (RF) emissions

In addition to reducing infrared and acoustic emissions, a stealth vehicle must avoid radiating any other detectable energy, such as from onboard radars, communications systems, or RF leakage from electronics enclosures. The F-117 uses passive infrared and low light level television sensor systems to aim its weapons and the F-22 Raptor has an advanced LPI radar which can illuminate enemy aircraft without triggering a radar warning receiver response.

48.3.3 Measuring the RCS

The size of a target's image on radar is measured by the radar cross section (RCS), often represented by the symbol σ and expressed in square meters. This does not equal geometric area. A perfectly conducting sphere of projected cross-sectional area 1 m^2 (i.e. a diameter of 1.13 m) will have an RCS of 1 m^2. Note that for radar wavelengths much less than the diameter of the sphere, RCS is independent of frequency. Conversely, a square flat plate of area 1 m^2 will have an RCS of σ=4π A2 / λ2 (where A=area, λ=wavelength), or 13,982 m^2 at 10 GHz if the radar is perpendicular to the flat surface. At off-normal incident angles, energy is reflected away from the receiver, reducing the RCS. Modern stealth aircraft are said to have an RCS comparable with small birds or large insects, though this varies widely depending on aircraft and radar.

If the RCS was directly related to the target's cross-sectional area, the only way to reduce it would be to make the physical profile smaller. Rather, by reflecting much of the radiation away or by absorbing it, the target achieves a smaller radar cross section.

48.4 STEALTH TECHNOLOGY FUTURE

Stealth technology is expected to advance in the future with the use of AI, advanced materials, and multi-domain integration:

(i) AI

Machine learning and other AI technologies can enhance situational awareness and help assets adapt to threats. For example, AI can help set flight paths and control ECMs. **(ii) Advanced materials**

New stealth materials can absorb radar and sonar, and can sense, process information, and identify themselves.

(ii) Multi-domain integration

Stealth assets can work with other platforms, like drones and satellites, to disrupt enemy detection networks.

(iii) Quantum stealth

Quantum stealth technologies may be developed to minimize visibility across all detection spectrums.

(iv) Metasurfaces

These thin layers of meta-material can redirect scattered waves without changing the target's geometry.

(v) Plasma stealth

Ionized gas can create a cloud around a platform to absorb or deflect radar.

(vi) Adaptive aeroelastic wings

These wings can change shape in flight to deflect airflow.

(Vii) Fluidics

Fluid injection can be used to control the direction of an aircraft.

48.4.1 Other developments in stealth technology include:

Active camouflage,

- Improved electronic warfare,
- Unmanned, surface, and underwater vehicles, and
- Mirrored surfaces in space.

48.5 NEXT-GENERATION STEALTH TECHNOLOGY FOR AIRCRAFT AND NAVAL VESSELS

Stealth has become the single biggest game changer in contemporary warfare by changing the way aircraft, naval ships, and most other platforms are designed and used on contemporary battlefields.

The aspects of stealth are very important because, through them, a mission remains a secret and there is a high chance of success. Speculating on stealth technology in avionics and naval vessels,

48.5.1 Evolution of Stealth Technology in Aircraft

This is in regards to the new advancements in stealth aircraft, which have revolutionized aerial combat. Truly stealth aircraft's progenitor initially included aircraft such as the 117 Nighthawk. The 117 Nighthawk became the first operational aircraft that was designed with stealth in mind and was given a shape and material that would reduce its radar signature.

After the 117 Nighthawk, the Air Force B2 Spirit was another memorable creation in terms of stealth. B2 Spirit employs what is known as a flying wing design; this aspect contributes to its small radar cross-section. This stealth airplane also has modern raw materials and paint to make it invisible to the radar systems of its enemies. The aspects of stealth provided by the B2 Spirit perform well in enabling it to get deep into the enemy airspace and hit intended targets with little chance of neutralization.

The B2 stealth bombers are considered to have the highest level of invisibility among modern aircraft. Both these bombers include concentrated composite materials, electronic counters, and low-observable technologies to comprehend a new degree of stealth. The B2 stealth fighter jet is another example of a lean or low silouhette with cohesive stealth technology. This also however holds combat versatility.

48.5.2 Naval Stealth Technology

The merchant navy has also had a sneak peak in technological development. Originally, the merchant navy was associated with the transport of goods and cargo at sea via ships, but presently it is involved with the condimentation of stealth features for safeguarding important cargo consignments and threats during the movement through the dangerous territories. That is why the use of stealth technology is not limited only to military affairs and is exemplified by the described trend.

48.6 STEALTH TECHNOLOGY USED IN INDIAN AIRFORCE

The Indian Air Force (IAF), recognizing the immense potential of stealth technology, has been integrating these advancements to bolster its air superiority and mission effectiveness. Here, we will explore the principles of stealth technology and examine the specific aircraft within the IAF that employ these cutting-edge features.

(i) LCA Tejas

The Light Combat Aircraft (LCA) Tejas, developed by Hindustan Aeronautics Limited (HAL), incorporates some stealth features such as reduced radar cross-section through its design and the use of radar-absorbent materials. While not a full-fledged stealth aircraft, Tejas represents a significant step in integrating stealth principles into indigenous aircraft.

Design Features:

The Tejas has a low frontal RCS due to its compact design, use of composite materials, and attention to detail in minimizing the aircraft's radar signature.

(ii) Advanced Medium Combat Aircraft (AMCA)

The AMCA is India's ambitious project aimed at developing a fifth-generation stealth fighter. With advanced stealth features, the AMCA will have a low radar cross-section, advanced avionics, and superior maneuverability.

Stealth Features:

The AMCA design includes an internal weapons bay, serpentine air intakes to shield engine compressors from radar, and extensive use of RAM. Its shape and structure are optimized for minimal radar reflection.

Expected Capabilities:

The AMCA is expected to perform air superiority, ground attack, and electronic warfare missions, providing a significant edge in modern aerial warfare.

48.7 FUTURE PROJECTS OF INDIAN AIR FORCE FOR STEALTH TECHNOLOGY

48.7.1 Collaborations with Global Leaders

India has been exploring collaborations with countries like the United States, Russia, and Israel to enhance its stealth technology capabilities. These collaborations aim to integrate cutting-edge stealth technologies into Indian platforms and develop indigenous solutions.

48.7.2 Research and Development

The Defence Research and Development Organisation (DRDO) is actively involved in researching advanced stealth technologies, including metamaterials and adaptive camouflage. These advancements are expected to further enhance the stealth capabilities of Indian aircraft.

48.8 APPLICATIONS OF STEALTH TECHNOLOGY IN THE INDIAN AIR FORCE

By integrating advanced design principles and cutting-edge materials, the IAF is able to maintain air superiority and execute complex missions with reduced risk of detection. Before we look into the specific applications of stealth technology within the IAF, let's explore the fundamental principles that make this technology so pivotal in modern warfare.

(i) Fighter Aircraft

Future stealth fighters like the AMCA will be designed for air superiority and strike missions, capable of penetrating enemy airspace and engaging targets without detection.

(ii) Unmanned Aerial Vehicles (UAVs)

Stealth UAVs are being developed for surveillance, reconnaissance, and strike missions, allowing for high-risk operations without risking pilot lives.

(iii) Advanced Training and Simulation

Incorporating stealth technology into training programs and simulation exercises ensures that Indian pilots are well-prepared to operate and counter stealth aircraft.

Stealth technology is transforming the capabilities of the Indian Air Force, providing a strategic edge in modern aerial warfare. With ongoing projects like the AMCA and collaborations with global leaders, India is poised to become a formidable force in stealth aviation. As detection technologies evolve, the Indian Air Force will continue to innovate and adapt, ensuring that they remain at the forefront of stealth technology and aerial combat.

48.9 IIT KANPUR UNVEILS ANALAKSHYA STEALTH TECH TO MAKE OBJECTS INVISIBLE TO RADAR

IIT Kanpur has introduced a groundbreaking innovation in stealth technology with its Analakshya Metamaterial Surface Cloaking System (MSCS). This cutting-edge system is designed to make objects almost invisible to radar, offering significant advancements in defence and national security.

Developed by a team of researchers and students from IIT Kanpur, the Analakshya MSCS is a textile-based broadband metamaterial microwave absorber. Its unique capability lies in its ability to absorb radar waves across a wide spectrum, rendering Synthetic Aperture Radar (SAR) imaging ineffective. This means better protection against radar-guided missiles and enhanced stealth capabilities for modern warfare.

The innovation reflects India's growing self-reliance in defence technology, with over 90% of its materials sourced domestically. After extensive testing from 2019 to 2024, the technology proved effective in various operational conditions.

It has been licensed to Meta Tattva Systems Pvt. Ltd. for manufacturing, marking a critical step towards its deployment by the Indian Armed Forces.

It strengthens India's operational capabilities and offers a significant strategic edge in maintaining national security. The system not only represents a leap in stealth technology but also highlights the collaboration between academia, industry, and the armed forces.

48.10 MORE ON FUTURE OF STEALTH TECHNOLOGY

Future developments in stealth technology are expected to advance more as engineers continue to work on attempts to make planes and naval ships even more invisible to radar. Active camouflage, advanced materials, and the improvement of electronic warfare are promising tools that will augment the stealth features of future platforms.

In the case of aircraft, one of the two dominant forms of AI seen in today's world, which is machine learning, is predicted to greatly contribute to improving stealth technology. These technologies can become useful to set flight paths, control ECMs, and react to threats in real-time, which makes the chances of being detected by stealth fighters and bombers even slimmer.

The evolving component in the naval domain is thus mentioned to be unmanned, surface, and underwater vehicles with stealth capabilities. The difference lies in the fact that these catalytic platforms can work for reconnaissance, surveillance, and sometimes for any combat operations that require less visibility of the operated equipment.

(i) Integration of Stealth Technology in Modern Military Strategies

Regarding the aircraft, stealth technologies have modified the nature of aerial combat, where the stealth fighters and bombers, such as the B2 stealth bombers, can engage the enemy with fewer chances of being attacked. In terms of striking capability, the most memorable one is the B2 Spirit aircraft, which can independently perform tactical strikes and deliver its missions without being identified and plays an important role in the air force combat power system.

(ii) Modern Military Strategies in Stealth Technology

Another application of stealth technology is in surveillance, especially by surveillance aircraft. Stealth airplanes, which are fitted with sensors, can sip intelligence in the hostile airspace; in other words, one can acquire vital information about the enemy that is not necessary to counterattack. For instance, the B2 stealth plane is capable of bombing and, at the same time, spying, bringing out the flexibility of stealth in current warfare.

(iii) Naval Stealth and the Merchant Navy

The merchant navy has also started implementing stealth technology to safeguard carrier/merchant routes and important consignments. This idea is a perfect reflection of modern tendencies in the development of the global economy since the contested nature of trade routes means that they have to stay invisible to potential threats in order to remain stable. Including stealth into the adaptability of the merchant navy brings extra security to locations that are deemed unsafe, allowing a constant supply of goods and services.

(iv) Technological Innovations and Future Prospects

That is why one can state that the further evolution of stealth technology promises even greater revolutionary changes. There are works on new products that have made them stealthier in radar and I.R. These products will be used in aircraft and naval ships. For example, modern developments in technology are focused on the creation of adaptive camouflaging systems, the integration of which might render stealth fighters and naval vessels virtually invisible.

Also, the use of unmanned systems integrated with stealth capability is increasing. These stealth-equipped UAVs and UUVs are capable of performing multiple operations, from surveillance to combat, with less threat. These automated platforms can augment manned B2 stealth bombers and stealth airplanes, thereby forming a versatile model for stealth actions.

CHAPTER 49

INTEGRATED MISSILE DEFENCE SYSTEM

49.1 INTEGRATED GUIDED MISSILE DEVELOPMENT PROGRAMME OF INDIA

The Integrated Guided Missile Development Programme (IGMDP) was an Indian Ministry of Defence Programme for the research and development of the comprehensive range of missiles. The programme was managed by the Defence Research and Development Organisation (DRDO) and Ordnance Factories Board. The project started in 1982–83 under the leadership of **Dr. APJ Abdul Kalam** who oversaw its ending in 2008 after these strategic missiles were successfully developed.

On 8 January 2008, the DRDO formally announced the successful rated guided missile programme was completed with its design objectives achieved since most of the missiles in the programme had been developed and inducted by the Indian armed forces.

By the start of the 1980s, the DRDL Hyderabad had developed competence and expertise in the fields of propulsion, navigation and manufacture of aerospace materials based on the Soviet rocketry technologies. Thus, India's political leadership, which included Prime Minister Indira Gandhi, Defence Minister R. Venkataraman, V.S. Arunachalam (Scientific Advisor to the Defence MInIster), decIded that all these technologies should be consolidated.

This led to the birth of the Integrated Guided Missile Development Programme with **Dr. Abdul Kalam,** who had previously been the project director for the SLV-3 programme at ISRO, was inducted as the DRDL Director in 1983 to conceive and lead it. While the scientists proposed the development of each missile consecutively, the Defence Minister R. Venkataraman asked them to reconsider and develop all the missiles simultaneously. Thus, four projects, to be pursued concurrently, were born under the IGMDP:

Dr. APJ Abdul Kalam started multiple projects simultaneously to develop the following types of Indian Guided Missiles:

(i) Short Range Surface to Surface Missile (SSM) **'Prithvi'**

(ii) Long Range Surface to Surface Missile (SSM) **'Agni'**

(iii) Medium Range Surface to Air Missile (SAM) **'Akash'**

(iv) Short Range Surface to Air Missile (SAM) **'Trishul'**

(v) Anti-tank Guided Missile (ATGM) **'Nag'**

The Agni missile was initially conceived in the IGMDP as a technology demonstrator project in the form of a re-entry vehicle, and was later upgraded to a ballistic missile with different ranges. As part of this program, the Interim Test Range at Balasore in Odisha was also developed for missile testing.

After India test-fired the first Prithvi missile in 1988, and the Agni missile in 1989, the Missile Technology Control Regime (then an informal grouping established in 1987 by Canada, France, Germany, Italy, Japan, the United Kingdom and the United States) decided to restrict access to any technology that would help India in its missile development program. To counter the MTCR, the IGMDP team formed a consortium of DRDO laboratories, industries and academic institutions to build these sub-systems, components and materials. Though this slowed down the progress of the program, India successfully developed indigenously all the restricted components denied to it by the MTCR.

The starting of India's missile program influenced Pakistan to scramble its resources to meet the challenge. Like India, Pakistan faced hurdles to operationalize its program since education on space sciences was never sought. It took Pakistan decades of expensive trial errors before their program became feasible for military deployment.

(i) Prithvi missile

The Prithvi missile is a family of tactical surface-to-surface short-range ballistic missiles (SRBM) and is India's first indigenously developed ballistic missile. Development of the Prithvi began in 1983, and it was first test-fired on 25 February 1988 from Sriharikota, SHAR Centre, Pottisreeramulu Nellore district, Andhra Pradesh. It has a range of up to 150 to 300 km. The land variant is called Prithvi while the naval operational variant of Prithvi I and Prithvi III class missiles are code named **Dhanush**. Both variants are used for surface targets.

The Prithvi is said to have its propulsion technology derived from the Soviet SA-2 surface-to-air missile. Variants make use of either liquid or both liquid and solid fuels. Developed as a battlefield missile, it could carry a nuclear warhead in its role as a tactical nuclear weapon.

The initial project framework of the IGMDP envisioned the Prithvi missile as a short-range ballistic missile with variants for the Indian Army, Indian Air Force and the Indian Navy. Over the years the Prithvi missile specifications have undergone a number of changes. The Prithvi I class of missiles were inducted into the Indian Army in 1994, and it is reported that Prithvi I missiles are being withdrawn from service, being replaced with Prahar missiles. Prithvi II missiles were inducted in 1996. Prithvi III class has a longer-range of 350 km, and was successfully test fired in 2004.

(ii) Agni Missile re-entry technology

A technology demonstrator for re-entry technology called Agni was added to IGMDP as Prithvi was unable to be converted to a longer ranged missile. The first flight of Agni with re-entry technology took place in 1989. The re-entry system used resins and carbon fibers in its construction and was able to withstand a temperature of up to 3000 °C. The technologies developed in this project were eventually used in the Agni series of missiles.

(iii) Trishul missile

Trishul is a short-range surface-to-air missile developed by India as a part of the Integrated Guided Missile Development Program. It has a range of 12 km and is fitted with a 5.5 kg warhead. Designed to be used against low-level (sea skimming) targets at short range, the system has been developed to defend naval vessels against missiles and also as a short-range surface-to-air missile on land. According to reports, the range of the missile is 12 km and is fitted with a 15 kg warhead. The weight of the missile is 130 kg. The length of the missile is 3.5 m. India officially shut down the project on 27 February 2008. In 2003, Defence Minister George Fernandes had indicated that the Trishul missile had been de-linked from user service and would be continued as a technology demonstrator.

(iv) Akash missile

Akash is a medium-range surface-to-air missile developed as part of India's Integrated Guided Missile Development Programme to achieve self-sufficiency in the area of surface-to-air missiles. It is the most expensive missile project ever undertaken by the Union government in the 20th century. Development costs skyrocketed to almost US $120 million, which is far more than other similar systems.

Akash is a medium-range surface-to-air missile with an intercept range of 30 km. It has a launch weight of 720 kg, a diameter of 35 cm and a length of 5.8 metres. Akash flies at supersonic speed, reaching around Mach 2.5. It can reach an altitude of 18 km. A digital proximity fuse is coupled with a 55 kg pre-fragmented warhead, while the safety arming and detonation mechanism enables a controlled detonation sequence. A self-destruct device is also integrated. It is propelled by a solid fueled booster stage. The missile has a terminal guidance system capable of working through electronic countermeasures. The entire Akash SAM system allows for attacking multiple targets (up to 4 per battery). The Akash missile's use of ramjet propulsion system allows it to maintain its speed without deceleration, unlike the Patriot missiles. The missile is supported by a multi-target and multi-function phased array fire control radar called the 'Rajendra' with a range of about 80 km in search, and 60 km in terms of engagement.

The missile is completely guided by the radar, without any active guidance of its own. This allows it greater capability against jamming as the aircraft self-protection jammer would have to work against the high-power Rajendra, and the aircraft being attacked is not alerted by any terminal seeker on the Akash itself.

Design of the missile is similar to that of the SA-6, with four long tube ramjet inlet ducts mounted mid-body between wings. For pitch/yaw control four clipped triangular moving wings are mounted on mid-body. For roll control four inline clipped delta fins with ailerons are mounted before the tail. However, internal schema shows a completely modernized layout, including an onboard computer with special optimized trajectories, and an all-digital proximity fuse.

The Akash system meant for the Army uses the T-72 tank chassis for its launcher and radar vehicles. The Rajendra derivative for the Army is called the Battery Level Radar-III. The Air Force version uses an Ashok Leyland truck platform to tow the missile launcher, while the Radar is on a BMP-2 chassis and is called the Battery Level Radar-II. In either case, the launchers carry three ready-to-fire Akash missiles each. The launchers are automated, autonomous and networked to a command post and the guidance radar. They are slewable in azimuth and elevation. The Akash system can be deployed by rail, road or air.

The first test flight of Akash missile was conducted in 1990, with development flights up to March 1997.

The IAF has initiated the process to induct the Akash surface-to-air missiles developed as a part of the Integrated Guided Missile Development Programme. The Multiple target handling capability of Akash weapon system was demonstrated by live firing in a C4I environment during the trials. Two Akash missiles intercepted two fast moving targets in simultaneous engagement mode in 2005 itself. The Akash System's 3-D central acquisition radar (3-D car) group mode performance was then fully established.

(v) Nag Missile

Nag is India's third-generation "Fire-and-forget" anti-tank missile. It is an all-weather, top attack missile with a range of 0.5 to 4 km.

The missile uses an 8 kg high-explosive anti-tank (HEAT) tandem warhead capable of defeating modern armor including Explosive Reactive Armour (ERA) and composite armor. Nag uses Imaging Infra-Red (IIR) guidance with day and night capability. Mode of launch for the IIR seeker is LOBL (lock-on before launch). Nag can be mounted on an infantry vehicle; a helicopter launched version will also be available with integration work being carried out with the HAL Dhruv.

Separate versions for the Army and the Air Force are being developed. For the Army, the missiles will be carried by specialist carrier vehicles (NAMICA-Nag Missile Carrier) equipped with a thermographic camera for target acquisition. NAMICA is a modified BMP-2 infantry fighting vehicle license produced as "Sarath" in India. The carriers are capable of carrying four ready-to-fire missiles in the observation/launch platform which can be elevated with more missiles available for reload within the carrier. For the Air Force, a nose-mounted thermal imaging system has been developed for guiding the missile's trajectory "Helina". The missile has a completely fiberglass structure and weighs around 42 kg.

Nag was test fired for the 45th time on 19 March 2005 from the Test Range at Ahmednagar (Maharashtra), signaling the completion of the developmental phase.

49.1.1 India Completes Successful Tests of Phase-II Ballistic Missile Defence System

On July 24, 2024, the Indian Defence Research and Development Organisation (DRDO) successfully conducted a flight test of its sophisticated Phase-II Ballistic Missile Defence System, featuring the AD-2 endo-atmospheric missile. This flight marks a critical advancement in India's preparedness for network-centric warfare.

The target missile, simulating an enemy ballistic missile, was launched at 16:20 from Launch Complex IV in Dhamra. Within minutes, the system's radars, strategically placed on land and sea, detected the threat, activating the Advanced Defence (AD) Interceptor system. The interceptor missile was subsequently launched from Launch Complex III at the Integrated Test Range (ITR) in Chandipur at 16:24.

This flight test, which achieved all set objectives, also demonstrated the prowess of India's indigenous defence technologies. The Phase-II AD endo-atmospheric missile, a two-stage, solid-propellant, ground-launched system, is designed to intercept and neutralize enemy ballistic missile threats in both endo and low exo-atmospheric regions.

Throughout the test, the missile's performance was meticulously monitored via a range of tracking instruments including Electro-Optical Systems, Radar, and Telemetry Stations, spread across various locations, including aboard ships.

An endo-atmospheric missile is designed to operate and intercept targets within the Earth's atmosphere. Unlike exo-atmospheric missiles, which function outside the atmosphere, endo-atmospheric missiles handle engagements within the dense atmospheric layers. These missiles are crucial for missile defense systems, as they are capable of intercepting and destroying incoming ballistic missiles or other airborne threats before they can reach their targets. Endo-atmospheric missiles, such as the AD-2 tested by the Indian Defence Research and Development Organisation (DRDO) on July 24, 2024, are characterized by their high-speed interception capabilities, advanced guidance systems, and robust aerodynamic design to withstand the pressures and temperatures of atmospheric flight. Powered by solid or liquid propellants, these missiles are engineered for rapid acceleration and precise maneuverability, making them essential for modern air and missile defense strategies.

This successful trial underscores the DRDO's capability to defend the nation against high-class ballistic missile threats, particularly those capable of reaching distances up to 5,000 km. The development of this missile system incorporates numerous state-of-the-art technologies crafted by various DRDO labs, highlighting the organization's growing proficiency in indigenous technology.

This flight test represents not only a significant technological leap but also strengthens India's position in modern warfare, ensuring a robust defence mechanism against potential threats.

49.2 INTEGRATED AIR AND MISSILE DEFENSE OF USA

In air and missile defense (AMD), the Integrated Air-and-Missile Defense system (IAMD) is an SMDC research program to augment the aging surface-to-air missile defense systems and to provide the United States Army with a low-cost, but effective complement to kinetic energy solutions to take out air threats. Brigade level higher energy lasers are used in truck mounted systems called HELMTT. At lower levels, the Army needs to develop interceptors that don't cost more than small, unmanned aircraft systems. In early research they have successfully used 5-kilowatt lasers on a Stryker combat vehicle. The Mobile Expeditionary High-Energy Laser (MEHEL) was used at MFIX at Fort Sill, Oklahoma, in the first half of April, 2017.

The United States Army Integrated Air and Missile Defense [IAMD] Battle Command System (IBCS) is a plug and fight network intended to let any defensive sensor (such as a radar) feed its data to any available weapon system (colloquially, "connect any sensor to any shooter"). The system is designed to shoot down short, medium, and intermediate range ballistic missiles in their terminal phase by intercepting with a hit-to-kill approach. IBCS has been developed since 2004, with the aim to replace Raytheon's Patriot missile (SAM) engagement control station (ECS), along with seven other forms of ABM defense command systems.

The IBCS program is part of the Army's Integrated Air and Missile Defense (IAMD) effort. IBCS aims to create an integrated network of air defense sensors, such as AN/MPQ-64 Sentinel and AN/TPS-80 G/ATOR, AN/MPQ-53, AN/MPQ-65A and GhostEye (LTAMDS) in **Patriot missile system,** GhostEye MR in NASAMS, AN/TPY-2 in Terminal High Altitude Area Defense (THAAD) and Ground-Based Midcourse Defense (GMD),[AN/SPY-1 and AN/SPY-6 in Aegis BMD, and AN/APG-81 in Lockeed Martin F-35 Lightning II, allowing them to interoperate with IBCS engagement control stations. IBCS engagement stations will be able to take fine control of army-fielded air-defense systems like Patriot and THAAD, directing radar positioning and suggesting recommended launchers; naval, aerial and Marine systems will only be able to share either radar tracks or raw radar data with the IBCS network. The Army requires all new missiles and air-defense systems to implement IBCS support.

Northrop Grumman was announced as the prime contractor in 2010; between 2009 and 2020, the Army had spent $2.7 billion on the program.

By May 2015, a first flight test integrated a networked S-280 engagement operations center with radar sensor and interceptor launchers. This test demonstrated a missile kill with the first interceptor. By Army doctrine, two interceptors were launched against that missile. By April 2016, IBCS tests demonstrated sensor fusion from disparate data streams, identification and tracking of targets, selection of appropriate kill vehicles, and interception of the targets, but the "IBCS software was 'neither mature nor stable'". On 1 May 2019 an Engagement Operations Center (EOC) for the Integrated Air and Missile Defense (IAMD) Battle Command System (IBCS) was delivered to the Army, at Huntsville, Alabama.

In July 2019, the TRADOC capability manager (TCM) for Strategic Missile Defense (SMD) has accepted the charter for DOTMLPF for the Space and Missile Defense Command (SMDC/ARSTRAT).

On 30 August 2019 at Reagan Test Site on Kwajalein atoll, THAAD Battery E-62 successfully intercepted a medium range ballistic missile (MRBM), using a radar which was well-separated from the interceptors; the next step tested Patriot missiles as interceptors while using THAAD radars as sensors; a THAAD radar has a longer detection range than a Patriot radar. THAAD Battery E-62 engaged the MRBM without knowledge of just when the medium range ballistic missile had launched.

IBCS' second limited user test was scheduled to take place in the fourth quarter of FY20.

In July 2020 a Limited user test (LUT) of IBCS was initiated at WSMR; the test ran until mid-September 2020. The LUT was originally scheduled for May but was delayed to handle the COVID-19 safety protocols. The first of several LUTs of IBCS, by an ADA battalion was successfully run in August 2020. IBCS successfully integrated data from two sensors (Sentinel and Patriot radars), and shot down two drones (cruise missile surrogates) with two Patriot missiles in the presence of jamming. In the week after, by 20 August 2020 two more disparate threats (cruise missile and ballistic missile) were launched and intercepted; the ADA battalion then ran hundreds of drills denoting hundreds of threats for the remainder of the IBCS tests (the increased effort occupied the entire unit); the real-world data serve as a sanity check for Monte Carlo simulations of an array of physical scenarios amounting to hundreds of thousands of cases. IBCS created a single uninterrupted composite track of each threat and handed off each threat for separate disposition by the air and missile defense's integrated fire control network (IFCN). In 2022, IBCS successfully completed initial operational test and evaluation (IOT&E).

In September 2020 a Joint exercise against cruise missiles demonstrated AI-based kill chains which can be formulated in seconds; One of the kills was by a "M109-based" tracked howitzer.

The ranges of the IAMD defensive radars, when operated as a system, are thousands of miles. Cross-domain information from ground, air, and space sensors was passed to a fire control system at Project Convergence 2021 (PC21), via IBCS, during one of the use case scenarios. At PC21 IBCS fused sensor data from an F-35, tracking the target, and passing that data to AFATDS (Army Field Artillery Tactical Data System). The F-35 then served as a spotter for artillery fire on ground target data.[56] More than 100 technologies were prototyped in experiments at PC21

By August 2020, a second Limited User Test (LUT) at White Sands Missile Range was able to detect, track, and intercept near-simultaneous low-altitude targets as well as a tactical ballistic missile, over several separate engagements. Army doctrine can now be updated to allow the launch of a single Patriot against a single target.

On 24 February 2022 THAAD radar and TFCC (THAAD Fire Control & Communication) demonstrated their interoperability with Patriot PAC-3 MSE missiles; in other words IBCS can engage targets using both THAAD and Patriot interceptors, freed of a siloed solution (THAAD-only / Patriot-only, etc.). For example, in a scenario where a THAAD system has to conserve its All-Up-Rounds, IBCS can calculate which targets are within the reach of its PAC-3 MSE interceptors, and instead fire the PAC-3 interceptors at those targets within range.

49.2.1 High Energy Laser Tactical Vehicle Demonstrator

A contract for the U.S. Army Space and Missile Defense Command/Army Forces Strategic Command's High Energy Laser Tactical Vehicle Demonstrator (HEL TVD) laser system, a 100-kilowatt laser demonstrator for use on the Family of Medium Tactical Vehicles, was awarded 15 May 2019 to Dynetics-Lockheed. A 300-kilowatt laser demonstrator (HEL-IFPC) effort supersedes the HEL TVD.

In July 2021 RCCTO conducted a combat shoot-off on just how to control pointing these high-energy lasers. Raytheon is providing the high energy laser (Directed Energy Maneuver-Short Range Air Defense system —DE M-SHORAD) for the Strykers in 2022.

RCCTO has awarded a contract to build a 300-kW high-energy laser (HEL) for the Army in FY2022, capable of defending against airborne threats, by acquiring, tracking, and maintaining the HEL's aimpoint on the threat until it goes down.

49.3 THE U.S. ARMED FORCES CARRIED OUT FOR THE FIRST TIME THE INTERCEPTION OF A BALLISTIC MISSILE FROM THE STRATEGIC ISLAND OF GUAM

The United States Armed Forces continue to strengthen their deterrence and response capabilities in the Pacific. The latest milestone in this effort has been a ballistic missile interception test conducted from the island of Guam, home to some of the most important U.S. naval and air installations in the region.

Designated as Flight Experiment Mission-02, the test involved intercepting a medium-range ballistic missile launched off the coast of Andersen Air Force Base in Guam. One of its objectives was to validate the capabilities of the defense systems deployed on the Western Pacific island.

Broadly speaking, these defense systems integrate the new AN/TPY-6 radar with a vertical launcher for SM-3 (Standard Missile-3) Block IIA missiles, in combination with the AEGIS combat management system. Once the test missile's launch was detected, the radar began tracking it and providing data for the interception systems, which successfully neutralized the ballistic missile.

According to statements from the Pentagon and the U.S. Missile Defense Agency, this test marked the first time a ballistic missile has been intercepted from the island of Guam. Officials highlighted that this test represents a significant step forward in strengthening Guam's defense systems: "Today's event marks a crucial step in the initiatives and partnerships for the defense of Guam, providing critical support to the overall concept, requirements validation, data collection, and model development for the future Guam Defense System (GDS). The focus remains on defending Guam and protecting forces against any potential missile threats in the region."

They added: "FEM-02 is the first demonstration conducted from Guam as part of the long-term initiative to defend the island and will contribute to broader efforts to develop, install, and operate the Guam Defense System (GDS). This system will consist of a combination of Department of Defense (DoD) service components working together to deliver an enhanced, layered, integrated air and missile defense system. Collectively, DoD components will develop and deploy an integrated air and missile defense capability on a persistent basis."

The significance of this demonstration from Guam cannot be overstated, as the U.S. has been advancing plans to deploy more military assets on the island and its facilities. For example, the island has recently hosted the deployment of strategic bombers for hypersonic missile launch tests, the first deployment of Virginia-class attack submarines, and plans for the expansion of Andersen Air Force Base.

"In the context of national defense, one of the Department of Defense's main priorities, Guam is also a strategic location for sustaining and maintaining U.S. military presence, deterring adversaries, responding to crises, and ensuring a free and open Indo-Pacific," the U.S. Missile Defense Agency stated in its December 10th release.

49.4 NATO INTEGRATED AIR AND MISSILE DEFENCE (NATO IAMD)

NATO Integrated Air and Missile Defence (NATO IAMD) is an essential and continuous mission in peacetime, crisis and conflict, safeguarding and protecting Alliance territory, populations and forces against any air or missile threat or attack. This mission is conducted with a 360-degree approach and tailored to address all air and missile threats, emanating from all strategic directions, and coming from both state and non-state actors. To that end, it incorporates all measures – such as 24/7 air policing and ballistic missile defence – to contribute to deterring any air and missile threat, or to nullify or reduce their effectiveness. NATO IAMD is an essential element of NATO's deterrence and defence posture, contributing to the Alliance's indivisible security and freedom of action, including NATO's ability to reinforce its deployments and to provide a strategic response.

NATO Integrated Air and Missile Defence provides a highly responsive, robust, time-critical and persistent capability. It is aimed at ensuring a desired level of control of the air, so that the Alliance is able to conduct the full range of its operations and missions in peacetime, crisis and conflict.

NATO IAMD is particularly crucial in the current strategic environment, which is characterized by a significant proliferation of various types of air and missile capabilities and their excessive use in conflicts, as demonstrated during Russia's war of aggression against Ukraine.

In response to Russia's war against Ukraine, Allies have deployed additional IAMD capabilities to NATO's eastern flank, demonstrating Allied solidarity and resolve.

Allies continue to strengthen NATO IAMD by improving its readiness, responsiveness and integration through various initiatives, such as the implementation of the IAMD Rotational Model across the Euro-Atlantic area, with an initial focus on the eastern flank. Allies remain committed to enhancing the effectiveness of IAMD and taking all necessary steps to respond to the security environment.

NATO IAMD is implemented through the NATO Integrated Air and Missile Defence System (NATINAMDS), a network of interconnected national and NATO systems comprised of sensors, command and control assets, and weapons systems.

NATINAMDS comes under the authority of NATO's Supreme Allied Commander Europe (SACEUR).

49.5 INDIA SHOWS ITS DETERRENT HOLDS CHINESE CITIES AT RISK

India's recent test of an *Agni*-V missile using a multiple independently targetable re-entry vehicle (MIRV) is a milestone in the country's long-held ambition to acquire an ICBM which could credibly inflict unacceptable damage on Chinese cities.

On 11 March 2024, India successfully tested its first MIRV using the under-development *Agni*-V intercontinental ballistic missile (ICBM) as a delivery vehicle. The successful launch indicates that, after well over a decade of testing, the *Agni*-V has begun its last development phase, that of integrating the proven missile with a reliable set of warheads.

According to the Ministry of Defence press release, the successful test was carried out from Dr. APJ Abdul Kalam Island in Odisha, India's primary missile-testing facility. The facility's location on India's eastern seaboard allows for test launches into the Bay of Bengal with minimal disruption to civilian air and maritime traffic. A low-definition screen capture of the test was released shortly after the press release, showing the missile's launch with two of its three stages visible. Partly obstructed by the exhaust plume, a yellow wheeled transporter-erector launcher (TEL) is also visible and closely resembles other launch vehicles seen at the Defence Research and Development Organisation (DRDO) missile complex near Shamirpet outside Hyderabad, in southern India. The TEL is also visible in satellite imagery of the site taken two days before the test was reported, suggesting it was very unlikely that a rail launcher was used, even though the facility provides this option and the Agni-V may possibly utilize a rail-launch option. In recent years India appears to have created some ambiguity around the launcher types used in tests, which likely underscores the strategic value of the system.

The missile launched during the recent test displayed a shorter, rounder and wider nose cone than those seen in previous tests and public displays, suggesting a payload larger than a single warhead. The number of dummy warheads carried by the missile could be as few as two, and is unlikely to have been more than three. The Agni-V is one of two Indian ballistic missiles designed to be equipped with MIRV technology, the other being the under-development and as yet not flight-tested Agni-VI system.

49.6 NIRBHAY CRUISE MISSILE ADVANCES SIGNAL INDIA'S GROWING DEFENSE CAPABILITIES

India is bolstering its cruise missile capabilities with domestic development of a turbofan engine recently tested on the subsonic Nirbhay cruise missile, which has a strike range of 1,000 kilometers. The new engine, known as the "Manik," paves the way for development of the long-range land attack cruise missile (LRLACM), slated for testing by 2028.

The Nirbhay is considered a powerful addition to India's missile arsenal alongside the supersonic BrahMos cruise missile, also domestically produced.

A Manik-powered Nirbhay conducted a successful test flight off the coast of the east Indian state of Odisha in mid-April 2024. Dubbed the indigenous technology cruise missile (ITCM), the enhanced Nirbhay achieved a low-altitude "sea-skimming" flight using waypoint navigation, India's Ministry of Defence (MOD) reported. Defence Minister Rajnath Singh called the test a "major milestone."

Waypoint navigation essentially incorporates coordinates specifying a location through which the missile must pass on its way to the target.

The recent trial also cleared the way for integrating the Manik engine into the LRLACM being developed by India's Defence Research and Development Organisation (DRDO), according to Janes, an intelligence analysis website. The LRLACM will operate from land, air and naval platforms, and is considered an eventual successor to the Nirbhay.

The test flight also aimed to validate the performance of enhanced radio frequency seekers and other subsystems. The ITCM program's primary goal is a 100% domestically developed and produced cruise missile. The DRDO is collaborating with local research and defense firms, including the Bengaluru-based Gas Turbine Research Establishment, developer of the Manik engine, the MOD stated.

The Nirbhay is "capable of deep penetration into adversary territory to strike high value targets with precision," the DRDO stated. "India is in the list of select few countries having capability to design and develop this class of cruise missiles."

Deployed from a land-based mobile launcher, the Nirbhay can carry a 450-kilogram payload and be armed with high explosives or a small nuclear warhead, according to the news magazine India Today. "The missile is also equipped with advanced avionics and software to ensure better and reliable performance," according to the MOD.

The cruise missile is expected to be made available to all three branches of the Indian Armed Forces, the Times of India newspaper reported in November 2023, and will significantly enhance the military's capabilities, providing commanders with versatile and potent options.

49.7 US STEALTH DESTROYER IS THE FIRST TO CARRY HYPERSONIC MISSILES

The future USS Zumwalt (DDG 1000) is underway for the first time conducting at-sea tests and trials on the Kennebeck River in the Atlantic Ocean. The Zumwalt is the largest destroyer ever built for the U.S. Navy.

The hypersonic missiles can travel at speeds of more than 3,830 miles per hour and typically travel at high altitudes of up to 50 miles above sea level. A U.S. Navy stealth ship is getting a major upgrade: USS Zumwalt is currently being fitted with an experimental hypersonic weapon system called Conventional Prompt Strike.

Hypersonic weapons are seen as the next stage in non-nuclear warfare, as they enable targets to be destroyed swiftly and accurately and from far greater distances than conventional weapons allow. In a briefing document leaked by a former Massachusetts Air National Guard member, it was revealed that China had tested its DF-27 intermediate-range hypersonic weapon system last year, according to AP News.

Hypersonic weapons travel at more than five times the speed of sound, or "Mach 5" — approximately 3,830 mph (6,160 km/h).

Conventional Prompt Strike launches in the same way as a ballistic missile but is then propelled by a hypersonic glider vehicle. The glider vehicle is capable of traveling at up to Mach 8 — approximately 6,140 mph (9,880 km/h). Each of the Zumwalt-class destroyers would be deployed with four weapon tubes, with each tube carrying three Conventional Prompt Strike missiles.

When firing at long range, missiles traveling at hypersonic speeds may be intercepted by advanced defense systems. After detecting an incoming missile, a defense system predicts the flight path and launches an interceptor missile. Hypersonic missiles typically have some maneuverability, making them harder to shoot down, although much of the development focus to date has been on maneuverability for accuracy, rather than for evasion. They generally tend to rely on being too fast for detection systems to react.

The degree of maneuverability also depends on where the missile is in its trajectory. Any object traveling at immense speed is already subject to high forces, including drag from wind resistance. Its own momentum, which keeps it on course, means you would need very high forces to achieve even a small turn. Maneuverability for evasion purposes is not of much benefit; any turn would cause the missile to lose speed, which would make subsequent interception easier.

To overcome some of the drag, hypersonic missiles typically travel at high altitudes of up to 50 miles (80 km) above sea level. For comparison, a typical passenger aircraft has an optimal cruising altitude between 5 and 7 miles (8 and 11 km). To achieve the required speed, hypersonic missiles need to minimize drag, and air density is much lower the higher you go.

The Zumwalt-class destroyers are already equipped with cutting-edge technology, including electric propulsion systems, wave-piercing hull, low radar detectability and automated damage control systems. And with the Conventional Prompt Strike system, it could gain the ability to strike farther than ever before.

49.8 INDIA CONDUCTS MAIDEN TESTS OF LONG RANGE HYPERSONIC AND SUBSONIC NAVAL MISSILES

On November 16, 2024, India's Defence Research and Development Organisation (DRDO) conducted a successful flight trial of India's first long-range hypersonic missile. A press release by the Ministry of Defence (MoD) stated that the missile is designed to carry various payloads for ranges greater than 1,500 kms for the Armed Forces. It also mentioned that the missile conducted successful terminal maneuvers and impacted with a high degree of accuracy.

India's test highlights the intensifying global race for hypersonic, including India's growing maturity in developing hypersonic systems, which it has invested in since the 2000s. It warrants further analysis of what this means for regional stability, particularly India's precarious relationships with China and Pakistan. Recent developments also indicate bilateral and regional approaches for the next U.S. administration to consider in an intensifying global race for hypersonic weapons.

The recent test represents a public culmination of India's multi-decade research into hypersonic systems. India's hypersonic ambitions began far before 2007 when it first took delivery of its BrahMos missile, a joint venture with Russia. In 2004, India first publicized its indigenous development of the Hypersonic Technology Demonstrator Vehicle, demonstrating its scramjet engine in a 2020 flight test.

The timing of India's test comes alongside rapid developments in the global hypersonic arsenals. The test occurred days after China showcased the GDF-600. India finds itself in a global environment marked by rapid developments in the hypersonic space. Most recently, the United States tested its hypersonic missile in the Pacific as part of its Long-Range Hypersonic Weapon program. Over the past year, Russia loaded its Sarmat intercontinental ballistic missile (ICBM) with nuclear-capable Avangard HGV and was alleged to have used its Zicron hypersonic missile in its ongoing war in Ukraine. North Korean state media reported a test of the Hwasong 16B hypersonic missile, described by President Kim Jong Un as a "key piece of the nuclear deterrent."

India's attention is on China, a neighbor with whom it shares a long history of border crisis and security concerns. Touted as the leader in hypersonic technology, U.S. Department of Defense (DOD) officials state that China surpasses Russia and the United States in the development and advancement of conventional and nuclear-capable hypersonic weapons. A recent intelligence leak revealed that the People's Liberation Army 2023 successfully tested its DF-27 intermediate-range ballistic missile with a hypersonic glide vehicle in 2023, allowing it to easily penetrate missile defense systems. As per the 2023 China Military Power Report, the DF-17 HGV-armed medium-range ballistic missile "will transform the PLA's missile force." While the DF-17 is meant for conventional missions, it can equip nuclear warheads.

Former commander of the U.S. Strategic Command General John Hyten has emphasized that hypersonic technology gives a country "response, long-range, strike options against distant, defended and/or time critical threats when other forces are unavailable." Hypersonic weapons offer India a capability combining extreme speed, maneuverability, and low-altitude flight, making them harder to track and detect. India's test launch indicates its move towards diversifying its strategic capabilities and strengthening its deterrent by enhancing traditional warfighting advantages. DRDO states that the missile carries "various payloads," prompting several questions on the specifications, such as its speed and the hypersonic delivery systems within the missile.

China and Pakistan are yet to provide an official response to India's hypersonic missile test. China is unlikely to respond, an approach that it has maintained in the past. However, China has responded to military developments, including the unprecedented deployment of a multiple independently targetable reentry vehicle (MIRV)–capable ICBM. The most recent response was to the Agni-V MIRV capable ICBM, showcasing its capability of reaching targets in China. Foreign Ministry spokesperson Zhao Lijian cited UN resolution 1172, which condemns India and Pakistan's respective nuclear tests in 1998, stating that "maintaining peace, security and stability in South Asia meets the common interests of all."

Given India's expanding capabilities, this may tempt Pakistan to respond by ramping up its hypersonic developments and potential reliance on China for hypersonic technology in the coming days. While Pakistan does not have an indigenous hypersonic program, the Pakistan Air Force recently claimed the development of a hypersonic-capable missile as part of a larger modernization effort to "counter evolving threats." A video released by the Pakistan Air Force featured a CM-400AKG anti-ship missile, a China-manufactured missile that allegedly travels at hypersonic speeds.

The recent test adds new urgency to India, China, and Pakistan's missile race. India's clear ambition to strengthen its hypersonic program and Pakistan's desire to pursue this capability, coupled with China's influence on the region, raise concerns about potential escalation risks. Hypersonic weapons can maneuver to different targets and present little warning time for defenders, complicating a tense nuclear balance. Moreover, the challenge of facing two nuclear peers—considered an emergent problem among U.S. strategists—represents India's current status quo. Hypersonic proliferation in the region will demand a deeper reckoning.

India's test launch also comes amid a complicated relationship with China. The test not only coincides with China's showcasing of new hypersonic missiles but comes weeks after India and China reached a new military patrol pact along the Line of Actual Control, four years after its deadly border clash began in 2020. The pact divides Indian experts on whether this is a strategic shift or more of a tactical move, and the implications of this new pact are yet to be seen, especially between two countries with a history of deep mistrust and border contentions. Tracking advancements in technological processes is critical moving forward for the India-China relationship.

CHAPTER 50

RADAR SYSTEMS IN WARFARE

50.1 RADAR IN MILITARY OPERATIONS AND ITS IMPORTANCE

The first early warning radar network in the world was developed by Britain during World War 2 to provide warning of approaching German bombers and was a major factor that helped save lives and win the war. Called the **Chain Home Radar**, it was cutting-edge technology at the time and has only developed since, with radar still being used for military operations today.

What is Radar?

A radar system transmits a high-frequency signal towards a target, which then bounces off the target and returns. The radar can then use the information that it has received from this to identify the target's position and speed. It can be airborne, ground-based or underwater. Because of this very subtle but accurate and helpful method, it is used by military operations across the world for both defence and offensive purposes.

50.1.1 Some Types of Radar

(i) Detective/search radars

Like the Chain Home Radar, these work by transmitting a short pulse of radio waves to search large regions. Technology has advanced so, modern versions of search radars are more accurate and they remain as important to military operations as they were during WW2. They can be used to determine the approximate range, velocity and angle of aircraft and ships and are key in protecting airspace and ocean boundaries as well as early warning detection and target acquisition of enemies in war zones.

(ii) Targeting radars

Similarly to detective radars, targeting radars search for targets but they are different in that they focus on small zones more frequently in order to lock onto a specific target, rather than searching for potential threats. This is vitally important to differentiate between moving targets, stationary targets and clutter so that military operations can be as efficient and informed as possible.

(iii) Instrumentation radars

Instrumentation radars are a form of tracking radars which are used by militaries across the world to measure metric data on equipment like aircraft, missiles, projectiles and satellites. This is essential for evaluating the performance of equipment at military test facilities so their effectiveness can be measured and the correct item chosen for use.

(iv) Weather-sensing radars

Weather radars use pulse doppler radar systems to bounce microwave signals off clouds and rain to measure weather conditions. This is used by militaries for air traffic control, in order to "see" what the weather is like ahead and safely know if the conditions will affect operations, both during the flight of an aircraft and at airports.

(v) Navigational radars

Like search radars, navigational radars use short wavelengths that bounce off targets. However, navigational radars target natural obstacles such as earth and stone, used so that ships and aircraft can avoid collision and know where they are going. This is essential for mobile military operations to prevent avoidable delays.

50.2 TOP 5 RADAR SYSTEMS WITH THE INDIAN ARMED FORCES

(i) The ROHINI [central acquisition radar (3D-CAR)]

- The central acquisition radar (3D-CAR) is a 3D radar developed by DRDO for use of tracking 150 targets al long distance.
- There are two variant of 3D-CAR radar namely- **ROHINI and REVATHI**
- The ROHINI is the Indian Air Force specific variant, whereas the REVATHI is for the Indian Navy.
- A third variant, known as the 3D Tactical Control Radar is developed for the Indian Army.
- It is a medium range radar with surveillance range of more than 180km and covers an elevation of 18km.
- Capable of detecting low-altitude targets, and also supersonic aircraft flying at over Mach 3 speed

(ii) The Swordfish [long range tracking radar]

- The Swordfish is an Indian active electronically scanned array (AESA) long-range tracking radar specifically developed to counter ballistic missile threat.
- The radar has a range of 600-800 km and is capable of spotting small objects also.
- Upgradation of this radar is in the process by the DRDO to increase its range.

(iii) The INDRA

- INDRA is an acronym for Indian Doppler Radar and is series of 2D radars developed by India's DRDO for the Army and Air Force.

- The INDRA-I is mobile surveillance radar for low-level target detection while the INDRA-II is for ground-controlled interception of targets.
- The INDRA II radar uses pulse compression for detection of low flying aircraft in heavy ground clutter with high range resolution and ECCM capabilities.
- This radar produced by Bharat Electronics Limited and is used by Indian Air Force and Army.
- The range of the radar is 90km with vertical coverage of 30-3000m.

(iv) The Rajendra

- This radar is a passive electronically scanned array radar developed by the DRDO.
- It is a multifunction radar, capable of surveillance, tracking and engaging low radar cross-section targets.
- It is the heart of the Akash surface-to-air missile system and is the primary fire-control sensor for an Akash battery.
- This radar uses 3D target detection along with multi-target tracking and multiple missile guidance.
- Capable of distinguishing between hostile and friendly targets, it automatically tracks up to 64 targets.
- Three hundred sixty degrees rotation at a moderate speed allows it to perform 360-degree surveillance.

(v) The Swathi Weapon Locating Radar (WLR)

- Swathi is mobile artillery locating phased-array radar developed by India.
- This counter-battery radar is designed to detect and track incoming artillery and rocket fire to determine the point of origin for Counter-battery fire.
- This radar has a range of 40km depending on the incoming target, i.e. missile, mortar.
- Capable of covering elevation of -5°-75° with accuracy.

50.3 NAVIGATING THE FUTURE: MILITARY RADAR SYSTEMS IN MODERN WARFARE

In the ever-evolving landscape of military technology, radar systems have emerged as a cornerstone of modern warfare. These sophisticated systems not only provide invaluable situational awareness but also play a crucial role in enhancing the effectiveness and safety of armed forces.

Military Radar Systems Market Size is anticipated to reach USD 27.3 Billion by 2030 and registering a CAGR of 7.20% by 2022-2030

50.3.1 The Importance of Military Radar Systems

(i) **Situational Awareness:** Radar systems serve as the eyes and ears of the military, providing real-time information about the surrounding airspace, land, and sea. This awareness is essential for threat detection, tracking, and response.

(ii) **Force Multiplier:** Radar systems are force multipliers, allowing military forces to operate more effectively with fewer resources. They enhance the ability to detect and engage targets, thereby increasing mission success rates.

(iii) **Navigation and Safety:** Radar aids in navigation, especially in challenging conditions like low visibility, adverse weather, and at night. It ensures safe airspace management and reduces the risk of collisions.

(iv) **Countermeasures:** Radar technology is essential for the development of electronic countermeasures (ECM) and electronic warfare (EW) systems. These capabilities are crucial for protecting military assets from enemy attacks.

50.3.2 Evolving Capabilities of Military Radar Systems

(i) **Stealth Detection:** Radar systems have evolved to detect stealthy aircraft and vessels that attempt to evade traditional radar. Advanced signal processing and multi-spectrum radar technologies enable better stealth detection.

(ii) **Long-Range Surveillance:** Modern radar systems offer extended detection ranges, enabling early warning of potential threats, including ballistic missiles and unmanned aerial vehicles (UAVs).

(iii) **Adaptive Beamforming:** Adaptive beamforming technology enhances radar performance by focusing energy on specific targets or areas of interest. This increases the accuracy of tracking and reduces interference.

(iv) **Integration with Other Systems:** Military radar systems are increasingly integrated with other sensors, such as infrared and electro-optical systems, to provide a comprehensive picture of the battlefield.

50.3.3 Key Players in the Radar Systems Industry

(i) **Lockheed Martin:** Known for its advanced radar systems, Lockheed Martin is a global leader in aerospace and defense technology. Their AN/TPQ-53 radar system is used for counter-mortar and counter-artillery missions.

(ii) **Raytheon Technologies:** Raytheon Technologies is renowned for its radar solutions, including the AN/SPY-6 air and missile defense radar, which is used in the Aegis combat system.

(iii) **Northrop Grumman:** Northrop Grumman specializes in radar systems, including the AN/APG-77 radar used in the F-22 Raptor and the AN/TPS-80 Ground/Air Task-Oriented Radar (G/ATOR).

(iv) **Thales Group:** Thales Group is a global player in defense technology and provides radar systems like the Ground Master 400 and Sea Fire naval radar.

50.4 RECENT ADVANCES OF RADAR TECHNOLOGY

Radar technology has made great progress in recent years. Higher resolution, more precise tracking capabilities and lower margin of error are at the beginning of the progress. At the same time, **artificial intelligence and data analytics** help radar systems become more-efficient and smarter. Future radars will be able to operate over a wider frequency range, perform better in complex environments, and be managed with more automation.

Modern radar systems are combining advanced materials, solid-state modules, digital signal processors, and complex A-D converters to give a better look to military and civilian users who need the best possible capability in small, compact, and efficient packages.

Modern radar systems are combining advanced materials, solid-state modules, digital signal processors, and complex A-D converters to give a better look to military and civilian users who need the best possible capability in small, compact, and efficient packages.

Modern radar systems often have imaging capability, can yield digitized signals quickly and easily for use with graphical overlays, can be networked together so the total system is greater than the sum of its parts, and can serve several different functions—such as wide-area search, target tracking, fire control, and weather monitoring—where previous generations of radar technology required separate systems to do the same jobs.

Most important, however, is the relative ease and speed with which modern analog radar signals **can be converted to digital information**. Not only does this open a wide variety of signal processing options, but it also enables radar information to be made available in real time or near real time on Internet-type networks for inclusion in the digital battlefield and Global Information Grid visions of the future.

(i) RF energy

Radar works essentially by bouncing radio waves off a target and detecting the return signals. Radar systems as early as World War II were simple, but were state-of-the-art for that time. Those early radar systems were tube based and mechanically steered, and involved RF transmitters, RF receivers, analog signal processing, and a video display.

Later, radar systems improved in sensitivity to where they could virtually detect and monitor waves on the ocean and insects in the air. Increasing sensitivity, however, compounded one of the biggest challenges that radar systems confront—so-called clutter, or reflected signals from objects that are not of interest. In World War II when radar technology was not nearly as sensitive as it is today, this was not a big problem. Large objects showed up as blips, while relatively small things of no interest didn't show up at all.

Increasingly sensitive radar systems were able to detect wave action on the ocean's surface, but they could have difficulty trying to find an enemy periscope in a rough ocean. Radar designers originally dealt with the clutter problem by finely tuning transmitted RF signals to match the return signals of targets of interest as closely as possible. Engineers tune transmit signals by altering the size of transmit antenna modules. Finely tuning transmit signals for targets of interest has helped reject clutter, but can make radar systems useful only for a narrow range of applications. Hence, users needed a separate radar system for wide-area surveillance, target tracking, fire control, and weather monitoring.

It has come from a simple concept that there was one radar per function. Each was tuned to its own frequency for that function. At the lower frequencies the radiating elements get bigger, and inside of those are transmitters. You can have a radar system the size of buildings, all the way up to an airborne radar that would fit on your table—yet always one radar had its own function.

Raytheon designs radar systems run the gamut, ranging from the huge Sea-Based X-Band (SBX) Radar that sits essentially on a mobile offshore oil rig to search for ballistic missile launches, all the way down to small panel-type radars on unmanned aerial vehicles and the active electronically scanned array (AESA) radar systems on the F/A-18 jet fighter bomber.

(ii) Digital radar signals

The first glimmer of what lay on the horizon in radar technology came with the invention of solid-state technology. We moved into solid-state radars with processing and RF chips, which enabled us to go to active electronically steered arrays.

Solid-state technology led to digital signal processing, ultrafast analog-to-digital (A-D) and digital-to-analog (D-A) converters, fast commercial off-the-shelf (COTS) microprocessors, and high-speed digital networking. Increasingly, radar systems handle all signal processing in the digital realm, rather than in analog, which offers increasingly fast and efficient processing, broad new opportunities for offering radar information on the tactical Internet, as well as innovative new ways to display radar information.

In essence, once radar signals are converted from analog to digital, they are limited only by the state of the art in digital processing, which today is not a serious limitation at all. It's not a processing problem, that new powerful generations of field-programmable gate arrays are available to process complex fast Fourier transform (FFT) algorithms on which radar processing depends.

In fact today's latest generations of FPGAs, digital signal processors (DSPs), powerful COTS microprocessors, and multiprocessing computing architectures are enabling radar systems designers to put radar-processing algorithms written decades ago into use. Sometimes we can use algorithms that have been on the shelf for 50 years by applying new technology for them.

With such digital computing power at their command, radar systems designers can even start thinking about doing without processes like digital down-conversion, and can begin designing systems able to sample signals immediately after they are digitized.

We can now use commercial parts for A-D converters that enable us to do direct digital sampling. We are seeing a trend where we remove analog components from systems.

Now we can almost digitize microwave signals without the down-convert. We need A-D converters with sufficient sampling rates to do that. Everyone wants higher efficiency and more dynamic range, and we can do that with COTS technologies and innovative architectures.

(iii) Transmit/receive efficiency

Improvement in radar systems, however, is not happening only on the digital side. New semiconductor materials, such as gallium arsenide (GaAs) and gallium nitride (GaN) are helping systems designers improve efficiency and shrink overall system size. Raytheon, for example, is involved in a project with backing from the U.S. Missile Defense Agency (MDA) called Next Generation Transmit/Receive Integrated Microwave Module—otherwise known as NGT—that seeks to use GaAs and GaN technologies for major radar systems upgrades.

As part of the program, Raytheon radar experts are demonstrating GaN-based monolithic microwave integrated circuit (MMIC) amplifiers with substantially higher RF power with greater efficiency than does current radar technology.

Raytheon is using GaN technology developed for the U.S. Defense Advanced Research Projects Agency in Arlington, Va., as part of the agency's Wide Bandgap Semiconductor program. GaN technology not only can extend radar ranges, sensitivity, and search capabilities, but also can help designers reduce the size of radar antennas.

The NGT program is important because it is the first significant government-funded contract to address the use of the more capable GaN semiconductors in a relevant environment. This recent demonstration shows that GaN technology performs better in transmit/receive modules representative of those used in modern radars.

In addition to GaN and GaAs semiconductors, complex A-D converters (CADCs) from companies like Analog Devices Inc. in Norwood, Mass., as well as mixed-signal integrated circuits from companies like TriQuint Semiconductor Inc. in Hillsboro, Ore., also are contributing to the improvement of next generation radar systems, says Lockheed Martin's Nespor.

(iv) Imaging radar

The enhanced transmit/receive and computer-processing technologies of modern radar systems also are giving rise to imaging radar systems—or those that produce high-resolution pictures from returning signals.

Signal-processing experts at Mercury are working on a technology called circular synthetic aperture radar—also known as circular SAR, or video SAR. The processor-intense part of radar is image formation, which requires significant numbers crunching.

A specialist in multiprocessing and other computer-intensive applications, Mercury is involved in synthetic aperture radar and ground-moving target indication.

Circular SAR involves flying a specialized radar system aboard an aircraft that files around an area of interest, and builds a 3D image based on successive layers of radar data. Going to things that are more processor intensive, such as circular SAR, is where the research is going.

Although imaging radar is not a new technology, until recently it involved recording volumes of radar data in the air and performing intensive processing later on the ground. We would just let the numbers crunch and spend days analyzing it. Today the imaging radars are doing image formation in real time. Processing time is less than aperture time, so it is nearly real-time image generation, or radar acting like a camera.

Mercury and other signal-processing experts are using cell processors to break radar image processing tasks down into separate, manageable tasks. SAR image formation, and the cell processor, are coming up with a tiling scheme that breaks the image down into little tiles. You are doing FFTs, but you have to glue stuff together. You look at the capability of the processor to handle the tile, so you keep chip count as low as possible.

(v) Multifunction Radar

Perhaps the most important development in radar technology in recent years is the shift to multifunction radar—or systems that can perform a variety of applications with the same system. The move to solid-state radar gave the first glimpse of performing several applications with the same radar.

The kinds of active electronically steered radar arrays that solid-state technology enabled, is where you get into the transmit/receive module, and that is how we get away from the single-purpose radar. It is a cost tradeoff. The customers are looking for as much functionality as possible, for the same price or less.

Multifunction radar systems can save on space, weight, and power by combining several stand-alone radars into one system, and also can help reduce demands for human operators by combining functions and systems. Today a lot of acquisitions are coming along where a given radar procurement will replace four or five radars current in the field, and that gives you a more cost-effective infrastructure.

The latest incarnations of the Raytheon Spy 1 and Spy 3 dual-band shipboard radar may replace as many as a dozen conventional antennas. You will start seeing the space, size, and weight allocated for antennas on ships shrinking, as well as their functionality increasing.

Sophisticated and rapid processing of digitized radar signals enable systems designers essentially to slice return signals from wideband radars into two or more portions, and use the separate portions for separate applications—such as wide-area search as well as target tracking. A lot of weather radars look to get Doppler, yet now we have weather radars looking down below 10,000 feet in range, and can predict tornadoes that occur in the lower atmosphere.

Today, operators cannot look at separate portions of the return signals all at the same time, although they can switch among signal portions quickly enough so that it appears to be near real time. As always, systems designers must determine the radar's priorities. You can choose to make the antenna wideband, or have multiple bands you switch in and out to design antennas and receivers with broader bandwidths. Those trades are what the radar engineer does today.

(vi) Sustainability and growth

Radar designers are not only seeking to improve system capability, however. In these budget-constrained days they also are looking into radar architectures designed for growth, technology insertion, and long-term sustainability.

They are using open architectures to design for growth. They use VME and other commercial standards at the backplane level. Software changes about as fast as the hardware, so they don't want to get locked into proprietary software. They are using a Linux-based operating system more than we used to, and they also are using Windows NT and are moving to Vista.

Embedded computer suppliers to Lockheed Martin and other large defense contractors also are keeping a close eye on sustainable technology. "We produce a variety of 'shop window' products—board-level hardware and software that is truly COTS," says Michael Stern, product manager for access multiprocessor AXIX at GE Fanuc Intelligent Platforms Embedded Systems in Towcester, England.

GE Fanuc aims its embedded technology at medium-sized radar systems that call for multi-board embedded computers with five or six quad processors, and as many as 70 or 80 central processing units per enclosure.

(vii) The future of radar

Placing real-time radar information on tactical networks and viewing the data through Internet-like interfaces may open new possibilities in disseminating radar to those who need it and blending information from one radar system with another or other kinds of sensors.

They are looking at XML techniques so that sensor information can look like a Web page.

This level of digitally processing and disseminating radar data could have its downside, warns Lockheed Martin's Nespor. "Wide-open sensor apertures could be wide-open doors to bad guys. Cyber warfare must deal with this."

Radar signal processing is one of the most demanding embedded computing applications known in deployed military and aerospace systems. Signal processing demands fall into two general categories—converting signals from analog to digital, and then back into analog, and making sense of signals once they are digitized.

Radar users today are dealing with an enemy amidst civilian buildings; they want to find him with precision before he moves.

Radar signals and frequency bands are getting wider—in the hundreds of gigahertz and beyond—to provide better resolution and precision in positional tracking. Wider-bandwidth radar signals create two burdens—digitizing that signal with A-D conversion, and the power of the signal processor that follows. You have to do the processing proportionally faster because you want to do the same kind of algorithm you did before, but at a higher bandwidth, so you need more horsepower.

It is computer horsepower that is the operative phrase when it comes to radar signal processing. The never-changing requirement from our radar customers is more performance.

It is the radar developers who are quickest to adopt the highest-performance computing capability. Radar guys are up against the highest performance and the highest risk. Effectively they have algorithms that consume an unlimited amount of compute power.

For Curtiss-Wright, the processing technology most in demand from radar systems integrators is the VPX high-speed serial switched fabric. Way over 50 percent of their radar processors are going with VPX. VPX is truly a sea change.

50.5 INDIA EYES 8,000-KM RANGE RADAR FROM RUSSIA TO COUNTER CHINESE AERIAL AND MISSILE THREATS

India and Russia are reportedly in advance stages of talks over a mega defence deal that could give a massive boost to India's air defence infrastructure. The deal, which is expected to be worth more than $4 billion, involves India purchasing an early warning radar system '**Voronezh**' from Russia with a range of over 8,000 kilometers.

This comes as Defence Minister Rajnath Singh is on a visit to Russia.

The advanced 'Voronezh' series radar is manufactured by Russia's Almaz-Antey Corporation, a specialist organization for manufacturing anti-aircraft missile systems and radars.

The Voronezh radar is an advanced long-range early warning system boasting a range exceeding 8,000 kilometers. It is designed to detect and track threats, including ballistic missiles, fighter jets, and intercontinental ballistic missiles (ICBMs).

If acquired by India, the radar will provide comprehensive coverage against aerial threats from China, South and Central Asia, and large parts of the Indian Ocean region.

Capable of tracking over 500 objects simultaneously, the Voronezh radar has a maximum range of 10,000 km, with a vertical range exceeding 8,000 km and a horizon range of over 6,000 km. Russia claims that the system can detect stealth aircraft and deliver detailed data on ICBMs, as well as near-Earth objects in space.

The negotiations between Indian and Russian officials regarding the deal have been going on for a while. Last month, a team from Almaz-Antey visited India to interact with the offset partners that will be involved in the project, the Sunday Guardian reported. The outlet noted, citing sources, that at least 60 per cent of the system will be manufactured by Indian partners, in line with the government's 'Make in India' initiative. If the deal is finalised, the radar system is likely to be installed in Karnataka's Chitradurga district, where a site has reportedly been surveyed. The district already hosts several of India's most advanced and top-secret aerospace facilities.

The acquisition of the radar system is expected to greatly enhance India's threat-detection and surveillance capabilities across Asia and the Indian Ocean Region (IOR), bolstering its air defence infrastructure to address regional and global challenges.

On the second day of his visit on Tuesday (10 December, 2024), Defence Minister Rajnath Singh co-chaired a high-level meeting with his Russian counterpart, Andrey Belousov, to discuss military-technical cooperation. The meeting underscored the special, strategic and privileged defence partnership between India and Russia. Defence Minister Singh also reviewed the status of the two remaining S-400 Triumph missile systems that India is awaiting from Russia.

The deal for these advanced missile systems was signed in 2018.

CHAPTER 51

LIDAR SYSTEMS IN WARFARE

51.1 STRATEGIC IMPORTANCE OF THE US ARMY'S LIDAR COLLECTION CAPABILITY

Light Detection and Ranging (LIDAR) is a remote sensing technology that utilizes laser pulses to measure distances and create detailed, high-resolution 3D maps of the Earth's surface. By emitting rapid laser pulses and measuring the time it takes for the light to bounce back, LIDAR generates precise spatial data that is invaluable for a variety of applications, particularly in military contexts. Unlike commercial satellites, which can provide useful imagery, LIDAR offers unparalleled depth perception, allowing for the identification of terrain features, infrastructure, and even vegetation in a three-dimensional context.

The capabilities provided by LIDAR are particularly essential during the preparation stages of operations. It offers detailed information on obstacles, potential cover and concealment options, and logistical routes, which is vital for mission planning. For instance, commanders can evaluate potential engagement zones, plan troop movements, and assess the vulnerability of infrastructure—all through a comprehensive 3D understanding of the battlefield. Such insights are crucial for effectively deploying forces, optimizing strategies, and minimizing risks during operations.

Since the inception of the U.S. Army Geospatial Center (AGC) HR3D BuckEye Support Team, (AGC has) satisfied over 185 Requests for Information (RFI), assisted in the mission planning of over 180 operations in the CENTCOM and AFRICOM AORs, and provided support to over 50 JSOC and non-JSOC Special Operations training exercises, to include two operations of national significance. Additionally, the HR3D program provided new CONUS collections for the past, present, and future training locations in excess of 20,000 square kilometers. This critical data helped to improve and develop new tactics, techniques and procedures for future operations.

The contributions of AGC Buckeye are invaluable to both humanitarian and military operations. These efforts significantly enhance the capabilities within AFRICOM.

Buckeye has been an integral part of the base defense apparatus as they provide specific imagery that can be collected on request in support of a dynamic environment. Buckeye's ability to create products at the UNCLASSIFIED level is unique and incredibly important on a base that exists as the sum total of a multitude of organizations, some of whom cannot access SIPR.

In the realm of large-scale combat operations, the integration of LIDAR into reconnaissance and planning phases is critical. When deployed on aircraft, the U.S. Army Geospatial Center's (AGC) LIDAR sensors can conduct broad-area collection missions, capturing extensive geographic areas with remarkable accuracy. This high-resolution geospatial data lays the groundwork for situational awareness, enabling military planners to visualize the terrain and make informed decisions during the competition phase prior to crises or conflict.

The rapid evolution of warfare, characterized by increased complexity and the integration of advanced technologies, necessitates a sophisticated approach to intelligence gathering. LIDAR not only enhances traditional reconnaissance methods but also complements emerging technologies like artificial intelligence and machine learning, which can analyze vast datasets for actionable insights. Without LIDAR, the Army would be at a strategic disadvantage, unable to fully harness the potential of these advanced systems.

Without the ability to collect LIDAR data, the impact to combat planning, training, and mission execution could be detrimental. The loss of this advanced capability would lead to a significant reduction in the military's situational awareness. Commanders would operate with limited understanding of the terrain, potentially jeopardizing operational effectiveness and increasing the risk to personnel. Tactical decisions based on outdated or incomplete data could result in misallocation of resources, failed missions, or even catastrophic losses.

51.2 GROWING IMPORTANCE OF LIDAR AND THE ACCOMPANYING THREAT POSED BY CHINA

Up to 2018, the global LiDAR market was dominated by US companies, but, now, one Chinese firm alone accounts for 47 percent of the global market share.

51.2.1 What is LiDAR?

LiDAR is an emerging, dual-use, remote sensing technology that uses light in the form of a pulsed laser to measure distances and map the surrounding environment. Unlike radar (Radio Detection and Ranging), which uses microwaves, and sonar (Sonic Navigation and Ranging), which uses sound waves, the use of reflected light in case of LiDAR provides greater speed, precision and resolution in mapping, though all three technologies use the same basic principle of emitting energy waves to detect and track objects. LiDAR also offers an advantage over cameras, since it can work in any lighting condition and has a better detection range.

LiDAR is an emerging, dual-use, remote sensing technology that uses light in the form of a pulsed laser to measure distances and map the surrounding environment.

Though the first commercially available LiDAR system came out in 2008, earlier devices were expensive, bulky and required frequent maintenance. It was the increasing interest in autonomous vehicles that spurred significant technological developments in the field and have now enabled it to become more powerful, efficient and cost-effective. The combination of LiDAR with artificial intelligence platforms has also allowed enhanced object recognition and classification, allowing the technology to evolve even further.

51.2.2 Applications of LiDAR

Advanced driver-assistance systems (ADAS) and autonomous vehicles like self-driving cars use 3D LiDAR map data to navigate roads and other environments. LiDAR also provides sensing capabilities for lane-keeping and collision avoidance. The automotive industry constitutes the bulk of the demand for the technology with global automotive LiDAR revenues estimated to be US $ 332 million in 2022. Tesla alone purchased over US $ 2 million worth of LiDAR equipment from manufacturer Luminar earlier this year.

LiDAR is used in agriculture to monitor crop conditions and soil health. It is used in weather forecasting to measure temperature, cloud cover, air density and other atmospheric parameters. It is also used in geology and mining for mapping and surveying landscapes. Smart cities pair LiDAR with other technologies to integrate sensor networks in utilities, transportation and infrastructure. Additionally, LiDAR also has applications in manufacturing, construction, forestry, aviation, bathymetry and energy.

In the military domain, LiDAR is being used to support autonomous navigation capabilities for uncrewed ground and aerial vehicles. These can be used to conduct battle damage assessment, thereby eliminating the need for military personnel to be physically present on the battlefield.

In the military domain, LiDAR is being used to support autonomous navigation capabilities for uncrewed ground and aerial vehicles. These can be used to conduct battle damage assessment, thereby eliminating the need for military personnel to be physically present on the battlefield. The US Army has stated that potential applications of LiDAR include "platform target identification, aim point selection, range instrumentation support, weapon defences and mapping."

51.2.3 The growing importance of LiDAR and the accompanying threat posed by China

The rapid advancements in automated systems and vehicles have led to a growing demand for increasingly sophisticated sensor technology. This has enabled another emerging technology to progress remarkably and come into prominence, namely, Light Detection and Ranging (LiDAR). While the demand for LiDAR systems is being primarily driven by autonomous vehicles, it is finding applications in a wide assortment of fields, both commercial and military. This has, however, inadvertently resulted in China acquiring a sizeable portion of the market in this domain, an issue which is now raising serious national security concerns around the world, particularly in the United States (US).

51.3 COUNTERING CHINA'S LIDAR THREAT BY U.S. MILITARY SYSTEMS

Today, Chinese companies are rapidly consolidating control over the global LiDAR market, with PRC-origin sensors now widely deployed across civilian and military networks worldwide, including in the United States. These sensors often serve as essential nodes within interconnected public safety, transportation, and utility systems, which is a clear benefit to the United States. However, Chinese LiDAR's system-wide integration also leaves its users vulnerable to espionage and sabotage, potentially enabling Beijing to access sensitive U.S. data or disrupt critical operations.

For decades, Beijing has used cyber operations to breach sensitive networks and infiltrate critical infrastructure in the U.S. China's military and intelligence services could leverage Chinese-made LiDAR systems for espionage purposes, much as they have exploited the compromised communication gear sold by Chinese telecommunications giant Huawei. The widespread adoption of Chinese-made

LiDAR technology also advances Xi's "comprehensive national security" concept, which merges the technological development with state security to enhance China's geopolitical advantage. In practice, Beijing could weaponize Western reliance on Chinese-made systems by manipulating or disrupting LiDAR supply chains as it has done repeatedly with rare earth elements to pressure other countries into accepting its strategic demands.

Countering Chinese LiDAR dominance requires integrating LiDAR into a comprehensive industrial policy that bolsters U.S. technological leadership and economic competitiveness. This approach must go beyond simply reducing reliance on untrusted vendors from foreign countries of concern. It should also focus on expanding domestic LiDAR production capacity and fostering trusted supply-chain partnerships with allied nations. Establishing and enforcing rigorous cybersecurity standards for LiDAR technology will also be essential to safeguarding critical infrastructure and ensuring LiDAR's secure integration into both civilian and military networks.

LiDAR's most well-known civilian application is in the autonomous vehicle (AV) industry, where LiDAR sensors enable real-time object detection and safe navigation in dynamic driving environments. LiDAR has also been integrated into drone, train, and airport transportation systems for similar purposes. Additionally, utility companies and critical infrastructure providers are increasingly using LiDAR to monitor pipelines, power lines, and rail networks, proactively identifying structural weaknesses and environmental hazards before they compromise system integrity.

Urban planners are also integrating LiDAR with artificial intelligence and machine learning to build so-called "safe cities." For example, many cities rely on LiDAR sensors to monitor traffic flows at major road intersections and automatically adjust traffic signals to reduce congestion. LiDAR-enabled alerts can also provide emergency responders with critical, real-time information about vehicle accidents and other safety hazards, enabling more efficient response times. Lastly, LiDAR can optimize city services, such as waste management and energy distribution, by delivering precise data on usage patterns and infrastructure conditions, streamlining resource allocation, and enhancing public safety.

LiDAR's applications are becoming essential to **modern military and defense systems**, too. U.S. reconnaissance and missile guidance systems rely on LiDAR to enhance target acquisition, terrain analysis, and navigation in hostile environments. Looking toward the future of warfare, the Department of Defense's Joint All-Domain Command and Control initiative aims to integrate data and technology across military platforms to improve battlefield awareness, which will include outfitting next-generation autonomous military vehicles and drones with fully integrated LiDAR suites. For instance, the U.S. Army's Future Vertical Lift program, focused on developing advanced helicopters, will rely on LiDAR for real-time terrain mapping and obstacle avoidance capabilities.

51.3.1 Military Applications of LiDAR

(i) **Battlefield Mapping:** Enables precise terrain analysis, including in urban environments.

(ii) **Enemy Detection:** Helps locate enemy positions and infrastructure.

(iii) **Navigation:** Enhances movement by providing detailed spatial awareness.

(iv) **Identifies the depth of underwater threats.**

(v) **Laser Weapon Support:** Predicts performance by analyzing atmospheric conditions.

(vi) **High-Altitude Surveillance:** Systems like DARPA's HALOE map large areas rapidly from the air, offering advanced data collection over 100,000 feet above ground.

China has also prioritized LiDAR as part of its ongoing military modernization. Under its "intelligentized warfare" concept, the People's Liberation Army (PLA) is integrating LiDAR into defense systems to boost battlefield awareness and precision-strike capabilities. LiDAR has reportedly been installed on autonomous Chinese military platforms, including an autonomous fighting vehicle developed by the PLA in partnership with UISEE Technology and Dongfeng Motors. Equipped with advanced LiDAR from Chinese manufacturers such as Hesai, these and other LiDAR-enabled platforms are poised to become a "trump card," or decisive advantage, in enhancing Chinese reconnaissance operations, according to Chinese state media.

Today's use cases aside, LiDAR's potential remains largely unrealized. The global LiDAR market is projected to exceed $2.8 billion by 2025, driven by demand for high-resolution mapping, autonomous systems, and automation technologies. North America accounted for 36.6 percent of LiDAR-related revenue in 2021, the highest percentage in the world.

While non-Chinese companies in the United States, Germany, Canada, and Israel are key competitors operating on a level playing field, Chinese LiDAR firms benefit from extensive state support, giving them an outsized advantage. This state-backing allows Chinese companies to dominate the market, especially in price-sensitive sectors, with Chinese firms accounting for more than 80 percent of global LiDAR sales. As LiDAR adoption grows, China's market dominance could provide Beijing with strategic leverage across other emerging industries, such as precision agriculture, renewable energy, and advanced robotics.

51.3.2 LiDAR Functionality and Known Risks

LiDAR's transformative potential is undeniable, but its rapid proliferation carries significant risks. The technology's ability to collect and transmit precise spatial data makes it a prime tool for espionage and sabotage, especially when these systems are manufactured or otherwise controlled by companies located in foreign countries of concern. These risks are set to intensify as LiDAR systems are increasingly deployed near critical infrastructure, transportation hubs, utility grids, and defense nodes across the United States.

China has long recognized LiDAR's strategic value. In 2018, China identified LiDAR as a critical "chokepoint technology," and in 2020, Beijing directed that LiDAR be integrated into military systems. Through its military-civil fusion strategy, which mandates that Chinese companies collaborate with China's defense and intelligence agencies, Beijing has blurred the lines between civilian and military technologies.

At least one Chinese LiDAR company, Hesai, has acknowledged in investment disclosures that the Chinese government "may influence or intervene" in its "operations at any time." Such disclosures highlight the inherent national security risks associated with embedding PRC-manufactured LiDAR systems in U.S. and foreign infrastructure. While Hesai denies any affiliation with China's military, its inclusion on the Defense Department's Section 1260H list, which designates companies linked to the Chinese military-industrial complex, underscores the U.S. government's concerns. Notably, after Hesai's legal challenge briefly led to its removal from the 1260H list in October 2024, the Defense Department promptly re-listed the company "based on the latest information."

Chinese law not only facilitates military-civil fusion, it mandates cooperation between Chinese companies and state security agencies. China's 2017 National Intelligence Law, 2021 Data Security Law, and recently revised Counter-Espionage Law require Chinese companies, including LiDAR manufacturers, to assist state intelligence operations. China's National Intelligence Law, for example, explicitly requires "all organizations and citizens" to support, assist, and cooperate with national intelligence efforts. This legal framework extends far beyond China's borders, potentially enabling the Chinese government to exploit PRC-manufactured LiDAR systems abroad. This jeopardizes the security of any nation that integrates Chinese-made LiDAR into its networks and critical infrastructure systems.

Despite these and other known risks, the methods that adversaries such as China could use to exploit LiDAR technology remain poorly understood. Understanding exactly how LiDAR works and how foreign adversaries could exploit its known vulnerabilities is key to preventing its misuse.

51.3.3 How LiDAR Works and Adversarial Risks

LiDAR sensors are equipped with high-speed optical transceivers, including laser emitters and photodetectors, which enable them to transmit and receive information through light pulses.

Modern LiDAR systems use coded light pulses, or "fingerprints," to distinguish their signals from background noise and reduce interference from other sensors. With each new generation of LiDAR, the sophistication of pulse encoding increases, potentially introducing more serious risks.

Specifically, as pulse codes become more complex, adversaries could develop new methods to manipulate these signal patterns, potentially using them to interfere with LiDAR systems and the broader networks to which they are connected. This could involve designing "fingerprints" that disrupt LiDAR sensor functionality, including the disabling of a sensor's safety features.

As LiDAR's signal processing capabilities advance, each sensor's supporting hardware is also becoming more sophisticated, further increasing the risk of exploitation. This vulnerability arises from several factors. First, LiDAR sensors utilize advanced embedded processors and non-volatile memory to store firmware, operational logs, and other information. The complexity of these advanced processors could allow for the introduction — and concealment — of malicious code or firmware backdoors that are difficult to detect and neutralize, even upon close inspection.

The incorporation of advanced custom silicon chips in many newer LiDAR sensors, especially those produced in China, presents another layer of risk. These custom chips can be specifically engineered to include hidden vulnerabilities, known as "hardware trojans," which could provide a LiDAR company or other hostile actor with unauthorized access or control over LiDAR devices.

Furthermore, LiDAR sensors often rely on embedded operating systems, such as Linux, and use Ethernet standards for data transmission. As a result, these sensor systems typically connect to the broader internet, significantly increasing their vulnerability to cyberattack. For instance, many LiDAR sensors include a built-in web interface, accessible via a standard browser and powered by an onboard web server. These web interfaces, if not properly secured, could be remotely accessed and manipulated by hackers or LiDAR companies located in foreign countries of concern to gain control over deployed LiDAR sensors and potentially LiDAR-enabled networks.

What is more, LiDAR systems frequently receive software updates over the internet, often affecting network settings, cybersecurity protocols, and core sensing algorithms. These over-the-internet updates, especially those originating in foreign countries of concern, could introduce additional cybersecurity vulnerabilities, making Chinese-made LiDAR systems highly vulnerable to manipulation.

51.3.4 Disruptive Cyberattacks

Like other connected technologies, LiDAR systems are inherently vulnerable to cyberattacks. However, these risks increase significantly when sensors are manufactured in, or by companies beholden to, foreign countries of concern. Foreign adversaries or the companies they control could potentially embed malicious code in the encrypted components found in most LiDAR sensors, making detection extremely difficult. These types of compromises raise serious cybersecurity risks, particularly because LiDAR's integration into supply chains and transportation networks creates single points of failure. Manipulating even a small number of sensors could have widespread repercussions.

LiDAR sensors, as networked devices often connected via Ethernet and linked to systems with internet access, are highly vulnerable to a specific type of disruptive cyberattack known as a network-based attack. These attacks commonly rely on malware, which can be embedded at multiple stages: during manufacturing, network integration, or through routine firmware updates. Once malware is installed, attackers can remotely access LiDAR sensors via internet-connected pathways (or access points), potentially disabling the sensors or manipulating them to malfunction or transmit false information.

For example, after gaining remote access to one of these compromised networks, cyber actors could instruct LiDAR sensors to send false fingerprints to other nearby sensors, causing them to malfunction. Such an attack could result in targeted disruptions or the disabling of select safety systems in these networks. In extreme cases, it could cause widespread failures of entire LiDAR-dependent systems, leading to large-scale disruptions in public safety and national defense operations, such as paralyzing autonomous vehicle networks or hindering critical infrastructure monitoring.

The threat of attacks that cause LiDAR sensors to transmit false information is equally severe. Compromised sensor activity could erode policymaker confidence in the reliability of LiDAR-dependent systems, including advanced weapons platforms. This potential loss of faith could undermine strategic decision-making and policymaker resolve to deploy these systems during future conflicts. During wartime, LiDAR-enabled weapons could fail when needed most. Similarly, sensor malfunctions could sow doubt about the integrity of LiDAR-dependent critical infrastructure systems. This extends to systems needed to support future force mobilization, such as domestic rail networks, potentially jeopardizing U.S. military readiness and responsiveness.

Another type of disruptive cyberattack targeting LiDAR systems is an optical attack, in which an adversary sends coded light pulses directly to the sensor's lens. Unlike network-based attacks, optical attacks do not require internet connectivity, making them difficult to detect. The only known mitigation for optical attacks is to use LiDAR sensors from a trusted source, though even this relies on the assumption that the sensors have not been tampered with during manufacturing or distribution.

An adversarial fingerprint could disable the sensor immediately or trigger it to stop functioning at a pre-determined point in the future. In many respects, these optical attacks resemble electronic warfare techniques such as GPS jamming. Unlike jamming, however, they do not require continuous interference. A single instance of interference could potentially disable thousands of LiDAR sensors, leading to widespread disruption across the systems that depend on them, such as those that monitor critical infrastructure.

The malware necessary to carry out an optical attack could be embedded in the LiDAR sensor during its manufacturing process or introduced later via routine firmware updates. For instance, malicious code could be hidden deep within the millions of transistors found in the custom silicon chips used in modern LiDAR receivers, making it nearly impossible to detect prior to or after deployment.

Moreover, given the extreme sensitivity of modern LiDAR receivers, these types of optical attacks could be delivered from ground-based and airborne systems. For example, an attack could be initiated by something as simple as an individual using a handheld laser device to send the adversarial "fingerprint" to nearby LiDAR sensors. To cover larger geographic areas, delivery would likely involve high-altitude balloons and/or aircraft equipped with ground-directed laser systems.

Importantly, the emitted signal does not need to strike the LiDAR sensor directly to achieve its intended effect, as LiDAR systems are designed to detect faint secondary reflections from distant objects. This means that even if the sensor is shielded from direct exposure, an optical attack could succeed by illuminating the general area within several hundred meters of a LiDAR sensor.

Satellite-based laser systems, similar to those already deployed on remote monitoring satellites, could also disable compromised LiDAR sensors in critical areas across the United States within seconds. For instance, a satellite emitting a laser beam with a 5-meter diameter at Earth's surface and a pulse repetition frequency of 10 megahertz could deliver an optical attack across a 60-square-kilometer area, the size of Manhattan, in just 0.3 seconds. An area the size of Washington, DC, (160 square kilometers) could be targeted in approximately 0.8 seconds.

China's advancements in satellite-based laser systems, such as those observed in recent orbital missions, heighten the risk of Beijing using space-based technologies to disable compromised LiDAR sensors. Last year, a Chinese satellite, the Daqi-1, was observed using its onboard laser system to take measurements around the Hawaiian Islands. Additionally, Chinese state media reported that a Chinese Changguang remote sensing satellite successfully demonstrated space-to-ground laser links. These examples indicate that Chinese-controlled satellite laser systems are already in orbit with the demonstrated capability to conduct optical attacks across wide geographic regions, including within the United States.

51.3.5 Unauthorized Data Exfiltration

While disruptive cyberattacks can occur through network-based and optical methods, data exfiltration usually requires an internet connection to covertly transmit sensitive information abroad. This is an area where China has long excelled. In 2018, Chinese hackers breached U.S. Navy defense contractors, stealing data related to undersea warfare and missile programs. Similarly, in 2021, Huawei was found to have transmitted sensitive data from surveillance systems installed at the African Union headquarters to servers in China for over five years.

These and other documented examples show how China leverages unauthorized data exfiltration to conduct both military and industrial espionage.

Other industries using LiDAR face similar vulnerabilities. The absence of standardized cybersecurity regulations across sectors leaves safeguarding efforts decentralized, with protection varying widely by industry. This sector-by-sector approach results in weak and inconsistent security measures, with many users lacking the knowledge or protocols to prevent data from being covertly transmitted to adversarial entities.

One notable example of such exploitation was the recent "Cloud Hopper" cyber-espionage campaign, attributed to a hacker group linked to China's Ministry of State Security. The campaign targeted foreign cloud service providers, exploiting vulnerabilities to steal intellectual property and sensitive government data from companies and organizations across several industries. This incident highlights how Chinese cyber actors can leverage cloud services, potentially including those tied to LiDAR manufacturers, to access vast amounts of sensitive information.

51.3.6 Limitations of Current Cybersecurity Protocols and Certifications

A LiDAR manufacturer under the influence of a foreign country of concern could embed malware in devices in ways that are nearly impossible for customers or third-party cybersecurity organizations to detect. While cybersecurity standards such as ISO/SAE 21434 and ISO/IEC 27001 provide best practices for automotive cybersecurity and information security management, they are not designed to safeguard products sourced from untrustworthy manufacturers.

Most cybersecurity certifications are essentially compliance checks that rely on information provided directly by the manufacturer and rarely include a review of source code. Auditors typically examine documentation about the manufacturer's development processes and cybersecurity protocols and then certify products based on whether this documentation meets established standards. These certifications presume the manufacturer is trustworthy and the information provided is accurate.

However, certifications alone cannot guarantee the security of a particular product. A LiDAR manufacturer located in or otherwise beholden to a foreign country of concern could easily program hidden malware into its devices that would evade typical detection techniques by cybersecurity auditors. Detecting the malware would require significant penetration testing, which is expensive, highly technical, and uncommon.

A recent event illustrates the limitations of existing cybersecurity protocols and certifications in detecting firmware errors or malware in Chinese LiDAR sensors. On March 1, 2024, at precisely 12:00

a.m. Coordinated Universal Time (UTC), two LiDAR sensor models produced by Chinese manufacturer Hesai were involved in a global, synchronized disruption. This disruption was caused by a firmware error that failed to account for 2024 being a leap year. The sensors incorrectly calculated timestamps after being synchronized to UTC, which is the standard used by LiDAR systems. The result was the grounding of autonomous vehicle fleets in both China and the United States due to timing mismatches between Hesai's LiDAR sensors and the vehicles themselves.

This event underscores how a malicious attack using an intentional time trigger could cause far more widespread disruption if significant numbers of compromised LiDAR systems from adversarial nations were deployed in consumer vehicles or critical infrastructure. If an unintentional error like this can go undetected by cybersecurity auditors and customers, intentionally hidden malware would likely prove equally elusive.

51.3.7 Policy Recommendations For LiDAR

LiDAR sensors are becoming crucial for automating and controlling key U.S. infrastructure, including dockyards, drawbridges, traffic signals, and autonomous vehicles. As LiDAR deployment accelerates — particularly in connected vehicles — millions of these systems will be operating across the United States in the coming years. LiDAR's rapid expansion heightens security risks, particularly when sensors are sourced from countries of concern, such as China. These vulnerabilities could severely compromise U.S. infrastructure and defense systems.

The U.S. government is beginning to address the security risks posed by Chinese technology. In late September 2024, the Department of Commerce's Bureau of Industry and Security proposed a rule to ban the sale or importation of connected vehicles using specific Vehicle Connectivity Systems (VCS) and Automated Driving Systems (ADS) linked to China or Russia. This includes telematics, cellular, and satellite modules in VCS and ADS software. However, this proposed rule does not explicitly cover LiDAR systems or other use cases beyond vehicles, leaving many broader vulnerabilities unaddressed.

To address the risks posed by adversary-influenced LiDAR systems, U.S. policymakers must prioritize LiDAR in America's broader industrial policy. This includes fostering domestic LiDAR competition and building strong public-private partnerships between American LiDAR manufacturers, the U.S. government, and key civilian stakeholders, such as utility and car companies. These and other proactive measures will be necessary to avoid costly "rip-and-replace" scenarios in the future, akin to ongoing efforts to remove untrusted Huawei telecommunications gear from U.S. networks.

51.3.8 Empowering Law Enforcement and Regulatory Agencies

(i) **Evaluate the Feasibility of DoT Procurement Bans:** Congress could consider legislation that would ban the DoT from procuring LiDAR sensors manufactured by companies based in foreign countries of concern. Such measures could also restrict the use of DoT funds, such as SMART Grants, to ensure they are not used for acquiring LiDAR systems produced by these entities.

(ii) **Increase and Enforce Section 301 Tariffs on Chinese LiDAR:** The U.S. Trade Representative should consider increasing the existing 25 percent tariff on LiDAR imports from China to 50 percent or higher to counter unfair competition caused by state subsidies and oversupply practices in China. Additionally, Customs and Border Protection should be tasked with investigating past LiDAR imports from Chinese companies to ensure proper enforcement of tariffs. This would help protect U.S. LiDAR manufacturers from predatory pricing and support a more competitive domestic market.

(iii) **Create a DHS Task Force on Emerging LiDAR Threats:** DHS, through CISA, should create an inter-agency task force focused on emerging threats posed by LiDAR systems, with a particular emphasis on vulnerabilities in transportation and critical infrastructure. The task force should release regular public reports on vulnerabilities, mitigation strategies, and incident response planning for cyberattacks targeting LiDAR systems.

(iv) **Expand the 'Covered List' to Include Chinese LiDAR Manufacturers:** The Federal Communications Commission (FCC) should add Chinese LiDAR manufacturers to its "Covered List" of banned entities, preventing U.S. companies from using federal subsidies to purchase these systems. Congress could also direct the FCC to investigate this issue through new legislation, if needed.

(v) **Conduct an ODNI-Led Technical Threat Assessments for Foreign LiDAR Systems:** The Office of the Director of National Intelligence (ODNI) should lead inter-agency efforts to conduct comprehensive technical threat assessments on LiDAR systems sourced from foreign adversaries. These assessments would evaluate cybersecurity vulnerabilities, espionage risks, and supply-chain concerns, providing crucial intelligence for protecting U.S. infrastructure and military capabilities.

(vi) **Require Transparency for Municipal LiDAR Deployments:** Municipal governments deploying LiDAR systems in public infrastructure projects (such as smart cities) should be required to notify the federal government about the source of the technology and whether it originates from a foreign country of concern. This transparency measure will allow communities to make informed decisions about the technology being integrated into their environment. Congress should require such disclosures as part of federal grant conditions, mandating that recipients of federal infrastructure funding disclose technology sources.

51.4 LIDAR IN INDIA: HOW IS IT USED ACROSS VARIOUS PROJECTS?

While Lidar and aerial photo capture may be enjoying a large and increasing international market, we don't see any company forthcoming to invest in LiDAR in India on a large scale for fear of burning its fingers; government policy on aerial data collection has been unclear. However, it seems signals are now turning green for a private sector eager to conduct an aerial survey here. India has always been interested in the processing of data and so it's logical that Indian companies have shown interest in the processing of the millions of laser points we call point cloud. A number of companies are now capable of classifying LiDAR data, creating terrain models etc. for various domains including cartography, mining, power and telecommunications network industries. We will see the growth of much smaller companies capable of providing processing services at very low margins and bring in competitive pressure.

For Indian geospatial companies, the path to a success story will then rely on their capacity to adapt by providing value added services to their partners/clients; and by positioning themselves as integrators of holistic solutions.

3D Lidar data might support major cities in India that lacked basic geospatial data for their management. Flooding problems in Mumbai, tsunami damage and flood-related heavy damage to oil refineries in Gujrat were pointed to as applications for which 3D data was crucial. The ambitious Indian Government project for urban renewal of 63 cities (JNURM) will cost Rs. 12,05,360 million and a fraction of this investment could help in providing data to guarantee the success of the mission. Therefore government concerns should be addressed within a geopolitical context and a workshop held at which the issue of security and high-resolution data might be discussed.

Further, airborne remote sensing needed to be deregulated and the process for obtaining survey permission made more efficient and transparent. The geospatial education system will need to look at LiDAR as an important remote sensing option and build it into the curriculum. Like we have seen in the other data processing sectors, once there is an abundance of well-trained professionals, the pricing strategies will be one of the next controlling factors. The private sector should come forward with investment and government should play the role of facilitator in encouraging this development.

51.4.1 Where is the usage of LiDAR in India?

These are the various projects, where we can see the usage of LiDAR in India from a geospatial perspective:

- Power line transmission and pipeline corridor planning
- Earthquake study and disaster management
- Space and planetary studies
- Urban modelling
- Oil and gas exploration surveys
- Transportation studies and surveys including road, rail, air and sea.
- Climatic study
- Flood risk mapping
- Forestry
- Study on airport exclusion zones
- High-resolution mapping
- 3-dimensional modeling of structures
- Engineering and construction surveys
- Real estate development
- Coastal zone mapping
- Rectification of satellite imagery

CHAPTER 52

UNMANNED AIR VEHICLES (UAVS)

52.1 INTRODUCTION TO UNMANNED AERIAL VEHICLE (UAV)

An unmanned aerial vehicle (UAV), or unmanned aircraft system (UAS), commonly known as a drone, is an aircraft with no human pilot, crew, or passengers on board. UAVs were originally developed through the twentieth century for military missions for humans, and by the twenty-first, they had become essential assets to most militaries. As control technologies improved and costs fell, their use expanded to many non-military applications. These include aerial photography, area coverage, precision agriculture, forest fire monitoring, river monitoring, environmental monitoring, weather observation, policing and surveillance, infrastructure inspections, smuggling, product deliveries, entertainment, and drone racing.

The term drone has been used from the early days of aviation, some being applied to remotely flown target aircraft used for practice firing of a battleship's guns, such as the 1920s Fairey Queen and 1930s de Havilland Queen Bee. Later examples included the Airspeed Queen Wasp and Miles Queen Martinet, before ultimate replacement by the GAF Jindivik. In addition to the software, autonomous drones also employ a host of advanced technologies that allow them to carry out their missions without human intervention, such as cloud computing, computer vision, artificial intelligence, machine learning, deep learning, and thermal sensors. For recreational uses, an aerial photography drone is an aircraft that has first-person video, autonomous capabilities, or both.

An unmanned aerial vehicle (UAV) is defined as a "powered, aerial vehicle that does not carry a human operator, uses aerodynamic forces to provide vehicle lift, can fly autonomously or be piloted remotely, can be expendable or recoverable, and can carry a lethal or nonlethal payload". UAV is a term that is commonly applied to military use cases. Missiles with warheads are generally not considered UAVs because the vehicle itself is a munition, but certain types of propeller-based missiles are often called "kamikaze drones" by the public and media. UAVs may or may not include remote-controlled model aircraft. Some jurisdictions base their definition on size or weight; however, the US FAA defines any unmanned flying craft as a UAV regardless of size. A similar term is remotely piloted aerial vehicle (RPAV).

UAVs or RPAVs can also be seen as a component of an unmanned aircraft system (UAS), which also includes a ground-based controller and a system of communications with the aircraft. The term UAS was adopted by the United States Department of Defense (DoD) and the United States Federal Aviation Administration (FAA) in 2005, according to their Unmanned Aircraft System Roadmap 2005–2030. The International Civil Aviation Organization (ICAO) and the British Civil Aviation Authority adopted this term, also used in the European Union's Single European Sky (SES) Air Traffic

Management (ATM) Research (SESAR Joint Undertaking) roadmap for 2020. This term emphasizes the importance of elements other than the aircraft. It includes elements such as ground control stations, data links and other support equipment. Similar terms are unmanned aircraft vehicle system (UAVS) and remotely piloted aircraft system (RPAS). Under new regulations which came into effect 1 June 2019, the term RPAS has been adopted by the Canadian Government to mean "a set of configurable elements consisting of a remotely piloted aircraft, its control station, the command-and-control links and any other system elements required during flight operation".

52.2 HOW ARE DRONES CHANGING MODERN WARFARE?

Many may not realise that the origin of Unmanned Aerial Vehicles (UAVs), commonly known as drones, dates back almost a century, while modern UAV technology has been rapidly developing in civilian and military contexts over the past four decades. Drones represent the intersection of two important trends in military technology – the precise nature of weapons and the rise of robotics which, when combined, are flown remotely at no risk to a pilot and capable of delivering a lethal payload. The UAVs used in modern warfare has altered the dynamics of military operations, offering unique tactical advantages and enhanced the operational efficiency in various combat scenarios. Drones have been used for various purposes, including surveillance, reconnaissance, and targeted strikes. Overall, uncrewed systems represent a transformative advance in military technology, reflecting significant investment and development worldwide. As nations continue developing and deploying these systems, it is crucial to understand their implications and integration into Australian tactics, techniques and procedures.

In Ukraine, drones have become an important weapon to gain an asymmetric edge over Russian forces. Their availability, rapid development, ease of deployment and use make them indispensable in the military operations on Ukrainian soil. While this technology has altered the character of modern warfare, the UAVs have not had a decisive effect on the adversary to date. One lesson from the war in Ukraine remains clear – innovations in uncrewed systems are disrupting the way modern warfare is being conducted. It is therefore crucial to understand both the opportunities and limitations that drones bring to inform future doctrine, training, planning, as well as future investments in these technologies that can have an asymmetric effect on the battlefield. This Land Power Forum article is part of a larger project - the first evidence-based research on the lessons learnt from the use of drones in the war in Ukraine – drawing on both Ukrainian and Australian expertise.

52.2.1 Historical Evolution of UAVs

Looking back to the 1960s, the British Commonwealth artilleries recognized the emerging drone capability as crucial for target acquisition. Initially used to extend the range of artillery observers, the focus on reconnaissance and target acquisition remained unchanged for decades. In the past two decades, the United States has extensively utilized drones in counterterrorism operations in regions such as Iraq, Syria, Afghanistan, Pakistan, and Yemen, highlighting their capability to execute precise strikes with reduced risk to friendly personnel. In 2020, another notable example was the use of drones in the Nagorno-Karabakh conflict, deployed for precision strikes and surveillance, significantly impacting the conflict's outcome by providing superior aerial intelligence and striking capabilities.

However, the war in Ukraine has seen the first large-scale deployment of smaller drones by both Ukrainian and Russian forces for real-time intelligence gathering and direct combat engagements, illustrating the tactical versatility of these systems in contemporary conflicts. This unprecedented use of drones on a large scale in Ukraine offers valuable lessons for Australia. Understanding these lessons is essential for developing future drone capabilities and tactics in warfare.

Technological advancements have allowed drones to become smaller, more agile, and more capable. A decade ago, drones used in Ukraine were rudimentary and primarily used for harassment. However, by 2023, the proliferation of small drones capable of strikes, equipped with small payloads, had become capable of disrupting Russian attacks. Additionally, the progression beyond electro-optical and infrared capabilities has been significant. Prior to 2015, large platforms like Global Hawk, Triton, Predator, and Reaper were needed for multiple payloads. Following the Russian invasion in Ukraine, however, larger drones proved unsurvivable in a modern, well-equipped air defence environment and, as such, were removed from the battlefield. Battlefield aviation (both rotary and fixed wing) was also forced from the frontline. Smaller drones rapidly developed locally – able to both survive below the targetable threshold and to thrive in it – have quickly filled the space. These small drones are now capable of being equipped with multiple payloads, such as electro-optical infrared (EOIR), electronic warfare (EW), or geospatial intelligence payloads. This proliferation of multi-use drones at lower costs represents a significant technological advance.

History demonstrates that technological advancements in a warfare often outpace ethical and regulatory frameworks. For instance, the use in World War I of gas, machine guns, landmines, and cluster munitions were all fielded before any formal ethical or regulatory controls were established. Similarly, UAV technology is advancing rapidly, necessitating proactive engagement to understand and regulate its deployment effectively. The swift integration of these evolving technologies into military operations highlights the need for vigilant oversight and ethical considerations.

52.2.2 Death of Large Drones

Drones are generally classified based on distinct characteristics, such as function, size, payload, geographical range, flight endurance, and altitude. According to the NATO classification, the drone dynamics in Ukraine have showcased class I and class III drones, where class I drones are less than 150kg, and class III drones are greater than 600kg. In the war in Ukraine, small military drones – both fixed-wing and rotary, integrated with the ground units – have been commonly used for surveillance, target acquisition, battle damage assessment, and information warfare.

Class III drones – large surveillance and strike drones – are useful for gathering information over the lengthy periods of time (i.e., between 12-26 hours, with a General Atomics MQ-9 Reaper now achieving up to 40 hours) and can execute remote strikes. However, without air superiority, they are vulnerable to air defenses, electronic countermeasures, and are expensive to replace ($5m for one Bayraktar TB2). The large drones did not survive once Russian air defence and electronic warfare systems were properly integrated in Ukraine; while other class III drones (i.e., Russia's Orlan-10 reconnaissance drone) faced obstacles in providing good quality intelligence due to unreliability of its systems.

The Turkish-built Bayraktar TB2One is a vivid example of the large (class III) armed military drones that captured Ukrainian headlines and the public imagination in the first days and weeks of the Russian full-scale invasion of Ukraine. It was used for ISR (intelligence, surveillance, and reconnaissance) and tactical strike missions. The acquisition of Bayraktar TB2 drones was presented as a deterrent against Russia's aggression, and provided just as important as the supply of Javelin Anti-Tank Guided Missiles. The TB2s supplemented manned Ukrainian jets and provided the Ukrainian Air Force a ground strike capability that did not endanger its pilots in the critical few days and weeks since the full-scale Russian invasion. They were used as decoys in the mission to sink the missile cruiser Moskva, the flagship of Russia's Black Sea Fleet. One of the last successful employments of TB2 by Ukrainian forces was seen in the battle to recapture Snake (Zmiinyi) Island in the Black Sea in May and June 2022 where TB2s directly targeted several navy vessels and served as a surveillance and targeting platform in the raid against Chornobaivka airport in Kherson region.

The Bayraktar TB2's low speed (i.e., 130 km/h) and size (i.e., 12m) made it vulnerable on a battlefield with dense anti-air defence systems and electronic warfare at high density. As such, the TB2 disappeared over the skies of Ukraine from September 2023 due to the evolving dynamic of the war. This contrasts with the effective performance of TB2 in Libya, Syria, and during the second Nagorno-Karabakh war in 2020, conflicts without a dense layer of air defenses. Similar scholars from security studies community argue that large drones have limited utility in high-intensity conflicts, and their success depends on the failure of an adversary's air defence.

52.2.3 Displacement of Human Roles & Reducing Human Risk

Drones displace rather than replace humans on the battlefield. This displacement reduces risk in dangerous situations and preserves human life. Drones can provide greater persistence over various missions, maintaining optimal performance without the limitations of human endurance. For example, drones, whether ground-based, aerial, or maritime, are still controlled by humans but displace their roles in the surveillance chain, removing them from frontline exposure and risk.

Drones can and have saved lives. This is particularly important in asymmetric warfare, where an asymmetry in the availability of personnel can mean putting fewer people in harm's way and ensuring a better return on investment in human resources over time. Ukrainian forces have removed the risk to humans by conducting surveillance and reconnaissance with longer range drones that are capable of moving at rapid speed and maintaining their presence for long periods of time. When equipped with multiple sensor packages, these drones have been essential in detecting enemy movements, locations and force compositions during the day or night. Small, class I drones, have changed the operational tempo of artillery, shortening time-critical targeting and firing cycles from about 30 minutes to 3-5 minutes, helping to increase precision and pace of artillery fires and keep soldiers safer in the observer role. Much like harassing and interdiction artillery fire, the presence of drones also disturbs and exhausts Russian forces. Additionally, UAVs can enhance the capacity of both human and machine teams, optimizing performance and effectiveness in various operations. The ability to blend human and machine organizations has led to improved outcomes and operational efficiencies for Ukrainian forces.

52.2.4 Multi-Purpose use of UAV

Overall, drones have been shown to provide lethality at range, low cost, and with economy of effort that can be used in the air, land and sea. The ability to see farther accurately, coupled with cost savings, has made UAVs indispensable for both Ukrainian defensive and offensive operations. Apart from the overwhelming use of drones in the air, the war in Ukraine unveiled successful use of the naval drones, and the infancy of the land drones for various purposes.

52.2.5 Counter-Drone Defenses

With the development of uncrewed systems, the measures to counter these technologies are also developing in a rapid innovation cycle. The initial strategy to counter an uncrewed aircraft (UA) involved shooting it down while airborne. However, preventing its launch by targeting its ground components can be also effective. This involves either detecting the ground stations visually or electronically and striking with artillery or drones.

A key countermeasure to prevent drones from reaching their target has been the EW (Electronic Warfare) – a broad range of activities related to controlling the electromagnetic spectrum – in the form of jammers, spoofers, deceiving enemy communication systems, radar, and other electronic devices. Jamming is conducted by directing a high amount of energy towards the UA to mask the radio frequency control or data signals. Jammers – used by both Ukraine and Russia – sent out powerful electromagnetic signals causing the target drone to fall down, veer off course, or turn around and attack its operator. For example, in Bakhmut drone operators could not fly further than 500 metres due to Russian jamming or were leaving messages on the adversary drone operators' screen – 'Slava Ukraini!' ('Glory to Ukraine!') before downing the drone.

Counter-uncrewed aerial systems (C-UAS) are specifically designed to detect, track, and neutralize UAVs. This can be achieved through detection systems (radar, electro-optical/infrared sensors, and acoustic sensors that identify UAVs in the airspace); jamming devices (portable or fixed); missile weapons (anti-missile systems, air to air weapons); kinetic weapons (shotguns, anti-air machine guns and cannons); **directed energy weapons (high-energy lasers or microwave beams** can disable UAVs by damaging their electronics or propulsion systems); nets and drone-on-drone technologies (physically intercepting or capturing UAVs).

Disrupting or neutralizing any links relied on by the UAS - such as launch and recovery units, ground control stations, communications equipment, logistics, and supporting systems - can significantly hinder the UAS's operational capabilities. The proliferation of cheap, expendable (class I) drones has created an unfavorable interception curve for traditional air defence platforms. Larger drones with a distinct radar cross-section are easy, slow-moving targets for air defence interceptors and artillery, while traditional air defence has some limitations against small systems. An emerging challenge of counter-drone defence is its cost-effectiveness, where the counter drone system must be cheaper than its target. Greater investments in EW will be a key part of any counter against the threat of drones.

UAVs enable relatively unsophisticated adversaries to significantly impact high-value assets, exemplifying their role in asymmetric warfare. Drones are cheap and easy to use, allowing adversaries to leverage limited resources to achieve disproportionate effects. For example, defeating expensive military equipment such as tanks with inexpensive drones demonstrates how UAVs can create credible threats without sophisticated technology. Most Western militaries consist of a small number of very expensive assets. For example, if approximately seventy $20 million Abrams tanks in Australia can be defeated with $700 drones, it means that there is no need to be a sophisticated player to pose a credible threat.

52.3 RAPID INNOVATION AND ADAPTATION OF DRONE IN MODERN WARFARE

The rapidly evolving nature of modern warfare in Ukraine necessitates an accelerated cycle of innovation, which currently ranges from a week to approximately three months. New solutions or significant modifications to existing technologies are continuously required to maintain a competitive edge over the adversary. This fast-paced environment underscores the importance of continuous improvement and quick iteration in developing military strategies and technologies. Ukraine innovates (FPV drones with lethal effects), Russia responds (use of 'cope cages'). Russia acts (the use of Iranian Shahed-136 drones to attack cities and cities infrastructure), Ukraine adapts (answered this attack with the German Gepard air-defence system).

The evolution of robotics and autonomous systems, in all domains, is a clear example of the adaptation battle. Some fail to anticipate the array of future threats or are unable to judge which threats are the most serious. A final cause for adaptive failure is that enemies actively seek to interfere with and degrade their opponent's ability to learn and adapt.

The rapid development and deployment of UAV technology requires constant adaptation, innovation and evolution in military doctrines. It is essential to learn from both sides in conflicts like Ukraine and integrate those lessons into tactics, techniques and procedures. The increasing use of uncrewed systems across different domains demands that countries stay up to date with technological advancements. This ongoing evolution impacts investments and preparedness for future conflicts. Any war will stimulate competition, and then, of course, stimulate the arms race. The moment the UAS were used, the counter-UAS systems emerged, and then the tactics, techniques, and procedures (TTPs) had to be adapted, etc. The TTPs that worked for the first six months of the war no longer work today – both the platforms used and the tactics that underline the use of those platforms. Hence, it is important to adapt to achieve missions and tasks with the resources at hand, despite the obstacles that the adversary may create, and this has to be done faster than the adversary's capability to respond – maintain the 'adaptation battle'.

52.4 FUTURE OF WARFARE: HOW DRONES ARE SHAPING COMBAT?

Drones have played a key role in defense and security operations for some time now, but it's only over the last few years – and particularly with the devastating conflict in Ukraine – that the true combat potential for drones has been realized. The use of drones in combat is now the norm, and **combined with advanced computers and artificial intelligence**, is creating a capability that will give forces a superior advantage on the future battlefield.

The last decade has seen a surge in the use of small and inexpensive UAS – including quadcopters and fixed-wing examples – that can carry out lethal attacks. The first notable example of this was during the fight against Islamic State (IS) in Syria and Iraq in the mid-2010s. IS pioneered the use of cheap off-the-shelf quadcopters that could drop hand grenades, as well as hobby RC planes that could carry explosives. While primitive, the armed drones were effective in stopping advancing forces that were attempting to retake IS-controlled cities such as Mosul.

The war in Ukraine has only served to accelerate and mature the technology and tactics of drone warfare, with much of this conflict now characterized by the use of commercial-off-the-shelf (COTS) drones for both surveillance and lethal strikes. In early 2024, Ukraine's Minister of Digital Transformation stated that Ukrainian drones had destroyed or damaged 73 tanks, 10 howitzers, and 369 Russian personnel, all in just one week.

Ukrainian forces have pioneered the use of cheap first-person-view (FPV) drones – initially developed for recreational activities such as filmmaking and drone racing – that are configured with explosives and flown directly into a target using skilled operators. A typical mission could see one drone deployed by an ISR team, which then passes off targeting data to a separate FPV team to complete the 'kill chain'.

Drones are now highly integrated into Ukraine's force structure with the Ukrainian President announcing in February a new strategic branch of the armed forces known as Unmanned Systems Forces.

52.4.1 Future technology developments for drones – the AI revolution

As computer processing advances further and artificial intelligence (AI) is deployed on more systems at the edge, including those pioneered by Systel, then the autonomous capabilities of drones will only advance further. One example of how autonomy and AI will underpin new drone tactics is **swarming**, where multiple UAS will work collaboratively together to perform a mission with limited human input.

A flavor of the devastating impact of swarming drone attacks has already been seen with Russia's bombardment of civilians in Ukraine, as well as Houthi attacks in the Red Sea and the organization's use of Iranian-made loitering munitions on naval and merchant shipping. These attacks show how multiple relatively cheap drones can overwhelm air defenses and quickly deplete defensive ammunition stocks.

In future, swarming drone attacks will be more intelligent and integrate various electronic warfare countermeasures that spoof and degrade air defense systems further.

The use of AI will also allow drones to be more survivable in degraded environments and, like swarming, carry out missions with limited human inputs. Currently, many UAS still rely on control inputs from a human operator and a ground control station – particularly the cheap commercial FPV drones used in Ukraine – which is challenging in environments where jamming exists and signals can be blocked.

An answer to this has been to leverage AI – including machine vision – to aid with autonomous navigation and target selection, especially in the terminal attack phase. This has been seen with

Ukraine's attacks on Russian infrastructure using long-range drones, as well as a new-generation of FPV drone that features an automated terminal attack phase that overcomes the jamming of control and data links.

This new reality of conflict has not gone unnoticed by the US military, including the US Army, which announced this year that it would "rebalance" its aviation modernization investments to include the phasing out of drone systems that are not survivable in modern conflicts, including the Shadow and Raven UAS.

The Army has said that it will increase investments in "cutting-edge, effective, capable and survivable" unmanned aerial reconnaissance capabilities, as well as procurement of commercial UAS. Sensors and weapons mounted on a variety of unmanned systems and in space are more ubiquitous, further reaching, and more inexpensive than ever before."

As part of its rebalance, the Army will accelerate efforts on the Future Tactical Unmanned Aircraft System (FTUAS), a program that Systel is playing a key role in, that will give US Army Brigade Combat Teams a rapidly deployable ISR capability. Another key program for the US Army is Launched Effects (LE), which will see small UAS launched from Black Hawk helicopters to provide key data for commanders during large-scale combat operations.

52.4.2 Modern rugged computers for drone warfare

As drones incorporate more intelligent features, particularly AI-enabled ISR and targeting capabilities, the requirement for more powerful edge computing will grow. These computers will have to be rugged enough to withstand the harsh environments experienced by UAVs, as well as reliable enough to survive the high tempo of major combat operations. This includes protection from environmental factors, as well as electromagnetic compatibility (EMC) considerations.

In addition to this, computers will have to be designed around the limited size, weight and power (SWaP) available on UAS platforms, which has been an engineering challenge for drone manufacturers ever since their inception. Advanced AI processing has traditionally required larger processors and more power, although this has been difficult to achieve on power-constrained platforms such as UAS, where space is also extremely tight.

Systel has been working on specialist rugged computers for UAS – including working alongside a number of leading drone manufacturers – for a number of years now, and has become a leading industry expert on ultra-lightweight, small-form factor rugged computers that have high processing capabilities.

One such example is Sparrow-Strike – introduced in 2023 – which is already being trialed on UAS platforms and is a gamechanger for drone platforms of the future. By incorporating the NVIDIA Jetson Orin NX edge AI system on module (SOM), the Sparrow-Strike is capable of next-generation autonomous and uncrewed mission-critical applications, and can be configured according to the end users' unique requirements in a fast-changing security environment.

52.5 PENTAGON STARES DOWN "DRONE SWARM" THREAT

Swarms of low-cost drones powered by AI — perhaps hundreds of them at the same time — could soon pose an existential threat to the war machinery.

Why it matters: Advanced militaries can largely fend off individual drones. But swarms of them deployed on a single target have the potential to reshape the global balance of power.

The big picture: Future battles will be fought with a connected mix of cheap weapons — not sophisticated weapons systems that only superpowers can afford.

Expensive aircraft carriers or stealth bombers, which require massive time commitments to develop and build, will become vulnerable to drones that almost every military and armed group can access. This moment hasn't yet arrived, but it is rushing to meet us.

Those are largely point-and-shoot weapons controlled remotely by individual combatants, and American soldiers and ships are typically capable of shooting them down.

But if they're deployed in a swarm, Iranian-supplied drones that cost thousands of dollars each could take on multibillion-dollar warships.

State of play: The Pentagon is racing to develop a defense against swarms while building AI-coordinated drone systems of its own.

The Defense Department has signed at least five deals with contractors that explicitly reference swarming, according to a report from the Center for Security and Emerging Technology that examined procurement contracts from 2020.

Army Gen. Erik Kurilla, commander of U.S. Central Command, told the Senate Armed Services Committee last week that cheap drones are "one of the top threats," and called swarms an even bigger concern. He said the U.S. should "continue to invest in things like high-powered microwave" weapons to defend against mass drone attacks.

52.6 INDIA'S AND PAKISTAN'S DEVELOPMENT OF DRONES

On 23 August, 2024, a small Indian drone was spotted across the Line of Control in the airspace over Pakistan-administered Kashmir, a reminder of the ever-present tensions between Islamabad and New Delhi. While Pakistan claims it shot down the drone, India asserts that the drone inadvertently strayed due to a malfunction. This episode, though, is emblematic of a larger and more concerning trend—the rapid militarization of drone technology in South Asia.

Similar sighting of drones from Pakistan have been reported by India multiple times in the Indian territory of Punjab. These drones have been accused to carry surveillance, drugs, small arms and propaganda material.

Both India and Pakistan have been heavily investing in drones, which are increasingly being viewed not only as tactical assets but also as strategic tools capable of influencing the balance of power in the region. As drones become integral to militaries around the world, the potential for miscalculation, unintended escalation, and destabilization in future confrontations grows.

In the evolving dynamics of India-Pakistan relations, where tensions remain high and the risk of miscalculation is ever-present, future-oriented confidence-building measures (CBMs) for drone technology will be vital. With the anticipated rise of more advanced **autonomous and swarm drones** operating near sensitive borders, the potential for rapid, unintended escalation grows. To prevent this, future CBMs. must prioritize real-time crisis communication systems, artificial intelligence (AI)-driven transparency in drone operations, and preemptive agreements on limiting disruptive drone technologies. These forward-looking measures will help bridge the trust gap and ensure that cutting-edge drone advancements enhance stability rather than heighten the risk of conflict.

52.6.1 India's Drone Development

India's drone program has seen rapid advancements in recent years, driven by a desire to modernize its military and address emerging security challenges, particularly along its borders with China and Pakistan. By mid-2024, India had inducted between 2,000 and 2,500 drones, with the total expenditures ranging from USD $361.45 million to $421.69 million. These drones, which are used for surveillance, cargo transport, and tactical operations, are an integral part of India's military modernization efforts.

In the evolving dynamics of India-Pakistan relations, where tensions remain high and the risk of miscalculation is ever-present, future-oriented confidence-building measures for drone technology will be vital.

One of the most notable developments in India's drone program has been the deployment **of swarm drones**—unmanned aerial vehicles (UAVs) that operate in coordinated groups to execute complex missions. These swarm drones are seen as being a core component of India's defense strategy, particularly in terms of neutralizing threats from Pakistan. In February 2023, the Indian Army began inducting its first heterogeneous swarm UAV system, developed by the India-based NewSpace Research & Technologies, showcasing India's commitment to integrating advanced drone technology into its military. India's swarm drones have both offensive and defensive uses. Offensively, these drones can overwhelm enemy defenses by attacking multiple targets simultaneously, potentially including high-value assets such as nuclear delivery vehicles. Their ability to operate in large, coordinated groups makes them particularly effective in saturating and disrupting adversary defenses, potentially neutralizing critical military infrastructure. The potential for swarm drone operations to be coordinated with India's precision missile systems could enable greater real-time accuracy for strikes – including pre-emptive attacks – on critical targets within Pakistan. Furthermore, AI-based surveillance systems can provide real-time data to support these operations, improving the accuracy and effectiveness of drone strikes. India's swarm drones can also support its air defense and nuclear alert systems, through intercepting rival drone-based attacks and surveilling Pakistani nuclear forces deployed near border areas.

India's strong industrial base also gives it a significant advantage in sustaining drone operations during a protracted conflict. India benefits from a dynamic ecosystem with over 200 startups dedicated to drone technology. This capability may allow India to quickly replace lost drones and maintain operational pressure over extended periods – an advantage that can prove crucial in modern warfare.

52.6.2 Pakistan's Drone Development

Recognizing the strategic value of drones, Pakistan has also made significant progress in developing and acquiring UAVs. Pakistan's drone fleet includes advanced platforms like the Bayraktar TB2 and Akinci from Turkey, as well as the Wing Loong II and CH-4 from China. These drones, along with indigenous systems like the Burraq and Shahpar, provide Pakistan with a versatile and capable force that can be used for a variety of missions, including precision strikes, surveillance, and electronic warfare.

The Pakistan Air Force (PAF) currently plays a crucial role in supporting drone operations, as seen in recent air-defense exercises. By providing cover for UAVs, the PAF may coordinate drone strikes while simultaneously conducting manned air operations. By using loitering munitions alongside drones, Pakistan can target India's high-value assets such as the S-400, Advanced Air Defense (AAD), and Prithvi Air Defense (PAD) systems, effectively disrupting Indian defenses.

In the coming years, Pakistan may coordinate operations between advanced drones with its next-generation missile forces, such as the Fatah-2 missile, to enhance its ability to penetrate India's air defenses. These drones would be capable of overwhelming India's radar and missile interception systems, creating openings for precision missile strikes on critical targets. This approach could help offset India's growing numerical and technological superiority in air defense, leveraging advanced drone-missile integration to challenge even the most sophisticated defenses.

52.6.3 Implications for Strategic Stability in India-Pakistan Region

The rapid proliferation of drones in South Asia threatens to negatively impact strategic stability. As drones become more integrated into military strategies, their potential to disrupt traditional deterrence dynamics becomes increasingly apparent. In nuclearized South Asia, the ability of drones to conduct precision strikes on strategic targets, including nuclear assets, raises the stakes in any conflict.

India's potential integration of swarm drones with its missile defense systems and precision strike capabilities could fundamentally shift the strategic balance in the region. While offensively these drones might be capable of targeting Pakistan's nuclear delivery vehicles, such a scenario would likely only occur in the event of an extreme conventional conflict. The real threat to strategic stability may lie in how these technologies are perceived as part of India's increasing counterforce options, alongside systems like multiple independently-targetable re-entry vehicles (MIRV)s and precision missiles such as the Agni-P.

In the nuclearized South Asia, the ability of drones to conduct precision strikes on strategic targets, including nuclear assets, raises the stakes in any conflict.

Although Pakistan has developed drones domestically, its drone supply is still largely dependent on foreign imports and its domestic production capacity is limited by its weak indigenous drone industry. This reliance on external sources could become a vulnerability in a protracted conflict, such as a full-scale war with India similar to the Russia-Ukraine conflict.

CHAPTER 53

UNMANNED GROUND VEHICLES (UGVS)

53.1 INTRODUCTION TO UNMANNED GROUND VEHICLES (UGVS)

An unmanned ground vehicle (UGV) is a vehicle that operates while in contact with the ground without an onboard human presence. UGVs can be used for many applications where it is inconvenient, dangerous, expensive, or impossible to use an onboard human operator. Typically, the vehicle has sensors to observe the environment, and it autonomously controls its behavior or uses a remote human operator to control the vehicle via teleoperation.

On March 29 2024 as part of the Eastern Ukraine Campaign in the Ukraine War, a platoon of Russian UGV's equipped with AGS-17 automatic grenade launchers was deployed for an assault near the town of Berdychi in Ukraine, marking the first ever use of UGV's for direct frontline assaults.

An autonomous UGV (AGV) is an autonomous robot that replaces the human controller with artificial intelligence technologies. The vehicle uses sensors to feed a model of the environment, which supports a control system that determines the next action. This eliminates the need for humans to watch over the vehicle.

An autonomous vehicle must have the ability to:

- Navigate based on maps that allow the vehicle to pick a route from origin to destination.
- Detect objects such as people and vehicles.
- Travel between waypoints without human assistance.
- Avoid harming people, property or itself, unless those are part of its mission.
- A robot may also be able to learn autonomously.

Autonomous learning includes the ability to:

- Gain capabilities.
- Adjust strategies based on the surroundings.
- Adapt to surroundings.
- Behave ethically in pursuit of mission goals.

Armed autonomous machines must distinguish between combatants and civilians. This is particularly true in conflicts where combatants intentionally disguise themselves as civilians to avoid detection. Even highly accurate but imperfect systems can pose unacceptable civilian losses.

53.2 SOME UNMANNED GROUND VEHICLES BEING DEVELOPED AROUND THE WORLD

Unmanned Ground Vehicles (UGVs), robot combat vehicles or autonomous vehicles, no matter what you call them, are increasingly being developed, prototyped and tested for future use in autonomous and teaming operations or scenarios with manned vehicles. In the U.S., the Army has expressed growing interest in UGVs, after recently selecting four companies to develop robotic combat vehicle prototypes for a variety of future use cases.

Interest in new UGV technology is also growing around the world. Last year, the U.K. Ministry of Defense (MOD) held its first ever trial of heavy UGVs, Estonia hosted a large international test of UGVs and Australia ended the year with several weeks of a major robotic combat vehicle exercise.

Following are the updates on some of the latest prototyping, development, purchasing and deployment of new military UGVs.

(i) General Dynamics Land Systems

TRX SHORAD

General Dynamics first unveiled the Tracked Robot 10-ton (TRX) Short Range Air Defense (SHORAD) variant in March 2023, prior to the Army selecting them as one of four companies/teams competing to win its Robot Combat Vehicle (RCV) contract that seeks to deploy UGVs teamed with manned vehicle units by 2028. The variant unveiled in March 2023 adds increased counter UAS (C-UAS) capability to the demonstrator that the company previously launched in October 2020. TRX SHORAD is also capable of autonomously launching and receiving small unmanned aircraft systems (UAS). The original non-SHORAD version of the TRX was selected for the first phase of the Army's RCV program, and GDLS is currently designing and building two prototypes of the modular TRX. GDLS describes the UGV as featuring "lightweight materials and a hybrid-electric propulsion system" that is being developed for several different use cases including indirect fire, autonomous resupply, complex obstacle breaching, C-UAS, and electronic warfare (EW) among others.

(ii) HDT Global

WOLF-X

WOLF-X is the prototype UGV from a team that includes HDT Global as the lead vehicle developer, BAE Systems as the armament and situational awareness provider, and McQ Inc. — a Virginia-based defense electronics manufacturer — as the prime contractor. The 8×8 wheeled UGV is capable of being transported in a CH-47F. Powered by a hybrid diesel-electric system, WOLF-X also features a lithium-ion battery and can be configured with an MK44 30 mm chain gun and enhanced armor. Team

HDT is one of the four developers selected to participate in the Army's RCV program, with a WOLF-X prototype scheduled for delivery to the Army for test and evaluation this year.

(iii) Textron Systems Corp.

RIPSAW M3

Textron Systems is the lead developer of the RIPSAW M3, and also one of the four prototype vehicles selected for the Army's RCV program. The company partnered with Howe & Howe and Teledyne FLIR on the development of M3, a variant of the RIPSAW M5 system first launched by Textron in 2019. The Army has completed over 2,000 miles durability testing with the RIPSAW family since 2020, and describes all RIPSAW variants as featuring a basic flat-top deck configuration with an open architecture design and common chassis. RIPSAW is also designed with a size that fits inside of a CH-47F, and has 85 feet of deck space for payload configuration, and a 1,200-horsepower hybrid electric powertrain according to details released by Textron Systems. During AUSA 2023, Textron exhibited a version of the M3 demonstrator featuring a Kongsberg RS6 weapons system and a Javelin missile launching system. The company is developing a RIPSAW prototype for delivery to the Army later this year.

(iv) Oshkosh Defense, LLC

Oshkosh RCV

Oshkosh Defense is collaborating with Pratt Miller Defense and QinetiQ on the development of two different prototype robotic combat vehicles as part of the Army's RCV program. The company displayed both variants during the 2023 AUSA Conference. One of the variants was equipped with a Kongsberg RS6 remote weapon station, L3 Smoke Obscuration Module, LW 30x113 mm cannon, and a Switchblade 300 loitering munitions/UAS launcher system from AeroVironment. The second variant featured the Kongsberg CROWS-J equipped with an M2. 50 caliber machine gun and the Hoverfly Tethered Unmanned Aerial System (TeUAS). Oshkosh is on track to deliver two prototypes for testing to the Army in August.

(v) Team Lynx

XM30

Team Lynx is led by American Rheinmetall Vehicles (ARV), and includes Alison Transmission, Anduril Industries, L3 Harris, Raytheon and Textron Systems. The group was selected to design in July 2023 the prototype XM30 Mechanized Infantry Combat Vehicle to serve as the U.S. Army's Optionally Manned Fighting Vehicle. XM30 will replace the M2 Bradley Fighting Vehicle and is the Army's first ground combat vehicle designed using new digital engineering tools and techniques. XM30 is not expected to be operational as a fully unmanned vehicle in its initial deployment, according to a June 2023 media briefing given by Army acquisition leader Doug Bush. According to details about the XM30 released by Raytheon, the vehicle is being designed to be operated with two crew members and a third "virtual crew member, who will scan an area, identify potential threats and notify the crew." The Army is now in the third phase of prototyping for the XM30 program and estimates its first unit could be equipped with a field-ready XM30 by 2029.

(vi) Milrem Robotics

THeMIS

Established in 2013, Milrem Robotics is a Tallinn, Estonia-based robotics supplier that first introduced its THeMIS UGV in 2019, when the prototype was subsequently acquired for testing by Thailand's Army and Defence Technology Institute. THeMIS is eight feet long and has a height of three feet with a chassis that can carry up to 1.3 tons. Milrem's power options for THeMIS include a diesel engine and electric generator as well as a lead acid or lithium-ion battery. Its run time on electric power is up to 1.5 hours, and Milrem has developed logistics, combat, ISR and EOD configurations of THeMIS. The U.S. is one of 16 total countries that have acquired the THeMIS UGV for testing and evaluation purposes. The company has also delivered 15 THeMIS UGVs to Ukraine to support logistics, casualty evacuation and route clearance operations.

(vii) Israel Aerospace Industries

Rex MK II

First unveiled by Israel Aerospace Industries (IAI) subsidiary ELTA Systems Ltd. (ELTA) in September 2021, REX MK II is an all-wheel drive hybrid electric powered UGV that can carry up to 1.3 tons. According to IAI, Rex MK II also features remotely controlled weapons systems, including a 7.62 mm machine gun and a. 50 caliber heavy machine gun. The multi-mission vehicle can operate completely autonomously with command and control or remotely controlled by an operator. IAI also notes that two Rex vehicles can be transported inside of a single V-22 Osprey. One of the most recent updates around Rex MK II came when it demonstrated manned-unmanned teaming capability in an exercise with BAE Systems last year. The U.K.'s MOD is one of several international customers that has purchased the Rex MK II.

(viii) Elbit Systems

ROBUST

Jointly developed as a technology demonstrator by Elbit Systems and the Israel Ministry of Defense, the Robotic Autonomous Sense and Strike (ROBUST) UGV was first unveiled in October 2022. The 6×6 vehicle is equipped with a 30 mm autonomous turret, an active protection system, a robotic autonomy kit, and a built-in robotic arm for receiving and launching drones. Last year, ROBUST began testing and demonstration exercises including a trial organized by the U.K. MOD's Defence Equipment and Support (DE&S) division for UGVs weighing over 5.5 tons. The vehicle can be remotely controlled and has a turret equipped with a 12.7 mm heavy machine gun and pintle-mounted 7.62 mm machine gun, according to details released by Elbit. No timeline has been provided by Elbit for when ROBUST will be ready for deployment and operation.

(ix) Rheinmetall

Autonomous Combat Warrior Wiesel

German aerospace and defense manufacturer Rheinmetall unveiled the Autonomous Combat Warrior variant of its Wiesel ground vehicle in December 2021. The vehicle can be remotely controlled by a tablet, and the company added the unmanned capability to its Wiesel family of ground vehicles using its "PATH A-Kit." According to Rheinmetall, the A-Kit is a navigation system that enables full autonomous navigation through a combination of advanced sensors, algorithms, and real-time data analysis. The vehicle's autonomous mode is enabled by programming waypoints for it to navigate between on a tablet. Rheinmetall has released very few details about the robotic vehicle since its 2021 launch, although it did participate in the U.K.'s 2023 heavy UGV trial.

(x) Hanwha

Arion-SMET

Korean aerospace and defense manufacturer Hanwha's Autonomous and Robotic Systems for Intelligence Off-road Navigation – Small Multi-purpose Equipment Transport (Arion-SMET) is a six-wheeled all-electric vehicle that can drive up to 62 miles and carry up to 1,200 pounds in payload. Hanwha developed Arion-SMET to support infantry operations, such as ammunition transport, medical evacuation, reconnaissance, and fire support. In December 2023, the U.S. Army, Marine Corps, and Ministry of National Defense of the Republic of Korea participated in a Foreign Comparative Test (FCT) to evaluate the Arion-SMET's ability to navigate over a variety of terrains, heavy equipment transportation, soldier following and remote-controlled driving. The Marine Corps has not yet fielded a UGV, and is currently defining its requirements for the technology.

53.3 UNMANNED GROUND VEHICLES SUCCESSFULLY DEMONSTRATED AT PNTAX -23

The Robotics for Engineer Operations (REO) team at the U.S. Army Engineer Research and Development Center (ERDC) successfully conducted demonstrations of their Unmanned Ground Vehicles (UGV) without the use of the Global Navigation Satellite System (GNSS) at the Position, Navigation, Timing Assessment Experiment (PNTAX).

PNTAX is part of Army Futures Command's campaign of persistent experimentation and continuous learning, where participants field-test their technology to understand its operational effectiveness in a denied and degraded environment. This experiment provided the opportunity to run tests focused on localization and freedom of movement of an UGV on challenging roads and rugged combat trails, while access to standard navigation was not available.

ERDC's REO team, comprised of researchers from the Information Technology Laboratory, Construction Engineering Research Laboratory, Geospatial Research Laboratory and Geotechnical and Structures Laboratory, all participated.

Chief Warrant Officer Charles Bashor (CW4) from the Construction Engineering Research Lab said events like this are crucial in the development of up-and-coming warfighting technology.

"Events like PNTAX provide the challenging, Army-relevant environments and Soldier touchpoints that allow ERDC to accurately test the capabilities they develop early in the development cycle and incorporate the feedback," Bashor explained. These demonstrations simultaneously help support AFC's mission, to transform the Army through innovation and ensure war-winning future readiness.

The field experiments took place over three day and three-night operations, providing the REO team a chance to experience varying lighting and environmental conditions, mimicking the challenging terrains these vehicles may encounter in future conflict. ERDC's REO team utilized their UGVs to perform test runs and obtain key performance metrics of their mapping capabilities. With a goal of mapping a specific area without using GNSS positioning, the experiments tested ERDC-developed algorithms for absolute and relative localization in environments with few distinctive features.

At PNTAX, the robotic platforms operated autonomously on combat trails, and the developed algorithms were minimally impacted by the active GPS jamming/spoofing signals present. With this success, ERDC Senior Research Scientist Dr. Anton Netchaev said their team is ecstatic about REO's progress.

"I believe that our team truly united under the challenging environments of PNTAX, and the performance of the capabilities delivered by REO has exceeded our initial expectations," said Netchaev. This shows ERDC's advancement in creating GNSS-denied operational capabilities for the Army, even in areas where key features are limited.

Operating fully autonomously, these UGVs allow Soldiers to perform their missions from a safe location, while providing the most up to date information about the battlespace, including terrain features, go/no-go zones, location and type of obstacles, etc. Through multiple Soldier touchpoints, technology is placed into the Soldier's hands, allowing them to experiment with the technology and offer direct end-user feedback.

Data from the experiment will also be utilized by AFC's Next Generation Combat Vehicles Cross-Functional Team, which is writing the requirements for the Army's Robotic Combat Vehicle program. As part of the Army's Human Machine Integrated Formations (HMIF) concept, RCVs will be fielded in Brigade Combat Teams and will deliver increased situational awareness, lethality, and tactical options for Army formations in support of multi-domain operations. Its operators will remotely control RCVs or task them to operate semi-autonomously. Variants will serve as "scouts" or "escorts" for manned fighting vehicles.

The ERDC REO team will build on their success at PNTAX by incorporating a variety of supporting elements and enhancing navigational capabilities for military off-road operations. "The team will continue to expand the UGV's capabilities by including other sensing modalities," said REO Project Lead Dr. Ahmet Soylemezoglu, "This will help provide the Engineers with a more complete picture."

To further their experimentation, the REO team will participate in the annual Maneuver Support, Protection Integration Experiments (MSPIX 24) by the Maneuver Support Center of Excellence Battle Lab.

53.4 INDIAN ARMY HAS INTEGRATED THE KRUSHNA UNMANNED GROUND VEHICLE (UGV)

The Indian Army has recently integrated the Krushna Unmanned Ground Vehicle (UGV) into its operational arsenal. This advanced UGV, created by Club First, a robotics manufacturing firm headquartered in Jaipur, is designed to significantly improve the Army's capabilities in various critical areas. The Krushna UGV is specifically engineered to boost efficiency in surveillance and reconnaissance missions, providing enhanced monitoring and data collection in challenging environments. By leveraging this cutting-edge technology, the Army aims to strengthen its operational readiness and effectiveness in gathering intelligence and assessing tactical situations.

The Krushna Unmanned Ground Vehicle (UGV) is equipped with advanced technologies designed to perform effectively in demanding environments.

The key features of this UGV include:

(i) **Silent Operation:** The UGV operates silently, making it ideal for stealth missions where minimizing noise is crucial for maintaining operational secrecy.

(ii) **Thermal Day/Night Camera:** It is outfitted with a thermal camera capable of operating both day and night, with a detection range of up to 4 kilometres. This feature significantly enhances its surveillance and reconnaissance capabilities, allowing for detailed monitoring under various conditions.

(iii) **Remote and Autonomous Navigation:** The Krushna UGV supports both remote operation and autonomous navigation through GPS waypoint technology. This dual capability enables it to traverse challenging terrains that are difficult for conventional vehicles to access, providing versatility in a range of operational scenarios.

The integration of the Krushna Unmanned Ground Vehicle (UGV) into the Indian Army's operations is part of a larger effort to incorporate unmanned systems into its strategic framework. This initiative is designed to enhance border security and improve surveillance capabilities, addressing key priorities in India's defence strategy. The Krushna UGV's robustness in harsh climatic conditions and challenging geographical terrains makes it an invaluable asset for the Army, offering significant advantages in maintaining security and conducting operational missions effectively.

The Krushna Unmanned Ground Vehicle (UGV) is set to transform battlefield support for the Indian Army with its **advanced artificial intelligence capabilities**. This innovative technology is designed to significantly boost operational efficiency and enhance soldier safety. By providing enhanced surveillance and logistical support, the Krushna UGV offers a substantial upgrade to current military operations, empowering the armed forces with cutting-edge tools for improved performance and strategic advantage.

CHAPTER 54

UNMANNED UNDERWATER VEHICLES (UUVS)

54.1 INTRODUCTION TO UNMANNED UNDERWATER VEHICLES

Unmanned underwater vehicles (UUVs), also known as uncrewed underwater vehicles and underwater drones, are submersible vehicles that can operate underwater without a human occupant. These vehicles may be divided into two categories: remotely operated underwater vehicles (ROUVs) and autonomous underwater vehicles (AUVs). ROUVs are remotely controlled by a human operator. AUVs are automated and operate independently of direct human input.

54.1.1 Classifications of Unmanned underwater vehicles

(i) Remotely operated underwater vehicle

Remotely Operated Underwater Vehicles (ROUVs) is a subclass of UUVs with the primary purpose of replacing humans for underwater tasks due to the difficult underwater conditions. ROUVs are designed to perform educational or industrial missions. They are manually controlled by an operator to perform tasks that include surveillance and patrolling. The structure of ROUVs disqualify it from being able to operate autonomously. In addition to a camera, actuators, and sensors, ROUVs often include a 'gripper' or something to grasp objects with. This may throw off the weight distribution of the vehicle, requiring manual assistance at all times. Sometimes ROUVs require additional assistance due to the importance of the task being performed. The US Navy developed a Submarine Rescue Diving Recompression System (SRDRS) that can save up to 16 people up to 2000 feet underwater at a time. Such a large vehicle with the primary role of saving lives requires an operator(s) to be present during its mission.

(ii) Autonomous underwater vehicle

Autonomous Underwater Vehicles (AUVs) are defined as underwater vehicles that can operate without a human operator. Sizes can range from just a few kilograms up to thousands of kilograms. The first AUV was created in 1957 with the purpose of performing research in the Arctic Waters for the Applied Physics Laboratory at the University of Washington. By the early 2000s, 10 different AUV had been developed such as screw driven AUVs, underwater gliders, and Bionic AUVs. The earliest models used screw propeller thrusters while more recent models utilized automatic buoyancy control. The earliest model, SPURV, weighed 484 kg, went as deep as 3650 meters, and could travel for up to 5.5 hours. One of the most recent models, Deepglider, weighs 62 kg, can go as deep as 6000 meters, and can travel up to 8500 km.

The US Navy began using UUVs in the 1990s to detect and disable underwater mines. UUVs were used by the US Navy during the Iraq War in the 2010s to remove mines around Umm Qasr, a port in southern Iraq.

The Chinese military uses UUVs for mostly data collection and reconnaissance purposes.

On December 20, 2020, a fisherman in Indonesia spotted a glider-shaped UUV near Selayar Island in South Sulawesi. Individuals from the Indonesian military have categorized the vehicle to be a Chinese Sea Wing (Haiyi), created for the purposes of collecting data including water temperature, salinity, turbidity, and oxygen levels that can help chart optimal submarine routes.

The navies of multiple countries, including the US, UK, France, **India,** Russia, and China are currently creating unmanned vehicles to be used in **oceanic warfare** to discover and terminate underwater mines. For instance, the REMUS is a three-foot long robot used to clear mines in one square mile within 16 hours. This is much more efficient, as a team of human divers would need upwards of 21 days to perform the same task.

A survey conducted by RAND Corporation for the US military analyzed the missions which unmanned underwater vehicles could perform, which included intelligence, reconnaissance, mine countermeasures, and submarine warfare.

In November 2022, the Eurasian Times reported that China's Harbin Engineering University has developed trans-medium **'flying submarine'** drones capable of both underwater and air travel, noting the potential military applications of the vehicles.

54.2 WHAT IS AN AUV?

"AUV" stands for Autonomous Underwater Vehicle; they are unmanned, untethered vehicles used to conduct underwater research.

AUVs are unmanned underwater robots akin to the Curiosity rover NASA uses on Mars. As their (autonomous) name suggests, AUVs operate independently of humans. Unlike remotely operated vehicles (ROVs), which are tethered to a service vessel, AUVs have no physical connection to their operator. Rather, AUVs are programmed or controlled by operators who may be on a vessel or even on shore, who tell an AUV where, when, and what should be sampled. AUVs carry a variety of equipment for sampling and surveying such as cameras, sonar, and depth sensors. Unlike ROVs, which transmit video via their tethers almost instantaneously to a control room on a ship, an AUV stores all data, including images and other sensor data, on onboard computers until it can be retrieved after the AUV is recovered at the end of a dive.

Motivations for using AUVs include such factors as ability to access otherwise-inaccessible regions, lower cost of operations, improved data quality, and the ability to acquire nearly synoptic observations of processes in the water column. An example of the first is operations under Arctic and Antarctic ice, an environment in which operations of human-occupied vehicles and tethered platforms are either difficult or impossible. AUVs are becoming the platform of choice for deep-water bathymetric surveying in the offshore oil industry because they are less expensive than towed platforms as well as produce higher-quality data (because they are decoupled from motion of the sea surface). Finally, the use of fleets of AUVs enables the rapid acquisition of distributed data sets over regions as large as 10 000 km^2.

AUVs are a new class of platform for the ocean sciences, and consequently are evolving rapidly. The Self-Propelled Underwater Research Vehicle (SPURV) AUV, built at the University of Washington Applied Physics

AUVs can range in weight from only a few hundred pounds up to several thousand pounds. They may glide from the sea surface to ocean depths and back or they may stop, hover, and move like blimps or helicopters do through the air.

Fully autonomous operations carry power onboard. Power enables propellers or thrusters to move an AUV through the water and is necessary to operate sensors on the AUV. Most AUVs use specialized batteries, although some AUVs have used fuel cells or rechargeable solar power. Certain AUVs, such as gliders, minimize energy demands by allowing gravity and buoyancy to propel them.

AUVs are attractive options for ocean-based research. They can reach shallower water than boats can and deeper water than human divers or many tethered vehicles can. Once deployed in underwater, AUVs are safe from bad weather and can stay underwater for extended periods of time. They are also scalable, or modular, meaning that scientists can choose which sensors to attach to them depending on their research objectives. AUVs are also less expensive than research vessels, but they can complete identical repeat surveys of an area.

Scientists use AUVs to create maps of the ocean floor, record environmental information, sense what humans have left behind, identify hazards to navigation, explore geologic formations, document shipwrecks, and more.

Autonomous underwater vehicles (AUVs) are untethered mobile platforms used for survey operations by ocean scientists, marine industry, and the military. AUVs are computer-controlled, and may have little or no interaction with a human operator while carrying out a mission. Being untethered, they must also store energy onboard, typically relying on batteries. Motivations for using AUVs include such factors as ability to access otherwise-inaccessible regions, lower cost of operations, improved data quality, and the ability to acquire nearly synoptic observations of processes in the water column. An example of the first is operations under Arctic and Antarctic ice, an environment in which operations of human-occupied vehicles and tethered platforms are either difficult or impossible. Illustrating the next two points, AUVs are becoming the platform of choice for deep-water bathymetric surveying in the offshore oil industry because they are less expensive than towed platforms as well as produce higher-quality data (because they are decoupled from motion of the sea surface). Finally, the use of fleets of AUVs enables the rapid acquisition of distributed data sets over regions as large as 10 000 km^2.

AUVs are a new class of platform for the ocean sciences, and consequently are evolving rapidly.

A class of AUVs uses multiple thrusters to provide capabilities similar to that of a helicopter or a ship with dynamic positioning. The additional thrusters enable maneuvers such as hovering, translating sideways, and moving vertically. These vehicles are used when maneuverability is needed, for example, when operation near a very rough bottom is a necessity. The disadvantage is that the additional thrusters reduce efficiency for moving large distances or at high speeds.

While the vehicles described above are representative of the most commonly used systems, a wide range of other vehicles are in development or are in limited use. AUVs that come to the surface and **use solar panels to recharge batteries** have been demonstrated in seagoing operations. Gliders which extract energy from thermal differences in the ocean for propulsion have also been tested. A hybrid vehicle is being developed for reaching the deepest portion of the ocean. The hybrid vehicle operates as a tethered platform via a disposable fiber optic link for tasks requiring human perception, in other words, as a remotely operated vehicle (ROV), but operates as an AUV when that link is severed. These are just a few of the diverse AUVs being developed to answer the needs of ocean science.

54.3 UNMANNED UNDERWATER VEHICLES: A STRATEGIC OPPORTUNITY

Today, the market for underwater drones such as Unmanned Underwater Vehicles (UUVs) is growing significantly, due to their strategic capabilities.

The missions UUVs can assist with are varied:

- Mine clearance (detection and neutralization),
- Intelligence,
- Surveillance,
- Reconnaissance, and
- Anti-submarine warfare (ASM).

However, the objective is to improve the capabilities of Unmanned Underwater Vehicles, into fully autonomous AUVs (Autonomous Underwater Vehicles). **Advances in Artificial Intelligence (AI) technology make this possible.**

The new technological developments in underwater technology thus focus on increasing:

- the autonomy,
- agility, and
- storage capacity of drones.

54.3.1 Underwater Cables and Pipes

Underwater cables carry the production of offshore gas and oil infrastructures and wind farms, as well as 99% of the data exchanged between the world's telecommunication networks (source). Most intercontinental communications, financial flows, and access to telestored data (from the cloud) depend on them. These cables include everything from personal communications to financial transactions a sensitive national security data. They are therefore important strategic levers for states.

So much so, that the data is described as the 'new oil', and the cables as veritable trans-oceanic pipelines. States are thus becoming aware of the increasing vulnerability of these submarine cables.

Therefore, the threat to undersea cables is multifaceted. Sabotage, espionage, and foreign adversaries could track their whereabouts to sabotage them and cut rivals off from communications. For example, there has long been suspicion that Moscow is actively targeting these cables for spying purposes. Further, The Nordstream underwater gas pipelines were sabotaged in September 2022, with no clear answer who was behind it.

Private players such as the Google, Apple, Facebook, Amazon, Microsoft (GAFAMs) are laying submarine cables with colossal budgets, and tourist underwater robots capable of descending to 200 metres are being sold for as little as 2000€. Also, from the bottom of the oceans, players can take advantage of technological progress and an environment conducive to concealment. Therefore, states have great interest in the development of sensors and weapons projects for the air-sea space (source). Finally, the development of artificial intelligence (AI) is driving the market for underwater exploration robots and drones.

54.3.2 Protection of underwater Infrastructure

Protection of underwater infrastructure;

- anti-submarine warfare;
- mine warfare;
- search and recovery of objects damaged at sea;
- development of hydrography and oceanography;
- exploitation and exploration of underwater resources;
- military ISR;
- concealment;
- scientific exploration and control of the abyss;

are all reasons for a state to have a military intervention capability on the seabed.

ROVs are useful for a variety of underwater inspection tasks such as:

(i) Explosive Ordnance Disposal (EOD)

Military clearance divers carry out dangerous tasks including deploying a shaped charge to disarm a mine or examining and disrupting a Waterborne Improvised Explosive Device (WBIED). These tasks are still done by humans because they require high fidelity perception and dexterous intervention capabilities not previously available through a remote solution. Recently, with the release of more dexterous, compact manipulator arms, military organizations are investigating ways for ROVs to start to do some of these more dangerous tasks.

(ii) Port Security

The passageways between countries and regions are getting busier every day. With more unknown vessels entering port, each one needs to be inspected for potential threats. For example, countries like Saudi Arabia and other Gulf States are increasingly worried about Irani incursions in its ports.

(iii) Mine Countermeasures

Underwater vehicles can scout underwater minefields and possibly disarm naval mines. They may reduce, but not eliminate, the need for specialized diver teams to reconnoitre, identify, and demine potential landing beaches for amphibious warfare operations. However, as technology develops the needs for frogmen such as the elite operators of DEVGRU might become obsolete.

(iv) Smuggling

It is possible to hide contraband in the bottom of a ship's hull or inside its propellers. But with an ROV, port security professionals can quickly receive live video from underneath a ship. Once again, this avoids putting a diver's life at risk.

(v) Search and Recovery

Drownings often occur in bodies of water that are also extremely dangerous for divers. The availability of an ROV for immediate deployment minimizes the risk to divers. An ROV can also work in large areas in combination with a dive team. When a person is located, divers can either follow the ROV lanyard to the victim, use the grabber to recover the victim or deploy a buoy. Recovery operations include scenarios such as recovering a downed aircraft or drowning victims. Teams of highly trained military or civilian divers have historically performed these tasks. Recently, ROVs have been able to assist and, in some cases, take over these high-risk tasks.

(vi) Training

Military training programmes currently use ROVs. Many forces have training exercises that require specific diving exercises or the execution of underwater maneuvers. An ROV allows all movements and interactions of the students to be recorded for later review and personal improvement exercises.

54.4 POSITIONS OF STAKEHOLDERS ON UNMANNED UNDERWATER VEHICLES

If we are witnessing a general remilitarization of the oceans and seas, this trend is being replicated at the bottom of the sea. The great powers are engaged in a deafening underwater competition: each is increasing its capacities and setting up operational systems to prepare a seabed warfare.

(i) United States

The US Navy is integrating Subsea and Seabed Warfare (SSW) into their Full Spectrum Undersea Warfare (FSUSW). American defence companies such as General Dynamics Mission Systems are proposing their innovations as part of an overall forward-looking vision for seabed warfare. From 2021 to 2025, the US Navy would like to spend $1.9 billion on UUVs. It is expected to spend $941 million on USVs (unmanned surface vehicles) and UUVs in 2021 alone, an increase of 129% over

2019. The US Navy hopes to develop a 'shadow fleet' of unmanned ships to make up for its numerical deficit in the face of the rising Chinese People Liberation Army Navy (PLAN).

(ii) Russia

Russia hopes to maintain a cost-effective oceanographic picture of the surrounding seas, particularly in the Arctic Ocean. Unmanned Underwater Vehicles also meet the need to defend submarines against NATO's sophisticated torpedoes and mines. Moreover, Russia appears to be the only country in the world with the capability to integrate nuclear turbine generators into unmanned aerial vehicles (UAVs) or small underwater installations. It is already deploying the Poseidon strategic intercontinental autonomous torpedo powered by a nuclear turbine generator. These small reactors can also supply electricity to charging stations for drones.

(iii) China

Although China's fleet of AUVs is still primarily research experiments and early-stage prototypes, PLA Navy scientific research documents (PLAN) indicate that China is very interested in these technologies. The PLAN primarily focuses on using AUVs for maritime surveillance and reconnaissance, mine warfare and countermeasures, submarine cable inspection and anti-submarine warfare (source).

Moreover, advances in unmanned vehicle research may also permit the PLAN to use AUVs to tap or sever undersea fiber-optic cables in a conflict, which concentrates near northern Taiwan. These cables are crucial not just for information dissemination in Taiwan, but also for the trans-Pacific data exchanges that facilitate global internet access.

In future, the Chinese military may use large AUV models for anti-submarine warfare (ASW) and operations near the seabed. Indeed, the control of strategic maritime areas for China, such as Jiangsu Province, may explain this desire to militarize the seabed. With a coastline of more than 1000 kilometres, Jiangsu province faces Japan and South Korea.

(iv) France

France, with the world's largest submarine domain, cannot remain aloof from this new strategic space. On the 14th of February 2022, the French Minister of the Armed Forces presented a new seabed strategy.

"The seabed is a new area of power relations that we must master in order to be ready to act, to defend ourselves and, if necessary, to take the initiative, or at least to retaliate".

The French Navy's 'Mercator' strategic plan confirms this as a priority within the French Military Programming Law 2019-2022, with a budget of €2.9 million. For France, the objective is to have a permanent large-scale operational capacity from 2025.

(v) United Kingdom

Like other nations, British defence needs a greater presence in the undersea battlespace. The Royal Navy needs fully autonomous systems that can sense, understand, and decide without human intervention. These systems will increase in the undersea battlespace and take over the boring, dirty, and dangerous tasks of manned platforms.

Britain is seeking to develop an autonomous undersea force (its "force mix"). As part of its Defence and Security Accelerator (DASA) program, the MOD awarded Marlin Submarines Ltd a procurement contract. This contract contains provisions to develop an autonomous version of an Extra Large Unmanned Underwater Vehicle (XLUUV). The XLUUV is a two-stage project with a total value of £2.4 million.

54.5 FUTURE PLANS ON UNMANNED UNDERWATER VEHICLES

(i) United States

The US Navy wants to develop and procure Extra-Large Unmanned Undersea Vehicles (XLUUVs). The US Navy established the XLUUV program, also known as the Orca program, to address a Joint Emergent Operational Need (JEON). The Navy defines XLUUVs as UUVs with a diameter of more than 84 inches, meaning that XLUUVs may be too large to launch from a manned Navy submarine.

(ii) Russia

According to the information published by Izvestia (a major Russian newspaper), in late 2021 the Russian Ministry of Defense approved a test plan for the new unmanned underwater vehicle (UUV) Klavesin-2R-PM in the Far East, after being tested earlier in the Crimean Sea. In addition, Russia is working on the air-independent propulsion unit on the order of the Foundation for Advanced Research.

The unmanned underwater vehicle Sarma, developed by the Lazurit design bureau, will be able to dive to a depth of 1 km. Similarly, the vehicle will carry out assigned missions autonomously for three months. The submersible will be able to cover a distance of more than 8,000 km. The Sarma can perform a range of tasks, including bottom survey, bottom topography, monitoring of long objects (gas pipelines), mapping, geological exploration, and maintenance and repair of underwater communications in the oil and gas sector. "Maintenance of under-ice monitoring and lighting systems" is also among the potential tasks of drones (source).

(iii) China

The Chinese People's Liberation Army (PLA) Navy has a project dubbed the "Great Undersea Wall". Intended to protect the South China Sea from underwater incursions, the Great Undersea Wall is similar to the Acoustic Surveillance System (SOSUS). SOSUS is a network of seabed hydrophone arrays built by the US to detect and monitor Soviet submarines during the Cold War. It consists of active and passive sensors, UUVs, and manned submersibles.

Beyond the three major Chinese AUV design centres, a growing number of research institutes and private companies are entering the Chinese AUV market. As of 2019, China hosts 159 AUV research projects at more than 40 Chinese universities. This is a significant increase from the 15 major universities that had built AUV research teams four years earlier.

In September 2021, Chinese researchers tested a new underwater drone in what appears to be its first open-water test. This holds significant implications for geopolitical stability, as the tests took place in the South China Sea. A university with close ties to the Chinese military created the drone using a bio-inspired design. The drone uses the shape of a manta ray to help it glide efficiently through the water (source).

(iv) France

France intends to develop collaborative combat in order to free itself from environments and to combine the effects of armies (multi-domain concept) with hyperconnectivity, which represents a real technological challenge for underwater vectors (source). To deal with the difficulties of underwater communication, research and development work is suggesting interesting avenues. However, no satisfactory solution has yet emerged.

(v) The UK

The UK Submarine Delivery Agency's Autonomy Unit has launched a tender worth up to £21.5 million for the procurement and testing of an 8–12-meter Extra Large Autonomous Underwater Vehicle (XL-AUV), known as the CETUS project.

The CETUS project involves the design and construction of a 'very large autonomous underwater vehicle' that could one day work alongside the British attack submarines. The resulting XL-AUV will help the UK Navy reduce the risks associated with the acquisition of future UUVs.

The UK Navy already has an Extra Large Unmanned Underwater Vehicle (XLUUV) which they use to test various payloads. The Project CETUS work follows on from XLUUV experimentation and will be the service's first purpose-built XL-AUV. Design work should be completed in 2022-23, with the delivery of the demonstrator to follow in 2023-24.

54.6 LIMITATIONS OF UUVS

While UUVs are a valuable addition to naval and submarine fleets, there are limitations to underwater communications. Military organizations must look at how to overcome such limitations for the persistent missions the military wishes to accomplish. The main problem with UUVs is communication. The only communication with these vehicles is snippets of information that come back acoustically. The vehicles are unable to send back any data unless they surface to do so.

To address the challenges of underwater communications, solutions are being explored such as small software-defined radio systems, advanced acoustic detection and communication systems, and artificial intelligence and machine learning. It's not just about putting UUVs in the water, it's about getting the "right data to the right decision-makers, in the right format" (source). Ultimately, this would mean that UUVs could facilitate decisions and provide states with a tactical advantage. However, **there are some solution like ULTRA that offers wireless communication underwater using lasers.**

54.6.1 Role of AI in UUVs

AUVs generally operate in a set pattern to map a large area of the ocean. **Artificial intelligence could allow AUVs to make decisions, modify their trajectory, and even decide whether to take offensive action.**

The difficulties of underwater communication thus make the autonomy of UUVs all the more necessary. They need to be able to make decisions, and that's where AI comes in. It is a matter of equipping the vehicles with AI and machine learning for automatic target recognition. The concept is to teach the AUV what a mine looks like, for example. So if it detects one it tags it, comes to the surface, quickly offloads the images and continues on its mission (source).

Thanks to AI, AUVs will be able to go beyond missions that require them to follow a predefined trajectory, moving from one landmark to another. With AI, the aim is to maximise the search parameters to identify objects of interest. The UUVs will be empowered to make decisions and share information with other resources, in a logic of interoperability and connected warfare.

A vehicle equipped with a sophisticated side-scanning system, for example:

- Could search for and locate a mine;
- then call a vehicle equipped with a high-quality camera nearby to get a better view;
- all without human intervention.

The AUV will also be able to select the data of interest to avoid overwhelming the operator.

Finally, from a military perspective, vehicles will have to go much deeper and stay longer for missions. This is because the missions require:

- High survivability,
- Higher energy density,
- Greater degree of autonomy,
- This will require **the use of Artificial Intelligence.**

54.7 AUV-150 OF INDIA

AUV (Autonomous Underwater Vehicle)-150 is an unmanned underwater vehicle (UUV) being developed by Central Mechanical Engineering Research Institute (CMERI) scientists in Durgapur in the Indian state of West Bengal. The project is sponsored by the India's Ministry of Earth Sciences and has technical assistance from IIT-Kharagpur.

The vehicle was built with the intent of coastal security like mine counter-measures, coastal monitoring and reconnaissance. AUV-150 can be used to study aquatic life, for mapping of sea-floor and minerals along with monitoring of environmental parameters, such as current, temperature, depth and salinity. It can also be useful in cable and pipeline surveys. It is built to operate 150 meters under the sea and have cruising speed of up to four knots.

(i) Structure

AUV-150 is cylindrical-shaped with streamlined fairing to reduce hydrodynamic drag. It is embedded with advanced power, propulsion, navigation, and control systems. The UUV includes a pressurized cabin which is necessary for the diving and flotation system to work properly; this also helps to increase its sealing power against water leakage into the cabin. The AUV 150 weighs 490 kg, is 4.8 meters long and has a diameter of 50 cm.

(ii) Control system

The AUV is autonomous, that is automatic and self-controlled. It has an onboard computer that can be pre-programmed to dive to preset depths, move along preset trajectories, and return to the base after completing the assigned tasks. A remote-control option is provided in order to perform special tasks.

(iii) Propulsion

It is propelled by water-jet propulsion which comprises thrusters for generating motion in different directions to control surge, sway, heave, pitch, and yaw, while preventing the vehicle from rolling. Two arrays of cross-fins have also been fixed at the two ends to provide additional stability to the AUV.

(iv) Navigation

The autonomous vehicle is equipped with a number of navigational equipment to locates its own geographical position Such as inertial navigation system, depth sonar, altimeter, doppler velocity log, global positioning system through ultra-short baseline system and forward-looking sonar to facilitate obstacle evasion and safe passage.

(v) Payload

The AUV is equipped with an underwater video camera that can send wireless video pictures from underwater to a monitor above water surface along with side scan sonar. The submarine is equipped with CTD or conductivity-temperature-depth recorder and several sensors that can measure orientation, current and speed.

(vi) Communication

For smooth communication and distant intervention, the vehicle is equipped with hybrid communication system: it uses radio frequency while on surface and acoustic under water.

(vii) Power

The vehicle uses a **Lithium polymer battery** and can operate up to depths of 150 metres at speeds of 2-4 knots.

54.7.1 Test trial of AUV-150

A full-scale prototype of AUV-150 was put to sheltered freshwater test in January 2010 under tight security. It was brought in a truck from Durgapur in West Bengal under secrecy inside a container up to Kulamavu reservoir at Idukki in Kerala and the vehicle was later taken to the middle of the reservoir on a boat. The choice fell on Idukki since Kulamavu has a centre associated with Indian Navy and the reservoir being very deep. The biggest arch dam of Asia, Idukki reservoir is located about 250 km from Kerala's capital Thiruvananthapuram, and its lake covers an area of 60 km^2. The preliminary test was conducted for seven to eight days and the results was reported to be very satisfactory. The final leg of the still-water trials was conducted in the reservoir between September and October 2010.

54.7.2 Sea trial of AUV-150

The first series of sea trials of AUV-150 was commenced from 13 July 2011 off the Chennai coast. From July 13 to July 16 the diving depth of the AUV-150 was increased in stages, it reached consecutive depths of 35.79m 79.86m and 119.95m, and finally on 17 July 2011 Auv-150 reached the specified depth of 150 m. Despite extreme rough sea environments (Sea-state of 4), the sea trial was satisfactory. Although minor problem was faced in recording video frames.

54.8 TAKING STOCK OF INDIA'S EVOLVING UNMANNED UNDERSEA CAPABILITIES

On July 28, 2023, the Garden Reach Shipbuilders and Engineers (GRSE), an Indian state-owned shipyard, launched an autonomous underwater vehicle (AUV) named ***Neerakshi***. A significant milestone in India's efforts to strengthen its underwater capabilities, it is a collaborative effort between GRSE and Aerospace Engineering Private Limited (AEPL). Officials said that the vehicle can be used for a variety of functions such as mine detection, mine disposal, and underwater survey.

Lauding the indigenous partnership behind the prototype's production, Defense Research and Development Organization (DRDO) Chairman Samir V. Kamat said that he is optimistic about the Indian defense industry's capacity to provide innovative and cutting-edge technologies in order to meet the ever-growing demands of the armed forces.

UUVs are built to perform very unique roles. Their main functions include mine countermeasures; intelligence, surveillance, and reconnaissance; anti-submarine warfare; search and rescue operations; anti-surface warfare; protection of critical infrastructure; monitoring checkpoints; oceanographic research; etc. As there are different types of UUVs, coming in several shapes and sizes, their operators are provided with a higher degree of flexibility.

Additionally, their abilities to inflict very high damage, especially when operated in groups of large numbers (also called swarms), as well as their lower implementation and operational cost, and improved endurance when compared to manned platforms, are expected to make them a preferred choice in the near future. Although UUVs are mostly used for peacetime operations today, they are also highly capable of offensive roles. Just as unmanned aerial systems have transformed land-air warfare, UUVs are also set to revolutionize naval warfare – completely changing the way naval forces perceive maritime operations.

54.8.1 India's Forays into the UUV Arena

For more than a decade, the DRDO and its cluster laboratories have been involved in the production of indigenous AUVs that are designed and developed fully in India. In 2010, the Indian Navy floated a tender to invite interest from state-owned and private defense companies regarding a requirement for at least 10 AUVs. These were expected to carry out surveillance, reconnaissance, and oceanographic survey missions and should be operable at a depth of 500 meters for a duration of 7 to 8 hours.

A few years later, a collaboration between DRDO and IIT Madras resulted in an AUV that completed its preliminary testing in the Bay of Bengal. This technology demonstrator prototype had a flat-fish configuration and was reportedly able to carry payloads weighing up to 500 kgs at a depth of 100 to 300 meters. At the same time, another variant – which was aimed to be used mainly for mine disposal activities – was also under development by the DRDO's Naval Science and Technology Laboratory (NSTL) in Vishakhapatnam and Electronics Corporation India Ltd (ECIL). Apart from these, India also built a low cost AUV called Samudra, which is being used for deep-sea exploration purposes.

In 2015, former Defense Minister Manohar Parrikar had addressed the Parliament regarding a feasibility study that was undertaken to assess the DRDO's capabilities to design and develop different kinds of UUVs. In the "Indian Naval Indigenization Plan (2015-2030)," a guideline document released to enunciate its most important requirements as well as self-reliant endeavors including the development and use of critical technologies such as artificial intelligence (AI), the Indian Navy discussed in detail the need to "enhance operational capabilities of naval forces in underwater warfare, reconnaissance and surveillance."

At the Naval Commanders' Conference in 2021, Defence Minister Rajnath Singh launched the "Integrated Unmanned Roadmap for Indian Navy" that is expected to guide the capability development of unmanned platforms in the Indian Navy from 2021 to 2030. In July 2022, an unclassified version of this publication was released for the benefit of industry partners so that their development of unmanned technologies and solutions will be synchronized with the Navy's requirements.

Along with state-owned entities, private companies are also equally involved in the development of indigenously built UUVs. At DefExpo 2020, the flagship event of the Ministry of Defense, Larsen and Tourbo (L&T) displayed its newly developed underwater drones, Adamya, Amogh, and Maya. Adamya in particular can be launched from both surface ships and submarines, and can last for 8 hours at a depth of 500 meters. Further, the company also signed an MoU with NewSpace Research and Technologies (NRT) to design and build UUVs.

While New Delhi has said that it prefers domestic procurements, this might not be possible in the short term. As it requires UUV platforms on an urgent basis, imports should be relied on until indigenous capabilities are well established. Private companies' contributions and research and development projects will become crucial in enhancing these critical technologies, and therefore, the government and the navy should push for an inclusive approach that will bring together all the important stakeholders.

Very soon, UUVs will become the norm in naval operations and manned platforms will be more mission-specific. Even though the drone industry in India is still in its infancy, one can definitely argue that a naval dimension has been added to it. As the strategic competition in the IOR intensifies and India attempts to keep a check on Beijing's presence in the region, it must push beyond surveillance and intelligence collection activities, and maintain a strong UUV capacity with adequate offensive capabilities.

CHAPTER 55

DIRECTED ENERGY WEAPONS (DEWS)

55.1 INTRODUCTION TO DIRECTED ENERGY WEAPONS (DEWS)

A directed-energy weapon (DEW) is a ranged weapon that damages its target with highly focused energy without a solid projectile, including lasers, microwaves, particle beams, and sound beams. Potential applications of this technology include weapons that target personnel, missiles, vehicles, and optical devices.

In the United States, the Pentagon, DARPA, the Air Force Research Laboratory, United States Army Armament Research Development and Engineering Center, and the Naval Research Laboratory are researching directed-energy weapons to counter ballistic missiles, hypersonic cruise missiles, and hypersonic glide vehicles. These systems of missile defense are expected to come online no sooner than the mid to late-2020s.

China, France, Germany, the United Kingdom, Russia, **India,** Israel, and Pakistan are also developing military-grade directed-energy weapons, while Iran and Turkey claim to have them in active service. The first use of directed-energy weapons in combat between military forces was claimed to have occurred in Libya in August 2019 by Turkey, which claimed to use the ALKA directed-energy weapon. After decades of research and development, most directed-energy weapons are still at the experimental stage and it remains to be seen if or when they will be deployed as practical, high-performance military weapons.

55.2 TYPES OF DIRECTED ENERGY WEAPONS (DEWS)

55.2.1 Microwaves

The microwave frequency is commonly defined as being between 300 MHz and 300 GHz (wavelengths of 1 meter to 1 millimeter), which is within the radiofrequency (RF) range. Some examples of weapons which have been publicized by the military are as follows:

(i) Active Denial System

Active Denial System is a **millimeter wave source** that heats the water in a human target's skin and thus causes incapacitating pain. It was developed by the U.S. Air Force Research Laboratory and Raytheon for riot-control duty. Though intended to cause severe pain while leaving no lasting damage, concern has been voiced as to whether the system could cause irreversible damage to the eyes. There has yet to be testing for long-term side effects of exposure to the microwave beam. It can also destroy unshielded electronics.

(ii) Vigilant Eagle

Vigilant Eagle is a ground-based airport defense system that directs high-frequency microwaves towards any projectile that is fired at an aircraft. It was announced by Raytheon in 2005, and the effectiveness of its waveforms was reported to have been demonstrated in field tests to be highly effective in defeating MANPADS missiles.

The system consists of a missile-detecting and tracking subsystem (MDT), a command-and-control system, and a scanning array. The MDT is a fixed grid of passive infrared (IR) cameras. The command-and-control system determines the missile launch point. The scanning array projects microwaves that disrupt the surface-to-air missile's guidance system, deflecting it from the aircraft. Vigilant Eagle was not mentioned on Raytheon's Web site in 2022.

(iii) Bofors HPM Blackout

Bofors HPM Blackout is a high-powered microwave weapon that is said to be able to destroy at short distance a wide variety of commercial off-the-shelf (COTS) electronic equipment and is purportedly non-lethal.

(iv) EL/M-2080 Green Pine | EL/M-2080 Green P

The effective radiated power (ERP) of the EL/M-2080 Green Pine radar makes it a hypothetical candidate for conversion into a directed-energy weapon, by focusing pulses of radar energy on target missiles. The energy spikes are tailored to enter missiles through antennas or sensor apertures where they can fool guidance systems, scramble computer memories or even burn out sensitive electronic components.

(v) Active Electronically Scanned Array (AESA)

AESA radars mounted on fighter aircraft have been slated as directed energy weapons against missiles, however, a senior US Air Force officer noted: "they aren't particularly suited to create weapons effects on missiles because of limited antenna size, power and field of view". Potentially lethal effects are produced only inside 100 meters range, and disruptive effects at distances on the order of one kilometer. Moreover, cheap countermeasures can be applied to existing missiles.

(vi) Anti-drone rifle

A weapon often described as an "anti-drone rifle" or "anti-drone gun" is a battery-powered electromagnetic pulse weapon held to an operator's shoulder, pointed at a flying target in a way similar to a rifle, and operated. While not a rifle or gun, it is so nicknamed as it is handled in the same way as a personal rifle. The device emits separate electromagnetic pulses to suppress navigation and transmission channels used to operate an aerial drone, terminating the drone's contact with its operator; the out-of-control drone then crashes. The Russian Stupor is reported to have a range of two kilometers, covering a 20-degree sector; it also suppresses the drone's cameras. Stupor is reported to have been used by Russian forces during the Russian military intervention in the Syrian civil war.

Both Russia and Ukraine are reported to use these devices during the 2022 Russian invasion of Ukraine. The Ukrainian army are reported to use the Ukrainian KVS G-6, with a 3.5 km range and able to operate continuously for 30 minutes. The manufacturer states that the weapon can disrupt remote control, the transmission of video at 2.4 and 5 GHz, and GPS and Glonass satellite navigation signals. Ukraine has also used the EDM4S anti-drone rifle to shoot down Russian Eleron-3 drones.

Due to the threat posed by drones in regard to terrorism, several police forces have carried anti-drone guns as part of their equipment. For example, during the policing of the Commonwealth Games in 2018, the Australian Queensland Police Service carried anti-drone guns with an effective range of 3 km (2 mi). In Myanmar, police have been equipped with anti-drone guns "ostensibly to defend VIPs".

Counter-electronics High Power Microwave Advanced Missile Project

The Counter-electronics High Power Microwave Advanced Missile Project (CHAMP) is a joint concept technology demonstration led by the Air Force Research Laboratory, Directed Energy Directorate at Kirtland Air Force Base to develop an air-launched directed-energy weapon capable of incapacitating or damaging electronic systems by means of an EMP (electromagnetic pulse).

THOR/Mjolnir

The Tactical High-power Operational Responder (THOR) is a high-power microwave directed energy weapon developed by the United States Air Force Research Laboratory (AFRL).

Radio Frequency Directed Energy Weapon

This UK-developed system was unveiled in May 2024 and uses radio waves to fry the electronic components of its targets, rendering them inoperable. It is capable of engaging multiple targets, including drone swarms, and reportedly costs less than 10 pence (13 cents) per shot, making it a cheaper alternative to traditional missile-based air defense systems.

55.2.2 Laser Weapons

A laser weapon is a directed-energy weapon based on lasers.

The U.S. Navy has deployed a ship-based high-energy laser to defend against small and fast-moving ocean surface vessels as well as missiles and drones. The Navy installed a 60-kilowatt laser weapon on the destroyer the USS Preble in August 2022.

The Air Force is developing high-energy lasers on aircraft for defensive and offensive missions. In 2010, the Air Force tested a megawatt laser mounted on a modified Boeing 747, hitting a ballistic missile as it was being launched. The Air Force is currently working on a smaller weapon system for fighter aircraft.

The Airborne High Energy Laser, or AHEL is a directed energy weapon system designed to be mounted on airborne vehicles for neutralizing ground emplacements, vehicles, or other airborne targets, and brings us closer to the **"star wars"** laser systems depicted in those 80s movies mentioned above.

Russia is also developing a ground-based high-energy laser to "blind" their adversaries' satellites.

Limitations of laser weapons

One key challenge for militaries using high-energy lasers is the high levels of power needed to create useful effects from afar. Unlike an industrial laser that may be just a few inches from its target, military operations involve significantly larger distances. To defend against an incoming threat, such as a mortar shell or a small boat, laser weapons need to engage their targets before they can inflict any damage.

However, to burn through materials at safe distances requires tens to hundreds of kilowatts of power in the laser beam. The smallest prototype laser weapon draws 10 kilowatts of power, roughly equivalent to an electric car. The latest high-power laser weapon under development draws 300 kilowatts of power, enough to power 30 households. And because high-energy lasers are only 50% efficient at best, they generate a tremendous amount of waste heat that has to be managed.

This means high-energy lasers require extensive power generation and cooling infrastructure that places limits on the types of effects that can be generated from different military platforms. Army trucks and Air Force fighter jets have the least amount of space for high-energy laser weapons, and so these systems are limited to targets that require relatively low power, such as downing drones or disabling missiles.

Ships and larger aircraft can accommodate larger high-energy lasers with the potential to burn holes in boats and ground vehicles. Permanent ground-based systems have the least constraints and therefore the highest power, making it potentially feasible to dazzle a distant satellite.

Another important limitation for platform-based high-energy laser weapons relates to the infinite magazine concept. Since the truck, ship or airplane must carry the power source for the laser, and that will limit the power source's capacity, the lasers can only be used for a limited amount of time before they need to recharge their batteries.

There are also fundamental limits to high-energy laser weapons, including diminished effectiveness in rain, fog and smoke, which scatter laser beams. The laser beams also need to remain locked onto their targets for several seconds in order to inflict damage. Current prototype laser weapons are also proving a challenge to maintain in combat zones.

In the future, high-energy laser weapons are likely to continue to evolve with increased power levels that will expand the range of targets they can be used against.

Emerging threats posed by low-cost, weaponized drones like those in use in conflicts in the Middle East and Ukraine make it more likely that high-energy lasers will also find nonmilitary applications such as defending the public against terrorist attacks.

55.2.3 Particle-beam weapons

Particle-beam weapons can use charged or neutral particles, and can be either endo-atmospheric or exo-atmospheric. Particle beams as beam weapons are theoretically possible, but practical weapons have not been demonstrated yet. Certain types of particle beams have the advantage of being self-focusing in the atmosphere.

Blooming is also a problem in particle-beam weapons. Energy that would otherwise be focused on the target spreads out and the beam becomes less effective:

Thermal blooming occurs in both charged and neutral particle beams, and occurs when particles bump into one another under the effects of thermal vibration, or bump into air molecules.

Electrical blooming occurs only in charged particle beams, as ions of like charge repel one another.

55.2.4 Plasma

Plasma weapons fire a beam, bolt, or stream of plasma, which is an excited state of matter consisting of atomic electrons and nuclei, and free electrons if ionized, or other particles if pinched.

The MARAUDER (Magnetically Accelerated Ring to Achieve Ultra-high Directed-Energy and Radiation) used the **Shiva Star project** (a high energy capacitor bank which provided the means to test weapons and other devices requiring brief and extremely large amounts of energy) to accelerate a toroid of plasma at a significant percentage of the speed of light.

Additionally, the Russian Federation claims to be developing various plasma weapons.

55.2.5 Sonic Weapons

Long-Range Acoustic Device (LRAD)

The Long-Range Acoustic Device (LRAD) is an acoustic hailing device developed by Genasys (formerly LRAD Corporation) to send messages and warning tones over longer distances or at higher volume than normal loudspeakers, and as a non-lethal directed-acoustic-energy weapon. LRAD systems are used for long-range communications in a variety of applications and as a means of non-lethal, non-projectile crowd control. They are also used on ships as an anti-piracy measure.

According to the manufacturer's specifications, the systems weigh from 15 to 320 pounds (7 to 145 kg) and can emit sound in a 30°- 60° beam at 2.5 kHz. They range in size from small, portable handheld units which can be strapped to a person's chest, to larger models which require a mount. The power of the sound beam which LRADs produce is sufficient to penetrate vehicles and buildings while retaining a high degree of fidelity, so that verbal messages can be conveyed clearly in some situations.

55.2.6 Strategic Defense Initiative

In the 1980s, U.S. President Ronald Reagan proposed the Strategic Defense Initiative (SDI) program, which was nicknamed **Star Wars**. It suggested that **lasers, perhaps space-based X-ray lasers,** could destroy ICBMs. in flight. Panel discussions on the role of high-power lasers in SDI took place at various laser conferences, during the 1980s, with the participation of noted physicists including Edward Teller.

A notable example of a directed energy system which came out of the SDI program is the **Neutral Particle Beam Accelerator** developed by Los Alamos National Laboratory. This system is officially described (on the Smithsonian Air and Space Museum website) as a low power neutral particle beam (NPB) accelerator, which was among several directed energy weapons examined by the Strategic Defense Initiative Organization for potential use in missile defense. In July 1989, the accelerator was launched from White Sands Missile Range as part of the Beam Experiment Aboard Rocket (BEAR) project, reaching an altitude of 200 kilometers (124 miles) and operating successfully in space before being recovered intact after reentry. The primary objectives of the test were to assess NPB propagation characteristics in space and gauge the effects on spacecraft components. Despite continued research into NPBs, no known weapon system utilizing this technology has been deployed.

Though the strategic missile defense concept has continued to the present under the Missile Defense Agency, most of the directed-energy weapon concepts were shelved. However, Boeing has been somewhat successful with the Boeing YAL-1 and Boeing NC-135, the first of which destroyed two missiles in February 2010. Funding has been cut to both of the programs.

55.3 RESEARCH ON HIGH POWER MICROWAVES (HPM) WEAPONS

HPM weapons create beams of electromagnetic energy over a broad spectrum of radio and microwave frequencies in both narrow-band (bandwidth ~< 1%) and wide-band (bandwidth >> 1%) with the intent of coupling/interacting with electronics within targeted systems either by causing damage or temporary disruption from which the system cannot self-recover in time to accomplish its mission.

Directed energy HPM provides the U.S. Navy many benefits including speed of light attack, deep magazines and modest electrical power from the host platform. Additional benefits include broad beams for wide area coverage with comparatively simple targeting, and the ability to be used in congested (urban) environments where kinetic weapons may have constricted use.

Present focus areas are twofold. First is to explore technologies, including effects mechanisms, which explore opportunities to engage targets at higher frequency (X through K bands) and, thus, greater directivity. Second is the reduction of required power on target through the exploration of novel HPM systems that provide rapid on-demand shot-to-shot waveform parameter adjustability. Enabling areas of research might include development of small sources with sufficient control over the waveform such that they can be arrayed. They might also include high power amplifiers and higher frequency variants of existing concepts, both beam driven and solid state.

Additional research focus areas should include research into the coupling of high-power RF to, its interaction with, and its effects upon electronic systems. The goal of this is to enable the development of predictive effects tools. It should also include the use of waveform parameter adjustability with the goal of maximizing effects on electronics. Finally, developments of any sort that lead to significant (> factor of 2) improvements in size, weight, power and cost of existing systems will be considered.

55.3.1 Research Challenges and Opportunities

Exploration of the capabilities and limitations of conventional HPM-beam-driven sources to optimize output power at X band through K band including not only consideration of (potentially novel solid state and vacuum electronic-based HPM sources, but also the supporting pulsed power, antennas and other subsystems.

Solid state HPM sources that provide shot-to-shot waveform parameter adjustability (bandwidth, frequency, pulse width, envelope, etc.) from L band through K band that are able to rapidly adjust not only to rapidly changing assessment of target vulnerabilities, but also to real-time sensing of optimal coupling paths.

Computer simulation and modeling codes to enable the next generation of HPM sources and amplifiers, both beam driven and solid state. Focus of this code development will be on modeling of bulk materials and material interfaces subjected to high power densities and plasma interactions.

Theory and analysis of HPM effects that merges approaches to assess efficient coupling with an understanding of the various mechanisms by which coupled RF energy can have an effect on target systems of interest. This could include RF coupling and modeling tools to capture both complex EM wave interactions with electronics and associated enclosures, and electronic component disruption.

55.4 FOCUS ON LASER WEAPONS INTENSIFIES

The robotic craft swoops in low, closing on its target. The enemy's sensors try to get a fix, as the planet surface races past below. In an instant, a beam reaches out from below at the speed of light, the high-powered laser burning through its target. This is not Star Wars: This is Scotland, in January 2024, where the UK Ministry of Defence and its industry partners conducted the first successful firing of their DragonFire laser weapon against an aerial target.

With this trial, the culmination of £100 million of investment to date, the United Kingdom joined other nations racing to develop and deploy what are known in military parlance as directed energy weapons (DEW). Though the technology is yet to mature, the United States has begun to deploy early laser weapons on several of its naval destroyers, as well as testing ground- and air-based versions.

Following the October 7 attacks by Hamas, **Israel** has sought to expedite development of its own Iron Beam laser weapon to help shoot down incoming rockets and drones, augmenting the kinetic interceptors of its Iron Dome missile defence system. China, Russia, France, India, Turkey, Iran, South Korea, Japan, and others are investing in their own national programmes, with varying degrees of progress.

Various technology programmes have sought to make DEW a reality. In the 1980s, the Reagan administration famously sought to develop more-powerful lasers to defend the United States against Soviet missiles as part of the Strategic Defense Initiative. Ultimately, this did not work—a costly and failed effort often derisively dubbed "Star Wars". Yet after decades of low-power devices being rolled out globally, recent years have seen increasing military investment and technological advances in high-energy lasers (HEL).

Competing nations are pouring so much investment into DEWs because, if the technology can be matured, such systems hold the potential to tip both the military and economic calculus of modern warfare in their users' favour. HEL and HPRF/HPM systems deliver an effect on target at the speed of light, drawing on an energy source rather than traditional munitions. Compared with traditional gun- or missile-based alternatives, these characteristics of DEW promise increased accuracy, speed of engagement, magazine depth, and flexibility to re-task the weapon against a variety of targets.

Currently, DEWs are comparatively large, relying on large power sources and stable platforms such as a ground battery or naval ship. But in future, miniaturized and more-efficient energy storage systems could enable their rollout across all domains—with the U.S. and European next-generation fighter programmes envisaging integrating such weapons into the fighter aircraft of the future. Reducing reliance on kinetic munitions that must be constantly replenished would similarly take pressure off military logistics and industrial production, enabling forward deployed forces to operate for longer without resupply of ammunition, so long as they had access to suitable energy sources.

Munitions production capacity is tightly constrained despite ongoing global efforts to ramp it up. Low-cost drones and rockets have swung the economic calculus of offence and defence in favour of those using large volumes of cheap unmanned systems and munitions to overwhelm more-sophisticated air and missile defenses. Maturing DEW technologies therefore promise more cost-efficient ways of engaging a variety of threats, especially these rockets and drones.

Directed energy weapons need to be further developed to become more mobile, reliable, and affordable.

55.5 DEWS: ULTRA-SHORT PULSE LASER AND ATMOSPHERIC CHARACTERIZATION

Ultra-Short Pulse Laser (USPL) and Atmospheric Characterization initiatives explore the scientific limitations of DEW. Of particular research interest are efforts that offer breakthroughs in precision dynamic engagements against multiple maneuvering targets with selectable effects (hard kill, sensing, non-lethal), and those offering deep magazines, low cost per shot and "lowest" to zero collateral damage. This challenge includes research into fundamental understanding of laser sources, adaptive optics compensation techniques, long range atmospheric propagation physics, and the characterization of laser-matter interactions. This program will develop scientific understanding, components and subsystems to enable a USPL-based DEW best suited for naval applications.

55.5.1 Research Challenges and Opportunities in DEWs

(i) **Atmospheric and Adaptive Optics:** Controlling atmospheric interactions (like filamentation) while maximizing intensities requires the characterization and control to minimize the losses of laser intensity at distance. Accomplishing this will require adaptive optics with advanced control systems capable of handling non-linear laser propagation against fast moving targets in a maritime, littoral or high turbulence environment. Also of interest are systems capable of compensation through the high turbulence conditions of low maritime altitudes. Interest also includes beacon vs non-beacon solutions for near real-time adjustment of adaptive optics in real-time atmospheric conditions, and the break-up of laser beams due to atmospheric lensing. This includes high repetition rate of high power USPLs against their propagation effects in the atmosphere in order to validate propagation models with controlled experiments at over one kilometer (>1 km).

(ii) **Materials Interaction and Response:** Understand the theory of laser pulse material interaction and the effects due to concentrated strong electric fields through experimentation and observations to understand, validate and document USPL physics. Examine whether or not greater pulse energy at longer duration causes more effect than less pulse energy at shorter pulse duration, including femtosecond pulses.

(iii) **Laser Source Development:** Research laser sources with high repetition rates to address weapon capabilities and capacity within platform size, weight, energy and power constraints. Additional interest includes USPL ruggedization, reductions in size, weight and power or cooling, USPL pulse compression and expansion, and novel compact long wavelength infrared USPL sources.

55.6 "KALI 5000" INDIA'S OWN DIRECTED ENERGY WEAPON

KALI (Kilo Ampere Linear Injector) is a linear electron accelerator being developed by DRDO (Defence Research and Development Organisation) and BARC (Bhabha Atomic Research Centre). KALI was developed by keeping an industrial mindset but later when the developers understood the great potential of KALI as a weapon, they researched further into the subject and now they are working towards fulfilling it. KALI 5000 is India's warfare weapon under development which can stop missiles, air crafts, enemy satellites anything with an electric circuit inside it. It is designed to work in such a way that if a missile is launched in India's direction, it will quickly emit powerful pulses of Relativistic Electrons Beams (REB).

The KALI series (KALI 80, KALI 200, KALI 1000, KALI 5000 and KALI 10000) of accelerators are described as "Single Shot Pulsed Gigawatt Electron Accelerators". They are single shot devices, using water-filled capacitors to build the charge energy. The discharge is in the range of 1GW. Initially starting with 0.4GW power, present accelerators are able to reach 40GW. Pulse time is about 60 ns.

The KALI-5000 is a pulsed accelerator of 1 MeV electron energy, 50-100 ns pulse time, 40kA Current and 40 GW Power level. The system is quite bulky as well, with the KALI-5000 weighing 10 tons, and the KALI-10000, weighing 26 tons. They are also very power hungry, and require a cooling tank of 12,000 liters of oil. Recharging time is also too long to make it a viable weapon in its present form.

The KALI has been put to various uses by the DRDO. The DRDO was involved in configuring the KALI for their use. It can be utilized as an Anti-satellite weapon by making it perfect with some medium range rocket and it very well might be an exact response to the Chinese hostile to satellite rocket.

The X-rays emitted are being used in Ballistics research as an illuminator for ultrahigh speed photography by the Defence Ballistics Research Institute (DBRL) in Chandigarh. The Microwave emissions are used for EM Research.

The microwave-producing version of Kali has also been used by the DRDO scientists for testing the vulnerability of the electronic systems of the Light Combat Aircraft (LCA), which was then under development.

It has also helped in designing electrostatic shields to "harden" the LCA and missiles from microwave attack by the enemy as well as protecting satellites against deadly Electromagnetic Impulses (EMI) generated by nuclear weapons and other cosmic disturbances, which "fry" and destroy electronic circuits. Electronic components currently used in missiles can withstand fields of approx. 300 V/cm, while the fields in case of EMI attack reach thousands of V/cm.

Its ability to emit powerful pulses of electrons and the conversion of electron energy into E.M. Radiation fueled the hopes that the KALI could be used in a High-Power Microwave gun.

However, application of KALI in defence field will take some time. The system is still under development, and efforts are being made to make it more compact, as well as improve its recharge time, which, at the present, makes it only a single use system. There are also issues with creating a complete system, which would require development of many more components.

India is believed to have already conducted a successful test. There had been unconfirmed reports blaming India for the Siachen glacier avalanche in 2012 which has caused the death of around 135 Pakistani soldiers which the sources claim as a result of KALI's successful test melting the hard ice sheets.

There have been reports of placing the weaponized KALI in an IL-76 aircraft as an airborne defense system.

55.7 DURGA-2: INDIA'S DIRECT ENERGY WEAPON TO BE TESTED SOON

The laboratory of DRDO primarily working on the DEW is CHESS (Centre for High Energy Systems and Science) located in Hyderabad.& Laser Science and Technology Centre (LASTEC) located in Del

There are many projects in progress related to DEWs under DRDO's sleeves. Some of the projects are Kilo Ampere Linear Injector (KALI), Project Aditya, Durga-II and Air Defence Dazzlers.

KALI has a very interesting story behind it and possible the first military use of DEW in the world. BEL has already been manufacturing the **Laser Dazzlers** and supplying to Indian navy. The laser dazzler is used as a 'non-lethal method' for stopping suspicious vehicles, aircraft, and unmanned aerial vehicles (UAVs) from approaching secured areas. In February 2022, The Chinese PLA Navy has again been accused of using laser beams against the P-8 Poseidon Maritime Reconnaissance Aircraft of Australia.

(i) Project Aditya: 100KW Laser

Aditya is a 100KW laser developed by DRDO. Unlike modern day laser, which is based solid state, Adita is a Gas Dynamic Laser & is not easily portable. It has a lot of complex plumbing too. Aditya uses 0.7m aperture telescope to cause damages at 0.8 km and 2.5 km distance. The beam delivery system has to simultaneously acquire and track the distant static and moving target in real time and point and focus the laser beam on the target. Aditya was an experimental test bed of DRDO to seed the critical technologies for future laser weapon programs of DRDO.

(ii) Project Durga (Directionally Unrestricted Ray-Gun Array)

In the financial year 2021-2022, DRDO has requested $100 million budget to produce a high-power laser weapon. The classified project is dubbed as **DURGA II** (Directionally Unrestricted Ray-Gun Array), aimed to develop a 100-kilowatt, lightweight directed-energy system. Once developed, it will be integrated with various land-, sea- and air-based platforms of military. While no official time frame has been definitively outlined, anonymous sources within the Indian Defence Ministry have claimed that the Indian Army will soon receive the DURGA II.

The DURGA project may have started way back in early 2000 but gained momentum only in 2017 when DRDO tested a 1kW truck-mounted laser weapon over a range of 250m in Chitradurga. LASTEC has reportedly developed a 25 KW laser weapon that can target a ballistic missile during its terminal phase at a max distance of 5Km. The laser experts are now working to enhance this range to 100 km or beyond.

While no official time frame has been definitively outlined, the financial express reported the full prototype of DURGA will be one of the most crucial projects to watch out for in the near future.

(iii) Directed Energy Laser Systems (DELS)

DRDO's anti-drone system is one example of application of DEW which employs a laser-based hard kill measure to destroy the drones & UAVs from a range of 1-2.5 km. The system deployed here is probably using the 2kW or 10kW laser developed by DRDO.

DRDO mobile Directed Energy Laser System (DELS) Laser beam combination technology has been used for generating desired power for shooting down drones and is being modified for shooting down rockets/mortars similar in capability to Israel's Iron Beam. It features an Electro-optical sensor to track the targets and dual laser beam is used to neutralize the target.

DRDO's Directed Energy Laser Systems (DELS) is 10kW proof-of-concept directed energy system against UAVs like target and establishment of critical technologies of precision tracking/pointing and laser beam combination. During the period (January, 2018 – March, 2019), tests for detection and tracking of mini drone (hexacopter) with Battle Field Surveillance Radar (BFSR) and cuing to 10 kW DELS were carried out at ATR, Chitradurga. Testing of radar interfacing with 10 kW DES was carried out with 10 kW DES radar interface simulator and Air Defence Fire Control Radar (ADFCR) '**Atulya**' at BEL, Bengaluru. Track acquisition, tracking and laser pointing of 10 kW Anti-UAV Laser Weapon (AULW) system were carried out on flying DRONEs at CHESS site up to the range of 0.7 km.

(iv) High Power Microwave (HPM) System

DRDO has also taken up a project to develop a High-Power Microwave (HPM) system of Radio Frequency whose power lies in the S-band of electromagnetic spectrum. The HPM will be able to neutralize and fry the electronics of the drones from a distance of 5 km. During June, 2018, trial the indigenous Marx generator was tested for up to 300 KV, 20 Hz and excellent performance was obtained. These generators are used to generate the High-power Microwaves. The military application for the compact HPM system is extensive given the potential threat posed by missiles. These systems can also be integrated with the land based mobile vehicles and larger transport aircrafts.

SECTION-VI

NON-TECHNOLOGICAL WARFARE

CHAPTER 56

CROSS-BORDER TERRORISM AND PROXY WAR

56.1 PROXY WAR IN JAMMU & KASHMIR, INDIA

Jammu & Kashmir, a region in the northern part of the Indian subcontinent, is globally known for the battle of ownership which has been going on for seven decades. This place of mesmerizing beauty, often referred to as heaven on earth, used to be one single entity before the partition of British India into Pakistan and India in 1947. Today, Jammu & Kashmir is separated into three politically different regions with one administered Pakistan, India and Bangladesh. Initially the struggle to win over the whole region started as a political one, mainly between India and Pakistan, however, the struggle has evolved dramatically over the course of time. Today the 'war' over Jammu & Kashmir is a multifaceted one which consists of political as well as social aspects, with the issue of religion being a major one. The State of Pakistan has used the religious sentiments of the people in the region, most of whom are Muslims, to instigate them against Indian rule.

56.1.1 The Partition and its effects on Jammu & Kashmir

Jammu & Kashmir was one of the princely states during the British rule of the Indian subcontinent. It was a unique region because of its religious composition, its geography and its ruler. It had geographical proximity to both India and newly born Pakistan and while being a Muslim majority region during that time, it had a Hindu ruler, Maharaja Hari Singh. Three distinct areas of the State had a noteworthy composition of population. The Ladakh region had a Buddhist majority population, the Jammu region had a Hindu majority and the Kashmir Valley, a Muslim majority. Due to these distinct features, unlike present-day Bangladesh, which was awarded to Pakistan during partition because of being a Muslim majority region, the ruler of Jammu & Kashmir was given the opportunity to choose the fate of Jammu & Kashmir. The ruler of Jammu & Kashmir could choose to unify with either Pakistan or India. In theory, the ruler could also choose to establish Jammu & Kashmir as an Independent State. However, the people of many of the princely states preferred to unify with an already established democratic state than remaining subjects of a monarchy which could mean the continuation of authoritarian rule. In addition, Sheikh Abdullah, a popular mass leader, was building a movement against the rule of the Maharaja. Considering the complexity of the situation, the Maharaja of Jammu & Kashmir remained neutral and delayed his decision. The Maharaja had difficulty making a decision because of the heterogeneous composition of the population. The British representatives visited the Maharaja attempting to advise him on the issue but failed to persuade him to make a decision. Before the neutrality of the Maharaja could be converted into a peaceful transition to the next step for Jammu & Kashmir, Pakistan invaded Kashmir and Jammu provinces from the north. The invaders comprised of tribesmen from Pakistan's North West Frontier Province (NWFP) and regulars from its army. The invaders were organized in company-level units and armed with lethal weapons. Houses were burnt, property looted and destroyed and large-scale rapes and abductions of women took place. The Maharaja and his government were unable to defeat the coalition of the Pakistani Army. Despite their failure, the Maharaja did not give into Pakistan's aggression; instead the Maharaja decided to accede to India and signed the Instrument of Accession. The Instrument of Accession gave India the power to take control of Defence, Foreign Affairs and Communication and the complete authority of the State was to be decided later. Regardless of the fact that it was the Maharaja's exclusive right and decision, subsequent to Pakistan's invasion, to accede to India and perhaps not a democratic decision, it translated into the fact that India had the legitimate authority to take control of Jammu & Kashmir.

56.1.2 The War of India and Pakistan in 1947-1949

The first war between India and Pakistan over Jammu & Kashmir broke out in 1947. The Maharaja of Jammu & Kashmir took refuge in India leaving the fate of the State to the war. Indian troops had a difficult time fighting against the Pakistani Army because of their lack of experience and expertise to combat in the mountains of Jammu & Kashmir. While Prime Minister Nehru had the legal authority to exercise the legality of the Instrument of Accession vis-à-vis the whole State of Jammu & Kashmir (including the parts which are currently under administration of Pakistan), it is unclear why he did not and chose to take the issue to the international theatre (United Nations Security Council). Some sources claim that Prime Minister Nehru hoped that the international community would recognize Pakistan's aggression and intervene to stop further bloodshed. The United Nations (UN) passed a Resolution on 13 August 1948, through which Pakistan was requested to withdraw its army from Jammu & Kashmir. The plan was to arrange a free and fair plebiscite after the withdrawal of the Pakistani Army to give the Kashmiri people the chance to choose their fate. Another condition for the plebiscite to take place was to restore the situation in a pre-1947 State. Following the UN Resolution, an emergency government was established on 30 October 1948 in Jammu & Kashmir in which the popular mass leader Sheikh Abdullah, who initiated a movement against the Maharaja's rule, became the Prime Minister. India and was hopeful of an outcome in their favour but Pakistan had no intention of withdrawal and firmly held onto its struggle to capture power in Jammu & Kashmir. Amidst this stubbornness, the UN finally managed an agreement between India and Pakistan for a ceasefire on 1 January 1949, which awarded India the control of a significant (65%) part of Jammu & Kashmir whereas the rest of the area remained in control of Pakistan. The ceasefire line established by the UN which was agreed upon as a temporary solution became the de-facto border of India and Pakistan in the Jammu & Kashmir region. The free and fair plebiscite which was supposed to take place never happened because Pakistan did not withdraw its troops.

56.1.3 The 1962-Sino-Indian war

The region of Jammu & Kashmir in control of China is called **Aksai Chin**. Aksai Chin is situated in the easternmost part of Indian Administered Jammu & Kashmir. The border issue in relation to this portion was never resolved and it remains disputed till date. In the post partition period, India claimed ownership of Aksai Chin as part of its ownership of Jammu & Kashmir while China claims ownership of this region due to historical reasons. China refused to give up its control and consequently China and

India fought a one-month long war of bloodshed, known as the Sino-Indian war of 1962. The Chinese retained control of Aksai Chin and in addition China also gained control of 5,180 sq km of Pakistan Administered Jammu & Kashmir through a border agreement between these two countries in 1963. The occupation of Aksai Chin was crucial for the Chinese to establish routes for transportation among Xinjiang and Tibet. With Aksai Chin amounting to almost 20% of Jammu & Kashmir, the share of the State under India's administration dropped to 45% and Pakistan's share dropped to 35%. The disputed border between Chinese Administered- and Indian Administered Jammu & Kashmir is known as the Line of Actual Control.

56.1.4 The 1965-war and the Tashkent agreement

A political resolution which left Pakistan with a mere portion of Jammu & Kashmir did not satisfy Pakistan's national interests. It was not long after the ceasefire of 1949, that Pakistan attempted to increase its stake of Jammu & Kashmir. In 1965, Pakistan attempted to win over Jammu & Kashmir through a secret mission, called 'Operation Gibraltar', which entailed a sudden attack in Indian Administered Jammu & Kashmir with a battalion of 30,000 armed soldiers. Experts on the Kashmir-issue believe that the brief struggle over Rann of Kuch earlier in 1965, as well as communal violence between Muslims and Hindus over a sacred relic of Muslims, encouraged Pakistan to plot this attack. Earlier in 1965, Pakistan was awarded 10% of Rann of Kuch which was originally in India's possession following a fight among border guards of Sindh of Pakistan and Kuch of India and a mediation by two British High Commissioners. The Pakistani leaders envisioned that if similar situations could be created in Kashmir then they could win more areas of Indian Administered Jammu & Kashmir through mediation. In addition, before the war of 1965, Indian Administered Jammu & Kashmir experienced outrage of local Muslims due to the theft of a holy relic from a local mosque which escalated into communal tension between Muslims and Hindus. Pakistan was hopeful of using this discontent of Muslims in Indian Administered Jammu & Kashmir in their attack against India. Pakistan misjudged the Muslims' anger towards the Indian Government as well as their 'allegiance' to Pakistan as none of the ideas that inspired Pakistan into this sudden attack helped Pakistan in the end. While Pakistan had political and military support from the US, India was supported by Russia. When China stepped into the game taking a hostile stand against India and supporting Pakistan, the British Prime Minister of that period, Harold Wilson, promised his support to India on behalf of both the United Kingdom and the US. India made significant progress against Pakistani aggression in the 1965 war but the war ended in another ceasefire due to diplomatic pressure from the international community. Following the UN mandated ceasefire, the Tashkent agreement was signed on 1 January 1966, and both countries were left with the territory they already administered pre the 1965 war, which in essence meant that the 1965 war had no impact on the territorial control of India and Pakistan. The countries did not gain anything rather than losing the lives of thousands of their soldiers. The 1965-war was a major strike against hopes of a peaceful resolution of the Jammu & Kashmir dispute. A second attack by Pakistan in such a short period of time reiterated the fact that Pakistan would continue to attack India to take control of Jammu & Kashmir. India and Pakistan fought another brief war in 1971, as India got involved in the Bangladesh Liberation War, following which the Simla agreement was signed, in which the ceasefire line, being monitored by the United Nations Military Observer Group in Pakistan and India (UNMOGIP), was renamed as the Line of Control (LoC) and the UN was requested to withdraw the UNMOGIP from the LoC.

56.1.5 Militancy in Indian Administered Jammu & Kashmir

Amidst the despair of conflict between India and Pakistan, the leadership of Sheikh Abdullah was able to bring peace in Indian Administered Jammu & Kashmir in the 1980s. His leadership also allowed Jammu & Kashmir to enjoy more autonomy and less supervision from the Indian Central Government. Even though there was pessimism among some of the people of Jammu & Kashmir regarding some of his political decisions and affiliations, the overall situation was comparatively peaceful. However, the peace was to last only until his death in September 1982, and with this charismatic and popular leader gone, the discontent of people began to rise to the surface again.

The succession of Sheikh Abdullah's leadership to his son Farooq Abdullah could not hold onto the peace he had contributed to. The election of 1987 in Jammu & Kashmir is a landmark year in the conflict history of Kashmir. A coalition of several Islamic parties fought against the coalition of Farooq Abdullah's party and the Congress party of India. The election was allegedly rigged to declare the latter coalition victorious which led to the disappointment and distrust of the Muslims in the state. In addition to this, Amanullah Khan, Chairman of the Jammu Kashmir Liberation Front (JKLF) in exile in Pakistan started an armed movement with the alleged support of Pakistan's premier intelligence agency, the Inter-Services Intelligence (ISI), against the authority in Indian Administered Jammu & Kashmir. The alleged rigging of the election, the operations of Amanullah Khan and some political decisions of Farooq Abdullah led to the beginning of the uprising of militancy in Indian Administered Jammu & Kashmir.

Acts of violence started from the year 1987, but 1989 marks the rise of the militants in Indian Administered Jammu & Kashmir as the impact of militancy in that year was massive. Communal violence among different religious groups such as Muslims, Hindus and Buddhists began to erupt throughout the State. The Kashmiri Pandits, an elite group among Hindus, became particular targets of the Muslim militants. In addition, the State was buried under strikes after strikes called by militant groups. According to estimates, the number of strikes were so many that they accounted for one third of that year's working days. Government officials such as police, intelligence officers, members and leaders of the National Conference, which was the leading political party in Jammu & Kashmir, were killed to breakdown the political system of the State. In addition to strikes and killings of government officials incidents such as kidnapping, bomb blasts, rapes and destruction of government properties became regular. The militants succeeded to intimidate local people as well as the local authority through their tactics. Following that year until today, numerous militant groups emerged in particularly the Valley of Jammu and Kashmir, and they also kept subscribing to various different ideologies. Some militant groups started off with a secular nature, but many harbored religious motivation. The Jammu & Kashmir Liberation Front propagated independence from India while some of the Islamic political parties which were part of the coalition of the Muslims United Front during the 1987 election also joined the independence movement, but at the same time also formed their own militant wings. On the other hand, militant groups with religious inclination such as the Hizbul-Mujahideen (HM) (militant wing of Jamaat-e-Islami), Hezbollah and Lashkar-e-Taiba (Let) had a pro-Pakistani ideology. The religious movement had a devastating impact on the peace building efforts in Jammu & Kashmir which forced around 200,000 Kashmiri Pandits to flee their homeland.

The armed struggle was responded with a heavy hand by the Indian authority, as it met the definitions of **cross-border terrorism** supported by Pakistan. The Indian Government decided to heavily militarize its administered part of Jammu & Kashmir and a federal paramilitary unit called the Central Reserve Police Force was sent to join the local police force to battle against the terrorists. Later in July 1990, a special act known as the Armed Forces Special Power Act (AFSPA) was passed by the Indian Government to deal with the increasing violent situation. Various sources, especially those which are against the militarization of Jammu & Kashmir suggest that 500,000 security forces are deployed in Indian Administered Jammu & Kashmir to eliminate the terrorists. However, the Indian authorities have not acknowledged the presence of this number of security forces. Nonetheless, Indian Administered Jammu & Kashmir remains one of the most densely militarized places in the world.

The role of AFSPA has been very crucial in the conflict history of Kashmir. The Act is said to provide immunity to Indian soldiers appointed in Jammu & Kashmir even when there are violations of human rights. The representatives of India have repeatedly denied the accusation that there has been frequent human right violations in their response to terrorism. The Indian authorities claim that any security force personnel guilty of human rights violations has been punished. Independent sources concur this claim.

56.1.6 The Role of Pakistan

Pakistan has always viewed the rule of India in Indian Administered Jammu & Kashmir as foreign occupation and it has continuously attempted to annex Jammu & Kashmir which it considers its rightful ownership. It should not come as a surprise that Pakistan played an active role in the religiously motivated militancy and in its continuation till date.

Active military efforts and negotiations failed several times for Pakistan to win over Jammu & Kashmir which made a covert armed militancy a strategy through which Pakistan could fight against India avoiding war and negotiations. In essence, this became **a proxy war against India** through which Pakistan attempted to impose a heavy political and economic burden on India. A rise in militancy would mean that India would have to invest resources in Jammu & Kashmir and in addition, it would also jeopardize the political authority of India in Jammu & Kashmir. The fact that the overwhelming majority of Indian Administered Jammu & Kashmir is Muslim, enabled Pakistan to religiously exploit the people of Jammu & Kashmir, especially in the Valley of Kashmir. Since the beginning of militancy, Pakistan publicly promised its moral and diplomatic support to the militants in Indian Administered Jammu & Kashmir. However, this promise was not limited to moral and diplomatic support only. Pakistan provided all kinds of support, including military support to the militants to bring down Indian authority. In the beginning their support was provided to all kinds of groups as long as the militants were fighting against Indian authority in Jammu & Kashmir. It gradually shifted its support towards pro-Pakistani militant groups which were willing to fight for the cause of annexation of Jammu & Kashmir with Pakistan. Pro-Pakistani militant groups were provided training, ammunition, shelter and other necessary support by the Pakistani military establishment in their war against Indian authority. Several camps of the militant groups, under the cover of refugee camps, were established in so-called Azad Kashmir, a part of Pakistan Administered Jammu & Kashmir. Pakistan acquired, allegedly through illegal means, modern weapons and ammunitions from the US while it joined the US forces in their fight against terrorism in Afghanistan. It also gained access to a large number of weapons following the withdrawal of Soviet forces in Afghanistan in 1989. Pakistan had an abundance of ammunition and weapons as well as experience to assist the militants to implement a guerrilla war in Jammu & Kashmir. On top of this, Pakistan received funds from sale of narcotics in Afghanistan, one of the largest opium producer of the world, and from donations of other Muslim organizations which financed the proxy war.

The presence of the religiously motivated militant groups changed the dynamics of the Kashmir-issue completely making it a religious struggle. The influence of Pakistan is to such an extent that Pakistanis as well as Afghans who had no personal connection with Kashmir joined Kashmiris in their fight against Indian rule solely because of their religious motivation. To explain, Pakistan labels this armed struggle against India as *'Jihad'* or as a religious mission of Muslim brotherhood, which provides the militants the motivation of religious rewards and therefore makes it difficult to resolve the issue politically. Due to this inclusion of Pakistanis and Afghans, the nature and composition of the militant groups changed significantly. To make things worse, some of the militant groups with members from Pakistan and Afghanistan joined the Jihadi organization of Osama Bin Laden, called International Islamic Front, which resulted in the fact that the insurgency in the Kashmir Valley soon became a terrorist movement. The tactics also changed because of affiliation with international terrorist organizations. The militants started attempting suicide attacks which were not common at the beginning of the struggle. The primary targets of the militants are usually stations of Indian security forces, especially the defence system at the LoC. Their targets also included bridges which disrupted communication networks among the security forces and public properties such as schools and temples. The militants established their bases in rural areas and in locations where it was difficult for the law enforcements agencies to track them or fight against them. Rural areas remain the most affected in terms of casualties caused by militants, especially casualties of minority communities such as the Kashmiri Pandits. This did not only assist the militants to create pressure on the Indian security forces but also resulted in creating a sense of communal hatred among local people.

While most of the militants joined the pro-Pakistan militant groups willingly because of religious motivation, Indian authorities claim that some people have been forced at the threat of their lives and with offers of financial benefits to join the militant groups in Pakistan Administered Jammu & Kashmir. Initially, the idea of Jihad gained momentum among the local Muslim Kashmiris, however time passed by and casualties of militants increased without any winning on their part, the local Kashmiris kept losing interest to join the militant groups. In this scenario, to continue the militancy in Kashmir, the Pakistani military establishment adopted backup strategies; It started recruiting militants from other Muslims countries using the banner of a religious struggle. Enthusiasts from other Muslim countries such as Saudi Arabia, Iraq, Egypt, Libya and Sudan joined hands with Pakistani and pro-Pakistani militants in their proxy war against India. This backup strategy of sending foreign militants did not achieve the desired outcomes as many of these militant were killed by the Indian security forces. Nonetheless, foreign militants continue to join the militants in Kashmir because of their faith in the religious cause.

As the Pakistani intelligence services kept changing their strategies regarding the continuation of militancy in Jammu & Kashmir, Indian authorities also kept improving their tactics continuously besides deploying a large number of troops, both national and local security forces. While the militants established bases and communication routes in intractable places such as hill tracts and deep forests, Indian authorities resorted to devices of latest technology to track down the militants. India also made successful attempts to incorporate local people in its efforts to track and battle the militants through the establishment of '**Village Defence Committees'** which are provided basic training, equipment of communications and rudimentary weapons.

The plan of **a proxy war was executed to avoid direct confrontation**, yet the proxy war resulted into a direct conflict in 1999, around ten years after militancy had started, following the infiltration of Pakistani troops **in Kargil**, in Indian Administered Jammu & Kashmir. Pakistan denied the fact that the infiltrators were Pakistani troops, and tried to blame local militants. This denial did not stop India from launching air strikes against Pakistan. India's progress in this conflict and pressure from the US forced the Pakistani Government to withdraw its troops. The war lasted for three months and forced thousands of people on both sides of the LoC to flee their homes. Except for the Kargil war in 1999, Pakistan has been careful to contain the conflict to an extent that it would not end up in a direct confrontation but would still affect India significantly.

Although Pakistan supports militancy in Indian Administered Jammu & Kashmir, it brutally crushes down any attempt of an independence movement which threatens the Pakistani authority in Pakistan Administered Jammu & Kashmir. In Pakistan's vision, the only definition of independence for Kashmiris is the accession of Jammu & Kashmir to Pakistan. Pakistan's attempts to weaken India's authority using religious emotions have only made India more vigilant in taking protective measures against militancy and made the issue more complex. India has suffered the loss of tremendous economic resources, especially on the budget for defence, as well as lives of security forces personnel and civilians due to the sponsored militancy. The militancy has crippled the prospective of social and economic development of Jammu & Kashmir and the tourism industry, which used to be the primary industry of revenue for Jammu & Kashmir, has lost its appeal due to decades long conflict and has collapsed miserably. The conflict has also affected other industries and income generating activities of local people. It is doubtful how long India can sustain such an exhausting process of investing in Jammu & Kashmir at the same level, and neutralize growing radicalism and terrorism in the State.

In addition to changing the nature of this conflict, the inability of the Governments of India and Pakistan to reach a consensus on the negotiating table makes the issue more difficult. One of the primary motives of Pakistan through fueling religiously motivated militancy was to attract international attention showing alleged human rights violations by India in Jammu & Kashmir which would force India to go through an international mediation which Pakistan believes will bring out an outcome in Pakistan's favor. This strategy has failed, as Pakistan has not been able to gather any support at the international level in favor of its debatable policies regarding Jammu & Kashmir.

56.1.7 The Role of India

International human rights organizations, such as Amnesty International have published reports on human right violations that have taken place in India's efforts to deal with the ongoing militancy.

Although these mishaps on the part of India degrade the civil-military relationship in India, the Indian Government has taken steps to stop and reduce these violations of human rights by imposing monitoring systems and education on human rights within the army, and to check and balance cases of human right violations. Various cases in which army personnel have been tried and convicted because of alleged human rights violations have also been reported by independent sources. In addition to this, the Indian Government has also taken steps to rehabilitate surrendered militants through changing their identity and providing them an opportunity to reintegrate into the society.

56.2 DEALING WITH PAKISTAN BY INDIA

India has since 2016 maintained a posture of 'no talks with terror' accompanied by one of 'zero tolerance' for terrorism. India's official policy remains that it seeks 'normal neighborly relations' with Pakistan in an atmosphere free of violence, terrorism and hostility. Hence, India had no major structured dialogue with Pakistan since 2016. This posture was reinforced after the Pulwama attack of 2019. Pakistan, on its part, effectively took on a posture in 2019 of 'no talks with Article 370', tying up its India policy in knots. Then PM Imran Khan's subsequent rhetoric and personal attacks on India's leadership reduced the space for diplomacy for his own government, his successors and even for the permanent establishment: the army.

India's policy has been largely effective, arguably with an assist from Pakistan's internal crisis, as also from global conditions. India has faced no spectacular terrorist attack since Pulwama 2019, cross-border infiltration has dropped significantly, and an LOC ceasefire has been held since February 2021. A strong counter-terror policy has also implied a short shelf life for Kashmir-focused terrorists within Kashmir and elsewhere.

56.2.1 The Terrorist Veto

With elections having been completed in both countries, the bilateral dynamics have changed. New governments should normally have led to fresh thinking and an impulse to stabilize borders. Pakistan's civilian leaders did make some overtures for stabilizing ties, but once again, the army may not be on board.

Even as India's Prime Minister was taking the oath of office on 9 June 2024, terrorists struck in the Jammu region killing nine civilians. Three more attacks followed in the next three days, establishing a clear pattern of **cross-border terrorist attacks** on a region perceived to be vulnerable. The gambit may be deliberate and based on several different tactical objectives.

Pakistan is perhaps signaling to Jammu & Kashmir that the recent spectacular success of the electoral process in terms of voter turnouts was not acceptable and must not be repeated. It may have stayed its hand prior to the elections because Pakistan's army believed that the Pulwama terrorism of 2019 helped the incumbent government in India win the elections that year and a pre-election attack may have had the same impact in 2024. Moreover, the effective counter-terrorism grid in the Kashmir Valley did not allow the terrorists to launch attacks there, and the scale of the attacks was deliberately below the perceived threshold of India's response. The upcoming Amarnath Yatra pilgrimage, record numbers (over 21 million in 2023) of tourists in the valley and the scheduled J&K

elections by September- all strong markers of normalcy-give reason for the sponsors of terrorism to attempt attacks wherever they could succeed. Pakistan may also be pandering to its primary global benefactor, China, to suggest that it was imposing costs on India for having reduced the density of troops in the Jammu sector of the LOC to bolster the LAC against China. Islamabad would also be keen to tell the restive parts of POK that all is not well with their Indian counterparts.

Perhaps, the most important signal from Pakistan's army was sent to the junior partner in the hybrid government, PML-N President Nawaz Sharif, whose message to India's PM was to replace 'hate with hope'. Nawaz Sharif would feel enormous déjà vu. Two of his past attempts at peace-making with India – in 1999 and in 2015 – were met with disapproval from the army and active encouragement to terrorism to stymie these moves. More worryingly for him, they also led to his premature ouster from office.

India is reading these signals clearly. A hardline approach to terrorism would remain India's primary posture. A verifiable halt to cross-border violence would remain the essential precondition for movement on other issues like trade. Cross-border terrorism will remain the deal breaker for any attempt at stabilizing the relationship.

As India goes about grappling with the central strategic challenge of China, it would need to deal with an unpredictable Pakistan, on a recalibrated basis and with strategic patience. It would need to avoid the pitfall of strategic negligence, of the kind Israel showed with Gaza in October 2023. Diplomatic and intelligence channels would need to remain open, for India to keep engaging both the army and civilians across the border, focusing on low hanging fruit but first seeking security guarantees for the future. Any stabilization process will be challenged by several threats: of escalated terrorism, the increasing China – Pakistan military collusion or the stresses to the Indus Waters treaty mechanism. With a focus on maintaining the security gains of the past decade, India would need to navigate its western front with a combination of strategic patience, calibrated engagement and proactive defence.

CHAPTER 57

PSYCHOLOGICAL WARFARE

57.1 INTRODUCTION TO PSYCHOLOGICAL WARFARE

Psychological warfare is the planned tactical use of propaganda, threats, and other non-combat techniques during wars, threats of war, or periods of geopolitical unrest to mislead, intimidate, demoralize, or otherwise influence the thinking or behavior of an enemy.

While all nations employ it, the U.S. Central Intelligence Agency (CIA) lists the tactical goals of psychological warfare (PSYWAR) or psychological operations (PSYOP) as:

- Assisting in overcoming an enemy's will to fight;
- Sustaining the morale and winning the alliance of friendly groups in countries occupied by the enemy;
- Influencing the morale and attitudes of people in friendly and neutral countries toward the United States.

To achieve their objectives, the planners of psychological warfare attempt to gain total knowledge of the beliefs, likes, dislikes, strengths, weaknesses, and vulnerabilities of the target population. According to the CIA, knowing what motivates the target is the key to a successful PSYOP.

57.1.1 A War of the Mind

As a non-lethal effort to capture "hearts and minds," psychological warfare typically employs propaganda to influence the values, beliefs, emotions, reasoning, motives, or behavior of its targets. The targets of such propaganda campaigns can include governments, political organizations, advocacy groups, military personnel, and civilian individuals.

Simply a form of cleverly "weaponized" information, PSYOP propaganda may be disseminated in any or all of several ways:

- Face-to-face verbal communication;
- Audiovisual media, like television and movies;
- Audio-only media including shortwave radio broadcasts;
- Purely visual media, like leaflets, newspapers, books, magazines, or posters;

- More important than how these weapons of propaganda are delivered is the message they carry and how well they influence or persuade the target audience.

57.1.2 Three Shades of Propaganda

In his 1949 book, Psychological Warfare Against Nazi Germany, former OSS (now the CIA) operative Daniel Lerner details the U.S. military's WWII Skyewar campaign. Lerner separates psychological warfare propaganda into three categories:

(i) **White propaganda:** The information is truthful and only moderately biased. The source of the information is cited.

(ii) **Grey propaganda:** The information is mostly truthful and contains no information that can be disproven. However, no sources are cited.

(iii) **Black propaganda:** Literally "fake news," the information is false or deceitful and is attributed to sources not responsible for its creation.

While grey and black propaganda campaigns often have the most immediate impact, they also carry the greatest risk. Sooner or later, the target population identifies the information as being false, thus discrediting the source. As Lerner wrote, "Credibility is a condition of persuasion. Before you can make a man do as you say, you must make him believe what you say."

57.1.3 PSYOP in Battle

On the actual battlefield, psychological warfare is used to obtain confessions, information, surrender, or defection by breaking the morale of enemy fighters.

Some typical **tactics** of battlefield PSYOP include:

- Distribution of pamphlets or flyers encouraging the enemy to surrender and giving instructions on how to surrender safely;
- The visual "shock and awe" of a massive attack employing vast numbers of troops or technologically advanced weapons;
- Sleep deprivation through the continual projection of loud, annoying music or sounds toward enemy troops;
- The threat, whether real or imaginary, of the use of chemical or biological weapons.

57.1.4 Early Psychological Warfare

History and folklore are rife with instances where tactics of Psychological Warfare were used as catalysts to attain a certain objective and even turn the tide of the battle. If anything, the days of the yore have taught us that morale and confidence are stronger than any other weapons in the arsenal. Even the holiest book of Hindus, The ***Shrimad Bhagavadgita*** was but a product of a psywar tactic used by Shri Krishna to remind his protégé Arjun of his karma i.e. his duties in this mortal world when the latter got cold feet at the prospect of battling and possibly killing his own kith and kin.

Even in the Kurukshetra battle, the Pandavas decided to bluff the killing of *Ashwatthama,* the son of the *Kaurava* army commander Guru Dronacharya who was bent on routing the Pandava forces by deafening announcements which led to the disheartened commander dropping his weapons and ultimately getting killed.

Heads of Tribes and even monarchs of vast empires propagated their "representative of God" image to ensure that public order prevailed and their autocratic rule went unquestioned.

While it might sound like a modern invention, psychological warfare is as old as war itself. When soldiers the mighty Roman Legions rhythmically beat their swords against their shields they were employing a tactic of shock and awe designed to induce terror in their opponents.

Chanakya's Arthshastra which is a master treatise for statecraft professes to wage psychological warfare against the enemy, it also encourages active surveillance and emphasizes on the secret services being directly under the king[2]. In this masterpiece, Chanakya suggests that the 4 route principle of Sama, Dana, Bheda and Danda be followed while dealing with an adversary wherein Sama involves conciliation by way of treaties, mutual commonalities, highlighting the symbiosis of the partnership and recognitions, Dana essentially involves bribery in form of one or multiple incentives, Bheda consists of exploiting any brewing dissent in the enemy ranks and breaking the cohesive structure through ways of a well-oiled espionage and misinformation system; this tactic has opted when the nemesis is stronger and Danda which is the use of military forces against the enemy.

As centuries rolled and entire dynasties went from masters of the subcontinent to rulers confined in their own walled fortress tactics of Psychological Warfare kept on evolving with almost every battle from the terrorizing tales of Genghis Khan's sea of barbaric marauding hordes which led to most enemies surrendering without a fight to army camp rumors floated by Chhatrapati Shivaji Maharaj's ambassador in front of the enemy creating a puny image of the Maratha king which ultimately led to the latter's defeat courtesy of lowering of guards which were the result of a systematic creation of a meek image of his master by the diplomat himself in official negotiations.

It is not a hidden fact that the British managed to subjugate the entire Indian populace despite the latter having superior numbers with carefully planned and executed Psychological Warfare tactics. The British accomplished this by winning kingdom after kingdom by putting Psychological Warfare mechanisms and place tactics like deception, trade restrictions and *'divide and rule'* were the bread and butter of the British colonialists. Indians failed to understand the import of 'Knowledge is Power' and that fact was exploited by the British when they systematically decimated the existing education system thus crippling the Indians and enslaving their minds long before they enslaved our bodies. For over a century during the British Raj, there was no systematic educational framework in the country until the Intellectual Charter of India was ushered and, as was typical of the Imperialists, the charter dictated exclusion of technical education to Indians to handicap the ability of Indians and make them just competent enough to serve as subordinates in government offices, in the interest of the British. This new Psy War tactic was put in place by Lord T.B Macaulay and is known for famously quoting "We must do our best to form a class who may be interpreters between us and the millions whom we govern, a class of persons, Indian in blood and colour, but English in taste, opinions, words and Intellect"

In the 525 B.C. Battle of Peluseium, Persian forces held cats as hostages in order to gain a psychological advantage over the Egyptians, who due to their religious beliefs, refused to harm cats.

To make the number of his troops seem larger than they actually were, 13th century A.D. leader of the Mongolian Empire Genghis Khan ordered each soldier to carry three lit torches at night. The Mighty Khan also designed arrows notched to whistle as they flew through the air, terrifying his enemies. And in perhaps the most extreme shock and awe tactic, Mongol armies would catapult severed human heads over the walls of enemy villages to frighten the residents.

During the American Revolution, British troops wore brightly colored uniforms in an attempt to intimidate the more plainly dressed troops of George Washington's Continental Army. This, however, proved to be a fatal mistake as the bright red uniforms made easy targets for Washington's even more demoralizing American snipers.

57.1.5 Modern Psychological Warfare

Modern psychological warfare tactics were first used during **World War I**. Technological advances in electronic and print media made it easier for governments to distribute propaganda through mass-circulation newspapers. On the battlefield, advances in aviation made it possible to drop leaflets behind enemy lines and special non-lethal artillery rounds were designed to deliver propaganda. Postcards dropped over German trenches by British pilots bore notes supposedly handwritten by German prisoners extolling their humane treatment by their British captors.

During World War II, both Axis and Allied powers regularly used PSYOPS. Adolf Hitler's rise to power in Germany was driven largely by propaganda designed to discredit his political opponents. His furious speeches mustered national pride while convincing the people to blame others for Germany's self-inflicted economic problems.

Use of radio broadcast PSYOP reached a peak in World War II. Japan's famous "Tokyo Rose" broadcast music with false information of Japanese military victories to discourage allied forces. Germany employed similar tactics through the radio broadcasts of "Axis Sally."

However, in perhaps the most impactful PSYOP in **WWII**, American commanders orchestrating the "leaking" of false orders leading the German high command to believe the allied D-Day invasion would be launched on the beaches of Calais, rather than Normandy, France.

The **Cold War** was all but ended when U.S. President Ronald Reagan publicly released detailed plans for a highly sophisticated **"Star Wars"** Strategic Defense Initiative (SDI) anti-ballistic missile system capable of destroying Soviet nuclear missiles before they re-entered the atmosphere. Whether any of Reagan's "Star Wars" systems could have really been built or not, Soviet president Mikhail Gorbachev believed they could. Faced with the realization that the costs of countering U.S. advances in nuclear weapons systems could bankrupt his government, Gorbachev agreed to reopen détente-era negotiations resulting in lasting nuclear arms control treaties.

More recently, the United States responded to the September 11, 2001 terror attacks by launching the Iraq War with a massive "shock and awe" campaign intended to break the Iraqi army's will to fight and to protect the country's dictatorial leader Saddam Hussein. The U.S. invasion began on March 19,

2003, with two days of non-stop bombing of Iraq's capital city of Baghdad. On April 5, U.S. and allied Coalition forces, facing only token opposition from Iraqi troops, took control of Baghdad. On April 14, less than a month after the shock and awe invasion began, the U.S. declared victory in the Iraq War.

In today's ongoing War on Terror, the Jihadist terrorist organization ISIS uses social media websites and other online sources to conduct psychological campaigns designed to recruit followers and fighters from around the world.

57.2 PSYCHOLOGICAL WARFARE STRATEGIES USED THROUGHOUT HISTORY

(i) War Elephants

As the tallest terrestrial animal on the planet, few creatures are as intimidating as an elephant. Add body armor and some blades or spikes to tusks, and you have a fearsome battle beast! Used in Africa and India, both African and Asian elephants were incorporated into armies. **In India, war elephants were so common that they were entire corps of militaries.** While not completely invincible, an elephant could easily sweep individual soldiers aside by swinging its massive head. Famously, horses were often intimidated by larger elephants and could refuse to charge into battle facing them.

However, the psychological warfare inflicted on opponents by war elephants could be countered. Famously, flaming pigs were used to terrify war elephants, which could turn around and trample their own soldiers in an attempt to escape. If an elephant panicked, it could cause almost as much damage to its own troops as the enemy! Thus, using war elephants was a high-risk strategy. Although the Romans successfully overcame war elephants–at tremendous cost–when fighting the Egyptians and Carthaginians, they came to adopt some for themselves, although mainly for entertainment and spectacle.

(ii) Mongol Deal-Making

Over a thousand years after war elephants terrified the Romans, the cavalry forces of the Mongol Empire terrified cities from the Pacific Ocean to present-day Ukraine. The Mongols used effective psychological warfare against targeted cities by offering a deal: surrender and pay tribute to the Mongol Empire or face total destruction. The offer of a relatively generous deal made the prospect of brutal combat against highly trained and disciplined Mongol forces that much more painful. As a result, many cities and fortresses chose to become vassal states of the Mongol Empire–taxation rather than combat. To help ensure that most decided to surrender, the Mongols made sure to leave some survivors of each massacre who could spread the word of how brutal the Mongols could be.

The Mongols expanded much more rapidly by being able to pacify potential foes with their deal-making. Similarly, cocaine kingpin Pablo Escobar in Colombia in the 1980s used the deal of platao-plomo (silver or lead) to convince many law enforcement officials to turn a blind eye to his illegal activities. By offering the proverbial carrot or stick, Escobar benefited when most potential opponents chose the carrot. Like the Mongols, this allowed Escobar to expand rapidly and avoid armed conflicts that could have taken him down sooner. But just like how the Mongol Empire eventually fell apart, so did Escobar's drug empire. The drug lord was shot dead on December 2, 1993, while fleeing police in Medellin, Colombia.

(iii) Vlad the Impaler's Showmanship

If challenged, the Mongols had the military might to conquer virtually any foe and could be plenty brutal about it. However, Vlad the Impaler takes the term "brutality" to another level entirely. In 1456, at about age 25, Vlad III of Walachia defeated his primary rival for leadership, Vladislav II, in hand-to-hand combat and became the leader of the Transylvania region of present-day Romania. He ruthlessly executed just about anyone he did not like, ranging from petty criminals to potential political rivals and their families. Famously, he used impaling as his favored punishment, possibly learned during his childhood with the Ottoman Turks.

Having victims impaled on vertical poles allowed Vlad to visually intimidate potential foes. Despite his brutality, Vlad was tolerated–and even celebrated–throughout Europe for militarily defeating and otherwise terrifying the Muslims of the Ottoman Empire. Famously, invading Ottoman sultan Mehmed II turned back when he encountered thousands of impaled bodies outside the city of Targoviste. Vlad's sickening showmanship prevented a bloody showdown...at least that time. Later, Vlad the Impaler's violent ways ended with his own violent death in 1476, when he was ambushed and beheaded on his way into battle.

(iv) Boer War & Siege of Mafeking

During the late 1890s, Britain wanted to unite all of South Africa into one colony. Inland, the Boer republics, of primarily Dutch heritage, resisted the influx of British people. In the autumn of 1899, war erupted between the British Empire and the much smaller, but highly skilled and determined Boer republics. The Boers laid siege to several British towns and garrisons and upset the military reinforcements who came to relieve them, which shocked the world. Using modern weapons and guerrilla tactics, the Boers were able to outmaneuver British forces that were used to fighting poorly armed natives.

One of the forts that the Boers besieged was Mafeking, where the roles were reversed. Here, a handful of British soldiers used clever deceptions to trick the surrounding Boers into thinking the garrison was more heavily defended. The Brits, including the future founder of the Scouting movement, Robert Baden-Powell, pretended to establish minefields and barbed-wire fences, which convinced the Boers not to attack. After 217 days, British reinforcements arrived and broke the siege in May 1900.

(v) World War I Propaganda & Leaflets

While most people are familiar with the use of propaganda to bolster support for a war on the home front, World War I saw the use of anti-war propaganda intended to sap enemy morale and convince them to surrender. Already suffering from the industrialized war that saw the wide-scale introduction of machine guns, modern artillery, poison gas warfare, and even the first armored tanks, German troops were bombarded with leaflets announcing their efforts were futile. Some German troops surrendered and asked for the rations promised in the leaflets, perhaps hastening the war's end.

Future wars saw the use of propaganda leaflets on both sides. They could be dropped by planes or released from artillery shells. World War II saw both Axis and Allied powers try to convince soldiers from the other side to surrender and that they were being used as pawns for the elites. In addition to leaflets, both Germany and Japan used English-speaking radio broadcasts as propaganda. Both Allied and Axis broadcasts (and leaflets) tried to reduce enemy morale by claiming that the war was going according to plan for their side.

(vi) Nazi Rallies Fake Military Might

After World War I, Germany was forced to disarm. In the early 1930s, Nazi Party leader Adolf Hitler became the chancellor of Germany and embarked on a policy of rearming the nation. Part of the Nazis' aesthetic was massive rallies that featured displays of strength and vigor, intended to both inspire Germans and intimidate potential foes. Famously, the rallies featured over a hundred powerful searchlights aimed at the sky. This cathedral of light used most of Germany's searchlights, but their use at political rallies tricked nations like France and Britain into believing that Germany must have had many more not in use.

The use of Nazi rallies and aggressive propaganda likely led to Britain and France not trying to check Germany's rearmament. Germany re-occupied the Rhineland and took control of Czechoslovakia in 1938. In retrospect, Germany was not militarily prepared to fight France and Britain in 1938, and the appeasement shown to Germany at the Munich Conference only led Europe further down the path to World War II. However, the Nazis' skillful propaganda during the 1930s convinced many that it was ready and willing to fight and win.

(vii) Ghost Armies vs. Nazi Saboteurs

While World War I saw largely static trench warfare for much of the conflict, especially on the Western Front in France, World War II was much more maneuverable and complex. After the D-Day invasion of France in June 1944, the US used ghost armies of lightweight, artificial equipment, including inflatable tanks, to fool the enemy. In addition to physical decoys, ghost army units also used fake radio chatter and sounds of military action on loudspeakers to convince the Germans that forces were elsewhere than they actually were...or much larger than they actually were. Believing they were facing large units up to 35-40 times their actual size, the Germans chose to disengage rather than fight, potentially saving tens of thousands of Allied troops.

However, the Germans had their own tools of psychological warfare. In late 1944, as Germany planned a final major offensive to re-take lost territory in France and Belgium, it enlisted commando leader Otto Skorzeny to run an ambitious sabotage operation. Skorzeny, famous for rescuing imprisoned Italian dictator Benito Mussolini in September 1943, was a feared opponent. Operation Greif was intended to sow fear and confusion in the American lines during the Ardennes Offensive by implanting German agents in American uniforms who were fluent in English. These agents could then destroy equipment, plant false information, and basically run amok. Discovery that this was occurring did lead to temporary panic among US forces, but fortunately had little effect on the military situation.

(viii) Nazi Wunderwaffe

After its unsuccessful Ardennes Offensive, known in the United States as the Battle of the Bulge, there was only one possible way for Germany to end the war in anything other than total defeat:

Wunderwaffe. These "wonder weapons" were high-tech aeronautical wonders that included the Me-262 fighter jet, Me-263 rocket fighter, V-1 jet-powered buzz bomb, and V-2 long-range rocket. From September 1944 until late in the war, the V-2 rocket inflicted painful gouges on the London landscape. The V-2 was terrifying because it was supersonic and unstoppable; it could not be heard coming and could not be intercepted.

Although the V-2 only killed some 2,700 people in Britain, it was feared that the V-2 could potentially be launched from ships in the Atlantic at American cities. While Germany undoubtedly hoped that fear of its Wunderwaffe would bring the Allies to the negotiating table, it likely only increased their resolve to push for unconditional surrender. Ultimately, the capture of Me-262 fighter jets and V-2 rockets at the end of World War II in Europe greatly advanced aeronautic technology in the United States.

(ix) Espionage Rings & Red Scare

While espionage has long been a part of war, few were more active in modern espionage than the Soviet Union. At the end of World War II, it was revealed that spies had helped the Soviet Union gain secrets from the Manhattan Project. Active Soviet spy rings helped the USSR develop its own atomic bomb by 1949, erasing the American "trump card" that it had held. It was revealed in late 1945 that Soviet spying was not limited to atomic secrets but general classified information as well. In 1952, it was discovered that a wooden carving of the Great Seal of the United States given to the US ambassador by the USSR contained a listening device.

The early Cold War saw a sweeping hysteria about communist infiltration of American society. This Second Red Scare of the late 1940s and early 1950s saw politicians investigate alleged communist links of fellow politicians, government employees, and media figures. The communist victory in the Chinese Civil War, resulting in the rise of Red China, only amplified tensions. When communist North Korea invaded South Korea a year later, resulting in the US leading a military response in the Korean War, fear of communism grew further. Fortunately, none of the espionage–real and suspected–led to war between the United States and the Soviet Union.

(x) Spooky Recordings vs. Booby Traps

After the Korean War, America next took up arms against communists in Vietnam a decade later. This time, the conflict mainly saw guerrilla warfare in a jungle environment rather than conventional warfare. Both the US military and the North Vietnamese military (and their Viet Cong guerrilla allies) looked for psychological warfare advantages to sap the morale of their enemies. The US used spooky tape recordings in Operation Wandering Soul to play on the superstitions of North Vietnamese and Viet Cong soldiers, hoping to get them to desert their positions. Success was mixed, with enemy soldiers sometimes discovering the ruse and firing on the speakers or the recordings, spooking South Vietnamese soldiers and civilians as well as the intended targets.

For their part, the North Vietnamese and Viet Cong also played on Americans' anxieties. They used deadly booby traps to sap soldiers' morale. Knowing that you could be maimed or killed anywhere, even with no sign of an enemy present, made many soldiers question the war effort. Between 1966 and 1971, US military morale plummeted as the Vietnam War dragged on, and little seemed to be accomplished. Some soldiers in Vietnam turned to illegal drug use to cope with continual anxiety and harsh conditions.

(xi) Loud Music Breaks Enemies' Will

While much of psychological warfare is intended to frighten or deceive an enemy, some is just intended to tire them out. In December 1989, the US invaded Panama to oust drug-dealing dictator Manuel Noriega, whose police forces had just brutalized and threatened Americans in the country. With 13,000 American soldiers already in Panama thanks to the American-controlled Panama Canal, the additional 13,000 troops brought in during the swift invasion had little trouble defeating Noriega's forces. But Noriega himself managed to flee to the Vatican City embassy in Panama City.

Storming a foreign embassy is a sociopolitical no-go, so the US had little choice but to wait and see. To break Noriega's will, the military blasted hard rock at full volume at the embassy. Sure enough, Noriega eventually surrendered. The US continued using loudspeaker psyops during the Gulf War (1990-91), using Humvee-mounted speakers to convince Iraqi soldiers to surrender. A notable success was a combined loudspeaker and leaflet effort convincing 1,400 Iraqi soldiers to surrender to a much smaller force of US Marines.

(xii) Shock and Awe Air Wars

In early 1991, Operation Desert Storm commenced with a massive US-led bombing campaign against Iraq. The mass coordination of computer- and satellite-guided smart weapons decimated Iraqi military targets. This swift campaign featuring the most modern military technology became known as "shock and awe," with enemy forces having little hope of defending themselves with obsolete, Soviet-era weapons. The speed, precision, and impact of expensive American air weapons convinced many Iraqi forces to surrender quickly when ground forces rolled in.

The US repeated its shock and awe air attacks in Afghanistan after September 11 and in Iraq again in early 2003. In both cases, the enemy surrendered quickly: the Taliban regime in Afghanistan put up little organized resistance, and Iraqi dictator Saddam Hussein saw his forces surrender en masse after the US invasion in March 2003. Shock and awe undoubtedly helped demoralize these enemy forces, both of which were touted beforehand as hardened fighters. Few things can frighten an enemy like showing you can strike fast and hard without being hit back in return.

57.3 PSYCHOLOGICAL WARFARE OF PAKISTAN WITH RESPECT TO J&K, INDIA

Pakistan in recent years has been an ardent user of Psychological warfare against India. ISI- the Inter-Services Intelligence has been on the forefront and has played an important role in executing several Pakistani proxy wars over the years. The role of the agency though was later expanded to suit and guard the interests of the politicians and favoring and fostering military rule. It has been keenly interested in the Kashmir region and has been consistently monitoring the developments of unrest and violence as well as instigating them unceasingly. The capabilities of the organization were mainly put to test when the Soviet Union invaded Afghanistan, which led to it becoming more prominent with the support from the CIA and Saudi intelligence.

For decades, the Jammu and Kashmir region has been a play board for Pakistan's psychological warfare and a major security issue for India for decades. As Jinnah in one of his speeches said 'Jammu and Kashmir is the jugular vein for Pakistan'. Pakistan has used various means to implement its psychological warfare.

57.3.1 Psychological Warfare by Pakistan on Social Media

Pakistan is taking advantage of the social media to further its propaganda against India and the twitter platform is being abused by Pakistan, considering the misinformation it has been spreading on twitter and if it continues, the trail of erroneous and inaccurate accusations it might end up losing its credibility along with its integrity. The twitter account operated by DG ISPR was set up in December 2016 and despite having been created in 2016, this account has 845 tweets as of 23 July 2020, and a following to follower ratio of 0 to 3.9 million, which shows the tremendous impact it has generated globally in just a span of four years.

The spurt in this platform is very visible and various agendas and hashtags to create misinformation have been doing the rounds on the social site some of which include #GobackModi- which have been in use since 2019, #ChaosInIndia-can be traced to April 20, 2020 and #Islamophobia_In_India- 23 March 2019. These hashtags have been doing the rounds since 2019 and have trended on twitter. Pakistan's ISPR is famous for spreading fake news and creating tensions, this was specially seen with respect to the abrogation of article 370 and was seen round the clock on DG ISPRs. twitter handle.

Thousands of youth of Pakistan have been trained and used by ISPR, an agency for spreading fake news and propaganda against India. They have also made them believe that the work they are doing is no less than that of a soldier and they are fighting a narrative war with India. To further boost the morale of the youth and get the work done, the ISPR awards them when an individual is successful in creating a buzz on the social media and the individual with the most retweets on social sites like Twitter and Facebook gets the award which includes jobs and contracts with Fauji Foundation. Several twitter handles also came in the limelight when the names of the account were changed to those of Arab royalty like @pak_fauj changed its name to @SayyidaMona who is an Omani Royalty and posted anti-India tweets about the ties between India and the Gulf.

Twitter is being used as a tool for building narrative wherein Pakistan is showcasing itself as a peaceful country that wishes to maintain peace in Kashmir and the region prominently whereas it is itself one of the major causes of instability. All the while, playing the victim card and accusing the Indian Army of targeting the civilians in Pakistan occupied Kashmir. The former Director-General Asif Ghafoor has been called out time and again for sharing misleading information on social media, but a change in narrative has been observed from Major General Asif Ghafoor (2016-2020) to the present DG ISPR Major General Babar Iftikhar. The former Director-General had been time and again lauded for expressing his viewpoint on pertinent and topical issues as well as his assertive stance on India, his tweets were accusing and direct. The present Director-General, on the other hand, has a different approach and is seen shedding light on how the civilians are being attacked and killed by the Indian troops.

According to a report released by The Freedom House in 2019, an international internet rights group, Pakistan is one of the most restricted countries in the world for internet laxity and was placed 26 out of 100 in the world and in the 3rd position regionally. The analysis of the report can be that majority of the public do not have access to internet thus making twitter inaccessible. Does this mean that the twitter account operated by the DG ISPR is focused more on creating an anti-India sentiment throughout the world than to constitute an impact internally?

The tweets by Pakistan against India can be seen from the following examples:

(i) There has been an overflow of tweets related to the unprovoked ceasefires by the Indian troops and the violation of the ceasefire agreement. Such tweets have become a common occurrence on their twitter, appearing almost every second day whereas tweets related to the Jammu and Kashmir region, see an increase during the time nearing their independence day. Moreover, Pakistan has always monotonously asserted that it is India that violates the ceasefire and targets civilians in their land. Such tweets have increased since the abrogation of Article 370. The Chief of the Indian Army has denied these allegations and stood their ground.

(ii) Blood-curdling screams and unnerving noises are being heard late at night in South Kashmir. Residents maintain that these distressful sounds are coordinated actions by Indian-occupying forces to terrorize and subdue residents as a part of a long legacy of psychological warfare in the region.

57.4 PSYCHOLOGICAL WARFARE BY INDIA (PAKISTAN'S NARRATIVE)

The Indian military has been conducting psychological warfare since the early days of the Kashmir liberation movement. This includes alarming and haunting noises emanating from villages late at night and rumored ghost and jinn sightings. Kashmiris, who grew up in the 1990s, when the armed revolt against Indian forces was at its peak, report Indian soldiers breaking into their homes late at night to damage their property or create a disturbance. Such tactics constitute a deliberate strategy to make Kashmiris feel unsafe in their homes and communities. A climate of fear dampens morale and discourages residents from leaving their homes, including for political assembly. Such a widespread inducement of fear also weakens people's links with Kashmiri land and territory, disrupting vital linkages with the landscape, which remains a significant obstacle to Indian occupation.

The discomforting noises being heard in Kashmir are said to exploit local folklore on "*raantas*". These are mythical creatures, witches with disheveled hair capable of producing terrifying reverberations. "Raantas" are sometimes lovingly invoked by elders to discourage children from roaming in the cold winter nights to protect them from falling sick. The ongoing Indian occupation is even impacting cultural stories, and Kashmiri's relationships with their childhood tales are taking on new militarized meanings.

Similar tactics were used during the Vietnam war. The United States (US) military broadcasted terrifying sounds to dissuade the National Liberation Front soldiers. The collage of sounds was designed to tap into Vietnamese beliefs that those not buried at home will roam in a state of distress in the afterlife. The uncomfortable mix of voices, music, and sounds was intended to encourage Vietnamese soldiers to abandon their posts and return home.

The continual use of PSYOP in Kashmir has widespread consequences. Fear, intimidation, and unpredictability have become the defining features of everyday life. It is estimated that more than 60% of the civilian population in Indian-occupied Kashmir suffers from PTSD, depression, or anxiety.

In 2017, there were widespread accounts of attacks on Kashmiri women throughout the valley. Several women were knocked unconscious by unknown assailants to find their braids cut upon regaining consciousness. This created another wave of fear, women were terrified of leaving home, and many girls even stopped attending school. Local vigilante groups were formed to protect one's neighbourhood from so-called braid choppers leading to further paranoia and animosity. Besides, social fragmentation is already a notable feature of everyday life in the valley. Many are accused of being informants for the Indian military or armed groups, and a general atmosphere of mistrust and suspicion permeates.

Other examples of psychological warfare in Kashmir include military humanitarianism or "heart warfare" to win the hearts and minds of selective communities. Also prominent is the public display of brutal force, particularly against Kashmiri men, midnight raids to arouse fear and anxiety, enforced disappearances, humiliating treatment at frequent checkpoints, and marking and maiming Kashmiri bodies through the use of pellet guns.

Existing structures of violence set in place by the Indian occupying forces work alongside targeted psychological warfare interventions, and in fact, each is imbricated in the other. While proponents of psychological warfare argue that PSYOP (psychological operations) can reduce the loss of life, they are always almost used alongside more overt forms of violence, not in lieu of them.

The continual use of PSYOP in Kashmir has widespread consequences. Fear, intimidation, and unpredictability have become the defining features of everyday life. It is estimated that more than 60% of the civilian population in Indian-occupied Kashmir suffers from PTSD, depression, or anxiety. According to a 2015-report by Médecins Sans Frontières India, over 11% of adults in the Kashmir Valley are taking benzodiazepines. Addiction to drugs such as heroin is also on the rise. While it is speculated that young Kashmiris, particularly men, turn to drugs to cope with the day-to-day uncertainty of living in a conflict zone, the Indian army has also been alleged to supply narcotics to Kashmiri youth. The effects of the "longue durée" of occupation and violence on the public psyche require a nuanced rendering.

Psychological warfare remains a controversial military strategy and has the potential to escape detection largely. As little or no physical devastation is left behind, psychological warfare does not fit into familiar frames of war and is difficult to prosecute under international law. The US military openly hires PSYOP soldiers, and most militaries in the world, including that of India, utilize such tactics. Due to a lack of transparent journalism and ever-increasing curbs on Kashmiri journalists, accounts of psychological warfare in the region remain notably absent from mainstream news reportage. Attention to these dimensions of the occupation is essential to understand the true extent of the violence at play and its lasting legacies.

Earlier this year, Pakistan and India reaffirmed the 2003 ceasefire across the Line of Control (LoC), the de-facto border that divides Kashmir amongst the two countries. The ceasefire was welcomed, given the loss of life on both sides of the LoC. While analysts have lauded the move, Kashmiri demands for de-militarization and sovereignty remain unmet. It remains to be seen whether positive bilateral relations between Pakistan and India would lead to tangible improvements in people's everyday lives in the former princely state.

57.5 INDIAN POLICIES AND RESPONSE TO PSYCHOLOGICAL WARFARE

The Indian Armed Forces have been equally effective in Psychological operations, instances in parts of Kashmir in the aftermath of flushing out terrorists where the commander of the Army explains to the villagers the importance of peace and discredits the terrorists portraying them to be the cause of their inconveniences have also surfaced. In Assam, where the Naxalite problem is menacing, the Army has resorted to the distribution of posters highlighting the dark side of militancy and has successfully managed to win the support of the society.

India used the first of its Psy-War tactics during the Bangladesh Liberation War where Field Marshal Sam Manekshaw warned about the retributive acts of the 'Mukti Bahini' to the defending Pakistani Army through his radio broadcasts.

lobbies may use the press to conduct strategic psychological warfare through key communicators, activists, non-governmental organizations, political parties, advocacy groups and media Indian media's unique nature makes it ideal for the same. All this is good news for an exponent Psychological warfare to use media in strategic, proxy warfare and low-intensity conflict scenarios.

Previously, not so vocal on responding to the Psychological Warfare machinery, the ADGPI has geared up for countering the narrative by the adversaries by aiding big curtain blockbusters and OTT platform web series which display the valour of the Armed Forces in pushing back the enemy, effectively highlighting the Uri surgical strike.

On the field presently, Psychological Warfare is conducted by commanders only at the battalion level and there is no concrete framework or a dedicated Psychological Warfare of the Indian Armed Forces.

Psychological war's weaponry is sensitive and double-edged, has ample scope for misuse or being counter-productive and thus needs to be used with caution. Since a conventional war remains a distinct possibility, it is these tactics which give the edge over adversaries, and hence an all-society inclusive approach should be used to effectively thwart attempts by the adversary in these futuristic form of warfare.

CHAPTER 58

ECONOMIC WARFARE

58.1 INTRODUCTION TO ECONOMIC WARFARE

Economic warfare is an economic strategy used by unfriendly states with the goal of weakening the economy of other states. This is primarily achieved by the use of economic blockades. Ravaging the crops of the enemy is a classic method, used for thousands of years.

In military operations, economic warfare may reflect economic policy followed as a part of open or covert operations, cyber operations, information operations during or preceding a war. Economic warfare aims to capture or otherwise to control the supply of critical economic resources so that the friendly military and intelligence agencies can use them and enemy forces cannot.

The concept of economic warfare is most applicable to total war, which involves not only the armed forces of enemy countries, but also mobilized war-economies. In such a situation, damage to an enemy's economy is damage to that enemy's ability to fight a war.

Economic warfare is the use of or threat to use strategies like trade embargoes, tariffs, boycotts, the freezing of capital assets and other means against another country to weaken its economy. These measures are typically coercive, aiming to impose costs on a nation in an attempt to force leaders to change their behavior.

Policies and measures in economic warfare may include blockade, blacklisting, preclusive purchasing, rewards and the capturing or the control of enemy assets or supply lines. Other policies, such as, tariff discrimination, sanctions, the suspension of aid, the freezing of capital assets, the prohibition of investment and other capital flows, expropriation, and debasing the target's currency by counterfeiting. even without armed military war, may constitute economic warfare.

The U.S. government largely regards economic warfare as wholly distinct from military tools and tactics even though in multiple historic cases, the military has played a significant role in economic warfare well beyond enforcing sanctions. Adversaries currently see economic warfare as an attractive option, affording means to gain asymmetrical advantage over the United States.

In the recent time, the economic warfare used by USA and some other allied countries **against Russia** in response to its invasion of Ukraine is unprecedented. These steps include:

- Ban on Russian oil and other energy imports.
- Blocking of Russia's largest public and private banks.
- Ban on luxury goods from being exported from the U.S. to Russia.

- Sanctions against dozens of defense companies.
- Sanctions against 328 members of the Duma legislative body and the chief executive of Sberbank.
- Suspension of information exchanges with Russia's tax authorities and the U.S. Internal Revenue Service.

The effect of these sanctions on U.S. consumers has also been particularly acute. President Joe Biden announced that the U.S. would ban the import of Russian energy products, and the average price of gas in the U.S. subsequently shot up to an all-time high of $4.331 per gallon, within 3 days after this announcement.

No longer fearful and sensing Western nations' vulnerability to economic shocks and perturbations, adversaries appreciate the potential asymmetric advantage created through offensive economic warfare. Economic warfare provides adversaries a grey-zone coercion tool short of direct action by conventional forces. Representative examples of adversaries embracing and employing economic warfare as an instrument of national power are rapidly emerging. Iran and China, in particular, are each upping their game on employing economic warfare as a key instrument in their struggle against U.S. power.

58.2 HISTORY OF ECONOMIC WARFARE

(i) American Civil War

Union forces in the American Civil War had the challenge of occupying and controlling the 11 states of the Confederacy, a vast area larger than Western Europe. The Confederate economy proved surprisingly vulnerable.

Guerrilla warfare in the American Civil War was supported by a large fraction of the Confederate population that provided food, horses, and hiding places for official and unofficial Confederate units. The Union response was to ravage the local economy, as in the Burning Raid of 1864. Before the war, most passenger and freight traffic moved by water through the river system or coastal ports. Confederate railroads were already inadequate and suffered much damage. Travel became far more difficult. The Union Navy took control of much of the seacoast and the main rivers such as the Mississippi River and the Tennessee River, using the Mississippi River Squadron of powerful small gunboats. Land transportation was contested, as Confederate supporters tried to block shipments of munitions, reinforcements and supplies through West Virginia, Kentucky, and Tennessee to Union forces to the south. Bridges were burned, railroad tracks torn up, and telegraph lines were cut. Both sides did the same and effectively ruined the infrastructure of the Confederacy.

The Confederacy in 1861 had 297 towns and cities with a total population of 835,000 people, 162 of which were at one point occupied by Union forces with a total population of 681,000 people. In practically every case, infrastructure was damaged, and trade and economic activity was disrupted for a while. Eleven cities were severely damaged by war action, including Atlanta, Charleston, Columbia, and Richmond. The rate of damage in smaller towns was much lower, with severe damage to 45 out of a total of 830.

Farms were in disrepair, and the prewar stock of horses, mules, and cattle was much depleted; 40% of the South's livestock had been killed. The South's farms were not highly mechanized, but the value of farm implements and machinery in the 1860 census was $81 million and had been reduced by 40% by 1870. The transportation infrastructure lay in ruins, with little railroad or riverboat service available to move crops and animals to market. Railroad mileage was located mostly in rural areas and over two thirds of the South's rails, bridges, rail yards, repair shops, and rolling stock were in areas reached by Union armies, which systematically destroyed what they could. Even in untouched areas, the lack of maintenance and repair, the absence of new equipment, the heavy overuse, and the relocation of equipment by the Confederacy from remote areas to the war zone ensured the system would be ruined at war's end.

The enormous cost of the Confederate war effort took a high toll on the South's economic infrastructure. The direct costs to the Confederacy in human capital, government expenditures, and physical destruction totaled perhaps $3.3 billion. By 1865, the Confederate dollar was worthless because of high inflation, and people in the South had to resort to bartering for goods or services to use scarce Union dollars. With the emancipation of the slaves, the entire economy of the South had to be rebuilt. Having lost their enormous investment in slaves, white planters had minimal capital to pay freedmen workers to bring in crops. As a result, a system of sharecropping was developed in which landowners broke up large plantations and rented small lots to the freedmen and their families. The main feature of the Southern economy changed from an elite minority of landed gentry slaveholders to a tenant farming agriculture system. The disruption of finance, trade, services, and transportation nodes severely disrupted the prewar agricultural system and forced Southerners to turn to barter. The entire region was impoverished for generations.

(ii) World War I

The British used their greatly-superior Royal Navy to cause a tight blockade of Germany and a close monitoring of shipments to neutral countries to prevent them from being transshipped to there. Germany could not find enough food since its younger farmers were all in the army, and the desperate Germans were eating turnips (a rounded, white root that is eaten cooked as a vegetable) by the winter of 1916–17. US shipping was sometimes seized, and Washington protested. The British paid monetary compensation so that the American protests would not escalate into serious trouble.

(iii) World War II

Clear examples of economic warfare occurred during World War II when the Allied powers followed such policies to deprive the Axis economies of critical resources. The British Royal Navy again blockaded Germany although with much more difficulty than in 1914. The US Navy, especially its submarines, cut off shipments of oil and food to Japan.

In turn, Germany attempted to damage the Allied war effort via submarine warfare: the sinking of transport ships carrying supplies, raw materials, and essential war-related items such as food and oil. As the Allied air forces grew, they mounted an Oil campaign of World War II to deprive Germany of fuel.

Neutral countries continued to trade with both sides. The Royal Navy could not stop land trade, so the allies made other effort to cut off sales to Germany of critical minerals such as tungsten, chromium, mercury and iron ore from Spain, Portugal, Turkey, Sweden and elsewhere. Germany wanted Spain to enter the war but they could not agree to terms. To keep Germany and Spain apart, Britain used a carrot-and-stick approach. Britain provided oil and closely monitored Spain's export trade. It outbid Germany for the wolfram, whose price soared, and by 1943, wolfram was Spain's biggest export-earner. Britain's cautious treatment of Spain brought conflict with the more aggressive American policy. In the Wolfram Crisis of 1944 Washington cut off oil supplies but then agreed with London's requests to resume oil shipments. Portugal feared a German-Spanish invasion, but when that became unlikely in 1944, it virtually joined the Allies.

(iv) Cold War

The Covenant of the League of Nations provided for military and economic sanctions against aggressor states, and the idea of economic sanctions was regarded as a great innovation. However, economic sanctions without military ones failed to dissuade Italy from conquering Abbysinia.

In 1973–1974, the oil-producing Arab states imposed an oil embargo against the United States, United Kingdom, Canada, South Africa, Japan, and other industrialized countries that supported Israel during the Yom Kippur War of October 1973. Results included the 1973 oil crisis and a sharp rise in prices but not an end to support for Israel.

58.3 RECENT EXAMPLES OF ECONOMIC WARFARE

(i) Iran's aggressive economic war on commercial assets

Iran's proxy units have pointedly attacked commercial shipping with conventional weapons in the Red Sea region since the October 7, 2023, Hamas attack on Israel. The Houthi forces' indiscriminate missile attacks on commercial shipping have already disrupted global supply chains to such an extent that Tesla shut down its Berlin manufacturing line due to crucial component shortages. Major commercial shipping lines, including Maersk, Hapag-Lloyd, and others, have paused or re-routed their ships around the risk area. Shipping rate costs have increased almost three times the normal rate, and material increases in transit times for nearly 20% or more of global maritime container shipments are impacting the global economy due to these attacks. Unarmed with economic warfare-waging capabilities, DoD planners would undoubtedly have benefited from a range of escalating non-kinetic response options in anticipating and reacting to the Houthi commercial shipping attacks. Options might range from interdicting and denying the flow of Iranian arms into Yemen, to digitally denying/degrading missile guidance systems, to leveraging counter-threat finance playbooks already learned in Iraq and Afghanistan (because the administration has re-designated the Houthis as terrorists). Non-kinetic economic warfare options and playbooks are therefore critically needed. Military operations other than war viewed through an economic warfare prism would arm war planners with a range of responses for dealing with Houthi attacks — as well as attacks from any other potential adversaries.

(ii) China' Economic War

The People's Liberation Army (PLA's) doctrine and its "Three Warfares" non-kinetic option envision economic warfare as a key element of Beijing's warfare strategy. The Chinese military's embrace of **economic warfare** is on full display worldwide, in daily news items. On January 31, 2024, FBI Director Christopher Wray publicly warned Congress that Chinese military cyber forces "are positioning on American infrastructure in preparation to wreak havoc and cause real-world harm to American citizens and communities, if and when China decides the time has come to strike." Economic warfare is already targeting American industry, economic capabilities, and resources waged by doctrinally organized, trained, and equipped Chinese military forces. This strategy is intended to create chaos and confusion and to sap the support of Americans in the event of a Chinese takeover of Taiwan. The FBI Director's latest warning builds on his 2020 warning regarding Chinese action as "a threat to our economic security, and by extension, to our national security."

China is leveraging all its economic resources toward creating strategic political and military advantages over the United States. China has thereby successfully leveraged economic means — predatory debt lending, bribery, and trade agreements — to achieve military objectives and advantages. For example, the Chinese gained a warm-water port in the Horn of Africa without kinetic action by foreclosing on debt. In 2023, two underwater fiber-optic communications cables were cut within a short period by suspected Chinese organizations, disrupting and degrading Taiwanese commerce. Finally, the Chinese government pressures other countries not to sign free trade agreements with Taiwan.

CHAPTER 59

DIPLOMATIC WARFARE

59.1 INTRODUCTION TO DIPLOMACY

Diplomacy comprises spoken or written communication by representatives of state, intergovernmental, or non-governmental institutions intended to influence events in the international system.

Diplomacy is the main instrument of foreign policy which represents the broader goals and strategies that guide a state's interactions with the rest of the world. International treaties, agreements, alliances, and other manifestations of international relations are usually the result of diplomatic negotiations and processes. Diplomats may also help shape a state by advising government officials.

Modern diplomatic methods, practices, and principles originated largely from 17th-century European customs. Beginning in the early 20th century, diplomacy became professionalized; the 1961 Vienna Convention on Diplomatic Relations, ratified by most of the world's sovereign states, provides a framework for diplomatic procedures, methods, and conduct. Most diplomacy is now conducted by accredited officials, such as envoys and ambassadors, through a dedicated foreign affairs office. Diplomats operate through diplomatic missions, most commonly consulates and embassies, and rely on a number of support staff; the term diplomat is thus sometimes applied broadly to diplomatic and consular personnel and foreign ministry officials.

59.1.1 Diplomacy in different countries and regions

(i) Ancient India

Ancient India, with its kingdoms and dynasties, had a long tradition of diplomacy. The oldest treatise on statecraft and diplomacy, *Arthashastra*, is attributed to *Kautilya* (also known as *Chanakya*), who was the principal adviser to Chandragupta Maurya, the founder of the Maurya dynasty who ruled in the 3rd century BC. It incorporates a theory of diplomacy, of how in a situation of mutually contesting kingdoms, the wise king builds alliances and tries to checkmate his adversaries. The envoys sent at the time to the courts of other kingdoms tended to reside for extended periods of time, and Arthashastra contains advice on the deportment of the envoy, including the trenchant suggestion that "he should sleep alone". The highest morality for the king is that his kingdom should prosper.

New analysis of Arthashastra brings out that hidden inside the 6,000 aphorisms of prose (sutras) are pioneering political and philosophic concepts. It covers the internal and external spheres of statecraft, politics and administration. The normative element is the political unification of the geopolitical and cultural subcontinent of India. This work comprehensively studies state governance; it urges non-injury to living creatures, or malice, as well as compassion, forbearance, truthfulness, and uprightness. It presents a raj mandala (grouping of states), a model that places the home state surrounded by twelve competing entities which can either be potential adversaries or latent allies, depending on how relations with them are managed. This is the essence of realpolitik. It also offers four upaya (policy approaches): conciliation, gifts, rupture or dissent, and force. It counsels that war is the last resort, as its outcome is always uncertain. This is the first expression of the raison d'etat doctrine, as also of humanitarian law; that conquered people must be treated fairly, and assimilated.

(ii) China

The first records of Chinese and Indian diplomacy date from the 1st millennium BCE. By the 8th century BCE the Chinese had leagues, missions, and an organized system of polite discourse between their many "warring states," including resident envoys who served as hostages to the good behavior of those who sent them. The sophistication of this tradition, which emphasized the practical virtues of ethical behavior in relations between states (no doubt in reaction to actual amorality), is well documented in the Chinese classics.

This tradition of equal diplomatic dealings between contending states within China was ended by the country's unification under the Qin emperor in 221 BCE and the consolidation of unity under the Han dynasty in 206 BCE. Under the Han and succeeding dynasties, China emerged as the largest, most populous, technologically most-advanced, and best-governed society in the world. The arguments of earlier Chinese philosophers, such as Mencius, prevailed; the best way for a state to exercise influence abroad, they had said, was to develop a moral society worthy of emulation by admiring foreigners and to wait confidently for them to come to China to learn.

Once each succeeding Chinese dynasty had consolidated its rule at home and established its borders with the non-Chinese world, its foreign relations with the outside world were typically limited to the defense of China's borders against foreign attacks or incursions, the reception of emissaries from neighboring states seeking to ingratiate themselves and to trade with the Chinese state, and the control of foreign merchants in specific ports designated for foreign trade. With rare exceptions (e.g., official missions to study and collect Buddhist scriptures in India in the 5th and 7th centuries and the famous voyages of discovery of the Ming admiral Zheng He in the early 15th century), Chinese leaders and diplomats waited at home for foreigners to pay their respects rather than venturing abroad themselves. This "tributary system" lasted until European colonialism overwhelmed it and introduced to Asia the European concepts of sovereignty, suzerainty, spheres of influence, and other diplomatic norms, traditions, and practices.

(iii) Greece

The tradition that ultimately inspired the birth of modern diplomacy in post-Renaissance Europe and that led to the present world system of international relations began in ancient Greece. The earliest evidence of Greek diplomacy can be found in its literature, notably in Homer's Iliad and Odyssey. Otherwise, the first traces of interstate relations concern the Olympic Games of 776 BCE. In the 6th century BCE the amphictyonic leagues maintained interstate assemblies with extraterritorial rights and permanent secretariats. Sparta was actively forming alliances in the mid-6th century BCE, and by 500 BCE it had created the Peloponnesian League. In the 5th century BCE, Athens led the Delian League during the Greco-Persian Wars.

Greek diplomacy took many forms. Heralds, references to whom can be found in prehistory, were the first diplomats and were protected by the gods with an immunity that other envoys lacked. Their protector was Hermes, the messenger of the gods, who became associated with all diplomacy. The herald of Zeus, Hermes was noted for persuasiveness and eloquence but also for knavery, shiftiness, and dishonesty, imparting to diplomacy a reputation that its practitioners still try to live down.

Because heralds were inviolable, they were the favored channels of contact in wartime. They preceded envoys to arrange for safe passage. Whereas heralds traveled alone, envoys journeyed in small groups, to ensure each other's loyalty. They usually were at least 50 years old and were politically prominent figures. Because they were expected to sway foreign assemblies, envoys were chosen for their oratorical skills. Although such missions were frequent, Greek diplomacy was episodic rather than continuous. Unlike modern ambassadors, heralds and envoys were short-term visitors in the city-states whose policies they sought to influence.

59.2 KASHMIR: CHINA MAY POSE BIGGER CHALLENGE THAN PAKISTAN

China is already occupying 20 per cent of land belonging to Jammu and Kashmir. While Pakistan has failed to find support over Kashmir move, China, on the other hand, may hold the key in this India-Pakistan **diplomatic warfare**.

China is a global superpower with a chequered diplomatic history particularly for India. It launched an attack on India at the height of Hindi-Chini bhai-bhai sentiment espoused by then Prime Minister Jawaharlal Nehru and occupied Aksai Chin in Jammu and Kashmir. In the globalized set up, for years it blocked tagging Masood Azhar a global terrorist when the rest of the world agreed with Indian viewpoint.

While Pakistan has been visibly aggressive and desperate in attempting to swing international opinion on Kashmir, especially after the scrapping of special status of Jammu and Kashmir by the Narendra Modi government, China has been more subtle and nuanced.

This makes China a serious player in diplomatic warfare between India and Pakistan over the status of Jammu and Kashmir. China is present in Jammu and Kashmir in two main pockets — both illegally acquired.

China occupies about 38,000 sq km of Aksai Chin, which it claims to be a part of the Hotan County, lying in the southwestern part of Xinjiang Autonomous Region. China has rejected the Simla Accord of 1914 signed between representatives of China, Tibet and British India.

According to this accord, Ladakh is part of Jammu and Kashmir. China does not recognize this agreement saying that it was signed by a government that did not represent the people of China. The current communist regime of China came to power in the country in 1949.

China continues to recognize Ladakh as "disputed" territory where Indo-China boundary is yet to be demarcated. This stand has helped China justify its occupation of Aksai Chin.

Besides, China has got an area of over 5,800 sq km in Shaksgam Valley of Pakistan-occupied Kashmir and renamed it as Trans-Karakoram Tract. It was originally part of Hunza-Gilgit region of the princely state of Jammu and Kashmir. In return, China has stood by Pakistan in all its dirty games against India.

The Pakistan-China boundary agreement of 1963, under which Pakistan gifted a part of Jammu and Kashmir's land, calls for re-settlement of boundary limits in the region once the Kashmir issue between India and Pakistan is resolved. This clause has a hidden message for China that it may get more land in strategically significant region.

After the Narendra Modi government ceased the operation of Article 370 giving special status to Jammu and Kashmir and also bifurcated the state into two Union Territories. Ladakh will be carved out as a Union Territory without a legislature. China responded to the move angrily saying India has violated sovereignty concerns of China. The obvious reference was to its claim over Ladakh.

"Recently India has continued to hurt Chinese sovereignty by unilaterally changing domestic law. This act is not acceptable and won't be in any sense binding," the Chinese foreign ministry said on August 6. Here, the choice of words by the Chinese foreign ministry spokeswoman Hua Chunying was curious.

She categorically termed the Article 370, a move by the Modi government, a matter relating to "domestic law" of India but asserted it is not "binding" on China with regard to Aksai Chin, which was referred to by Union Home Minister Amit Shah when he moved the motion to make changes in the status of Jammu and Kashmir.

Ten days later, China extended unhindered support to Pakistan over its Kashmir policy. "China will continue to firmly support Pakistan in safeguarding its legitimate rights and interests and continue to preside over justice for Pakistan on the international stage," China's Foreign Minister Wang Yi said after hosting his counterpart from Pakistan.

This commitment came just ahead of Wang's meeting with Foreign Minister S Jaishankar, who visited Beijing as part of India's diplomatic outreach to international community in the aftermath of Kashmir move.

China then pressed for a closed-door deliberation of the United Nations Security Council over the Kashmir move and pushed for a statement after the meeting. But none of the other permanent members agreed to the proposal leaving both China and Pakistan utterly disappointed.

Both Pakistan and China have over the past week snoozed their comments over Kashmir sensing that the world opinion is not on their side. This was also evident during Wang Yi's return visit to Pakistan, where on Sunday, he reiterated China's support to Pakistan but a joint statement called for bilateral settlement of the dispute.

"Parties need to settle disputes and issues in the region through dialogue on the basis of mutual respect and equality," the joint statement said about Kashmir issue.

The China-Pakistan deliberation took place ahead of UNHRC session in Geneva where Pakistan presented a 145-page document to stress its claim that India has violated human rights in Jammu and Kashmir. India was well prepared to counter Pakistan.

But there could be more in the offing. China and Pakistan may again make a concerted effort at the UN General Assembly later this month to raise the Kashmir bogey at the annual session.

There is another development that may see more Chinese involvement in Kashmir. When special status of Jammu and Kashmir was scrapped, China was battling to save its image in view of massive demonstrations in Hong Kong that drew global attention. Now Hong Kong situation is largely under control and China is said to be ready for poking India diplomatically over Kashmir.

59.3 FOREIGN SECRETARY MISRI CONVEYS INDIA'S 'CONCERNS' ABOUT SECURITY OF MINORITY COMMUNITIES IN BANGLADESH

Foreign Secretary Vikram Misri on December 9, 2024, conveyed India's "concerns" over the safety and security of minorities in Bangladesh and urged the interim government there to follow a "constructive approach".

Mr. Misri made the remarks while holding a Foreign Office Consultation with his Bangladeshi counterpart, Mohammad Jashim Uddin, in Dhaka. He also called on Chief Adviser Muhammad Yunus, who urged India to "join his initiative to revive the SAARC (South Asian Association for Regional Cooperation)".

Mr. Misri is the first high-ranking Indian official to visit the neighboring country after the student-people uprising forced Prime Minister Sheikh Hasina to flee to India in August. Bilateral relations have nosedived since the interim government took charge.

Responding to Mr. Misri's remarks, Mr. Jashim Uddin expressed Dhaka's concerns over the recent breach of security at the Assistant High Commission of Bangladesh in Agartala and asserted that the safety of the minorities was an "internal matter of Bangladesh".

India-Bangladesh ties have been maintained despite political changes, can't be reduced to 'single issue', says Indian envoy

While meeting Mr. Misri, Mr. Yunus said students, workers, and the people came together to change the political system in Bangladesh. "Our job is to keep their dreams alive. It is a new Bangladesh," he said.

"I have underlined India's desire to work closely with the interim Government of Bangladesh authorities. At the same time, we also had the opportunity to discuss certain recent developments and issues, and I conveyed our concerns, including those related to the safety and welfare of minorities," Mr. Misri said in a statement at the end of the consultation with the Bangladesh Foreign Secretary.

59.4 NOW ISRAEL WANTS TO LEARN JAISHANKAR'S DIPLOMATIC TRICK TO WIN THE WAR OF NARRATIVES

In December 2024, Israeli Ambassador Reuven Azar highlighted Israel's need to bolster soft power, citing India's success in blending diplomatic and narrative-building strategies. He underscored Israel's efforts to counter Hamas in Gaza, manage differences with India on UN votes, and address challenges around Palestinian sovereignty. Discussing the broader regional implications, Azar outlined concerns over Syria's stability and the shifting balance of power in West Asia.

Israel's reliance on hard power during conflicts has led to challenges in shaping global narratives, especially with powerful media platforms like Al Jazeera and TRT dominating discourse. Israeli Ambassador to India Reuven Azar, during an interaction with The Times of India editors, acknowledged this gap and pointed to India's success in combining hard and soft power.

Israel's current military offensive in Gaza has drawn both support and criticism globally, with a notable wave of backing for Palestinians. As the only Jewish state in the region, Azar noted, Israel often finds itself outnumbered in geopolitical and narrative battles.

CHAPTER 60

POLITICAL WARFARE

60.1 WHAT IS POLITICAL WARFARE?

Political warfare is the use of hostile political means to compel an opponent to do one's will. The term political describes the calculated interaction between a government and a target audience, including another state's government, military, and/or general population. Governments use a variety of techniques to coerce certain actions, thereby gaining relative advantage over an opponent. The techniques include propaganda and psychological operations ("PsyOps"), which service national and military objectives, respectively. Propaganda has many aspects for a hostile and coercive political purpose. Psychological operations are for strategic and tactical military objectives and may be intended for hostile military and civilian populations.

Political warfare's coercive nature leads to weakening or destroying an opponent's political, social, or societal will, and forcing a course of action favorable to a state's interest. Political war may be combined with violence, economic pressure, subversion, and diplomacy, but its chief aspect is "the use of words, images and ideas". The creation, deployment, and continuation of these coercive methods are a function of statecraft for nations and serve as a potential substitute for more direct military action. For instance, methods like economic sanctions or embargoes are intended to inflict the necessary economic damage to force political change. The utilized methods and techniques in political war depend on the state's political vision and composition. Conduct will differ according to whether the state is totalitarian, authoritarian, or democratic.

The ultimate goal of political warfare is to alter an opponent's opinions and actions in favour of one state's interests without utilizing military power. This type of organized persuasion or coercion also has the practical purpose of saving lives through eschewing the use of violence in order to further political goals. Thus, political warfare also involves "the art of heartening friends and disheartening enemies, of gaining help for one's cause and causing the abandonment of the enemies'". Generally, political warfare is distinguished by its hostile intent and through potential escalation; but the loss of life is an accepted consequence.

60.2 TOOLS OF POLITICAL WARFARE

(i) Peaceful

Political warfare utilizes all instruments short of war available to a nation to achieve its national objectives. The best tool of political warfare is effective policy forcefully explained, or more directly, overt policy forcefully backed. But political warfare is used when public relations statements and gentle, public diplomacy-style persuasion—the policies of 'soft power'—fail to win the needed sentiments and actions.

The major way political warfare is waged is through propaganda. The essence of these operations can be either overt or covert.

White propaganda is maximally overt: there is attribution to a promoter; the attributed promoter and the actual promoter are one and the same; and no attempt is made to hide the fact that a viewpoint or "line" is being promoted. Most television advertisements are white propaganda turned to commercial ends.

Grey propaganda ranges in overtness from maximal to a slightly lesser degree: as in white propaganda, there is honest attribution to a source; but it differs from white propaganda by being less forthright either about the link between the source and the line being promoted or about its status as propaganda in the first place. Grey propaganda has alternatively been defined as the "semi-official amplification of a government's voice"; guerilla advertising uses the tools of grey propaganda to sell products and services, while in public service. Examples include Radio Free Europe and Radio Liberty (established during Cold War I).

Black propaganda is covert: this may be limited to the anonymous dissemination of internally consistent talking points (differing from white propaganda only in its lack of credited authorship), but it may also be the impersonation of a widely trusted organization (through the use of its branding, corporate style guide, distinctive turns of phrase, etc.) under a false flag, or a strategy as complex as the use of a botnet to inundate a social network with self-contradictory disinformation as if through a firehose of falsehood, amplifying the botnet's posts with so-called Likes and Retweets, and frustrating genuine users' bona fide searches for pertinent information by diminishing the signal-to-noise ratio. What unifies these disparate strategies implementing black propaganda is that, in all cases, it appears to come from a disinterested source when, in fact it, does not.

There are channels which can be used to transmit propaganda. Sophisticated use of technology allows an organization to disseminate information to a vast number of people. The most basic channel is the spoken word. This can include live speeches or radio and television broadcasts. Overt or covert radio broadcasting can be an especially useful tool. The printed word is also very powerful, including pamphlets, leaflets, books, magazines, political cartoons, and planted newspaper articles (clandestine or otherwise). Subversion, agents of influence, spies, journalists, and "useful idiots" can all be used as powerful tools in political warfare.

(ii) Aggressive

Political warfare also includes aggressive activities by one actor to offensively gain relative advantage or control over another. Between nation states, it can end in the seizure of power or in the open assimilation of the victimized state into the political system or power complex of the aggressor. This aggressor-victim relationship has also been seen between rivals within a state and may involve tactics like assassination, paramilitary activity, sabotage, coup d'état, insurgency, revolution, guerrilla warfare, and civil war.

Foreign infiltration or liberation occurs when a government is overthrown by foreign military or diplomatic intervention, or through covert means. The campaign's ultimate purpose is to gain control over another nation's political and social structure. The campaign could be led by the aggressor's national forces or by a political faction favorable to the aggressor within the other state.

There are three stages involved in the extension of control by the aggressor over the victim:

- **Penetration or infiltration:** the deliberate infiltration of political and social groups within a victim state by the aggressor with the ultimate purpose of extending influence and control. The aggressor conceals its endgame, which goes beyond the normal influential nature of diplomacy and involves espionage.

- **Forced disintegration or atomization:** is the breakdown of the political and social structure of the victim until the fabric of national morale disintegrates and the state is unable to resist further intervention. The aggressor may exploit the inevitable internal tensions between political, class, ethnic, religious, racial, and other groups. This concept is a similar strategy to 'divide and conquer'.

- **Subversion and defection:** Subversion is the undermining or detachment of the loyalties of significant political and social groups within the victimized state, and their transference to the political or ideological causes of the aggressor. In lieu of total and direct transference, the aggressor may accept intermediate states that still meets its objectives, such as the favor of politically significant individuals. Furthermore, the formation of a counter-elite, made up of influential individuals and key leaders, within the victim state establishes the legitimacy and permanency of a new regime. Defection is the transference of allegiance of key individuals and leaders to the aggressor's camp. The individual could relocate or stay-in-place in the victim country, continually influencing local issues and events in the aggressor's favor. Defectors also provide insider information to the aggressor.

- **Coup d'état,** is the overthrowing of a government through the infiltration of the political, military, and social groups by a small segment of the state apparatus. The small segment exists within the state and targets the critical political levers of power within a government to neutralize opposition to the coup and post-coup governing force. Several pre-existing factors are necessary for a coup: political participation being limited to a small portion of the population, independence from foreign power influence and control, and power and decision-making authority concentrated within a political center and not diffused between regional authorities, businesses, or other groups.

A coup utilizes political resources to gain support within the existing state and neutralize or immobilize those who are capable of rallying against the coup. A successful coup occurs rapidly and after taking over the government, stabilizes the situation by controlling communications and mobility. Furthermore, a new government must gain acceptance from the public and military and administrative structures, by reducing the sense of insecurity. Ultimately, the new government will seek legitimacy in the eyes of its own people as well as seek foreign recognition. The coup d'état can be led by national forces or involve foreign influence, similar to foreign liberation or infiltration.

- **Paramilitary Operations:** transitional political warfare ranging from small-scale use of violence with primitive organizational structure (e.g. sabotage) to full-scale conventional war. The transition and escalation includes a series of stages and depends on tactical and strategic objectives. Paramilitary activities include infiltration and subversion as well as small group operations, insurrection, and civil war.
- **Insurgency:** an organized, protracted political warfare tool designed to weaken the control and eliminate the legitimacy of an established government, occupying power, or other political authority. An insurgency is an internal conflict, and the primary struggle is to mobilize local populations for political control and gain popular support towards the insurgents' cause. Insurgencies include political and military objectives, with the end goal of establishing a legitimate, rival state structure. Insurgencies are unconventional military conflicts which incorporate a variety of methods, ranging from coercive tools like intimidation and assassination, to political tools like propaganda and social services. An insurgency's approaches and objectives could involve perpetual disorder and violence demonstrating the government's inability to provide security for the populace, weakening the government and killing or intimidating any effective opposition against the government, intimidating the population and discouraging its participation in – or support for – political or legal processes, controlling or intimidating police and military forces (which limits their ability to respond to insurgent attacks) or by creating government repression by provoking over-reactions by security or military forces.

60.3 EXAMPLES OF POLITICAL WARFARE

(i) China

China's political leaders during this century have had to first create a nation before they could proceed to contend with other national actors in the international arena. Consequently, insofar as both the Chinese Communist Party and the Kuomintang subscribed to a political warfare concept during their formative years of struggle; the concept was as much concerned with creating national identity and defeating domestic adversaries as it was with China's ability to compete in the world. Since the founding of the PRC in 1949 they have centered much of their political warfare efforts within the United Front Work Department. The Chinese conception of political warfare includes the "three types of warfare": public opinion warfare, psychological warfare, and legal warfare, among others. Political warfare encompasses influence operations such as the doctrine of China's peaceful rise.

Taiwan remains a major target of PRC political warfare efforts. China's political warfare campaign aims to isolate Taiwan from the international community and interfere in Taiwan's democratic system and institutions. **India has also become a target of increasing importance for PRC influence operations.**

(ii) Israel

The Israeli Defense Force was an early adopter of social media platforms to promote a positive image about itself by posting sexy images of female Israeli soldiers with weapons, ASMR media involving firearms, as well as more traditional content about humanitarianism and national glory. It operates multiple accounts that have a large following and actively recruits influencers on Facebook, Instagram, Telegram, TikTok, Twitter, and YouTube. The IDF's online presence in English is about twice as large as its activity in Hebrew and mainly targets young Jewish people in the United States. Besides positive self-portrayals, other objectives included shaping the narrative during its media blockade of the Gaza War (2008–2009), preparing information ahead IDF operations in order to be the first one out with the desired story, and quickly responding to unexpected incidents.

(iii) Soviet Union and Russia

Throughout the Cold War, the Soviet Union was committed to political warfare on classic totalitarian lines and continued to utilize propaganda towards internal and external audiences. "Active measures" was a Russian term to describe its political warfare activities both at home and abroad in support of Soviet domestic and foreign policy. Soviet efforts took many forms, ranging from propaganda, forgeries, and general disinformation to assassinations. The measures aimed to damage the enemy's image, create confusion, mould public opinion, and to exploit existing strains in international relations. The Soviet Union dedicated vast resources and attention to these active measures, believing that mass production of active measures would have a significant cumulative effect over a period of several decades. Soviet active measures were notorious for targeting intended audience's public attitudes, to include prejudices, beliefs, and suspicions deeply rooted in the local history. Soviet campaigns fed disinformation that was psychologically consistent with the audience.

Examples of Soviet active measures include:

- **Operation Trust:** was a counterintelligence operation conducted by Soviet intelligence against domestic and foreign adversaries. The operation, which ran from 1921 to 1929, set up a fake anti-Bolshevik underground organization, "Monarchist Union of Central Russia", which claimed to plan a conspiracy to overthrow the Soviet government. The Trust aimed to create the view that communism was over in Russia and the Soviet Union would abandon its revolutionary ways. Western intelligence services supported the fake anti-Bolshevik dissidents who provided false intelligence reports. The operation's purpose was to identify real dissidents and anti-Bolsheviks, internally and abroad. The operation resulted in the arrests and executions of Russian exile leaders and the general demoralization of anti-Soviet efforts.

- **The "Rumor" Campaign:** In October 1985, a Soviet weekly Literaturnaya Gazeta drew attention to a story in an obscure Indian paper, The Patriot, which alleged that the U.S. government engineered the AIDS virus during biological warfare research in the U.S., and that it was being spread throughout the world by U.S. servicemen who had been used as guinea pigs for the experiment. The story was broadcast by Moscow's Radio Peace and Progress in English and Turkish broadcasts to Asian countries, some of which had important U.S. military bases. The "Rumor" campaign was highly effective in the 1980s and continues to resurface today throughout the world.

Communist strategy and tactics continually focused on revolutionary objectives, for them the real war is the political warfare waged daily under the guise of peace; the purpose of which was to disorient and disarm the opposition, to induce the desire to surrender in opposing peoples...to corrode the entire moral, political, and economic infrastructure of a nation. Lenin's mastery of "politics and struggle", remained objectives for the Soviet Union and other global communist regimes, such as the People's Republic of China.

Soviet political warfare in Afghanistan

The Soviet Union remains a comprehensive example of an aggressive nation which expanded its empire through covert infiltration and direct military involvement. Following World War II, the Soviet Union believed European economies would disintegrate, leaving social and economic chaos and allowing for Soviet expansion into new territories. The Soviets quickly deployed organizational weapons such as non-political front groups, sponsored 'spontaneous' mass appeals, and puppet politicians. While many of these countries' political and social structures were in post-war disarray, the Soviet Union's proxy communist parties were well-organized and able to take control of these weak, newly formed Eastern European governments. Moreover, the clandestine operations of the Soviet intelligence services and the occupying forces of the Soviet military further infiltrated the political and social spheres of the new satellites. Conversely, in 1979, the Soviet Union was unable to successfully penetrate the Afghan society after supporting a coup which brought a new Marxist government to power. While Soviet units were already in Kabul, Afghanistan at the time of the coup, additional Soviet troops arrived to reinforce the units and seize important provincial cities, bringing the total of Soviet troops inside Afghanistan to 125,000–140,000. The Soviets were unprepared for the Afghan resistance which included classic guerrilla tactics with foreign support. In 1989, Soviet forces withdrew from Afghanistan, having been unable to infiltrate the Afghan society or immobilize the resistance.

(iv) Taiwan

The Republic of China Government in Taiwan recognized that its Communist adversary astutely employed political warfare to capitalize upon Kuomintang weaknesses over the years since Sun Yat-sen first mounted his revolution in the 1920s, and Chiang Kai-shek's regime had come to embrace a political warfare philosophy as both a defensive necessity and as the best foundation for consolidating its power in hope of their optimistic goal of "retaking the mainland". Both the Nationalist and Communist Chinese political warfare doctrines stem from the same historical antecedents at the Whampoa Military Academy in 1924 under Soviet tutelage.

The Nationalist Chinese experience with political warfare can be treated in a much more tangible way than merely tracing doctrinal development. In the Taiwan of the 1970s, the concept was virtually synonymous with the General Political Warfare Department of the Ministry of National Defense, which authored political doctrine and was the culmination of a series of organizational manifestations of its application.

(v) United States

American foreign policy demonstrates a tendency to move towards political warfare in times of tension and perceived threat, and toward public diplomacy in times of improved relations and peace. American use of political warfare depends on its central political vision of the world and its subsequent foreign policy objectives.

After World War II, the threat of Soviet expansion brought two new aims for American political warfare:

- To restore Western Europe through military, economic, and political support, and
- To weaken the Soviet hold on Eastern Europe through propaganda.

(vi) Pakistan (against India)

Pakistan is making serious attempts to unleash political warfare to create political, communal and economic instability in India. There was a deliberate attempt, by the proxies and corroborators, to discredit entire democratic process of India by questioning the credibility of the Election Commission and the electronic voting system. Similarly, if functioning of financial institutes and banking system is discredited, it can lead to collapse of economy and financial viability of a state. Kashmir is witnessing a very sophisticated political warfare, where endeavor is made to discredit the institutions of governance. Election boycott and projection of complete lockdown, or forced shut down under coercion, are some of the facets of ongoing political war in the Valley.

Mechanism to Fightback Political Warfare (India-specific)

Political warfare requires heavy investment in intelligence to detect it at an early stage of manifestation. The governments must strengthen institutions of governance, to make them robust and credible, which are able to deflect repeated attacks through the tools of political warfare. One must be mindful that there are multiple tools of directing political warfare and an adversary will not use the same tool time-and-again, and will surprise by opening different fronts to achieve success. The situation also requires democratic nations to develop the capacity to react proportionately to achieve a deterrent effect. This can be achieved through the development of specifically crafted practical "grey-zone" response option – which doesn't mean engaging in retaliatory subversion, but instead utilizing the value-based argument, amongst other tactics, to win the narrative war. The political warfare is directed to manipulate perception of the people and institutions of governance. Therefore, developing citizens as warriors to expose false narrative by adversaries is one of the best ways to fight back. Cyber spear and cyber shield is another potent offensive and defensive tool to fight political warfare. Cyber spear must be used to decode impersonation, voice modulation, expose falsehood and discredit the aggressor for falsehood. Cyber shield is imperative to prevent hacking of systems and putting up impenetrable fire walls. Non-state actors can conduct political warfare with unprecedented reach because they are faceless and amorphous, thus use of Artificial Intelligence (AI) to decode the identity of the non-state actors and their place of origin is imperative. The information arena is an increasingly important battleground, where perceptions of success can be determinative. Ethnic and religious harmony act as shield against political warfare. There is a need for a strategy to fight political warfare. It requires the institutions of democracy: credible political system that is able to give stable governance, political leaders capable of delivery of governance, a bureaucracy capable of implementing that governance, and civil society groups able to provide support and stability to those institutions.

In nutshell, political war can be dealt with effectively by approach of government. To formulate a doctrine and concept of operations, it is important to develop understanding of this new age warfare. The military component that should be spearheading the response to political warfare is Special Operation Forces, cyber and information warriors. The role of military is vital for countering and launching political warfare. First on the ground, in a target country, ideally should be Special Forces to coordinate and galvanize public support and once the stage is set, thereafter, cyber and information warriors, and intelligence wings of Special Forces must oversee coordination and direction of operations. This warfare may be whole of the government approach but the operations must be executed with utmost secrecy. During the initial stage of the operations, activities must appear benign and over exposure or over-reaction could compromise response or retribution. The only way India can make Pakistan pay for the price of cross border terrorism is political warfare. In fact, it may be a monumental mistake to attempt to wrest Pakistan occupied Jammu and Kashmir (POJK) militarily, however, there is a window of opportunity to not only make Pakistan pay the price for its cross-border terrorism but also make it unsustainable for it to hold on to POJK. Conclusion Political warfare can generate unintended consequences which can, at times, spiral into major challenge for the adversaries. Two important characteristics of political war are; one that it is difficult to predict when this war begins and when it terminates; second that it is ethical denunciation of formal rules of war. War is a contest of wills, and the digital information age has created a scenario in which political warfare is only going to become more lethal and amorphous. Increasingly it will be about a contest of narratives below the threshold of war. Key to succeed is to deny exposed flanks or fractured society and to maintain deniability and surprise while targeting an adversary. Offence is the best option to deter the adversary and thus, conceptualization and deeper understanding of political warfare is essential.

60.4 MODERN POLITICAL WARFARE

In the Cold War, Russia routinely employed active measures to subvert Western-allied governments, and in recent years it has pursued an array of destabilizing activities in the Baltics, including espionage, military pressure, and economic pressure. Iran, for its part, has used an array of proxies, as well as soft power (the use of economic, sociopolitical, and cultural influence), to gain influence in Iraq and Syria through religious, cultural, and economic means and by supplying training, equipment, and advisory services to a variety of partners.

Political warfare consists of the intentional use of one or more of the implements of power — diplomatic/political, information/cyber, military/intelligence, and economic — to affect the political composition or decision-making in a state. As an example, the political warfare tactic of economic subversion can be seen in the overlap of the diplomatic/political (routine diplomacy) and economic (trade) spheres. Political warfare is often — but not necessarily — carried out covertly, but it must be undertaken outside the context of conventional war.

60.4.1 What Are the Characteristics of Modern Political Warfare?

Researchers focused on three case studies — two state actors (Russia and Iran) and one nonstate actor (the Islamic State of Iraq and the Levant [ISIL]) — to derive the common characteristics of modern political warfare. Each country employs its particular advantages or strong suits to gain leverage. In Estonia, Russia capitalized on the sentiments of its Russian minority to fan protests over the decision by the Estonian government to move a Soviet monument — the Bronze Soldier incident — which escalated into protests, a sustained cyberattack, and then sanctions and threats. In subsequent years, the Russian government has maintained its hostile stance to destabilize Estonia and other Baltic states, including contesting the legitimacy of their independence from the former Soviet Union.

(i) Example of Russia

- Views its activities as defensive in reaction to the United States;
- Sees democracy promotion and free press as threat;
- In Estonia, used opportunistic approach, capitalizing on crises;
- Used shaping operations (e.g., propaganda directed at Russian speakers) to prepare the ground;
- Uses "New Generation Warfare" innovations in economic leverage, social proxies, and media penetration
- Uses propaganda for obfuscation rather than persuasion.

(ii) Example of Iran

- Heavily based soft power strategy on cultural, political, and religious influence versus differentiated approach to Shi'a, pan-Arab, and pan-Islamic audiences;
- Uses a worldwide network of cultural, informational, and influence organizations, backed by material support, including religious tactics, such as funding junior clerics and mass pilgrimages;
- Offers political and economic support to foreign political parties and leaders to install and influence governments;
- Uses Arab proxies in Syria (including Iraqi militias and Lebanese Hezbollah paramilitaries) that become political actors and spawn new proxies;
- Has well-developed financial and cyber tools.

(iii) Example of ISIL

- Acquires or invents quasi-state tools, including governance, tax, economic resource control, and management;
- Uses combined arms, innovated weaponry, and tactics;

- Uses sophisticated information operations to recruit, inspire, plan, and execute;
- Systematically indoctrinates new recruits to strip them of their old identities and prevent them from straying;
- Has powerful brand and is a unified and sustained organization;
- Has moved away from a broadcast model to a dispersed and resilient form of communication that relies on peer-to-peer sharing and redundancy across platforms;
- Targets different audiences with different messages — uses violence and emotive language liberally in Arabic-media productions to mobilize rank-and-file members.

Synthesizing the case study characteristics in the table yielded a list of key attributes that broadly describe how this form of warfare is conducted today.

Political warfare

- employs diverse elements of power, including a preponderance of nonmilitary means;
- relies heavily on unattributed forces and means;
- is increasingly waged in the information arena, where success can be determined by perception rather than outright victory;
- uses information warfare, which works by amplifying, obfuscating, and, at times, persuading;
- is employed with cyber tools to accelerate and compound effects;
- increasingly relies on economic leverage as the preferred tool of the strong;
- often exploits shared ethnic or religious bonds, as well as social divisions or other internal seams;
- extends, rather than replaces, traditional conflict and can achieve effects at lower cost;
- is also conducted by empowered nonstate or quasi-state actors; and
- requires heavy investment in intelligence resources to detect it in its early stages.

60.4.2 Where are the Gaps in U.S. Information Capabilities and Practices?

One of the key attributes of modern political warfare that emerges from the synthesis is the importance of the information space and the ability to operate effectively within that space.

Because this area can profoundly affect all other lines of effort, it must be considered and managed at the highest levels of government. Moreover, the revolution in communications and information technology has transformed the information space, thus requiring new models and new capabilities to compete effectively in this arena.

In this light, researchers identified several gaps in U.S. government information capabilities and practices:

- Strategic-level communications are high-profile and bureaucratically risky — characteristics that militate against speed and initiative.
- Interagency coordination and National Security Council guidance pertaining to message themes remain lacking.
- The new Global Engagement Center, established by presidential executive order and located at the U.S. Department of State (DoS), focuses on third-party validators or influencers from the bottom up, but it has encountered various limitations.
- U.S. military information support operations are challenged by significant manpower and funding shortages and limited new media training.
- U.S. Central Command is at the forefront of U.S. Department of Defense (DoD) social media communications, but other combatant commands are lagging. The U.S. Special Operations Command Joint Military Information Support Operations Web Operations Center is nascent.
- Unattributed communications may have counterproductive effects that should be anticipated and mitigated.

60.4.3 What are the Effective Measures to Confront Political Warfare?

Developing an integrated response to threats short of war includes (1) the need for strategy, (2) the need for a whole-of-government approach to statecraft led by an appropriately enabled DoS, and (3) the formulation and coordination of responses with and through other sovereign governments, allies, and partners. The authors also include recommendations for improving military contributions to such an integrated approach.

In terms of the **need for strategy**, the general requirement for a cost-effective approach to national defense suggests that early and effective nonmilitary responses — and nonlethal uses of the military element of national power — may provide the United States with valuable tools to deter adversaries, prevent conflicts from escalating, or mitigate their effects. In some cases, these approaches may effectively reduce or remove the incipient threats.

As for the **need for a whole-of-government approach**, DoS is the designated lead for conducting U.S. foreign policy and represents such foreign policy interests abroad. Thus, it is the logical entity to lead a whole-of-government response in this primarily political and diplomatic realm and to coordinate other agencies if given such policy guidance from the President. However, despite the deep country and regional expertise at DoS, this research highlighted significant gaps in organizational and operational capabilities and practices that should be remedied to enable DoS to effectively plan, coordinate, and execute interagency responses continuously, if so directed by the President.

U.S. plans and activities must necessarily be **coordinated with the governments** of those countries where the aggression, subversion, coercion, or destabilization is occurring, along with other partners or allies who are willing and able to contribute their resources and efforts in a common effort.

Eight **recommendations** are relevant to improving the practices and capabilities of the U.S. military — and special operations forces (SOF) in particular — to work with state and nonmilitary entities to combat nonconventional warfare through expanded deterrence, enhanced resilience, and preparations for national resistance, among other means.

Recommendations:

(i) To improve whole-of-government synergy, U.S. military commands — including deployed headquarters — should routinely involve civilian departmental representatives to understand, coordinate with, and support DoS and other civilian program execution.

(ii) DoD and SOF in particular should incentivize and improve the selection and training for military advisers serving at DoS headquarters, U.S. embassies, and other diplomatic posts to increase their effectiveness.

(iii) DoD and SOF should offer military planners to DoS as it builds its own cadre of planners and integrates regional and functional bureau plans; doing so will enable DoS to play a lead role in responses to political warfare.

(iv) Military commanders should develop and maintain collaborative relationships with their civilian counterparts through regular visits and frequent communications to develop common understanding of and approaches to political-military conflict.

(v) DoD should routinely seek to incorporate DoS knowledge and the current insights of the U.S. country team into military plans to develop effective responses to political-military threats.

(vi) The special operations community should make it a high priority to improve and implement fully resourced, innovative, and collaborative information operations.

(vii) Military commanders and DoS should identify critical information requirements for gray zone threats, and the intelligence community should increase its collection and analysis capabilities dedicated to detecting incipient subversion, coercion, and other emerging threats short of conventional warfare.

(viii) DoD and DoS should support the deployment of SOF as an early and persistent presence to provide assessments and develop timely and viable options for countering measures short of war.

BIBLIOGRAPHY

Alberts, David S., John J. Garstka, and Frederick P. Stein. "Network Centric Warfare: Developing and Leveraging Information Superiority", 2nd Edition (Revised), CCRP Publication Series, August 1999.

Ali, Salahudin. "Coming to a Battlefield Near You: Quantum Computing, Artificial Intelligence, & Machine Learning's Impact on Proportionality", Learning's Impact on Proportionality, 18 Santa Clara Int. Law, Vol. 18, Issue 1, 2020.

Ashraf, Nageen and Dr. Saima Ashraf Kayani. "India's Cyber Warfare Capabilities: Repercussions for Pakistan's National Security", NDU Journal 2023, 34-45.

Basu, Arindrajit. "India's International Cyber Operations: Tracing National Doctrine and Capabilities", www.unidir.org | © UNIDIR 2022.

Berendsen, Maj. René G. "The Weaponization of Quantum Mechanics: Quantum Technology in Future Warfare", Monograph, School of Advanced Military Studies US Army Command and General Staff College Fort Leavenworth, 2019.

Dudeja, Jai Paul. "Mind-Reading and Artificial Intelligence: Past, Present And Future (Science, Technology, Applications, Risks and Regulations)", Notionpress.com, 2024.

Dudeja, Jai Paul. "Climate Change A Global Challenge (Causes, Adverse Effects, Monitoring, Adaptation, And Mitigation)", Notionpress.com, 2024.

Dudeja, Jai Paul. "Lidar Techniques for the Remote Sensing and Monitoring of Toxic Agents in the Atmosphere". Presented as an "Invited Talk" and Published on page 36 in the Proceedings of "National Conference on Optoelectronics and MEMS Technologies", CSIO Chandigarh, Apr 16-17, 2004.

Dudeja, Jai Paul and S. Veerabuthiran. "Simulation Study of a Multiwavelength DIAL System for the Range-Resolved Concentration Measurements of Spectrally Overlapping Chemical Agents Simultaneously Present in the Atmosphere". Presented and Published in the Proceedings of "International Conference on Optics and Optoelectronics (ICOL-2005)", IRDE Dehradun, Dec 12-15, 2005.

Dudeja, Jai Paul, S. Veerabuthiran and Shilpi Roy. "Theoretical Analysis of a Multiwavelength Differential Absorption Lidar System for the Range-Resolved Detection and Concentration Measurement of Some Pollutants in the Atmosphere"; Ind. J. Environmental Protection, 25, 1084, 2005.

Dudeja, Jai Paul. Mukesh Kumar Jindal and S Veerabuthiran. "Lidar-Based Remote Explosives Detection Systems (LIBREDS): An Overview". Laser Horizon, Vol. 8, No. 2. pp 3-8, Dec 2006.

Dudeja, Jai Paul. "Selection of an Averaging Technique by Simulation Study of a DIAL System for Toxic Agents Monitoring". Accepted for Presentation at the "SPIE Europe Remote Sensing Conference" Florence, Italy during Sep 17-21, 2007.

Dudeja, Jai Paul. "Integrated Sensors for Laser-Based Standoff Detection of Explosive Materials". Accepted for Presentation as an "Invited Talk" at New Orleans, USA at the "First North American Symposium on Laser Induced Breakdown Spectroscopy (NASLIBS 2007)" organized by Mississippi State University, USA during Oct 8-10, 2007.

Dudeja, Jai Paul. "Differential Absorption Lidar (DIAL) for Monitoring of Pollution in the Ambient Air"; Proceedings of First National Conference on "Trends and Applications in Laser Technology & Optoelectronics", TALTO-1, Amity University Haryana, Gurgaon. Apr 04, 2013.

Dudeja, Jai Paul. "Laser-based Hybrid Sensors for Standoff Detection of Explosive Materials", Proceedings of First National Conference on "Trends and Applications in Laser Technology & Optoelectronics", TALTO-1, Amity University Haryana, Gurgaon. Apr 04, 2013.

Dudeja, Jai Paul. "Fiber-Optic Fire Sensors". Laser Horizon, 6, 22, June 2002.

Fakron, Malik. "Environmental warfare operation principles", Applied Sciences Research Periodicals, Vol. 2, No. 2, pp. 10-15, February 2024.

Ford, Matthew. "From innovation to participation: connectivity and the conduct of contemporary warfare", International Affairs, Volume 100, Issue 4, July 2024, Pages 1531–1549, https://doi.org/10.1093/ia/iiae061.

Furqan Ali, Mohammad; Dushantha Nalin K. Jayakody; Yury Alexandrovich Chursin; Soféine Affes; and Sonkin Dmitry. "Recent Advances and Future Directions on Underwater Wireless Communications", Archives of Computational Methods in Engineering (2020) 27:1379–1412 https://doi.org/10.1007/s11831-019-09354-8.

Gupta, Ashish. "Internet of Things": A New Paradigm for Military Operations", Air Power Journal, Vol. 10, No. 2, April-June 2015. Pp 37-56.

Jesse, W. J. Hamel, Maj. "Adaptive Airpower: Arming America for the Future through 4D Printing", Report submitted to Air University for the degree of Master of Operational Arts and Sciences, May 2015.

Luddy, John. "The Challenge and Promise of Network-centric Warfare", Lexington Institute, February 2005.

Mrazova, Maria. "Advanced composite materials of the future in aerospace industry", INCAS BULLETIN, Volume 5, Issue 3, 2013, pp. 139 – 150.

Neumann, Niels M. P. Maran P. P. van Heesch1, and Patrick de Graaf. "Quantum Communication for Military Applications", November 2020, DOI:10.48550/arXiv.2011.04989

Pandey, Gp Capt (Dr) Dinesh Kumar (retd). "Space Warfare: The Final Frontier of 'War In Space', Synergy – Volume 3, Issue 1, February 2024, pp 14-30.

Pant, Atul. "Internet of Things Centricity of Future Military Operations", J. Defence Studies, Vol.13, No. 12, April-June 2019, pp. 25-58.

Sen, Prof. Gautam. "Hybrid Warfare: Concept & Implications for India", PPF - Centre for Radicalisation and Security Studies July 2020.

Singh, Lt. General Dushyant (Retd), "Swarm Drones - New Frontier of Warfare", SP's Land Forces, Issue: Aero India 2021 Special.

Stan, Andreea-Maria. " Environmental Warfare and the Meaning of Long-lasting Widespread and Severe Effects in the Law of Armed Conflict" Diploma Thesis, Johanness Keppler University Linz, June 2021.

Tomar, Sanjiv. "Nanotechnology" The Remerging Field for Future Military Applications", IDSA Monograph Series No. 48 October 2015.

www.ingramcontent.com/pod-product-compliance
Ingram Content Group UK Ltd.
Pitfield, Milton Keynes, MK11 3LW, UK
UKHW062007290726
14090UKWH00022B/1437

9 798897 242443